BROWN'S BOUNDARY CONTROL AND LEGAL PRINCIPLES

BROWN'S BOUNDARY CONTROL AND LEGAL PRINCIPLES

Eighth Edition

DONALD A. WILSON, LLS, PLS, RPF
President of Donald Wilson Consulting, LLC
Newfields, NH, USA

C.A. "TONY" NETTLEMAN III, PLS, Esq.
President and CEO of Nettleman Land Consultants, Inc.
Sparta, NC, USA

WALTER G. ROBILLARD, PLS, Esq.
Retired Surveyor
Atlanta, GA, USA

For general information on our other products and services or for technical support, please contact our Customer Care Department within the United States at (800) 762-2974, outside the United States at (317) 572-3993 or fax (317) 572-4002.

Wiley also publishes its books in a variety of electronic formats. Some content that appears in print may not be available in electronic formats. For more information about Wiley products, visit our web site at www.wiley.com.

Library of Congress Cataloging-in-Publication Data Applied for:

ISBN Hardback: 9781119911708

Cover Design: Wiley
Cover Image: © Sanhanat/Adobe Stock Photos

Set in 10/12pt TimesLTStd by Straive, Chennai, India

SKY10073272_041724

CONTENTS

PREFACE

In the seventh edition of this text, we emphasized the advances and improvements in technology affecting the surveying profession and stressed how the law, in contrast, had remained fairly constant, with few exceptions. Ten years have now elapsed since that was written, and in the typical fashion of progression, some astounding changes have been made within the court system and in the surveying profession as a whole. Since people in general often tend to be litigious, several new title and rights issues have found their way through the system, and a few dramatic changes have taken place. Law evolves slowly, yet eventually meets the needs of society and, hopefully, the majority opinion is the most fair.

At the federal level, some Native American rights have been clarified, some right of way issues have been decided differently from what most people believed, and the courts have cleaned up a number of questionable areas. Lower courts have been catching up on a regular basis, and we have attempted to bring important survey issues to the reader.

At the state level, many new court decisions have been added to the new edition, a revised list of survey right-of-entry laws were added, and state-specific practices were added concerning Florida, Texas, along with several other states. While this book is written for a national audience, the new edition also realizes that surveying laws can vary significantly from state to state.

Curtis Brown's visions of a textbook-type treatment of our beloved profession have enjoyed a wide acceptance, and we find that there are very few surveyors who do not credit this text for providing genuine learning experiences and ultimately use Brown's to pass their exams as well as a ready reference well into their professional careers. The current authors would like to think, and believe, that we have taken Curt solid foundation and tweaked it such that it is still current and useful as both a learning and reference aid.

Authorship has changed somewhat since the previous edition and dramatically since the first edition. As we strive to produce a richer text, but not necessarily a larger one, or a more costly one, dealing with advances in both technology and law requires more discussion, references, and citations. Textbooks, like people and our surroundings, change and sometimes get more complex. The praise we have received from various levels of the court system is encouraging and appreciated, knowing that our efforts to produce a valuable work product have been recognized and found to be of value.

DONALD A. WILSON, LLS, PLS, RPF

President of Donald Wilson Consulting, LLC
Newfields, NH, USA
September 2023

C.A. "Tony" NETTLEMAN III, PLS, Esq.

President and CEO of Nettleman Land Consultants, Inc.
Sparta, NC, USA
September 2023

WALTER G. ROBILLARD, PLS, Esq.

Retired Surveyor
Atlanta, GA, USA
September 2023

CHAPTER 1

HISTORY AND CONCEPT OF BOUNDARIES

1.1 INTRODUCTION

The history and location of boundaries are steeped in the history of the world from the time before records were kept to today. These boundaries are a result of actions of individuals and nations and law. Boundaries can be related to the areas of history, politics, surveying, and law. Both boundaries of an international nature and those between individuals have caused problems that have been fought, are still being fought, and will continue to be fought in the future over their locations between nations, states, and individual parcels of land within "Happy Acres" subdivisions. In recent years, both local and international judicial tribunals have had to apply old, proven doctrines and have created new legal doctrines to resolve boundary issues. One cannot pick up a newspaper or a magazine without reading about some individual or nation with a boundary issue that is new or that has been festering for many years.

Wars have been fought both on an international scale and in local neighborhoods, and people have been killed over boundary disputes of an inconsequential nature involving pieces of land that have ranged from hundreds of miles to a fraction of a foot or meter. Boundaries are personal in nature, and people have been and will continue to be protective about the misidentification or misalignment of a known or perceived boundary infringement. The surveyor may become the common factor in a boundary problem, as a result of, for example, preparing an erroneous map showing the boundary between two or more nations or the erroneous depiction of a single line between two landowners.

Brown's Boundary Control and Legal Principles, Eighth Edition.
Donald A. Wilson, C.A. "Tony" Nettleman III, and Walter G. Robillard.
© 2024 John Wiley & Sons, Inc. Published 2024 by John Wiley & Sons, Inc.

Even after modern boundary issues have been, seemingly, resolved, members of both the legal profession and the surveying profession may question the results, asking incredulously, "How could the court do that?" Both the trial attorney and the testifying survey expert could not believe the court disregarded the case law on the subject.

In the primeval forest, particularly in the plant kingdom, there are no known boundaries between living things. Although some horticulturists dispute this, we accept the fact that plants do not create boundaries to separate themselves. Animals—especially humans—do create boundaries. We like to think that only humans create and appreciate boundaries, but it has been observed in nature that most mammals, some reptiles, and a few fish create, identify, mark, and defend boundaries.

In this book, we discuss the creation, identification, description, and recovery of boundaries among people. We do not include the recovery and interpretation of the evidence of once-created boundaries; rather, we examine how boundaries are created, how they are described, and the technical legal and ethical ramifications of such boundaries that separate rights, both real and perceived, in real property.

Some boundaries are created in a random manner, whereas others are created according to preconceived plans, identified by any manner of a written description(s), and then litigated according to common law, case law, or statute law[1]. Although it is not our intent in this book to dwell on the creation of boundaries by the lower forms of animal life, their actions in creating boundaries should be examined, because certain principles are similar. Many of these boundaries humans create remain for generations and, when they are retraced by modern methods and with a modern approach, may cause technical and legal problems for today's surveyors and courts.

Field examinations and studies by naturalists have revealed that most animals really do not create boundaries per se. However, it is recognized that they usually create terminal points (corners) and then identify the boundaries between these points,[2] although lower forms of animals may create boundaries that are not necessarily of a permanent nature.

Humans usually create boundaries in several ways. For the sake of simplicity, these may be placed in the following categories:

1. *By action*. Physical acts create a line and points on the ground. This is followed by placing actual monuments at the corner points and identifying these points (corners) and line objects. The lines and objects are then described and may be identified on plats or in field notes. This evidence created and left "on the ground" becomes the proof of the original work and lines and becomes the legal controlling factor in conducting *retracements*.

2. *By writings*. The written word becomes the method of creation when a person describes corners and/or lines in a deed and then conveys to these described lines, prior to the completion of a survey. The problem is created when what the surveyor places on the ground is and then fails to create a solid paper trail.

3. *By law*. Ancient common and modern statutes are relied on to create, modify, and relocate many modern boundaries.

The following principles are introduced in this chapter and discussed in detail in later chapters:

PRINCIPLE 1. Boundaries enjoy a long history in both mythology and Judaic-Christian history.

PRINCIPLE 2. A surveyor creates land boundaries. These created lines, which are separate and distinct from property lines, are determined by legal principles and law.

PRINCIPLE 3. A described closed boundary identifies a claim of right to any property interest for which any person can make a claim of possession through a claim of title. These boundaries may be either macro or micro in nature.

PRINCIPLE 4. A person or landowner can legally convey only the quality and quantity of interest in land to which he or she has title.

PRINCIPLE 5. In most instances, there are no federal laws describing real property rights.

PRINCIPLE 6. Although there are no federal laws of real property, property rights are identified by the state laws and are protected under the US Constitution.

PRINCIPLE 7. Real property rights are determined according to the laws in effect in the particular locale where the land is located. English common law is the predominant law, and it is described as the *lex loci.*

PRINCIPLE 8. Once boundary lines are created, the contiguous lines may, by law or by the actions of landowners who have vested rights, be changed or altered.

PRINCIPLE 9. Law does not provide for two original descriptions of the same parcel.

PRINCIPLE 10. Multiple boundary descriptions may exist for the same parcel, but only one is controlling.

PRINCIPLE 11. There can be only one original boundary survey and description; all subsequent ones are retracements.

PRINCIPLE 12. A resurvey can be conducted only by the entity who conducted the original survey. The law provides for resurveys of parcels, but only on a limited basis and under certain restrictions, the main one being that the bona fide property rights granted under the previous survey are not jeopardized. Two classes of resurveys are recognized: dependent resurveys and independent resurveys.

1.2 SIGNIFICANCE OF BOUNDARIES

The description of property by surveys and landmarks and by reference to boundaries is very ancient. Basically, property interests are separated by boundaries. From precolonial times in the United States, many wars, both local and regional, have been fought and people have been killed as a result of disputed boundaries. This problem was probably inherited from the European continent when the United States adopted English common law as the basis of its common law.

In Great Britain and in Europe, territorial boundaries have, for the most part, generally been stable because the lines were etched in antiquity. Once parish boundaries were established in England—many during Roman times—they formed invisible webs or lines around families and bound them into communities, and ultimately separated communities from one another. This historical background was passed on to the United States, and these distinctions exist today as a result of this influence.

Stories abound in both the United States and Great Britain in which boundaries have affected people's lives. Individuals and groups go to extremes over boundaries, for a boundary can have political ramifications in areas such as citizenship and jurisdiction in legal matters. A tale from colonial times tells of the decision of surveyors who were engaged to run the boundary line between Kentucky and Tennessee to place a jog in the line when a landowner placed a jug of rum near his property and told the surveyors that it was theirs if they found it to be in Kentucky. They did. Naturally, the line has a jog in it. One of the authors of this book, Walt Robillard, remembers that when he was a young boy growing up near the Canadian border, his grandfather would take him to a tavern that straddled the US–Canadian border. On the US side of the bar, the serving of drinks stopped at midnight and was "never on Sunday"; however, on the Canadian side, the drinking continued. At the stroke of midnight and on Sundays, all drinks were served on the Canadian side. The bar patrons would move physically from the United States into Canada.

In 1870, the Reverend Francis Kilvert, an Anglican priest in Wales, related how one of his parishioners occupied a house that straddled the border in Wales on the edge of Brilly Parish. It was suggested that it would be more desirable for this parishioner to give birth to her child in his parish. The line between the parishes was indicated by a notch on the chimney. To ensure that the child would be born in the proper parish, the midwife had the mother give birth standing up in a corner on the appropriate side of the parish line.

People take boundaries seriously. Yet what they really are saying is, "I want the rights that I am entitled to in this property" or "I want those rights in that parcel of land." Boundaries do not determine rights in land, but they identify the limits of any rights a person or group of people may have created or identified and now claim.

1.3 BOUNDARY REFERENCES

Principle 1. *Boundaries enjoy a long history in both mythology and Judaic-Christian history.*

Historically, the English language, using actual occurrences, developed certain terms that depicted and/or identified boundary problems. Until the advent of published maps, boundary identification and the resulting problems and discrepancies were passed from generation to generation by word of mouth.

It was not until mapping became a part of everyday living that boundaries were identified to such a degree of certainty that they no longer relied on the spoken word.

In all probability, many of the boundaries on modern maps were placed there based on the testimony of people who identified them at an earlier time. There are many place names that indicate evidence of boundaries. The Old English term maere translates to "boundary." An examination of modern British Ordnance Survey maps or maps produced by the US Geological Survey indicates names like "Merebrook" and "Merebeck," indicating that certain streams were considered boundaries.

Once boundaries were established and identified, they would be of no value if society could not ensure them with a degree of certainty. Once again, the gods and society were called on for guidance. The ancient Greeks ensured that boundaries would be sacrosanct. They "appointed" the goddess Terminus to be the protector of boundaries. This system was inherited by the Normans and Saxons in England in two ways: first, by the manner in which boundary stones were originally marked, and second, by the practice of beating the bounds.

The historic practice of beating the bounds consisted of the ritual of selecting children from the locality (usually boys), who, accompanied by a member of the town, a clergyman, and the parties to the land transfer, would walk, or perambulate, the boundaries. At each corner, one of the boys would be suspended by his feet and his head would strike the monument. Then, in the event of a future dispute, the boy would go to the corner marking the boundary and point out its location, as he remembered it.

For centuries, surveyors have marked boundary stones (corners) by cutting crosses into rock monuments (see Figure 1.1). This practice was probably brought to America by early English surveyors, who used the same practice in their home country. An examination of early survey and mapping practices indicates that early English surveyors would cut a cross into the monument as protection or to indicate the bounds

Figure 1.1 Boundary stone marked with a + of medieval origin. (Source: Courtesy of Prof. Angus Winchester.)

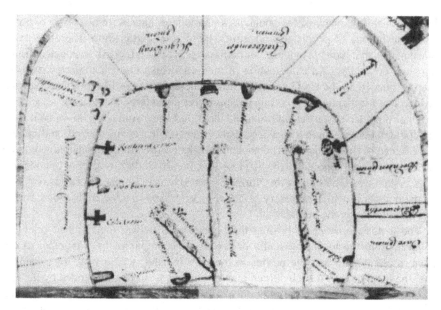

Figure 1.2 1675 Map of Exmoor Forest, Devon. Note crosses at some corners. (Source: Courtesy of Public Records Office, London.)

of a religious holding. They then indicated these beacons (monuments) on maps in the form of crosses (see Figure 1.2). In all probability, these crosses were cut into the stone and then shown on maps in hopes that the Christian God would protect them as Terminus had protected Greek boundary stones.

1.4 TERMINUS: THE GOD (OR GODDESS) OF BOUNDARIES

Following the Greeks, Terminus was designated by the ancient Romans as the god of boundaries. Some believe that this god evolved from the ancient Greek goddess Terminus. Today, surveyors, real estate attorneys, and judges who must make legal determination on land matters should consult the wisdom of this ancient god(dess). There are numerous references in the Old and New Testaments concerning boundary stones, markers, landmarks, and boundaries. Ovid, the Roman poet, wrote: "O Terminus, whether thou art a stone or a stump buried in the field, thou hast been deified from days of yore … thou dost set bounds to peoples and cities and vast kingdoms; without thee every field would be a root of wrangling. Thou courtest no favour, thou art bribed by no gold; the lands entrusted to thee thou dost guard in loyal good faith."[3]

To show faith in such a god and with the hopes that a favorable response from the god would bring peace to a community and stability to its boundaries, a festival called Terminalia was held on February 23. During this annual festival, landowners would meet at their common boundary stones. Each would place a garland of flowers, and the ceremony would culminate with a minor feast of cakes and honey and toasting with wine. Then an animal, usually a pig or a lamb, would be sacrificed and the bones and blood deposited near the site.

Titus Livy wrote in his *History of Rome* that the Romans showed such favor to Terminus that at Rome's founding a temple was erected to the god on one of the seven hills, and his domain was never questioned. To show that all of the gods of Rome looked to Terminus, Livy wrote: "The gods are said to have exerted their power to show the magnitude of this mighty empire.... The fact that the seat of Terminus was not moved, and that of all the gods he alone was not called away from that place consecrated to him, meant that the whole kingdom would be firm and steadfast."[4]

1.5 DISPUTES AND BOUNDARIES

Principle 2. *A surveyor creates land boundaries. These created lines, which are separate and distinct from property lines, are determined by legal principles and law.*

Disputes as to boundary location and/or boundary line identification predate recorded history. Until the development of modern maps at scales that permit adequate and positive identification of boundaries, individuals and communities depended on the spoken word to "seal" the location of boundaries and possession to maintain them. One historical method of identifying boundaries is beating the bounds. This practice was possibly a vestigial reminder of what had been a quasi-religious practice first used to identify parish boundaries between religious orders (see Figure 1.3). Today, this method is still referred to as "beating the bounds" and is still practiced on a very limited scale in a few states.

Figure 1.3 Beating the bounds of Edgmond Parish, Shropshire, about 1933. (Source: Courtesy of Local Studies Department, Shropshire County Library.)

Disputes over boundaries were frequent between communities and between church lands. The ancient ritual of beating the bounds was usually carried out during Rogation Week, the period between the fifth Sunday after Easter and Ascension Sunday. On the day selected, the parson, the constable of the townships, and the steward of the court (clerk) of the manors, accompanied by townspeople both young and old, would take ample supplies of food and drink and perambulate the boundaries to be identified. In this manner, they sealed in the memories of the townspeople boundaries that had never been committed to writing or placed on a map. To enhance the memory, young boys were selected and given an unforgettable experience at each of the beacons (corner monuments). Trials over disputed boundaries and depositions in many shire (county) courts have left us with excellent accounts of some of the rituals that helped the young people remember the disputed boundaries. Some are related as follows.

In 1687, an elderly man, William Gregory, testified that, as a child of seven in 1601, in a boundary dispute over a line in Exmore (Somerset), he had assisted in a perambulation of Exford Parish. As the group passed one of the boundary stones, one of the older gentlemen called to the boy, "William, put your finger on the meerestone, for it is soe hot it would scald him." William related "that in doing so he layd hold on my hand and did wring one of my fingers sorely so that for the present it did greive me very much." William then recalled that the person stated: "Remember that this is a boundary stone and it is a boundary to the parish of Exford."

Not to be outdone, in 1635, Robert Fidler testified in the matter of a boundary dispute that as a boy he "had his eares pulled and was set on his head upon a mearestone neere to a newe ditch of Ormisirke Moore and had his head knocked to the said stone to the end to make him better remember that the same stone was a boundary stone."

The ritual of perambulation, or beating the bounds, can still be found in some communities. Although antiquated, it nevertheless incorporates sound legal purposes and principles. During colonial times, it was required that owners of adjoining land walk and inspect their common boundaries yearly. The law also stipulated a penalty for those who failed to comply. More recently, in New England, municipalities are required to inspect and renew their bounds periodically. This remains the law in several states, although most do not carry out the "letter of the law." Some towns still undertake this job faithfully by surveying and marking their boundaries, and with new technology they are placing coordinates and global positioning system (GPS) values on monuments and corners. This ancient practice ensures landowners and others of the "true and correct" bounds and helps relieve surveyors of possible future surveying costs that are necessary to determine such lines when they are coincident with private boundaries. Yet disputes still arise when surveyors apply modern technology to ancient boundary descriptions. The practice of beating the bounds was replaced in 1677 by the enactment of the *Statute to Prevent Fraud* (*Statute of Frauds*) that required a written document be presented in order to have a cause of action for certain legal problems and to transfer an interest in land.

An examination of many early English maps and names reveals that some disputes were centuries old when William the Conqueror arrived to turn the Anglo-Saxon world into turmoil. Here is a selection of some of the names on present-day maps in the United Kingdom:

- calenge (Middle English): challenge, dispute
- ceast (Old English): strife, contention
- erioch (Gaelic): boundary
- devise (Old French): division, boundary
- flit (Old English): strife, dispute
- fyn (Welsh): end, boundary
- grima (Old Norse): marker boundary blaze on a tree
- ra' (Old Norse): landmark boundary, settlement on a boundary
- skial (Old Danish): boundary, boundary creek
- terfyn (Welsh): boundary
- threap (Old English): dispute

Few of these names were adopted in the United States or carried into American English and indicated on our maps when English common law was accepted, but we have developed our own words to describe the problems that result from boundary disputes.

The historical result of this ancient practice is that today in England the number of land surveyors who practice land boundary surveys to settle boundary disputes is probably less than 100 for the entire country.

1.6 ROLE OF THE SURVEYOR IN BOUNDARIES

Principle 3. *A described closed boundary identifies a claim of right to any property interest for which any person can make a claim of possession through a claim of title. These boundaries may be either macro or micro in nature.*

The surveyor should be able to make a distinction between the types or classes of boundaries that may be encountered. There are boundaries—and then there are boundaries. One will find both macro and micro boundaries. A *macro boundary* may be an international boundary between nations or between subdivisions of nations; a micro boundary is a boundary on a local level, such as between land grants or, possibly, between individual parcels of land. Surveyors can become involved with boundaries in two separate and distinct ways: those that represent major proportions or areas, which are macro boundaries, and others that are smaller and parochial in nature, which are micro boundaries (see Section 3.2). Few surveyors have the opportunity to create or retrace macro boundaries, but most surveyors are intimately involved with micro boundaries. Few retracing surveyors are asked to create or retrace international boundaries, state boundaries, or country boundaries that may be disputed; however, many of the boundaries they create or retrace will be of small parcels of a single lot or subdivision. A list of possible distinctions between macro and micro boundaries is provided in Section 3.2.

Although the methodology of creating macro and micro boundaries may be similar, the application of law may be entirely different when retracing these boundaries. The surveyor creates these invisible boundaries, which are a product of the work of the instruments used, the capabilities of the surveyor, and the methods employed in conducting the work.

Usually, the original surveyor creates the boundaries of land parcels through actions and/or words and according to the law. Once an original boundary is created and described, that description remains in effect forever, legally. According to federal statutes as well as common/case law, those lines remain fixed in perpetuity, from the time the first property rights are conveyed in reliance on the lines and corners described. Subsequently, the same surveyor or another surveyor or surveyors are the individuals who retrace the boundaries that were originally created and who may create new evidence for future surveyors to search for.

The first belief that any surveyor should have when entering the area of boundaries is that any boundary dispute can be resolved with the help of knowledgeable experts and with reasonable people as clients. The only problem one may encounter is that some disputes may take longer to resolve than others. One person stated that it required the death of the original parties to resolve a boundary dispute. Some disputes may be prolonged for generations, even to the point that they are identified on maps and become sealed in history. At that point, the origins of disputes become lost in history. One of the longest historical disputes in America was the Hatfield and McCoy feud, which crossed the state line between West Virginia and Ohio. What started as a simple timber trespass by Hatfields on McCoy lands, over the state lines, escalated to killing pigs and then scores of people in the two clans. The dispute lasted for over 100 years.

In examining British Ordnance Survey maps, one sees names such as Threapwood and Threapmuir. Threapwood is found in Wales near Wrexham, a tract of disputed land that belonged to no county, parish, or township. The residents were found to be paying no taxes and were subject to no local courts. It was a true no-man's-land. The boundaries had been disputed for centuries, and no county had ever gained authority over the people and the land. Similar situations exist in all US states, in both public land surveys and state-surveyed areas. As recently as 1994, surveyors in Louisiana discovered a "lost" strip of land between two federal townships that resulted in many legal problems for many occupants who had been residing and transferring the land without benefit of an original patent; in 2002, two parishes were still disputing their common boundary.

Many other macro boundaries are being disputed between countries and counties. Today, the states of Connecticut and Rhode Island are disputing a common boundary that was created in the 1700s. Recently, what should have been a simple described line between the states of Georgia and South Carolina was settled by the US Supreme Court when it determined that the original definitive boundary—namely, the center of the river, with all of the islands belonging to Georgia—had been changed by estoppel. The most recent boundary problem faced by Georgia is the north boundary with the state of Tennessee. The original charter calls for the "parallel 35 degrees." In 1811, the creating surveyor told the commissioners that his equipment could obtain a precision of only plus or minus 1 mi. The line was run and monumented. Recent precise measurements place the run boundary 1 mi south of the parallel. Georgia, facing a serious water shortage, now wants to change the boundary 1 mile north so that it hits the Tennessee River, which would give the state the ability to extract water from the river.

In a major boundary problem between England and Scotland, the dispute over the Threpelands was settled in 1552 by digging a ditch and giving half to each of the disputing parties. The ditch, called Scots Dyke, is still in existence. Here in the United States we do not have that flexibility.

The British left us with a legacy of boundary disputes but also with one of attempting to make permanent those important markers that identify land boundaries. A reference to today's Ordnance Survey maps will indicate such boundary landmarks as the Navelin Stone, which was established in 1200. This stone, also called the Avellan Stone, is identified in the charter established in 1210, depicting the boundaries of Cumberland in England (see Figures 1.4 and 1.5).

In other early attempts to resolve boundary disputes by legislative and legal means, the English tried to rely on boundaries identified by the centerlines of roads. To aid in maintenance and care, local governments were given authority to modify boundaries and give each governmental unit half the length of the road and responsibility for its entire care. The philosophy adopted was that all stones and markers placed along boundaries give tangible substance to those boundaries. In many instances, when the boundary stones were erected, proper names were given. Today, one can find such

Figure 1.4 The Navelin Stone, marking the boundary between Brisco and Cleator, Cumberland, established in 1200. (Source: Courtesy of Prof. Angus Winchester.)

Figure 1.5 The Navelin Stone depicted on a boundary map drawn in 1750, 550 years later. (Source: Courtesy of Prof. Angus Winchester.)

names as Kingstone, Earlestone, Sir Steven's Stone, Sargeant's Stone, and Attorney's Stone, all recording a long-forgotten history.

It must be remembered that even though boundary stones were very important, they did not eliminate other forms of boundaries, including natural boundaries. Such boundaries, or natural objects, are discussed to some extent in Section 4.19 because many judges, attorneys, and surveyors misunderstand their significance as controlling elements in boundaries.

Clients should expect surveyors to be expert measurers and collectors of data and evidence of boundaries. The work is not necessarily limited to land boundaries but could include boundaries above and below the surface of the Earth. In the event of a dispute, the surveyor's purpose becomes that of presenting these measurements, and the evidence recovered, to the court and jury for their deliberation and consideration. Hence, their skills and knowledge of the science of these measurements should be positive and should never be deficient.

Surveyors are the nexus of boundary issues in that they create them, they may describe them, and then they may be asked to retrace them or relocate them. The time frame between any two of these phases may be hundreds of years.

Surveyors create evidence and describe the evidence created in their own words; then they recover the evidence so as to have juries interpret it; and finally, to have courts apply the proper laws of evidence, they apply meaning to, and determine the intent of, legal documents that land surveyors and attorneys use to describe and locate land and boundaries of rights and interests, which we describe generally as *landownership*.

In this book, it is assumed that surveyors possess the mechanical measurement skills that are necessary and essential to create and locate boundaries correctly. Yet in today's modern technological world, new areas are evolving with which the student must become familiar: for example, geographic information systems (GISs), GPSs, and many other areas of pseudo-measurements that some wish to substitute for measurements. To understand boundaries fully, the student must first understand that measurements, actions, and words are the foundation for boundaries.

Measurements that create boundaries, measurements that are used as evidence of boundaries, and the words used to describe boundaries are all important elements and become controlling elements for the surveyor. The person who specializes in boundaries should realize that a dual responsibility is placed on surveyors.

First, a boundary between two individuals (estates) could not exist without being created. The boundary created not only can describe a parcel of land but can also be used to describe multiple interests within a boundary of the same parcel of land. Second, the boundary created must be relocated and identified at some time. In this phase, the surveyor will be required to take the description and, using those words, locate the parcel on the ground. This may require the surveyor to disagree with his or her peers as to what the words actually mean or what the evidence indicates. It is in this phase that disputes seem to arise, for no two persons see evidence in exactly the same light.

Unlike in other countries, surveyors in the United States do not have the authority to locate legal boundaries that are binding on all the parties involved. Their responsibilities lie in the area of interpreting legal descriptions and then placing these descriptions on the ground by conducting surveys to recover evidence of prior work or surveys. In addition to locating these title boundaries, surveyors may be called on to:

1. Locate the limits of possession.
2. Locate the limits of the claim of ownership, either under color of title or not under color of title.
3. Locate improvements on property.
4. Locate and describe rights and interests in land.

1.7 WHAT IS BEING CREATED? WHAT IS BEING LOCATED?

Principle 4.   *A person or landowner can legally convey only the quality and quantity of interest in land to which he or she has title.*

The surveyor who creates a boundary line creates nothing but an invisible line that is only described by numbers, yet no line can exist without its endpoints, the corners; in order to be a definite line, there must be a corner at each end of the boundary line. This principle was first identified by William Leybourn in his historic survey book, *The Compleat Surveyor*, in which he described a line as being "created by moving out of a point from one place to another so a line is thereby created, whether straight or crooked. And of the three kinds of Magnitudes in Geometry, viz. Length, Breadth, and Thickness, a Line is the first, consisting of Length only, and therefore the Line A B, is capable of division in Length only . . . .[5]

The surveyor who creates boundaries locates or creates nothing more than invisible lines and corner points that exist as legal fictions between property rights. Boundary lines without any other support have only legal dimensions; they have no physical dimensions until fences are constructed on the invisible lines; these fences are called for and identified in documents that are in the chain of title, or trees are marked and identified in field notes as line tree or tree on line and then that tree or trees become a reference indicating what it was that the creating surveyor intended. A boundary exists because the law permits it to exist, yet one cannot feel it, touch it, or see it; it is not manifested in any way by a dimension of width, only length. However, once it has been created, it has legal authority. One neighbor cannot cross over a neighbor's invisible boundary without being in trespass, and possibly being responsible for damages.

Regardless of the position of the surveyor, the responsibility that is assumed is that of creating or identifying rights and interests in land. Rights and ownership are related and are often confused, but they are not the same. The ownership of a land parcel carries with it responsibilities and liabilities, whereas rights will give a person, whether or not a landowner, certain legal rights that can be addressed in the courts.

Usually, to have a boundary created, that boundary must have terminal points, or corners. Each boundary line is controlled on each end by a corner, which may or may not be monumented. But in the event that the controlling corners are unmonumented, those corners have the same legal dignity as monumented corners.

In most instances, once a boundary is created, there is no need to resurvey that originally created boundary as long as the land is in the original ownership. A resurvey is required only when the originally created evidence becomes so questionable that a retracing surveyor is unable to find it or when a transfer of the property is contemplated. Boundaries are usually retraced in the event that the parcel is sold or a dispute arises when an adjoining boundary line is ascertained. As such, a new survey will identify the existing conditions of the boundary lines at the time of the recent conveyance but written in terms of the original description. Such a survey should identify the condition of the original corner monuments and should also redefine the definition of the courses (bearings and distances) in more modern terms. This is defined as a *retracement*. The practicing surveyor must be able to make the distinction between original surveys and retracements—this skill is absolutely required of the modern land surveyor.

In the process of attempting to understand boundaries and the various aspects of boundaries, the original boundaries of any parcel are created by the original survey; once an original boundary is created, it can never be re-created once rights are granted to the boundary. From that point on, unless the original creator conducts either a dependent or an independent resurvey, subsequent surveyors can only conduct retracements. Title and boundaries are symbiotic.

Types of surveys are generally categorized and defined as follows:

Original Survey This category occupies a special place in the law; since defined in 1809, the requirement in a retracement has been to "follow the footsteps" of the original surveyor, stated as "existing lines are to govern."[6] It cannot be overstressed how important this concept and requirement are, since the original surveyor created the original boundaries that defined the title to the property.[7] This rule has been followed in all courts and jurisdictions when it applies.

Resurvey A resurvey is the reestablishment or restoration of land boundaries and subdivisions by re-running and re-marking the lines represented in the original field notes and on the plat of an official survey.

A Dependent Resurvey is a retracement and reestablishment of the lines of the original survey or of a prior resurvey in their true original positions according to the best available evidence of the positions of the original corners.[8]

An Independent Resurvey is a retracement and reestablishment in reliance on evidence of the original survey in order to give official recognition and respect to all alienated lands within its scope, and where applicable, it also includes the establishment of new section lines and often new township lines, independent of and without reference to the corners of the original survey.[9]

A Retracement Survey is a survey that is made to ascertain the direction and length of lines, and to identify the monuments, and other marks of an established prior survey.

A First Survey is a survey made to locate, for the first time, the directions, distances, and monuments of a line or lines previously created without the benefit of a survey. In other words, title has previously been created and established, but the property has not been located or measured on the ground.

An Accurate Survey is a map reflecting the course and distance measurements, boundaries, and contents of a territory for a specified purpose. The map must reach the desired level of precision consistent with the purposes of the survey.[10]

1.8 ORIGINAL WRITTEN TITLE

The concept of title to land is unique to English law. The migration of Europeans to the New World resulted in a basic conflict with Native American beliefs and practice.

The European concept is primarily an English concept of landownership and its use. Native Americans had no known concept of written title, and at times uses by different tribes overlapped. Although tribes did recognize areas of specific claim or use, they held the belief that no individual or individuals could own land. Individuals only had the right to *use* land, and land was composed of certain rights of usage.

The English brought with them the concept of written title. Title as we know it was unknown to Native Americans. Possession was paramount. Today, we assume that most boundaries are defined in some sort of title document: a deed or a will, for example. Yet this is not entirely true. The law provides for and permits boundaries by several other means. These are discussed later in this book.

A very simple explanation of *land title*, which is separate and distinct from a land description, holds that it usually is a written document or legal instrument by which one can claim ownership to a separate and distinct identifiable parcel of land or property.

As the courts have recognized it, title may be considered as originating from varied sources, including: (1) conquest, (2) royal grants from a foreign power, (3) grants of original crown lands from one of the original states or from another state, (4) grants or patents from the US government from land considered originally as being in the public domain, and (5) lands in the form of newly created lands.

To claim lawful possession, one must have a claim of ownership, which usually is exhibited as a document purporting to give ownership. Regardless of how title to a person's property originated, potential problems might be uncovered by the surveyor that could cause problems in the location or relocation of boundaries. Several of the 13 original states not only granted lands within their original boundaries but also granted, with or without authority, lands outside these boundaries, usually under the terms of their original grants as they were and are interpreted. This situation happened between Tennessee and North Carolina; Virginia and West Virginia; Connecticut and Massachusetts; Virginia and Ohio; and New Hampshire, New York, Georgia, South Carolina, and Vermont.

Today, in any particular situation, the surveyor or attorney must consider whether the question is one of title—who owns it and how much—or one of boundary—what and where the boundaries are. This permits a court to determine what a person owns or who has better title to a parcel of land, even if it is considered as being unsurveyable or unlocatable.

1.9 RIGHTS AND INTERESTS IN LAND ARE COMPOSED OF A BUNDLE OF RIGHTS

Rights and ownership are related but are not the same. When a person owns a parcel, usually that person has the right to timber, water, minerals, and possession. Each right may be described, identified, and conveyed, and the owner may convey all of the rights, yet retain the right to pay taxes.

The original surveyor creates the boundaries between individual rights, and, once these rights are created and the original owner relies on these boundaries, no one, other than those who are beneficiaries, can change these boundaries. To understand what is being created, the surveyor, the attorney, the real estate agent, and especially the courts must understand the distinction between title, property, and rights or interest.

Title is the means or vehicle, usually documents, by which one acquires an estate. Rights, such as the right to take minerals, are attributes that a person may hold by being a landowner. A person holding a lien on land may have an interest but not a title, depending on the respective state where the property is situated. Interest and title are not synonymous.

Property may be considered as corporeal, meaning some right that describes a tangible element: a house, trees, a fence. The primary function of the surveyor has been in regard to rights, and interest in land has been to define these elements. The title to property is the exclusive domain of attorneys. However, as surveyors' responsibilities, identities, and capabilities have been changed and redefined by the courts, in many jurisdictions, the surveyor is no longer prohibited from giving an opinion about who holds the title to a piece of real property.

> The surveyor must be able to understand title but should refrain from giving opinions as to title issues.

Whatever rights or interests a person may have in land today are controlled and regulated by the laws of the state in which the land is located. The federal government has control over the public domain, Native American lands, lands involved in bankruptcy, state boundaries, navigation, lands seaward of state boundaries, and air rights crossing state boundaries (violations of air quality). Although no "law school" list of specific rights that attach to a person's land exists, certain "common-law" rights are recognized:

1. The right to dispose of property, not inconsistent with the law.
2. The right to have land free from interference.
3. The right to support of property, both subjacent and lateral.
4. The right to use waters that flow through or on property.
5. The right to any waters that flow through or touch property.
6. The right to all space above and below surface boundary lines.
7. The right to possess the property.
8. The right to convey or gift the property to second parties.
9. The right to use legal methods to protect these rights.

As stated, land is composed of assorted rights, both corporeal and incorporeal, which are held together by ownership or title. These rights were identified very early in English law. Jurists first identified the bundle of rights as extending from the center of the Earth to the accolades in the heavens. Some of these rights included, but were not limited to, timber rights; air rights; mineral rights; the right of possession; water rights; and the right of ingress, egress, and regress. These rights can become confusing and conflicting, as in Alaska, where Congress granted surface rights to one group of individuals and subsurface rights to another group.

Each right may be independent of the exterior boundaries of the parcel and may have boundary lines that are separate, distinct, and independent of the exterior boundaries of the parent parcel. In this situation, a landowner may ultimately convey numerous rights, and the surveyor may ultimately survey numerous boundaries within the parent parcel. This bundle of rights has been described as and likened to numerous straws, held together by the belt of landownership, each right with its boundary identified by its name and having its own separate description, boundaries, and rights to the holder. According to this concept, separate owners can each hold one or more of the straws in the bundle, and, as such, each owner would have a vested interest in the entire parcel. A surveyor may be asked to determine the source of a nuisance that violates a right, or he or she may be the cause of a nuisance in the form of trespass by the surveyor or his or her assistants.

Basic US real property law has its foundation in English land law, originating in feudalism. After CE 1066, all lands under English rule were considered as if owned in total by the reigning king, William I. William made conditional land grants to his followers, to specific English barons, and to certain individuals who submitted to his control. The grantees became holders of the land, or tenants. Because all grants were made in return for services by the tenant, the terms of the holdings under which each tenant held were free or unfree tenure.

Free tenure was divided into (1) military or knight service, wherein each knight was required to give a number of days each year in defense of the king; (2) spiritual or Frankalmoigne tenure, which required prayers or spiritual duties for the king; (3) socage tenure, which consisted of nonmilitary duties such as providing the king with crops or cattle; and (4) serjeanty, wherein personal services were provided to the king. Once each person provided the identified and specific tenure, his or her remaining time was his or her own, and the person could use the property freely in any manner so chosen. These four forms of tenure later became known as freeholds or freehold estates.

Serfs held their land under unfree tenure, in that they were bound to the land and could not use any of it for their own purposes. The main point in the feudal system is that the person who held tenure was also in possession or seisin of the property. Upon the collapse of the feudal system, only socage remained, from which we retain freehold estates.

Under early common law, the people who possessed seisin in reality owned a collection of rights incident to the land. The tenant holder's rights became known as his estate. Estates differed primarily in the length of time for which they might exist. In modern legal and technical senses, an estate is the degree, quality, extent, and nature of the interest that a person has in real property. Some find the term estate confusing in that it may define the corpus ("body"; either real or personal property or both), as in "all of my estate," or the rem ("thing"), as in "fee simple estate." An estate may be either absolute or conditional. In most instances, *fee simple* connotes absolute. Today, all estates are classified as freehold or non-freehold (less than freehold). Freehold estates are divided into fee simple, fee tail, and life estates.

A fee simple estate is the highest and greatest estate in land that one can obtain. Those who possess a fee simple or fee simple absolute estate are, for all purposes, the owners of the land. The words fee simple absolute in reality are not a single term; each distinct word carries a meaning that explains the entire term. Fee denotes that the estate is one that can be inherited or devised by a will or other documents. Simple denotes that the estate is not a fee tail estate, wherein the estate must be inherited by a specific person. Absolute means that there are no conditions or limitations so far as time is concerned on the estate, and this estate may continue forever (not like a fee or an estate that may be determinable upon the happening of an event), either precedent or subsequent.

Fee tail or estate tail is an early English type of estate, which in all probability was borrowed from the Romans. It is a true freehold estate limited by the grantor to the heirs of the grantee's body or to a special class of people, either male or female (e.g., the eldest, the youngest, or other). If the conditions are breached (no male or female heirs are produced by the grantee), the estate reverts to the grantor or his heirs at law or the sovereign. In the United States, individual states have determined that this type of estate, if adopted as part of English common law, that a conditional fee was present, and that upon the birth of a child it was converted to a fee simple estate,[11] or that statutes eliminated the estate and any reference to fee tail; *it then is converted to* connotes fee simple absolute.[12]

A life estate is considered a freehold estate because it can be conveyed to a third party, yet its duration is measured by some life. In essence, the life estate lasts only for the life of some person. An ordinary life estate is normally worded "to Jones for life and then to Brown in fee simple." An estate per autre vie has a measured life other than that of the holder of the estate and may be worded "to Jones for the life of Brown and then to Smith in fee simple." This estate and the right of legal possession terminates with the death of Brown.

Life estates may be created by expressed provisions or words in a will or deed or by contract between heirs or parties in interest. If a question exists as to the creation of a life estate, the courts will look at the precise words used. Terminology such as "to Mary Jones as long as she remains single" has been interpreted as creating both a life estate in Mary Jones and a fee simple estate.[13] A person who receives a gift of "rents and profits for his life"[14] or "the right to occupy the land as long as he may live"[15] has been construed to receive a life estate. A life estate may also be created by an operation of law wherein a surviving spouse is granted *dower* (for a widow) or a *curtesy* (for a widower). The law varies from state to state, and the prevailing statutes must be consulted.

The holder of a life estate has the right of possession, as with all freehold estates, but not all the rights of a fee simple absolute holder. He or she may not commit waste of the estate by reducing the value of the estate or by making unreasonable use of the estate proper (e.g., cutting timber, destruction or removal of minerals or structures, or improper placement of improvements). The holder of the life estate is obligated to pay taxes, to make all necessary repairs, and in some instances to provide the necessary insurance to protect the property.

A life tenant may convey any and all interests possessed but cannot encumber the property beyond the life conveyance terms. All conveyances purporting to convey any part of the remainder estate, without the prior approval of the remainderman, are void.

> A surveyor should never go beyond the words in the document.

In Pennsylvania, some land was sold with a quitrent; that is, the purchaser had to pay the seller a regular quitrent (a payment in money at specified intervals, theoretically, forever). This right was voided by the courts.

Whenever an estate is in question, the court attempts to convey a perfect estate or as large an estate as possible. Although the final decision is a legal one, it is important that surveyors or other people who research land records have knowledge of possible legal implications. In areas where surveyors commonly research legal records needed to conduct a survey, they should be able to recognize wording that could affect the interests of their client. If a researcher neglects to call attention to a possible problem and the client suffers damages, the researcher could be held liable and responsible.

Anything mentioned in a deed is part of the deed. Anything not mentioned, is not part of the deed. A good example is a fence; it may be evidence and ultimately a property line, but if it is not part of the document, it cannot control elements in the deed.

Everything in a deed should be honored if at all possible.[16]

Descriptions are not to identify land, but to furnish the means of identification.

Every deed[17], otherwise valid, will be considered to have intended to convey an estate of some nature.[18] Therefore, every attempt should be made to uphold the deed whenever possible. Where descriptions set forth in deeds are not ambiguous, they must be followed.[19] When, and only when, the meaning of a deed is not clear, or is ambiguous or uncertain, will a court of law or equity resort to established rules of construction to aid in the ascertainment of the grantor's intention by artificial means where such intention cannot otherwise be determined. It is permissible to bring in extrinsic evidence in the case of an ambiguity, so long as it does not add to, subtract from, modify, or otherwise conflict with the language of the deed.[20]

A deed must be read as a whole and every part thereof given effect if possible in order to arrive at the true meaning of the parties, and until each rule has been exhausted, resort should not be made to artificial and arbitrary rules of construction.[21]

1.10   ROLE OF THE COURT

All aspects of real property rights are protected and litigated in the respective state in which the land is situated by the Fifth Amendment of the US Constitution. The Constitution does not address real property rights as such except in the instance of the annexation of property for the public good. Thus, for any real property issues associated with the federal government, the law of the respective states in which the land is located applies, *lex loci*.

Since the titles to land in the United States originated from various foreign sources, and the Constitution recognized all prior valid rights in land, approximately 20 states are legally recognized as metes and bounds states, basing their real property and survey system on English common law, and that law applies in situations regarding land title and boundaries. In states that base their titles and surveys on the Public Land Survey System that originated under several public land laws, those laws apply under which the lands were surveyed and patented.

The role of the court in title and/or boundary questions is much different from that of the surveyor or the attorney. The surveyor's responsibility is to collect evidence of past boundaries described in documents, to collect evidence of possession and use, and to create new evidence to be left for future surveyors to recover. In questions of title or boundaries, the surveyor can then be called on to testify and give opinions to help the court or the jury understand complicated areas. Usually, an expert is not required if the facts are within the capabilities of the jury to understand. Surveyors should not be considered as advocates for a particular client or position.

Attorneys, on the other hand, are the means by which legal questions are presented to the courts. They are advocates, espousing the position of their clients, right or wrong. At times it may seem that surveyors are advocates, but one must differentiate between honest differences of opinion among surveyors and the advocacy of a surveyor who may seem to be an advocate.

> *What* boundaries are is a question of law; *where* boundaries are is a question of fact.

The courts are present to apply the various laws, both statute and common, to the facts presented. If there is a question as to the facts, it is in the province of the jury to decide what facts to believe and to apply. In most states:

In actual practice, the surveyor may encounter numerous attorneys and judges who do not understand this principle and maxim.

In applying this statement, courts will attempt to ascertain the application of common-law doctrines, such as adverse possession, estoppel, and agreement to boundaries, whereas juries will determine which of the two surveyors is to be trusted in testimony and how much weight should be given to any evidence and the resulting facts. Surveyors will ascertain the interpretation of words in a description that is contained in a deed, and the jury will determine which of the two is correct, whereas the courts and the judge will determine whether the deed meets the requirements for legality and sufficiency. A court or legislature cannot bestow this authority on any person or agency.

Because of the court's exclusive right to determine the meaning of words contained in a conveyance that is being questioned and then to determine where that parcel is located according to the description, it is necessary for surveyors to know and understand how courts interpret these meanings and what order of importance to place on them.

1.11 REAL AND PERSONAL PROPERTY

In most instances, the surveyor's concern as to the differences between real and personal property (see Figure 1.6) is of minimal interest, but to the client, these differences may be of extreme value. Real property is fixed, immovable, and permanent,

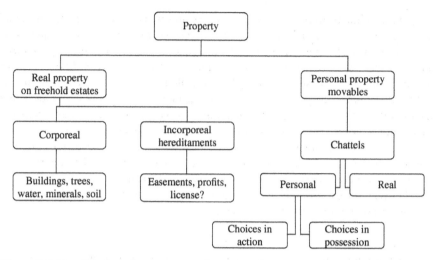

Figure 1.6 Classification of property as real or personal.

whereas personal property is consumable, can be destroyed, or is movable at will. At times, the distinction between the two may not be clear, yet it must be made.

Standing timber is realty; cut logs may or may not be personalty. A surveyor's failure to report cut logs observed in hot or cold decks in the course of a survey may result in liability if the client suffers damages by the surveyor's failure to do so.

Crops are considered personalty, but title may pass with the land. Perennial shrubs are realty, but nursery stock is not. Tree seedlings in heeling beds are personalty, but the exact same seedlings planted are realty. Fence posts in a pile are personalty, but the same posts in the ground may be realty. A surveyor's failure to identify pine seedlings in heeling beds behind a barn resulted in the client being sued for a large sum of money because the seedlings were considered as realty by the lessee of the vendor. The actual determination of classification (real or personal) rests on fact (evidence) and not necessarily on law. Under our judicial system, questions of evidence are determined by a jury. The classification is important because if property is realty it is governed by the Statute of Frauds, whereas property that is personalty comes under the Uniform Commercial Code.

Local laws or state laws determine questions of deed construction, boundaries, sufficiency of descriptions, rights of inheritance, and any other questions dealing with land. Thus, surveyors should be knowledgeable about these areas.

1.12 WHAT CONSTITUTES REAL PROPERTY

Principle 5. *A person or landowner can legally convey only the quality and quantity of interest to which he or she has title.*

Title to land does not constitute real property. Title in real property law is the right[22] or means by which one can claim just or legal possession to a parcel of land.

In the true sense, to determine boundaries, the surveyor generally uses the same documents as an attorney uses to inspect and comment on title.

Land is the solid material of the earth (soil, rock, clay, sand, lava, minerals, and so on). Real property has become synonymous with land. In its broadest description, real property is four-dimensional. The average surveyor, attorney, and court visualize real property or land in only two dimensions, length and width. Yet visionary surveyors are able to see that land or real property is also composed of depth (including height) and time. The third element, depth, in addition to surface rights, has been interpreted by the courts to include subsurface rights in minerals, waters, and passage, as well as aerial rights and the space above the land itself. English common law and the courts in the United States also recognize a fourth dimension: time. Time can describe or indicate the duration of the legal rights in real property that the vendor has to convey. Since ancient times, land has been considered as immovable or fixed in position. Hence, the courts and the resulting case law treated land as unique and considered the location of each parcel as definitive and distinguishable.

The common-law rule wherein the owner of the surface fee owned everything above the surface has been modified by case law, primarily because of air navigation. Today, the general rule is that the owner of the fee owns whatever airspace he or she can control or use, but the theories and case law on the use and control of airspace are not uniform from state to state. An action in trespass lies against anyone who violates a person's airspace, surface, or subsurface. Of course, this is in theory only and is not practical. Technically, airwaves, sound waves, air flights, or other violation of airspace is trespass, and under common law a surveyor would be prohibited from using electronic distance equipment across a person's property. It is doubtful that this concept could be enforced in court today.

With the advent of high-rise condominiums, parties buy and use rights to airspace within rooms. At times, surveyors have been asked to define, locate, and describe certain air rights in relation to signs, elevated walkways, and scenic or visual parameters.

In ancient law, underground waters were considered as part of the soil itself; thus, ownership was in the name of the surface owner.

Some states have legislatively separated water rights from the bundle of rights and treat these rights in a different manner, to the point that when a fee simple deed is drafted, water rights are not included in the conveyance and a separate deed is required to convey water.

In recent years, several states have modified this concept because they now consider waters as migratory in nature and as being owned in common; that is, the surface owner owns the waters in common with other landowners and has the right to use a portion of the waters for his or her benefit. Government may regulate the rate of pumping according to prior rights granted and for the common good. The courts may hold a party responsible for a decrease in water quality, as well as for detrimental damages to adjacent property caused by altering the flow or course of the runoff. In these instances, the surveyor may be called on to map or to identify water courses or to prepare topographic maps that will identify certain aspects of water.

Mineral ownership includes ores, coal, gravel, sand, oil, and gas. Even though oil and gas are classified as minerals, they are migratory in nature and are incapable of absolute ownership as a thing in place. The surface owner has had an exclusive right to drill, produce, and retain all oil and gas brought to the surface. In recent years, the tendency has been to change the law so as to require unit development of oil fields.

With respect to the ownership of lands adjoining tidewaters, lakes, or rivers, riparian rights attach to the land. However, the nature of riparian rights is still being determined by our courts. This complex subject is reviewed in Chapter 9—Riparian and Littoral Boundaries.

Vines, trees, crops, shrubs, and other growth (fructus naturales) are considered for most purposes as part of the land to which the roots are attached, yet there may be certain circumstances in which they are considered as personal property. A failure by the property surveyor or topographic surveyor to identify clearly the condition or state of certain fructus naturales may result in liability upon him or her. As a general rule, unless crops are reserved in writing, they pass with the sale of the land, and usually the requirements are clearly described in the Statute of Frauds. We do not discuss fixtures on property in this book because their treatment varies from state to state.

Principle 6.   *In most instances, there are no federal laws describing real property rights.*

One of the basic aspects of the US Constitution is that any areas not reserved by the federal government are delegated to the respective states. Since, in the formation of the Constitution, real property rights were not retained by the Framers, by default the respective states had virtual authority over individual and real property rights.

Thus, when any agency of the federal government either sues or is sued over aspects of real property, the law of the location of the property, or lex loci, is the respective law to apply.

The US Constitution is a creation of reserved rights and not delegated rights. What the Framers decided to delegate to the states they failed to mention in the Constitution.

Principle 7.   *Although there are no federal laws of real property, property rights are identified by state laws and are protected by the US Constitution.*

When decisions relating to real property rights are litigated, even with the federal government as a party, state laws control the litigation. This may not be so if a boundary problem is at issue. Control over real property falling within all state borders were not retained by the federal government. As states entered the union the fourteenth and the fifth Amendments became controlling.

Principle 8.   *Real property rights are determined according to the laws in effect in the particular locale where the land is located. English common law is the predominant law, and it is described as the lex loci.*

Although there is no federal law of real property, there exists federal survey law that is applicable to those lands that originated from the General Land Office (GLO) system of surveys. The first federal law enacted, still in effect today, was the Land Act of 1785. It was this act that created the entire GLO survey system. The act was modified and supplemented with subsequent federal laws, which are still in effect.

Although few states have enacted statutes to direct and control state surveys, most states apply common-law principles to the location of boundaries. Although several states, including Georgia and Texas, enacted surveying statutes to control surveys of their lands, most of the states in this category have relied on common law. In the modern era, many states have enacted statutes to control the creation of boundaries for subdivisions and other surveys of large parcels.

1.13 NATURE OF MODERN ESTATES

Although most surveyors are not involved in having to determine the effect of an estate on the survey, the modern surveyor should be familiar with what constitutes an estate. This is important in that the modern surveyor usually recovers more and much older documents than does the attorney. The surveyor may uncover documents that may have a great legal effect on the final determination of the case.

By law, an estate is the interest that a person has in real or personal property. The word is sometimes used to mean the property or assets of a person, such as in "the estate of John Doe." In general, real estates are classified by the time of enjoyment, and they are identified as follows: (1) estate in fee, (2) estate for life, (3) estate for years, and (4) estate at will.

An estate in fee, sometimes called an estate in fee simple, is the most absolute interest a person can have in land. It is of indefinite duration and is freely transferable and inheritable. The duration in fee could be considered as being "in perpetuity." More than one fee can be held on a given parcel of land; for example, one person may have the fee to minerals and another the fee to the land, excepting the minerals. A defeasible fee simple estate is one in which a future event must be met. The title is conveyed on the condition that certain things will be done within a time limit or that certain things will never be done. A fee may pass on the condition that a storm drain is installed or that the property is never used to sell alcoholic beverages. Restrictions of various natures have been placed, even to the point of prohibiting the use of property for "immoral" purposes.

An estate for life is an estate limited to the life of the person or persons holding it.

An estate for years is usually created by a lease between two parties whose relationship is that of landlord and tenant, such as a lease to use a parcel for 10 years, conditioned on payment of a given amount of money or other consideration.

An estate at will may be terminated at any time as described by law or contract.

1.14 TAXES ON LAND AND TAX MAPS

One of the obligations and responsibilities of landownership is the requirement to pay taxes on any property titled in the landowner's name. A landowner will find that there are numerous governmental agencies to which taxes will be owed. Taxes for the operation of governmental services, taxes for police, taxes for waste disposal, school taxes, and even income taxes can affect the ownership of property.

Usually, land parcels are located on tax maps and identified by a land parcel identifier that shows a *jigsaw* compilation of land parcels that are identified by the names of the record holders of the individual parcels. In most instances, little or no effort is given to accurately locate the individual parcels in relation to each other. Courts have held that tax maps cannot be used as evidence to ascertain boundaries; they can only be used to identify who is paying the taxes on the parcel.[23]

One of the police powers enjoyed by governmental bodies is the right to appropriate any lands for the nonpayment of taxes. If and when this is done to a parcel of land, different legal principles may apply concerning property rights and encumbrance on that parcel of land.

Some surveyors are prone to use tax maps to identify land boundaries. Courts are hesitant to permit this use because this type of map usually is not sealed by a surveyor and does not meet minimum standards of map preparation. They will permit the limited use to identify to whom the tax bill was sent and who paid the taxes. Courts have held that a tax map is admissible for limited purposes. The Vermont case of *Bull* v. *Pinkham Engineering Assocs. Inc.* states that "tax maps are not intended to be used in establishing boundaries on any survey, and it would not be in accordance with professional surveying standards to do so."[24]

The New Jersey court had this to say about maps: "[Z]oning involves more than looking at a map and drawing what appears to be logical or natural boundaries. It is more than an exercise in plane geometry. The mere observation of a plat map, tax map, or zoning map will lead to misleading abstract deduction and error."[25]

1.15 EASEMENTS AND LICENSES

Although of a legal nature, servitudes, restrictions, covenants, and conditions coincident with landownership should be understood by the practicing surveyor. At times, the surveyor may be asked to create such elements as a result of the work, or the surveyor may be asked to ascertain the extent, location, or possible effects that such elements may have on a parcel that is being surveyed.

When creating or relocating boundaries of servitudes, of which easements are but one area, a surveyor will be working in three dimensions. An easement may be on the surface of the ground or it may be located in the air as well as beneath the surface of the land, which may include waters. A company may require an easement or a right-of-way for an underground pipeline. In relation to any single line (e.g., a rapid transit line), a portion may be located on the surface, another underground, and, finally, some part may be elevated. These all require boundaries, which may be transition boundaries going from subsurface to surface to above ground. A surveyor locating mineral rights may find that the minerals located are thousands of feet below the surface of the Earth. The simple construction of a building or the building of a wall on or near the boundary line may require both an accurate and a precise survey, described adequately and legally and monumented sufficiently.

Surveyors, attorneys, and the courts often confuse easements and licenses, but they are actually different in many ways. The student should learn to recognize the basic differences, because often the boundary lines become important as to location, creation, identification, and relocation. One of the more important aspects of easements is location, as well as permitted uses and, still more important, the duration.

An easement is one of the many bundles of rights that enjoy a boundary or boundaries. It is a type of interest that one person has in the land of another. The other types of interest—covenants and servitudes—are discussed in Section 1.16. Courts, attorneys, surveyors, and even landowners should recognize that an easement is not a possessory right, it is a nonpossessory right that permits the holder of the easement the right to only the nonpossessory use within the boundaries described.

An affirmative or positive easement permits the possessor of the easement to do some physical act on, under, or over the lands of another party. The land that benefits is called the dominant estate, and the land to which the easement is attached is called the servient estate.

Negative easements are those in which the holder of the dominant estate can prevent the servient estate holder from some use of property. These may include easements for light, air, or scenic value.

An appurtenant easement benefits the dominant estate or its holder and attaches to the parcel of land, not to the holder. For example, an easement acquired by the owner of a landlocked parcel for the purpose of gaining access to a road is appurtenant to the land. The easement passes automatically with the sale of the land, whether or not it is mentioned in the conveyance; it is attached to the land. The converse of an appurtenant easement is an *easement in gross*. Easements in gross usually are to individuals and are for a specific purpose. In most instances, easements in gross are for the period of time the individual has the easement.

An easement may have either precise and definite or indefinite or imprecise boundaries depending on their method of creation.

An easement in gross attaches to a person, not to a particular parcel of land. For example, an easement granted to a railroad for a right-of-way, or to a person for the right to fish or hunt on a parcel of land, or for access and use of a swimming pool on a nearby parcel, are all easements in gross that attach to a person. In one instance, a person gained free airline passage for a lifetime by granting a right-of-way.

Easements are usually limited to the use cited. An easement for ingress and egress may not be enlarged to include surface or underground utilities. Easements may be created by reference in a deed or will, by a separate document, by implication, by necessity, and by prescription. In most states, an easement in writing must meet the same requirements as those for signature and recording as a deed. Courts usually make a narrow interpretation in deciding whether an instrument conveyed a fee title or an easement. Unless expressly identified, courts usually create an easement.

The most common method of creating an easement is through an express conveyance such as a deed or will. Those deeds that describe and convey an easement strictly without deeding a tract of land in fee are known as deeds of easement or easement deeds. The deed must describe correctly the interest conveyed and must comply with all the formalities required for the transfer of an interest in land. Easements may also be created by express reservation or exception in a deed of conveyance. By exception, the grantor is not conveying that which normally would be included and is retaining that which is being excepted.

According to the general rule of law, when the owner of a tract of land conveys part of it to another, the owner is said to grant with it, by implication, all easements that are apparent and obvious and that are reasonably necessary for the fair enjoyment of the land granted. Implied easements are often sought in litigation when there is no apparent right-of-way to a landlocked parcel of land. In theory, it is not possible to landlock a parcel of land through ordinary conveyancing. Implied easements over landlocked parcels are of two types. When an owner sells off a rear portion of a lot, it is implied that he or she must furnish an easement to it. If an owner retains the rear portion of a lot and fails to provide himself or herself an easement to his or her portion, in many but not all states, he or she has no implied easement.

Easements can be created by estoppel. In legal terms, an estoppel is a legal bar raised by the law which precludes a person, because of his or her conduct, from asserting rights that he or she might otherwise have. For example, if Jones, by her conduct, causes Brown to believe that he has an easement and in reliance to that conduct Brown erects improvements on a landlocked parcel, Brown may have an easement by estoppel. Because Jones's conduct or lies led Brown to do something that he would otherwise not have done, Jones is denied the right (estopped) to tell the truth.

If the language of a document does not create a specific easement, the courts will examine the entire transaction or document conveying the parcel and will determine whether an easement was implied at the time of the conveyance. For an implied easement, it is necessary that a common owner of the two parcels be identified, that the common owner retain one parcel, that the use for which the easement is necessary exists, and that the easement is reasonably necessary for enjoyment of the dominant estate.

Easements may be granted because of necessity. The courts have held that if two or more parcels are so situated that an easement over one or more is strictly "necessary" for the enjoyment of the other parcel(s), they will find that an easement exists of necessity. To effect an easement of necessity, the courts examine for a "common grantor" and whether alternative means of ingress or egress are available. Only availability, not cost or convenience, is considered. An easement of necessity exists only for the period of time during which the necessity exists. Once an alternative means is available, the necessity and easement cease to exist.

The requirements for creation of an easement by prescription vary from state to state but are usually the same as those for adverse possession. They are:

1. Adverse use without permission
2. Open and notorious use
3. Continuous use
4. Use for a specific purpose
5. Use for the statutory period

Easements may be terminated or extinguished in a number of ways. As with any other interest in land, an easement may potentially be unlimited in duration or may be created to last for a limited period of time, in which case the easement expires according to its own terms. Easements created for a specific purpose expire when the purpose has been accomplished.

Easements may always be extinguished through a release agreed upon by the parties involved. In addition, easements are terminated by merger. Because a person can never have an easement over his or her own land, if he or she acquires a parcel of land over which he or she has an easement, the easement ceases to exist. Stated in legal terms, when the dominant and servient estates are united in one entity, easements are extinguished and are not revived by a later separation.

Destruction of the servient estate, as by loss of land due to erosion, can terminate an easement. Destruction of a building existing on an easement may also terminate an easement. The holder of an easement may lose or terminate the easement by increased burden or use. When an increased burden inconsistent with the original use is placed on a property, such as placing underground utilities on an easement intended for access, the courts may void the entire easement.

The owner of an easement, under certain circumstances, may terminate the easement by abandonment. Abandonment means more than mere nonuse by the easement owner; it means conduct indicating an intention never again to exercise the right. Although nonuse is some evidence, it must be joined by an affirmative action by the holder of the easement, indicating an intention to forgo the right forever. As easements can be gained by either prescription or the process of estoppel, they can also be lost by the same processes.

In summary, an easement has these characteristics:

1. It is an interest in land (incorporeal hereditaments) and must be created by grant or agreement, expressed or implied.
2. The interest must be in the land of another.
3. The easement is nonpossessory because the owner of the easement can only prevent interference with his or her interest.
4. The privilege to use an easement must be capable of creation (e.g., the right of a ticket holder to view an event cannot be an easement).
5. The easement should be described by definite accurate boundaries and precise surveys.

The perfect easement is unique with respect to boundaries. It has multiple boundaries, with at least three boundaries being identified: boundaries of the dominant parcel, boundaries of the servient parcel, and boundary of the easement(s) itself.

For most land surveys, the surveyor is asked to certify that there are no encroachments on the land and/or to show easements. Liability can result from failure to identify or recognize the use of roads, underground pipelines, power lines, and the like.

A license is a personal, revocable, and usually unassignable permission or authority to do acts on the land of another without possession of an interest in the land.

The main distinguishing feature between an easement and a license is the license's revocability. An oral agreement to permit passage across an adjoiner's land without documents as required by the Statute of Frauds is nothing more than an oral license.

Many licenses never become easements. A ticket to a sporting event and a ticket to park a car are licenses. A license may develop into a contractual agreement between two persons such that if one is denied permission to enter upon the property, he or she may seek legal relief on the basis of breach of contract.

A surveyor or survey crew passing over a person's property with permission has a license. Without permission, the surveyor or crew are trespassers and, as such, assume liabilities above that of a licensee.

1.16 SERVITUDES, RESTRICTIONS, COVENANTS, AND CONDITIONS

A servitude is a restriction or a limited real right over another person's property that entitles the holder of the servitude to certain powers of use and enjoyment or prohibitions of use in relation to that property. In the true legal sense, the freedom of ownership of a parcel of land is restricted by a person other than the owner, who has a direct interest in the property and thus is "served." The condition of the ownership of the restricted property is termed a servitude, and the land itself has a burden placed upon it.[26] Under Roman law the number of servitudes was limited, but under modern law a large number of rights may be placed under servitude.

By law, servitudes may be classified as real (predial) or personal. A real servitude, such as an easement, is a right or rights established in favor of a parcel of land (the dominant tenement) over a second parcel (the servient tenement). A predial servitude is established for the benefit of a particular estate of land and is held for the benefit of the estate and not the individual.[27] As with easements, servitudes may be classified as positive or negative.

Although modern real servitudes may be unlimited in number, public policy restricts or limits the number, for it is against public policy that land be unduly burdened with restrictions. Servitudes have traditionally been classified as either rural or urban. Rural servitudes are concerned primarily with the land (e.g., grazing, rights-of-way, and water) and are usually positive. Urban servitudes are concerned with residential, commercial, or industrial property (e.g., drainage, support, party walls, light and air, sewer, and views) and are usually negative.

A covenant is an agreement between persons or parties that restricts the use of a freehold property. It is enforceable not only between the original parties who were privy to the agreement but by all parties who become assignees of the encumbered land. To be enforceable, a covenant must "touch and concern the land." Courts have held that it would be inequitable to permit a person to purchase a parcel at a price kept low because of a restrictive covenant and then allow him or her to sell it at a much higher price free of the restrictions.[28]

A covenant contained in a conveyance instrument is an agreement usually to restrain from doing certain acts and may take on certain forms:

1. Covenants contained in an agreement between more than one party, usually in the same neighborhood or subdivision.
2. Covenants contained in the deed of a single parcel.
3. Covenants, conditions, or restrictions contained in the deeds on the plat of a subdivision owner, binding on all owners or purchasers of lots within that subdivision.

A developer or owner may convey parcels with specific restrictions recited as he or she so desires or determines are for his or her benefit or that of the subdivision, and such restrictions will be binding on all future owners as long as they are not unlawful (racial or religious) or contrary to established rules of public policy. These restrictions will be binding on all future owners whether they have actual notice or not.

1.17 ACTIONS ON BOUNDARIES AND EASEMENTS

Principle 9.   *Once boundary lines are created, the contiguous lines may, by law or by the actions of landowners who have vested rights, be changed or altered.*

Most legal actions relative to land can be placed into one of the following three categories:

1. A question of title. This is primarily the lawyer's domain.
2. A question of boundary. This should be the surveyor's domain.
3. A combination of 1 and 2. This is the team's domain.

Category 2 can be subdivided into the following:

1. A question of what is the boundary. (This is a legal question.)
2. A question of where is the boundary. (This is a question of fact or survey.)

Boundaries are unique in that the invisible lines are subject to modification or alteration by various methods. In many instances, legal doctrines will legalize actions or inactions by landowners. Although the law is adamant that original lines will control, in an effort to provide some flexibility courts will recognize that under certain circumstances or conditions boundaries, once created, can be changed. Although the names may be different from state to state, the results are the same. One conflicting problem that attorneys fail to recognize is that most of these doctrines do not affect or create title. The major doctrines recognized for boundary modification or changing are:

1. Agreement, written and oral
2. Estoppel
3. Acquiescence

4. Adverse possession

5. Judicial action

These doctrines will be fully discussed in Chapter 12.

1.18   ONE UNIQUE PARCEL OR BOUNDARY

Principle 10.   Law does not provide for two original descriptions of the same parcel.

The responsibility of a land surveyor can be in one of two areas: as the creating surveyor or as the retracing surveyor. The surveyor cannot be both at the same time. Both have their strengths and their weaknesses. The creating surveyor is cloaked with the law, in that whatever boundaries are created have no error (legally).

The reasoning for this is that the two weakest elements of a description, the bearing (course) and the distance, are subject to change; as technology improves, the initial instrumentation of the original surveyor must remain in the controlling elements. The original surveyor created the footsteps for the retracing surveyor to find from the evidence.

On the other hand, the retracing surveyor is given the responsibility for the evidence of the boundaries that was left by the creating surveyor, or as the courts have related to the footsteps. The retracing surveyor only does that when he correctly does so.

Misunderstanding this principle has led to surveyor problems in that it is the surveyor's responsibility to rely on the most current description as well as the original creating description for the boundary research and retracement.

Principle 11.   Multiple boundary descriptions may exist for the same parcel, but only one is controlling.

Land is described as a bundle of rights, where the composite picture of land includes such elements as minerals, soil, timber, and water, with each of these elements capable of severances and sale by the fee holder to separate individuals under separate documents and at different times and in different interests. If the fee holder decides to sell the timber rights to one person and the minerals to a second person, then two separate documents should be created, each with its own individual boundary description.

This does not preclude that all of the interior boundaries are "held together" by the exterior boundary description that is separate and distinct in itself.

These boundaries may exist independently of each other.

1.19   THE ORIGINAL BOUNDARIES ARE SACRED

Principle 12.   There can be only one original boundary survey and description; all subsequent ones are retracements.

Along with the concept that the original surveyor creates the boundaries and that these boundaries have no error legally comes the recognition that the responsibilities of each of the two kinds of surveyors are different and distinct. The creating surveyor must first conduct the fieldwork and prepare the subsequent record, whether it be field notes or plat, to a degree of sufficiency that any subsequent surveyor will have little trouble and few problems in retracing the original work on the ground, using today's technology.

Historically, courts have made the distinction that surveys conducted to describe boundaries of a parcel of land and the accompanying interests have the following attributes:

1. They are without error and defined in the decision *Cragin* v. *Powell*[29] by the US Supreme Court.
2. They are *unassailable* through the courts.
3. They are permanent, even if no evidence is found to identify them.
4. They can be retraced and redefined by retracements, but never changed.

Principle 13.   *A resurvey can be conducted only by the entity who conducted the original survey. The law provides for resurveys of parcels, but only on a limited basis and under certain restrictions, the main one being that the bona fide property rights granted under the previous survey are not jeopardized. Two classes of resurveys are recognized: dependent resurveys and independent resurveys.*

Subsequent to the original survey that created the boundaries, when certain requirements or conditions are met, resurveys may be ordered. Two classes of resurveys are recognized: dependent and independent resurveys.

1.20 CONCLUSIONS

In this chapter, the concepts of land boundaries, landownership, and related topics were discussed. The reader should glean that boundaries are created and that until people erect physical evidence, including monuments as well as fences, trees, fields, etc., on the invisible lines that were created, they exist only by law. These lines describe rights to claims to possession of real property. We also noted that land is a collection of property rights that are freely assignable and that can be divested one at a time, and which can be lost or gained through legal doctrines. With this background, the reader should have gained an appreciation of the complexity of real property that boundaries separate. The practicing surveyor, the student of surveying, the courts, and landowners should recognize that the concept of boundaries is complex, controversial, and confusing.

The retracing surveyor should be versed in the technical aspects of boundaries. Although the surveyors must have an understanding of the laws regarding boundaries, they should realize that the making of any legal conclusions is the responsibility of the attorney.

BIBLIOGRAPHY

Brown, C., W. Robillard, and D. Wilson. *Evidence and Procedures for Boundary Location.* New York: Wiley, 1994.

Cazier, Lola. *Surveys and Surveyors of the Public Domain.* Washington, DC: U.S. Government Printing Office, 1975.

Colvin, Verplank. Annual Reports 1–7, *Topographical Survey of the Adirondack Region of New York.* Albany, NY: Weed Parsons Co., 1873–1879.

Creteau, Paul G. *Maine Real Estate Law.* Portland, ME: Castle Publishing, 1969.

Dean, Darrell, and John McEntyre. *Law and Surveying*, I.S.P.L.S. Surveying Series 3. West Lafayette, IN: Purdue University, 1975.

Hodgman, F. *A Manual for Land Surveying.* Published by Author, Climax, MI, 1903.

Kaplan, Johny, and Jon Waltz. *Gilbert Law Summaries: Evidence.* Gardena, CA: Law Distributors, 1979.

Love, John. *Geodaesia.* Reprint by Walter G. Robillard, Atlanta, GA, 2000.

Northrop, Elliott. *The Law of Real Property.* Boston: Little, Brown, 1919.

Padfield, Colin F. *Law Made Simple.* London: W.H. Allen, 1978.

Robillard, W. G., and L. Bouman. *Clark on Surveying and Boundaries.* Indianapolis, IN: Bobbs-Merrill, 1993.

Skelton, Ray H. *The Legal Elements of Boundaries and Adjacent Properties.* Indianapolis, IN: Bobbs-Merrill, 1930.

Uzes, François D. *Chaining the Land.* Sacramento, CA: Landmark Enterprises, 1977.

White, C. Albert. *A History of the Rectangular Survey System.* Washington, DC: U.S. Government Printing Office, n.d.

Wilson, Donald A. *Handbook for Maine Land Surveyors*, Vol. 11. Orono, ME: University of Maine at Orono, 1974.

NOTES

1. *Sawyer* v. *Grey*, D.C. Wash. (1913). USDCA.
2. Farley Mowat, *Never Cry Wolf* (Back Bay Books, Benica: CA, 2001).
3. William Leybourn, *The Compleat Surveyor* (London, 1657, reprinted).
4. Quoted in "On the Origin of Property: The Myth of Property," *Probate and Real Property Journal*, (Chicago: American Bar Association, 1992), 23.
5. Leybourn, *The Compleat Surveyor.*
6. *Bryan, & c. v. Beckley*, 16 Ky (Litt Sel Cas 91), (1809).
7. For additional discussion, refer to *Rivers v.* Lozeau, Fla. App. 5th Dist., 539 So.2d 1147 (Fla., 1989).
8. Section 5.1, Manual of Surveying Instructions, U.S.D.I., BLM, 2009.

9. Reference *CSX Hotels, Inc. dba the Greenbrier Resort, a West Virginia Corporation, v. City of White Sulphur Springs, West Virginia, et al., Frederick W. Kretzer, et al., v. City of White Sulphur Springs, West Virginia, et. al.* 217 W.Va 238, 617 S.E.2d 785, W.Va. 2005.

10. 5. Reference *CSX Hotels, Inc. dba the Greenbrier Resort, a West Virginia Corporation, v. City of White Sulphur Springs, West Virginia, et al., Frederick W. Kretzer, et al., v. City of White Sulphur Springs, West Virginia, et. al.* 217 W.Va 238, 617 S.E.2d 785, W.Va. 2005.

11. *Antley* v. *Antley*, 132 S.C. 306 (1925).

12. Ibid.

13. Official Code of Georgia Annotated (O.C.G.A.), Sec. 44-6-4 (1992).

14. *Cochran* v. *Groover*, 156 Ga. 323, 118 S.E. 865 (1925).

15. *Armstrong Junior College* v. *Livesey*, 189 Ga. 825, 7 S.E.2d 678 (1940).

16. *Lyford* v. *City of Laconia*, 72 A.1085, 75 N.H. 220 (1909).

17. *City of North Mankato v. Carlstrom, 2 N.W.2d 130, 212 Minn. 32 (1942).*

18. *Penienskice v. Short*, 194 A. 409, 38 Del. 526 (1937).

19. *Gawrylak v. Cowie*, 86 N.W.2d 809, 350 Mich. 679 (1957).

20. *Peacher v. Strauss, 47 Miss. 353 (1872).*

21. *Burchfield v. Hodges*, 197 S.W.2d 815, 29 Tenn. App. 488 (1946).

22. *Allen* v. *Lindsay*, 139 Ga. 648, 77 S.E. 1054 (1913).

23. *Resseau v. Bland*, 268 Ga. 634, 491 S.E.2d 809 (1997).

24. 107 Vt. 450, 752 A.2d 26 (2000).

25. *Bogert* v. *Washington Tp.,* 25 N.J. 57, 135 A.2d 1 (1957).

26. *Horney* v. *Price,* 189 N.C. 820 (1925).

27. *Black's Law Dictionary*, 5th ed., (St. Paul, MN: West Publishing, 1982), 1535.

28. *Frost-Johnson Lumber Co.* v. *Salling's Heirs*, 150 La. 756, 91 So. 207, 245 (1921). *Tulk* v. *Moxhay*, 1 H. & T. 105.

29. *Cragin* v. *Powell*, 128 U.S. 691 (La. 1888).

CHAPTER 2

HOW BOUNDARIES ARE CREATED

2.1 INTRODUCTION

To understand the legal fiction of boundaries, the surveyor must be able to make a distinction between the various elements of which boundaries are composed. In most instances, but not all, a boundary must have a corner at each end of the invisible boundary line. A monument may be placed at each corner point, or the corner may exist by reference to a document. The boundary line to be identified by the surveyor in conducting his or her retracement may be a deed line, a possession line, or a claim of ownership line. The deed line is factual, the possession line is legal, and the claim line is mental. Corners, monuments, boundary lines, and property lines should all mean different things to the surveyor; they should be addressed differently by the surveyor who creates them than by the surveyor who must ultimately retrace them and who may redescribe them in subsequent documents and then probably will have to defend their location in court. Unfortunately, these principles are sometimes misunderstood by students, practicing boundary surveyors, attorneys, and even the courts.

A deed line is factual; a property line is legal.

Neither the student nor the surveyor nor the court should confuse the distinction between the creation of existing boundaries, changing boundaries already in existence, and identifying the location of a boundary that is to be retraced. The changing of boundary lines already created will be discussed later in this book; for the most part, the principles are legal in nature.

Brown's Boundary Control and Legal Principles, Eighth Edition.
Donald A. Wilson, C.A. "Tony" Nettleman III, and Walter G. Robillard.
© 2024 John Wiley & Sons, Inc. Published 2024 by John Wiley & Sons, Inc.

The following principles are discussed in this chapter and, in most instances, are addressed by attorneys, but they should be understood by the courts and especially by the retracing surveyor.

PRINCIPLE 1. A landowner may divide a parcel of land in any manner not inconsistent with the law.

PRINCIPLE 2. There can be only one original survey that creates the boundaries of a parcel.

PRINCIPLE 3. The "footsteps" of the original survey are the evidence left on the ground as well as the documentation identifying the ground evidence that the original surveyor created.

PRINCIPLE 4. The retracing surveyor must be able to locate and then make a nexus between the ground evidence and the written evidence.

PRINCIPLE 5. Once a boundary or boundaries are created, no alterations or modifications are permitted in any manner by either the landowner or any surveyor once property rights have been granted or distributed according to the boundaries created.

PRINCIPLE 6. The original surveyor creates boundaries. It is the retracing surveyor who ascertains or identifies boundaries from the original evidence.

PRINCIPLE 7. Once created and approved, the original boundaries created are legally without error and are the exact dimensions as indicated by the creating surveyor.

PRINCIPLE 8. No surveyor or court has the legal authority to alter or modify a boundary line once it is created. It can be interpreted only from the evidence of where that boundary is located based on the "footsteps" in Principle 3.

PRINCIPLE 9. A retracing surveyor relates previously created and recovered evidence to a current survey, being mindful that the current survey is always subject to collateral attack by other surveyors.

PRINCIPLE 10. A modern survey may describe bearings and distances to a greater degree of precision using modern technology, but it can be no more precise than the original measurements that created the original lines.

PRINCIPLE 11. When there is a scarcity of written evidence that indicates no lines were created on the ground, any survey conducted subsequently may be held as the first survey. The courts may consider the first survey to be the original survey, or they may consider the original survey to be the first survey.

2.2 DEFINITIONS

Several books give academic and/or legal definitions of words that are important to the practicing surveyor. These are the definitions that should be used if one were to testify in court. But in many instances, they fail to meet the everyday practitioner's requirement for definitions that are understood. In an effort to introduce a degree of practicality, the definitions given here are those that the practicing surveyor, student, and landowner can appreciate without legal or academic confusion.

Any list of definitions can be short or extensive. The definitions presented here are only those basic ones that are needed to furnish a foundation for the understanding and the study of boundaries. Corners, lines, monuments, property lines, and boundary lines may have several definitions. The following are those that the authors use to make distinctions:

A *line* in surveying, which divides interests in real property, means, prima facie, a mathematical line (invisible) without breadth; however, this theoretical idea of a line may be explained, by the facts referred to and connected with the division, to mean a wall, a ditch, a crooked fence, or a hedge—that is, a line having breadth.

Usually, a *corner* is at the end of a boundary line or at a change in direction of a boundary line. Applying this concept, we see why an endpoint is also called the *terminal point* of a line, named after Terminus, a deity to the ancient Greeks and Romans. A corner may also be placed along a line where a third party may tie in or reference a senior line. To define a line, there must be two corners, or termini, one at each end of the line. The corners, to be controlling over other descriptive elements found in descriptions, must be created by the survey and must be called for in a legal document relating to the specific parcel.

In many instances, corners may be referenced in a description as *points*. Historically, points were described in early surveying textbooks. As early as 1657, William Leybourn referenced a point as being "without unity."[1] Then, in 1687, John Love, in *Geodaesia*, referred to a point as "that hath neither Length nor Breadth, the least thing which can be imagined, and which cannot be divided...."[2]

Not until some person places a physical object, the monument, at the *point* does the point become identifiable and retraceable.

Corners can be located or placed on the exterior boundaries of the parent parcel as well as inside the parcel itself. Usually, an interior corner will not control an exterior corner. To be controlling and legal, a corner does not have to be monumented. A corner has no physical dimensions of length, width, or depth. It has only a legal dimension, in that an original corner found legally identifies the point of the survey, regardless of whether it is monumented. This applies only if a conflict exists in a description; otherwise, it can control even if it is not called for, but this is predicated on other applications of law. When in a retracement a corner that was originally described as a *point* is identified, with no reference to having been monumented, it is difficult, if not impossible to identify with certainty in a subsequent survey. The original "point" was a product of the equipment and methodology used at that time, and these two elements cannot be totally replicated. The placement of that original "point" is unique. The certainty of identification and position becomes recoverable only when the original point is monumented and then reduced to writing

To be legally controlling, a *monument* is a physical manifestation set at or near a corner. A corner does not have to be monumented; it just has to be identified, with clarity, and called for in the survey or document. However, the law has set certain requirements for corners. As we will see, a call for a distance, more or less without the call for a corner or monument, is legally insufficient. Yet we can have monuments placed on the corner point or near the corner, either as a substitute for the corner point or to define possible multiple locations of the same point. Surveyors and landowners have problems when a monument is not placed at the corner point but near the point where the corner is located. One of the critical requirements for a monument to be controlling is that the *monument* recovered in a retracement be called for by words and the recovery description, not differing appreciably from the monument set at the time of the original survey, when the corner was established and a monument was set to identify that specific point.

Monuments may be classified as falling into two areas or descriptive categories: natural and artificial. Courts usually make a legal distinction between the two and hold that when there is a conflict between or question about a natural monument and an artificial monument, the natural monument usually will control legally.

Courts have held that streams, rivers, mountains, roads, and trees are supposed to identify a point that represents a corner—the legal point. But how can a stream that is 20 feet wide and 600 feet long be considered a monument? This discrepancy cannot be explained. It is even more difficult to explain how you can have a boundary line without having terminal corners. The reader will find that this is but one of many possible conflicts that may be encountered by the professional surveyor in retracement work. Research of early US case law reveals that many of the cases called these points *natural boundaries* and *artificial boundaries*.

Artificial monuments are those monuments that usually are placed at corner points by landowners, surveyors, engineers, and others. They may be referred to by various terms on maps, in descriptions, and in field books. One may find them referenced as iron pins or pipes (IPs), stakes, and, at times, trees, concrete monuments (CMs), nails, or whatever the surveyor decided to use or was available from the back of the survey truck. Courts have made the distinction that natural monuments control over artificial monuments, because their identification is more certain and they are less likely to be disturbed. (This is discussed in Section 5.10.)

The surveyor must make the distinction between a *boundary line* and a *property line*. Boundary lines between parcels are created in several ways, yet until written documents or legal principles attach, property lines are nonexistent. In theory, a boundary line remains fixed forever where it was located initially, but a property line may change by means of legal principles, including estoppel, agreement, adverse possession, or riparian rights.

When a surveyor establishes a survey line in the field to create a boundary, corners are usually set and then described and identified. Lines are then marked, and monuments are set at the corners. By principle, each line must have two corners, one on each end, to be fixed in place. Although, when placed, the monuments become visible to the surveyor and landowners, and the monuments that are set identify the position of the corners, the connecting lines between the corners remain invisible. However, if the creating surveyor blazed trees or placed posts or stakes along that line, these physical objects become the identification of that once-invisible line. Thus, the objects marked *at the time of the survey* become the location and identification of the boundary line created. The recovery of any of the evidence of this survey at a later date becomes the recovery of the original line, and the lines run in the field to create the original boundary controls.

Once a boundary line is created in the field or on paper, common law and statute law permit it to be modified, changed, and altered. When a line is thus changed, altered, or modified, according to law, it becomes a property line. Because surveyors can create boundary lines, and because a property line is a result of law, only the courts can certify or "sanctify" property lines or boundaries. Surveyors in the United States have not been given the authority to determine legal property lines. Usually, when two surveyors disagree on the survey location of a line, the parties resort to the legal system to make the final determination. This is often referred to as "the battle of the surveyors."

Once a boundary line is created and identified, it maintains its legality and position even if the corners on its extremities become lost. This may be difficult to understand, but the law recognizes a boundary line as being a legal entity, and, once it is created, as long as evidence of that line exists, the line is controlling.

On the other hand, although a true boundary is controlling, surveyors, in retracing these true boundaries, may consider it necessary and prudent to evaluate property boundaries. In many instances, the true boundary and the property boundary are one and the same; that is, they are superimposed on each other. The fence line or hedge line between the two adjacent parcels is superimposed on the deed line that was surveyed on the ground. However, the common law and statute law of most states permit people to modify these deed boundaries through legal doctrines, some of which are centuries old. Although they may be referred to by different terms in different states, for the sake of identification we will call them *agreement*, *estoppel*, *acquiescence*, and *adverse possession*. These doctrines require a landowner to complete some act in order to perfect the change of the boundary. Changing a boundary through a riparian change must be through, or a result of, an act of nature.

2.3 CLASSIFICATION OF BOUNDARIES

When a surveyor or an attorney looks at a boundary, he or she usually does not consider the magnitude or the complexity of that boundary. One may be asked either to create or to retrace a boundary of an entire parcel, a boundary of a portion of a larger parcel, or the entire boundary of a single minute parcel. The boundary may be composed of natural elements, artificial elements, words only, or a combination of the three. It may be either a complete boundary or a portion of a larger boundary.

These *macro* and *micro boundaries* enjoy the same attributes of definition in that they commence at a common point in the center of the Earth and extend to the accolades in the heavens. This was tested when the former Soviet Union made a claim extending its boundaries to the North Pole, at a distance of 20,000 feet under the Arctic Ocean. As an afterthought: when macro boundaries are disputed, nations bring out the nuclear weapons; when micro boundaries are disputed, the individuals involved resort to guns, knives, and fists.

As discussed in Chapter 1—History and Concept of Boundaries, all boundaries may be considered as either macro boundaries or micro boundaries. The distinction is not necessarily the manner in which the boundary was created but the rights or interests that were being separated and identified.

Since *macro boundaries* usually separate major interests, boundaries between nations, boundaries between political subdivisions of a nation, and major private or governmental and subdivisions may be placed in that category. *Micro boundaries* include smaller subdivisions within macro boundaries. Using this distinction, a boundary may be a macro as well as a micro boundary.

A township and range within a state may be a micro boundary within the state, but in relation to the respective sections identified within its boundaries, the township is a macro boundary for the respective sections (Section 3.2).

The law of retracements may apply differently to township lines (macro boundaries) than it does to section lines (micro boundaries). Consequently, if a subdivision of homes (micro boundaries) is placed within the boundaries of the section (macro boundaries), entirely new rules of retracement apply. When there are conflicts between boundaries, the problem must be approached in an orderly and systematic manner, using logic, analysis, and experience to arrive at a possible and suitable solution.

Having classified boundaries as macro or micro, they may be further classified as to the authority for their creation into two broad categories:

1. Rectangular
    a. Created by federal statutes
        1. Rectangular
        2. Metes and bounds
    b. Created by state statutes
        1. Rectangular
        2. Metes and bounds
    c. Created by private entities
2. Metes and bounds

Regardless of how the original boundaries were created or what category the retracing surveyor may place them in, the proper rules of retracements should be followed.

These categories are discussed in detail in Section 5.20. It will suffice at this time to recognize that the various methods of creation will help practicing surveyors understand that the methods and laws by which boundaries are created affect the methods of retracement.

2.4 METHODS OF BOUNDARY CREATION

In acknowledging the fact that a boundary is a factual invisible division line between two interests or estates in land, one must look at both common law and statute law to surmise how courts and legislative bodies have and will determine the various methods by which boundaries can be created. In applying the methods of boundary location, the surveyor can identify multiple methods by which the surveyors, the courts, or a landowner can create a boundary. It should be remembered that one does not have to be a surveyor to create a boundary. Courts can adjudicate and dictate a boundary; two or more contiguous landowners can create mutual boundaries; a single landowner can create a boundary by personal actions or words; and surveyors can create boundaries by their actions or words.

For study purposes, boundaries can be determined or created as follows:

1. *By action.* This is the actual creation of a boundary on the ground by survey, using the instrumentation and the words that are current at that time, by individuals or by actions. Corners are created and identified, monuments are established, and notes are made of the survey or actions; possibly, a plat will be prepared and lines marked on the ground. These actions are then incorporated into the resulting descriptions.

2. *By words.* Exterior and interior boundaries of tracts may be created by words. A deed describing "the south ¼" or "the north 10 acres" creates boundaries, as does a metes and bounds description having courses (bearings and distances) and corners. A major problem is that a "south ¼" description has no calls for monumented corners. If the description is the result of a survey, original lines are created, and the surveyor is an original surveyor. If the description is a words-only document, there is no original surveyor, only a first surveyor who locates the lines described from the words; as such, this survey is always subject to future collateral attack.

3. *By law.*

   a. *Statute law.* Boundaries can be a result of statutes. These boundaries may also be identified under point 1. The difference is that, whereas under method 1 the actions are performed as a result of a mutual desire to locate the boundary, under this section a governmental body, either federal or state, enacts legislation duly authorizing certain surveyors to survey large tracts of federal- or state-owned lands.

   The most noted and extensive surveys were those conducted by the federal government, now known as the Public Land Survey System (PLSS) or General Land Office (GLO) system. This system was created when several states, Georgia and Texas in particular, enacted legislation directing how their public lands would be surveyed for disposal to their citizens. Although the state legislation was not as extensive or as detailed as the federal legislation, it did identify such areas as how the respective parcels would be surveyed and the type of equipment to be used by the surveyors. More particularly, it directed how they would be identified and distributed. Some also included the qualifications that were required of surveyors. The federal laws that created the public land system (GLO) were supplemented with a subsequent act (Land Act of February 11, 1805) that basically set the dignity of the original surveys.

b. *Common law*. Although the common law has recognized that individual landowners can, under certain situations, establish their own unilateral or mutual boundaries, many states have identified, by statute, the specific requirements necessary to meet these historical methods.

Accepting the fact that the laws recognize the creation of such boundaries, the names often become confused, misapplied, misidentified, and generally muddled by landowners, attorneys, and, especially, the courts. In the founding days of our country, many lawyers and judges were also schooled in surveying, title examination, and land principles. This cannot be said of many of the younger professionals today. Not to detract from their capabilities, but today's young legal professionals have not been schooled in basic land principles; they usually lack a basic "land ethic." On occasion, the principles of agreement, estoppel, acquiescence, and the age-old doctrine of adverse possession to change, alter, and reidentify boundaries have to be identified, understood, and then accepted.

Creation of Boundaries by Running Lines

The foremost method of boundary creation, and the one best understood by the courts, is that of running on the ground by the surveyors and then reducing this running to words by creating field notes and plats, and possibly descriptions.

It is well recognized by common law and case law that a person may create boundaries of parcels, lots, and even subdivisions by their actions. A basic underlying principle is that as long as the person does not go beyond the exterior boundaries of the title lines, a parcel may be subdivided in any manner the law permits by running and establishing lines on the ground. This is not in contradiction to establishment by law, but it could be considered as establishment by common-law principles.

In the past, landowners were permitted to subdivide their parcels into lots of any size without any restrictions. Today, numerous subdivision laws have been enacted that limit the size of parcels and their potential uses. However, the retracing surveyor should realize that this does not resolve the issue of those many parcels that were created and sold with no restrictions—in many instances, the subdivision plats were not made a matter of public knowledge by registration.

If a person owns a parcel and wishes to subdivide it into smaller segments, this can be done by physically running survey lines, creating corners, either setting or not setting monuments, marking lines, creating field notes, and possibly preparing a plat of the subdivision. The landowner then conveys the individual lots as they were created. The description by which the lots were conveyed may or may not refer to the survey. If it does, the true intent is identified; that is, the survey would control the lines, regardless of any latent errors.

If a survey is not referred to, the rule is not clear. Courts have been inconsistent: some courts have permitted reliance only on circumstantial evidence that the original survey controls; others have relied on direct evidence. Some have held that if a plat is part of a description, all elements on the plat become part of that description, including the survey that created it, which includes the description of the lines run and the monuments set. Others have permitted surveyors to testify as to what was done and made their testimony controlling. To understand what local courts permit, each surveyor should become familiar with local legal decisions regarding boundaries as well as accepted universal legal principles.

Creation of Boundaries by Verbal Actions

If we were to consider maps, plats, and descriptions as verbal actions, it is well recognized at law that a landowner can create a boundary without benefit of survey. This may also include actions by landowners walking out lines, establishing corners, and setting monuments, and then writing their own deeds and descriptions to which rights will be conveyed.

Many times, a surveyor will be asked to locate a parcel described as "north ½," "east ¼," or the "south 10 acres" from a description prepared by the landowner. In analyzing such descriptions, there is one certain principle that surveyors, attorneys, and the courts must consider: in some states it is not unlawful for a landowner to convey a portion of the parent parcel by a description without the benefit of a survey. This occurs when a person owns a 10-acre parcel and conveys out the "south 5 acres" or the "south ½." These two descriptions may or may not be inconsistent. Yet once the deed is delivered or signed, an invisible boundary is created between the "north 5 acres and the south 5 acres" or the "north ½ and the south ½," even though no survey was ever conducted. The responsibility of the surveyor is to be able to place this invisible line on the ground by survey. This resulting survey is not an original survey that a surveyor creates when he or she runs the line first before making the plat. It should be considered as the first survey. The unique situation this first survey is placed in is that it is not controlling, as the original survey would be. A first survey by the initial surveyor is no more controlling than would be any other subsequent survey. This first survey is but an interpretation or an opinion by the surveyor or surveyors of the written description from which the surveyor works.

Creation of Boundaries by Survey

The attributes necessary for the surveyor who is given the responsibility of creating boundaries is entirely different from the attributes of the surveyor who is given the responsibility of finding the boundaries originally created or when conducting a subsequent retracement. We do not have the flexibility to pick and choose our surveyors for certain jobs, for registration laws hold that, once a surveyor is registered, that surveyor is legally and professionally capable of performing any of the actions that the law permits. *Being legally capable is not the same as being technically capable.* The surveyor who undertakes the responsibility either to create or to retrace boundaries will be assumed to be totally qualified to conduct either job once it is started.

Failure to follow the basic survey principles in creating the original surveys and the basic principles in conducting retracements may be tantamount *to negligence per se.*[3]

2.5 WHO MAY CREATE BOUNDARIES?

Principle 1. *A landowner may divide a parcel of land in any manner not inconsistent with the law.*

Notwithstanding laws to the contrary, landowners may subdivide land in any manner to suit their needs and desires. If questioned, it may well be considered a constitutional right to do so. Under the common law, one of the benefits of landownership is the ability of a landowner to sell property and describe this property in any manner the property owner wishes. At common law, there are no requirements that a landowner conduct a survey as a prerequisite for conveying a parcel of land or a property interest. There are distinctions as to the type of description that may be legal as far as the courts are concerned and adequate as far as a surveyor is concerned.

In many instances, when the landowner conducts his or her own survey or creates his or her own description and deed, a substantial monetary savings to the parties usually results, only to be offset later by the possible costs of legal fees and litigation between the original parties or future parties in interest. Surveyors should discourage and not aid landowners who practice this authorized but not recommended activity.

The property owner, unless prohibited or restricted by law, is empowered with the authority to subdivide the area in any manner he or she desires. Usually, there are no restrictions on how the original lines are marked or described.

Unknowingly, some landowners create a future opportunity for confusion and possible litigation by changing a boundary description from one method to another. Land that is created by an aliquot description will be redescribed to another form. For example, an original description that says "the S ½ SE ¼ of Section 7" will be redescribed as "the South 80 acres" or it may even be presented as a metes and bounds description with or without the call for monuments.

A landowner may even take a large parcel of land and create his or her own rectangular system within the parcel.

The retracing surveyor must be "tuned in" when this happens and properly apply the appropriate retracement techniques and rules of descriptions.

Principle 2. *There can be only one original survey that creates the boundaries of a parcel.*

Individual surveyors and attorneys, as well as judges, confuse what an original survey actually is. When having to decide whether the survey being retraced is an original survey, the retracing surveyor should look at the original titles to the property. The original survey should be correlated to the boundaries described in the original title. Because of the acceptance of the doctrines of estoppel, agreement, and acquiescence possession lines may affect the written deed boundaries.

Once approval and title(s) have been granted in reference to that survey, the creator of the parcels cannot alter any boundaries created without the approval of all interested parties.

When boundaries are referenced to one survey and the titles to the parcels are referenced and conveyed to the described boundaries, they become unalterable without legal action.

Principle 3. *The "footsteps" of the original survey are the evidence left on the ground as well as the documentation identifying the ground evidence that the original surveyor created.*

Evidence of the original boundaries can be misidentified by professionals who must conduct retracements with the hope of identifying the original locations of the originally created lines and the resultant established corners.

In order to conduct an adequate retracement, courts have held that, to have the totality of the evidence, the surveyor conducting the research should consider that the instructions given to the creating surveyors, the fieldwork, the trees marked, the monuments set, the plats prepared, and any reports created are all part of the original survey and should be obtained and considered.

Principle 4. *The retracing surveyor must be able to locate and then make a nexus between the ground evidence and the written evidence.*

A found monument, supposedly placed by some unknown person at some unknown time, should not be used to represent a described corner in order to be used in a retracement or to prove the original location of a corner. This can be said about any evidence the retracing surveyor wants to use to prove the existence of any survey, including an original survey. The retracing surveyor must have the documentary evidence that described the corners and their monument in order to consider the monument as being a controlling element for one termination of a boundary line.

> A found monument with no written history of its placement has no legal or survey value indicating the boundary.

Some surveyors and courts accept found monuments as primary evidence to solve boundary issues, but a found monument without a historical written connection is of little value without some other species of legal and survey support. In the novel *Uncle Tom's Cabin*, Little Topsy was asked the question; "Topsy, just where did you come from?" Her response was "I don't know. I just growed." If you are unable to determine a found monument's origin, it is better not to use it as a controlling element in your survey.

Principle 5. *Once a boundary or boundaries are created, no alterations or modifications are permitted in any manner by either the landowner or any surveyor once property rights have been granted or distributed according to the boundaries created.*

This principle may cause confusion and subsequent problems for landowners, surveyors, and attorneys, and possibly neighboring landowners. One of the guarantees of the US Constitution is that of being able to use land without government control or intervention. In modern times, that has been modified by zoning laws, ordinances, covenants, and so on. But one may survey and convey property in any manner not inconsistent with the law. This permits the landowner to survey and create boundaries for conveyancing. However, under the law, once these boundaries are created, the landowner can modify the boundaries in any manner desired as long as no property rights have been conveyed to third parties. But once a single lot or interest is conveyed according to these micro boundaries, all micro boundaries within that macro boundary become legally fixed and cannot be altered or modified without approval of all persons who have vested property rights.[4]

This is a difficult concept for landowners to comprehend while they still own the parent parcel, but it should be understood that the conveyance of a single lot in reference to the plat or subdivision line "seals" the location of all lines and corners referenced and identified on the plat. (Not recording the plat will not affect this principle.)

At common law, there are no restrictions as to how the landowner can convey these lands. Choices must be made as to whether a survey is to be made to identify the lines or the landowner is simply to convey the land using words.

Principle 6. *The original surveyor creates boundaries. It is the retracing surveyor who ascertains or identifies boundaries from the original evidence.*

In keeping with Principle 1, the landowner can create or establish boundaries on his or her own property, and the survey is the method by which these boundaries are actually created on the ground. A parcel of raw land has no boundaries. But once the surveyor runs and then identifies these lines, the boundaries are created and can never be altered by any subsequent surveyor.[5] Once these created lines are used to identify parcels. The elements of the original and all related information created from or identified from the original survey cannot be altered in subsequent surveys. The distances recited are those distances indicated; a subsequent surveyor only retraces the original survey, and the remeasured lines are only redescribed in relation to more modern and possibly more precise measurement(s). Because boundaries are created, the original actions cannot be altered by subsequent surveyors or landowners. It is subsequent surveyors who ascertain the location of the original boundaries by conducting retracements[6] based on the recovery of the original evidence and then relating this evidence to the description of the original boundaries.

The subsequent surveyor conducts a retracement that differs from the original survey both legally and in terms of permanency. Whereas the law has deemed the original corners established by the original survey to be legally errorless and unalterable, the subsequent retracement has no permanency. This was discussed in the Florida decision *Rivers* v. *Lozeau.*[7]

... [A] surveyor can be retained to locate on the ground a boundary line which has theretofore been established. When he does this, he "traces the footsteps" of the "original surveyor" in locating existing boundaries. Correctly stated, this is a "retracement" survey, not a resurvey, and in performing this function, the second and each succeeding surveyor is a "following" or "tracing" surveyor and his sole duty, function and power is to locate on the ground the boundaries, corners and boundary line or lines established by the original survey; he cannot establish a new corner or new line terminal point, nor may he correct errors of the original surveyor. *He must only track the footsteps of the original surveyor. The following surveyor, rather than being the creator of the boundary line, is only its discoverer and is only that when he correctly locates it.* (Emphasis added.)

2.6 SANCTITY OF THE ORIGINAL SURVEY

Principle 7. *Once created and approved, the original boundaries created are legally without error and are the exact dimensions as indicated by the creating surveyor.*

Although this principle is codified so far as federal lands are concerned in federal law in the Land Act of February 11, 1805, it is also accepted in common law for the metes and bounds states. When it becomes necessary to understand the full power of the original lines, it then becomes important for the creating surveyor, the retracing surveyor, and the courts to understand and appreciate the dignity of a found, identified original survey. The creating surveyor should strive to use the most precise means to run the surveys that create the boundary lines as accurately as possible and then record the actions definitively. The retracing surveyor should realize that modern methods of measurement will not duplicate the measurements that created the original boundaries. The courts must then realize that these original lines cannot be altered by judicial determination in that they are unassailable through any court, including the US Supreme Court. The retracing surveyor, as part of the totality of the evidence that is recovered, then should correlate his or her retracement measurements to the original measurements. A problem may present itself when the surveyor's research indicates that the original measurements are in one unit and his or her remeasurements are in a totally different unit (e.g., the original description is in poles and the remeasurements are in feet).

Historical correlation of measurements, angles, and distances in full degrees and poles and retraced distances should be in the same units.

The responsibility of surveyors is to examine evidence that is presented to determine where the true and correct locations of these lines are located. When attempting to locate these original lines, subsequent surveyors should always make their analyses in terms of the original units of measurements.

2.7 ORIGINAL LINES REMAIN FIXED

Principle 8. *No surveyor or court has the legal authority to alter or modify a boundary line once it is created. It can be interpreted only from the evidence of where that boundary is located. It can be interpreted only from the evidence of where that boundary is located based on the "footsteps" in Principle 3.*

This basic principle was created by the US Supreme Court in the decision *Cragin v. Powell*[8] when it wrote that original surveys are *unassailable through the courts*.

Retracing surveyors will encounter a minority of surveyors who, when finding an "error" in the original survey, believe it is their responsibility to "correct" the error and make the original bearings and distances as they should have been had they been surveyed correctly. These surveyors have no concept that, once the lines have been created, no subsequent surveyor has authority to re-create the original lines. When a creating surveyor indicates a distance or an angle, these are the original measurements, according to the creating surveyor's methodology and errors. By law, they are free of error, even though, in fact, we do realize that the creating surveyor made mistakes. This concept also extends to the presumption that the survey is correct[9] and that no subsequent surveyor or surveyors can correct the original lines; they can only redefine them. It is not the job or responsibility of subsequent surveyors to correct the originals. It is their job to report any discrepancies found. Differences do occur because of the different methods and equipment used in obtaining the original measurements and the subsequent measurements, as well as differences among the people who did the work and changes in the circumstances or conditions under which it was done.

2.8 DISTINCTIONS BETWEEN THE ORIGINAL BOUNDARY SURVEY, THE RETRACEMENT SURVEY, AND THE FIRST SURVEY

People, including many surveyors and attorneys, use words freely. When a person needs a word to explain a situation or any evidence and a word is not readily available, they are likely to create a name, which may become fixed in its use. This relates to the following three terms:

Original survey

Retracement

First survey

Such words as *love* and *friend*, as used in everyday speech, often have meanings that are different from their true meanings. Surveyors use the term *property boundary* very freely on maps, in reports, and in everyday conversation, yet no surveyor has the authority to identify and locate property boundaries, and few states define one of the attributes of surveying as "locating property boundaries." Without judicial authority, no surveyor has authority to locate or identify property boundaries.[10] Interestingly, though, several states have now added, as a distinct category of surveying, giving testimony in court as to boundaries.

Principle 9.   *A retracing surveyor relates previously created and recovered evidence to a current survey, being mindful that the current survey is always subject to collateral attack by other surveyors.*

Having seen that only the original entity that created the original boundaries can change boundaries after having secured the sanctity of vested property rights, the surveyor in the private sector can only conduct a retracement of an original survey.

The concept of the survey is best defined in *The Manual of Instructions*.[11] The *Manual* defines the areas of resurveys and retracements in the following areas (see Chapter 1—History and Concept of Boundaries):

1. Original surveys
   a. Retracements
   b. First survey
2. Resurveys
   a. Dependent
   b. Independent

Since an original survey may be conducted of both public and private lands, both independent and dependent resurveys may be conducted only by the entity that conducted the original survey. On the other hand, a retracement is a survey that is made to ascertain or redefine the direction and length of lines and to identify monuments that were established at corner positions by a prior original survey. A retracement is usually conducted within a macro boundary (e.g., a township) to ascertain micro boundaries (e.g., sections). It may be necessary to retrace several miles of lines within a macro boundary from found, proven, and original corners to set or relocate a lost corner or corners on a micro boundary.

A retracement may be used to recover new and additional evidence to ascertain the quality of an earlier survey. In a retracement, survey-found proven corners are rehabilitated, but any lost corners that are resurveyed have no finality of position.

The place of a retracement (resurvey) is best explained by the following excerpt from *Cragin* v. *Powell*:[12]

> The making of resurveys or corrective surveys of townships once proclaimed for sale is always at the hazard of interfering with private rights, and thereby introducing new complications. A resurvey, properly considered, is but a retracing, with a view to determine and establish lines and boundaries of an original survey … but the principle of retracing has been frequently departed from, where a resurvey (so called) has been made and new lines and boundaries have often been introduced, mischievously conflicting with the old, and thereby affecting the areas of tracts which the United States had previously sold and otherwise disposed of.

> In the first survey, lines are difficult to ascertain, but in most instances, they do not control as do the original lines.

2.9 ORIGINAL TECHNOLOGICAL METHODS OF BOUNDARY CREATION NOT RELATABLE TO MODERN METHODS

Principle 10. *A modern survey may describe bearings and distances to a greater degree of precision using modern technology, but it can be no more precise than the original measurements that created the original lines.*

In attempting to analyze a historic description of a boundary for retracing, an analysis should be made that compares the methods and equipment used for those that created the boundaries to the methodology that will be used to take the historical description and retrace that description today using the most modern of surveying equipment.

In many instances, ancient bearings and distances are the references for the modern survey. Although the more modern equipment may be more precise than the original equipment, the newly retraced boundaries can never replace the accuracy of the original surveys. Today, the surveyor is asked to retrace bearings and distances that may have been created hundreds of years ago and then relate them to references that are no longer used in descriptions. Until recent years, the universal instruments used to create bearings and their companion instances were the compass and the 2-pole chain, consisting of 50 links totaling 33 feet. A compass initially was graduated to a full degree with an estimated reading of ¼ of a degree (*not* 15 minutes). Usually, distances were measured with a 2-pole chain consisting of 50 links of 0.66 feet. The individual link could be estimated to probably 0.2 of a link (*not* 0.132 feet).

Today, we take these as absolutes and make a conversion into absolute reality, and then, using these absolute conversions, we use modern equipment to place on the ground these inexact references. In the process, we fail to consider the significant number relationships in the conversions.

If the retracing were to take the converted bearings and distances as absolutes, determining relationships would be made difficult, if not impossible.

2.10 ORIGINAL LINES MAY BE REDESCRIBED AS A RESULT OF A RETRACEMENT

Applying this principle with wisdom and foresight, American jurisprudence (the courts) anticipated that technology would probably change in the future, realizing that there has to be a finality to measurements because technology has refined the original measurements more precisely.

The courts have made the distinction between accuracy and precision. The major decisions held that the original measurements were without error in the location of the monuments, when found undisturbed, and thus they controlled. The basic premise is that the angles and distances cited in the field notes and then identified on the plats are the exact measurements created by the original surveyor using the equipment and methodology of that time, but any subsequent "survey" will use different equipment and different personnel for the subsequent work, and thus it is not and cannot be a duplication of the original work.

All measurements have a positional tolerance of location at the endpoints.

2.11 CONCLUSIONS

In application of this chapter to students and practicing surveyors, as well as attorneys and courts, we hope they will gain an appreciation of the distinct relationship among boundaries, boundary creation, and boundary retracement. When the term *boundary* is referred to in this sense, it is land boundaries that are intended, not property boundaries, in that a surveyor may conduct a property survey but that should be considered as a special-request survey that is specific and not universal in nature. A boundary line may be altered and become a property line boundary by special doctrines that are primarily legal in nature. Estoppel, acquiescence, adverse possession, and agreement are legal doctrines that may relegate deed or surveyed boundary lines to property boundary lines. These are legal dicta and should not be addressed by a registered surveyor. They should be understood, however.

There are two particular decisions that each boundary survey should take notice of: *Kerr* v. *Fee*[13] and *Rivers* v. *Lozeau*.[14] Rather than express the authors' personal views, it is better to quote what the courts expressed in these two decisions. In *Kerr*, the court stated as follows: "To survey land means to ascertain the corners, boundaries and divisions, with distances and directions, and not necessarily to compute areas included in defined boundaries. Knowing these, any competent mathematician can ascertain the areas."

Then in 1989, the Florida Supreme Court, in *Rivers* v. *Lozeau*, identified the responsibilities of the modern surveyor with these words:

First: "The definition of a legally sufficient real property description is one that can be located on the ground by a surveyor."

Second: "In a retracement survey, not a resurvey, ... each succeeding surveyor is a following or tracing surveyor, and his sole duty, function and power is to locate on the ground the boundaries corners and boundary lines established by the original survey...."

Principle 11.   *When there is a scarcity of written evidence that indicates no lines were created on the ground, any survey conducted subsequently may be held as the first survey.*

Whether a survey is considered as an original survey or a first survey may be a matter of being able to correlate the field evidence as well as the documentary evidence as to the time of creation. Some courts have held that the documentary evidence and the field evidence had to be created at the same time. But on the other hand, other judges have determined that if the field survey was conducted near the time of the documentation, that would suffice to meet the requirements.

The object of an original survey is to create boundary lines.
The object of a retracement is to find the evidence (footsteps) of the original surveyor and then redescribe the footsteps by today's standards.

NOTES

1. William Leybourn, *The Compleat Surveyer* (1657). Reprinted by Walter G. Robillard.
2. John Love, *Geodaesia* (London: 1687). Reprinted by Walter G. Robillard.
3. *Spainhour* v. *Huffman*, 377 SE 2 D 615 (Va. 1989).
4. *Kelsey* v. *Lake Childes Co.*, 112 So. 887 (Fla. 1927).
5. *Cox* v. *Hart*, 43 S. Ct. 154, 260 U.S. 427 (Calif. 1922).
6. *Cragin* v. *Powell*, 128 U.S. 691 (La. 1888).
7. *Rivers* v. *Lozeau*, 539 So. 2d 1147 Fla. Sup. (1889).
8. *Cragin* v. *Powell*, 128 U.S. 691 (La. 1888).
9. *Camp* v. *Winegar*, 210 P. 64 (Colo. 1922).
10. The authors question the authority of a state to enact a law that prohibits qualified surveyors from testifying in states in which they are not licensed. It is possible that an administrative agency is setting the legal standards for judicial proceedings.
11. Bureau of Land Management, *Manual of Instructions for the Survey of Public Lands* (Washington, DC: U.S. Department of the Interior, 1973), Sections 6.1 to 6.8.
12. *Cragin* v. *Powell*, 128 U.S. 691 (La. 1888).
13. *Kerr* v. *Fee*, 161 N.W. 545 (Iowa 1917).
14. *Rivers* v. *Lozeau*, 539 So. 2d 1147 (Fla. 1989).

CHAPTER 3

OWNERSHIP, TRANSFER, AND DESCRIPTION OF REAL PROPERTY AND ACCOMPANYING RIGHTS

3.1 CONCEPTS OF BOUNDARIES, LAND OWNERSHIP, AND LAND DESCRIPTIONS

The historical concepts of ownership, transfer, and the historical relationship of boundary description of real property are complex and are not subject to simple categorization, definition, or understanding. Nor can any of the principles recited or discussed be considered absolute. Usually, these areas are not the concern of the original surveyor or the retracing surveyor. Although the government has authority to regulate, tax, and limit land use through such controls as zoning, the courts in turn can determine title and quality and validity of descriptions conveying any interests in land. Courts can also adjudicate possession and its relationship to title and descriptions. Without an adequate and legal description of a parcel of land that can cover and include a single property interest or a multitude of interests on, below, or above the Earth's surface, the respective boundaries cannot be surveyed because of the ambiguous boundary descriptions that created them.

> Written title to land is the person's legal claim to possession.

What is described in a description? A description is similar to a basket full of apples. The basket that holds the apples is the title to the property that describes the

Brown's Boundary Control and Legal Principles, Eighth Edition.
Donald A. Wilson, C.A. "Tony" Nettleman III, and Walter G. Robillard.
© 2024 John Wiley & Sons, Inc. Published 2024 by John Wiley & Sons, Inc.

rights, both corporeal and incorporeal, that usually are described in the deeds that contain the descriptions. These were first described in English common law by some legal scholars and jurists as the *bundle of rights*. Land, being unique, is composed of a multitude of legal rights: the right of possession; timber rights; mineral rights; rights of ingress, egress, and regress; air rights; water rights; surface rights; subsurface rights; and a multitude of other rights. A new right is now recognized by the law: the right to have natural wind blow across your property. All of these rights are the individual apples in the basket, and, to be conveyable, each right must be contained within the boundaries of the parent parcel. Each of these rights may or may not have boundaries and may or may not coincide with the boundaries of the parent parcel. Thus, a parcel of land that is described by defining its perimeter may also have descriptions of rights falling within the parent description. If the owner conveyed the mineral rights to a coal company, the coal company would have the right to enter the property to extract the minerals located within the boundaries of the mineral deed, which is located within the boundaries of the parent parcel. This may require two boundary descriptions and possibly two boundary surveys, one for the exterior boundaries of the parent parcel and a second for the mineral rights. The fee owner may then convey the timber rights, which may have separate and distinct boundaries from the mineral rights, but may also enjoy overlapping boundaries of these rights.

Conflicts can occur in both interests and in descriptions. Conflicts in "ownership" may occur when two or more people have title or claim to the same property or property interest, when one person has title and a second has possession, or when the description that describes the interests is ambiguous or in conflict with a second description.

The aspects of land are unique.

The following principles are discussed in this chapter:

PRINCIPLE 1. One who grants title to property to a second person, that grantor, can legally grant no more interest than that which is owned at the time of the conveyance.

PRINCIPLE 2. Title insurance states only that if a defect is found in the title that is insured, the insurer will either defend the title in court or make good the value of the loss.

PRINCIPLE 3. A land description in a deed conveys only the interest described by the wording, terminology, and survey measurements that are depicted at the time the survey was conducted and the description was written.

PRINCIPLE 4. To be absolutely controlling in a resurvey or retracement, a corner and its monument must be called for in the original description.

PRINCIPLE 5. The unit of measurement indicated in the description is the unit of measurement supposedly used at the time of the survey or when the description was written and should be the reference measurement to which the surveyor should refer.

PRINCIPLE 6. Distances cited in modern descriptions are presumed to be along a horizontal straight line. Distances in early descriptions may or may not be presumed to be along a straight line. The contrary must be proved.

PRINCIPLE 7. One can presume that when a magnetic bearing is indicated in a description, the declination to be applied for correction to a true bearing is the declination in effect at the time the description was written or when the survey was made.

PRINCIPLE 8. The trial team should determine the exact nature of the problem, whether it is one of boundary, title, or equity.

PRINCIPLE 9. The complaint, representing the plaintiff and with input from the surveyor, should raise as many issues as possible.

PRINCIPLE 10. The attorneys representing the defendant, with input from the surveyor, should raise as many defenses in the answer as possible.

PRINCIPLE 11. To have a positive line capable of being positively located, the original must create a corner at each end and the retracing surveyor must find them.

PRINCIPLE 12. Boundaries and their controlling corners are invisible until the creating surveyor places objects that must be described and referred to in the original documents.

The two principal responsibilities of a boundary surveyor are as follows:

1. To create and describe the original boundaries to a degree of satisfaction.
2. To conduct a retracement to "discover" and redescribe the boundary evidence left by the original surveyor.

3.2 OVERVIEW OF BOUNDARIES

Although it is difficult to categorize boundaries, we suggest for study purposes that the student consider two broad categories of boundaries. This classification is based on the magnitude of the boundary itself and the possible impact, both presently and in the future. The two broad classifications are *macro boundaries* and *micro boundaries*.

1. Macro boundaries
   a. International boundaries
   b. Boundaries between subdivisions of a nation
      1. States
      2. Counties
      3. Parishes in Louisiana
      4. Cantons in Switzerland
   c. Historic land grants by foreign nations

     d. Conveyance of lands by major land treaties

     e. Major land purchases

     f. Major land subdivisions

2. Micro boundaries

     a. Township surveys

     b. Section surveys

     c. Individual lot surveys

For discussion purposes, the average surveyor rarely is concerned with some of the classes of macro boundaries. Perhaps the practicing surveyor will locate some form of one of the macro boundaries once or twice. Other forms of macro boundaries may constitute a major portion of the surveyor's work.

Although the practicing surveyor may not make a distinction or consider the importance of this classification, it may be important *legally*. If a surveyor were to become involved in performing a survey for a tract of land whose common boundary was also a boundary between two states and the line ran along a boundary line between these two states, or even between two counties within a state, macro boundaries would become important. The surveyor's location of the state or county line would have no lasting or binding effect on its final location, because any disputed state line boundary can only be adjudicated by the US Supreme Court, which has exclusive jurisdiction over boundaries between states. A state legislature has jurisdiction to determine the boundaries between counties. The results of the surveyor are evidentiary only.

Following this reasoning, let us consider the following scenario for both a metes and bounds parcel of land and a General Land Office (GLO) parcel of land.

Metes and Bounds Creation

1. Boundary of land grants from the king of England to North Carolina and Georgia:

     a. Macro boundaries for grant lines, micro boundaries for colony lines. Recent archaeological investigations in Georgia indicate that there may be some vestigial Spanish land grants that could possibly have an impact in the future.

2. In Georgia, subdivision of lands into districts:

     b. Macro boundaries for state, micro boundaries for districts and lots.

3. Sale of a Georgia lot into fractions of a lot:

     c. Macro boundaries for lot, micro boundaries for fraction subdivision.

4. Sale of a fractional lot into a neighborhood subdivision:

     d. Macro boundaries for subdivisions, micro boundaries for lots in subdivisions.

The same basic breakdown of boundaries can be followed for a GLO creation.

GLO Creation

1. Establishment of a principal meridian and the respective townships
   a. Macro boundaries for the principal meridian, micro boundaries for the townships.
2. Establishment of the townships and the sections:
   b. Macro boundaries for the townships, micro boundaries for the sections.
3. Division of the sections:
   c. Macro boundaries for the sections, micro boundaries for the section subdivisions.
4. Sale of a single lot within a fractional subdivision:
   d. Macro boundaries for the section subdivisions, micro boundaries for the lot that the surveyor has been asked to survey.

Throughout this entire exercise, the surveyor may be required to identify and survey several macro and micro boundaries until finally getting to survey "John's Little Acre."

Once the macro boundaries were created, the governmental bodies that received title and into whose ownership the lands were placed had full and complete authority and responsibility as to how these large areas would be surveyed, described, and disposed of. Two basic methods evolved, which can best be described as rectangular and metes and bounds. Of the two broad categories, the metes and bounds system may be older, although some early Roman land divisions were determined to be rectangular in nature.

Following is a breakdown of the types of land boundaries that a surveyor could encounter in his or her career.

1. Rectangular division (creation into squares or rectangles)
   a. Federally created parcels (GLO)
      1. Created by federal laws
         a. Land Act 1785
         b. Land Act 1796
         c. Land Act 1805
      2. Surveyed before disposal
      3. Set rules for creation and retracement
      4. Field notes and plats
   b. State-created parcels
      1. Created under state laws
      2. May or may not have been surveyed before disposal
      3. No set rules for retracement
      4. May have copied the federal rules or may have been different

2. Metes and bounds surveys

    a. Usually created by actual survey

    b. May be created by words alone

    c. May or may not have notes and plats

    d. Retracement usually determined by case law

3.3 PUBLIC AND PRIVATE LANDS

Distinctions should be made between public and private lands, because this difference becomes important when title is obtained or transferred, either to or from the governmental body, and in gaining and losing property rights and locating and modifying boundary lines. The meaning of the designation *public lands* has changed over the years. Initially, the term meant what is referred to now as the *public domain*. In today's thinking, public lands are any lands in the ownership of the federal government of which the public domain is a part. There are many thousands of square miles of publicly owned land that have been exchanged, purchased, and confiscated under federal laws, including the Racketeer Influenced and Corrupt Organizations (RICOs) statutes, by numerous federal agencies. Military bases, national forests, national wildlife refuges, and other federal areas are also considered public lands.

The original public lands and all remaining public lands were and are under the control of Congress, with the respective federal agencies acting in a custodial capacity as managers. Today, some federal agencies have authority to dispose of their public lands, whereas others are restricted from doing so. Throughout the years, Congress has passed various acts to divest the United States of its public lands; the Homestead Act of 1861 (since repealed), the Swamp Land Act of 1849, the Desert Land Entries Act of 1877, the Timber and Stone Act of 1878, the Taylor Grazing Act of 1934, and the Small Tracts Act of 1964 and 1983 are but a few. Many of these still exist.

The basis of title from the United States to its original public domain is a patent that was signed by the president, granting title to private citizens. The patent acted like a quitclaim deed and carried with it the same guarantees, or lack of guarantees, as does a quitclaim deed. Today, when a federal agency divests itself of land other than public domain, a federal officer given the authority to do so usually signs a quitclaim deed to the grantee. Basically, the federal government makes no guarantees of title to the grantee. In conveying an original patent of public domain, certain premises are recognized. The conveyance is of the last official survey of record, the doctrine of after-acquired title does not apply, and there are no warranties. Also, the description is that according to the survey accepted most recently.

Private lands are all the remaining lands in the ownership of all but the federal government. This includes lands owned by private citizens, state and local governments, corporations, and, where permitted by law, lands owned by foreign governments. These private lands originated from many sources: the original public domain; state grants; foreign grants, prior to the American Revolution; and by various Indian and other treaties. Regardless of how these lands were obtained, the owners originally had to show their source of title. Title is the basis of legal possession.

3.4 SOURCES OF TITLE

Title is the means or authority by which one justifies legal possession of property; it is not ownership, only evidence of ownership. Title can be obtained in many ways. Some of these are:

1. Title by conquest or by war and the ultimate peace treaty.
2. Title by patent from the US government, a form of grant equivalent to a quit-claim deed.
3. Title by state patent or grant.
4. Title by deed or private grant.
5. Title by descent from a deceased party.
6. Title by will from a deceased person.
7. Title by involuntary alienation (bankruptcy or foreclosure).
8. Title by adverse possession.
9. Title by eminent domain (public taking for public benefit).
10. Title by escheat (reverting to the state).
11. Title by dedication (dedicating to public or private use).
12. At law (courts granting title to accreted lands).
13. Title by parol granting under certain strict requirements.
14. Title by creation: a person creates the land or personal objects.
15. Title by custom.
16. Title by prior appropriation (e.g., the use of moving water when a boundary).
17. Title by historical discovery and ultimate possession.

Principle 1.   *One who grants title to property to a second person, that grantor, can legally grant no more interest than that which is owned at the time of the conveyance.*

Although not of primary interest to the surveyor, there should be an understanding that a person who grants title to real property or to a real property interest can grant no more interest than that owned originally. If one owns a half-interest in a parcel, no more than a half-interest can be granted; that is, the person can sell the entire parcel, but the purchaser does not obtain the remaining half-interest in the document because it simply was not there to be granted. The document by which the interest was granted is very important. If a quitclaim deed was used, the grantee still got the half-interest. But if a warranty deed was used, the grantee can litigate for the remaining half-interest or its value.

3.5 VOLUNTARY TRANSFER OF REAL PROPERTY

In the United States, early land titles were originally obtained from various sovereigns of foreign nations. Some of the early pioneers would purchase the "rights" to the land being settled from the original indigenous peoples. This one act led to many future problems, in that the Native Americans from whom the land was bought only thought they were selling an easement for use because ownership of land was totally foreign to them. The titles were given either as gifts, for services rendered, or for a minimal payment of money, and there have been instances where the same land was given to different parties, either by the same or by different individuals. Title may have been given to individuals, companies, or various entities in the form of royal charters. In some instances, the grants were given and then withdrawn. In other instances, overlapping grants and charters were granted that caused serious future title problems. As far as public lands are concerned, written land titles were first acquired from the controlling government in the form of a quitclaim deed called a *patent*. Under present law, once a patent is issued to a party, a valid voluntary transfer of title to real property by that party can be made only by a written instrument. Unless the document indicates otherwise, a conveyance of land is presumed to be a fee simple conveyance. After a title transfer has been made, most instruments affecting title to real property are recorded in a public repository so that subsequent purchasers and others dealing with the property in good faith may rely on the record to disclose evidence of title to the property.

Many early deeds contain restrictions on transfer: for example, prohibitions of sale to certain classes or ethnic groups, or prohibitions against using a parcel for a specific purpose. For the most part, all prohibitions of sale to ethnic groups have been ruled unconstitutional, but restrictions for specific uses have been ruled constitutional. In many early English colonies, many deeds were *entail*, in that usually only the eldest son could inherit. The tail estate relative to intestate inheritance has been terminated by statute in all of the states, but that does not rule out leaving to the eldest son by will.

Although the surveyor usually is not considered an expert on title matters, a knowledge of title is necessary when having to make a distinction as to the crux of the problem being investigated or researched. If the problem is a question of boundary, then putting one's efforts into and financing unnecessary title issues is a waste of effort and the client's funds. The surveyor should have sufficient training and experience in title matters to be able to advise the attorney as to whether the problem is one of title or one of boundary.

Since the judge at trial has the ultimate decision as to the qualification of witness, the surveyor should possess a sufficient knowledge of title and title matters to answer the question, if asked by the court, "Who do you think owns the property?"

3.6 CHAIN OF TITLE

Except in the Torrens system, in the United States and other countries, the *evidence* of title is recorded, not the ownership of land. Because only the evidence of ownership is recorded, to prove ownership one must show an adequate, legal, and continuous title record relating back to the first conveyance that described the parcel. This compilation of all title owners is known as a *chain of title* or *chain of record*.

Within each recorder's office, a grantee–grantor index and grantor–grantee index are usually maintained, and every property owner is charged with constructive notice of the content of those indexes, whether or not he or she has read them. Because of this burden, most property owners engage third parties to conduct title research back to the title's creation. Title companies, for a fee, issue chain-of-title insurance to cover omissions. The necessity of maintaining a chain of title can be eliminated by litigation or judicial decree, which provides the land description and declares who has superior title. After the statutory time for appeal has expired, the court's findings are final. Fires in aged courthouses and several wars have added to the destruction of many original plats, maps, deeds, and documents relating to land. Early in the history of converting many of the original documents from paper originals to new technological recording systems, such as microfilm, the systems had not been perfected to the point that one could totally rely upon them. With the assurances of some "experts" that the new systems were foolproof, the original written documents were destroyed. But the new systems were not as foolproof as originally thought, and the disintegration of the new record systems wreaked havoc on the recorded documents and the ability to recover historical records. In some areas today, researchers must contend with this problem.

The surveyor should realize that title is important in two ways. First, the original title description is predicated on the description that the surveyor creates as a result of the original survey. Second, whereas the attorney may have to research only a certain number of years (usually, 35 or 30–60 years, depending upon the standard of the particular state) according to title standards to perfect title, the surveyor must research the original title to perfect the description of the boundaries to their original creation.

Some state minimum standards of practice require that a description be traced back in time to its origin.

The Maryland court stated in the case of *Ski Roundtop Inc.* v. *Wagerman* et al.[1] that a "conveyance of public land by State is requisite for valid title to real property, and absent such conveyance, one purporting to transfer ownership interest in such property transfers nothing and no quantity of successive transfers by deed nor mere passage of time will metamorphose good title from void title." In this case, the two parties were in litigation over whose title a parcel belonged with, only to find that, after tracing the titles back to the sovereign, neither chain of title contained the parcel in question.[2]

The Pennsylvania court, in the case of *Roth* v. *Halberstadt*, stated that "the primary function of a court faced with a boundary dispute is to ascertain and effectuate the intent of the parties at the time of the original subdivision."[3]

3.7 TORRENS TITLE SYSTEM

In the Torrens system, the *owner* of the title is registered. In several states, an attempt has been made to establish the Torrens system, and it has been partially success-ful in Minnesota, Massachusetts, and the city of Chicago. To start a Torrens system and register the owner and location of each parcel, a court finding that eliminates the necessity for a chain of title and declares the land's location must take place. The survey and court costs for such findings have been a deterrent to widespread use of the system. In Chicago, a fire destroyed all the public title records, and it became necessary to have a court determination of who had ownership. In this instance, it was relatively easy to start the Torrens system. In Massachusetts, a Land Court was estab-lished to try all land boundary cases, to declare who had ownership, and to guarantee title to the person listed as owner. In essence, the Massachusetts Land Court system of recording who has ownership of land is a Torrens title system.

In the Torrens system, the law declares that no one shall acquire a right by prescrip-tive means; that is, adverse rights cannot ripen into fee title. One problem associated with the Torrens system is that the state must guarantee the right of ownership; thus, there must be an assurance fund to pay the costs for any error on the part of the court in its initial determination of who has the right of ownership. If the fund is insufficient, those loaning money and using the property as a lien will refuse to do so. In several states, this was one cause of disuse of the Torrens system.

Although the Torrens title system will certify title, it does not eliminate bound-ary disputes to the registered land. It only isolates or simplifies the issues, in that if registered parcels are involved, title does not have to be proven by the parties; thus, it becomes a simple issue of boundary location only. One must examine the specific registration statutes that created the title system, because some do not offer protection in that there can be no adverse possession of registered land.

3.8 UNWRITTEN RIGHTS OR TITLE TO LAND

Except for those states that recognize the Torrens system, every state recognizes that land can be transferred involuntarily from one person to another person under the law and that land boundaries can be altered by the acts of parties without written docu-ments and without altering title. Prolonged occupancy, agreement, and several other doctrines recognize acts, behaviors, and actions of adjoining landowners. Usually, the surveyor is not involved in determining the validity of adverse possession, agreement, or other doctrines, but the surveyor must have an acute awareness that these doctrines exist in the law and in surveying. The surveyor must also realize that it is the evidence that is recovered in the survey process that ultimately determines the validity of these doctrines.

Title to property or a property interest must be in writing to meet the legal requirements required by the Statute of Frauds.

Depending on the principle accepted by the respective jurisdictions, one does *not* have to initiate litigation to get title by adverse possession, for title is created or passes instantly at the exact moment when all the requirements are met. One will find deeds stating that "title is based on adverse possession"; it may be necessary to ask the court to perfect the title through litigation in order to have the court grant the claimant a "marketable title" based on the adverse claim, but that is incidental. This often is accomplished with an action to quiet title.

No legal doctrine actually conveys title. It has been held, though, that adverse possession creates a new title, but there is no written title until perfected and awarded by legal proceedings. There must be a writing to have proof of a good and marketable title.

3.9 METHODS OF VOLUNTARY TRANSFER OF TITLE

Title to real property may be transferred by voluntary means, involuntary means, or inheritance. In most instances, voluntary transfers must be in writing to meet the requirements of the Statute of Frauds. Involuntary transfers can result from foreclosures, failure to pay taxes, escheat, adverse rights, estoppel, liens, and bankruptcy. At one time, the transfer of ownership of freehold estates could be done without written documents. The process, called *feoffment*, merely required *delivery of possession*, evidenced by a ceremony known as *livery of seisin*, where in the presence of witnesses, a twig, a piece of turf, or some symbolic item of land was delivered to the new owner. Easements and other encumbrances incapable of possession had to be transferred by writing. In England, the Statute of Frauds required land or an interest in land to be transferred in writing, and in the United States, similar statutes were adopted by each colony and each state.

The most common methods of transferring title to real estate voluntarily are by grant deed, quitclaim deed, warranty deed, easement deed, and will. A *deed* is a conveyance of realty whereby title to or an interest in real property is transferred from one party to another. In a *grant deed*, unless otherwise stated, it is implied that previous to the time of execution of the conveyance, the grantor had not conveyed the same estate to another, and that the estate was free of encumbrances (e.g., mortgages), unless otherwise stated.

A *quitclaim deed* operates by way of a release; the grantor merely passes to the grantee whatever title, right or interest he or she may have had at the moment of signing the conveyance and upon delivery of the document. Under some state laws, the grantor warrants that neither the grantor nor anyone claiming under him or her has encumbered the property.

In a *warranty deed*, the grantor assures the grantee that he or she has good title and can be held liable for deficiencies that are not declared. The grantor warrants to defend or make compensation if the title is not valid; the choice is usually the grantor's.

An *easement deed* grants to another the right to a particular use of a property. Such a deed only grants one of the bundles of rights, not title to the entire parcel. This term is considered by some courts to be ambiguous.

3.10 DEED OR DESCRIPTION

The surveyor should not confuse the term *deed* with a description. A *deed* is the instrument or document by which a property interest is conveyed. It must meet certain requirements. A *description* is an intimate part of the deed, necessary to define the interest or land conveyed. Although there are common requisites of a deed, state statutes prescribe the necessary elements. There must be sufficient writings showing the names of the grantor and grantee, operative words of conveyance, and a sufficient description of the property. There must be competent parties; that is, parties must have the legal capacity to convey and receive. The description of the property must make possible location by *any* competent person. The deed must have been properly executed, delivered, and accepted. In some states it is not necessary to have a consideration, but in most states a consideration must be mentioned. It does not have to be the correct amount of the value of the parcel, but a reason such as "love and affection" may be questionable.

Case law has repeatedly held that the purpose of a description is to provide the "key" to finding or locating the land in question. To this should be added the reason for the description and when it is sufficient. The description is but one element of a deed. Courts have held that a deed may be void, voidable, or valid. This is for a judge to determine after hearing the evidence. To be a *valid deed*, the description must be sufficient and must describe a parcel of land that can be identified.[4]

> Whether a description is valid, void, or voidable is usually a question of law.

Every now and then, a decision will be rendered that exhibits wisdom from which a practicing surveyor or attorney can learn. The Florida Supreme Court set forth pearls of wisdom with regard to ascertaining the validity of a description.[5] Relative to a possibly ambiguous description, the court wrote:

"Although title attorneys and others who regularly work with them [descriptions] develop expertise as to land descriptions, the only professional authorized to locate land lines on the ground is a registered land surveyor . . . . *In fact, the definition of a legally sufficient real property description is one that can be located on the ground by a surveyor.*" (Emphasis added.)

Depending on the expertise, training, experience, education, and knowledge of the retracing surveyor, under this scenario four different surveyors could all meet and have four different situations.

1. Surveyor A can state: "This is a valid description because I can survey it and locate it with a great degree of certainty."
2. Surveyor B can state: "No surveyor, including me, can survey this description with any certainty." Then the description should be considered as void.
3. Surveyor C can state: "I am unable to survey this description because I do not have the experience to do so."
4. Surveyor D can state: "If I am to get more information, then I can give you an answer."

> An attorney relies on a deed to ascertain the quality of a title; the surveyor relies on the deed to ascertain the quality of the description.

3.11 TITLE OR LIEN

Often, there seems to be confusion as to what constitutes a lien. A *lien* is a charge, security, or encumbrance on a property to secure a debt. A lien may result by operation of law, as a result of trial determination, taxes, and so on. It permits the holder to recover the value of the lien against the property.

Perfect title is composed of two parts: equitable title and legal title. The formula is

> Legal Title + Equitable Title = Perfect Title

Unlike a lien, when a landowner gives a security deed or deed of trust to a parcel, the equitable title is granted for the money. The landowner still retains legal title, and it is registered in the titleholder's name. But the security deed is registered, indicating that the lienholder actually owns equitable title to the property.

The area of liens is really not the responsibility of the surveyor, who is retained to ascertain the quality of the boundaries. If in the course of work it is determined that a lien does exist, the surveyor has the obligation to inform the client of its existence.

3.12 DEED OF TRUST

In a mortgage, there are two parties: the *mortgagor* (the person in debt) and the *mortgagee* (the party lending money). In a deed of trust, there are usually three parties. The first party deeds the land to a *trustee* (the second party), who holds the deed in trust for the third party under the condition that the third party performs his or her obligations. Trust deeds generally came into being to make it easier for the mortgagee to foreclose on a debt.

3.13 MORTGAGE

This section is added to remind the surveyor that this area is primarily legal in nature. A document may look like a deed, read like a deed, smell like a deed, and even have the "magic" words of a deed, but it may not be a deed. That will be for the judge to decide.

A deed of trust is used in many states as a substitute for a common-law mortgage, where the legal title to real property is given to trustees either to repay a sum of money or for other considerations. It differs from a mortgage [death pledge] because it is security, usually for money.

A *mortgage* is a written interest in land created to secure the performance of a duty or payment of a debt. The term *mort* means death, and *gage* means pledge. For failure to pay off the mortgage, the land gave no return to the person in debt and was "dead." A mortgage is nothing more than a lien on land; in the event of failure to meet an obligation, it gives the person holding the mortgage the right to go to court and sue for title and possession of the land described in the mortgage.

3.14 ESCROW

Certain rigid laws must be complied with in the transfer of real estate. The seller cannot be expected to give up the title prior to receiving consideration, nor can the purchaser release the consideration until he or she has a clear title. This makes it necessary to have a trusted third party who oversees all conditions of a sale. In an *escrow*, "a grant may be deposited by the grantor with a third person, to be delivered on performance of a condition, and, on delivery by the depositary, it will take effect. While in the possession of the third person, and subject to conditions, it is called an escrow."[6]

3.15 TITLE ASSURANCE AND TITLE INSURANCE

Title assurance and title insurance are not the same. Because only evidence of title is recorded and because most purchasers are not qualified to determine whether the seller has good title, certain systems of title assurance have evolved. In general, we have abstractor opinions, attorney opinions, and title insurance. Title insurance is to real property what life insurance is to a person. For a fee, a guarantee of title, of location, or against all risks, including survey, can be obtained. Of course, the fee charged varies with the extent of the risk and the value of the property.

In the standard title insurance policy, the liability usually docs not include the following:

1. Losses arising from defects of title known to the insured to exist at the date of the policy;
2. Easements and liens that are not shown by the public records;
3. Facts, rights, interests, or claims that are not shown by public records, which impart constructive notice but that could be ascertained by inspecting the land, by making inquiry of persons in possession thereof, or by a correct survey;
4. Mining claims, reservations in patents, water rights or claims, or title to water, whether or not of record;
5. Acts or regulations of any government agency regulating occupancy (zoning) or use;
6. Items outside the period of search; and
7. Today, possibly, hazards contamination situations.

An *extended coverage policy* (ALTA) will rule out or eliminate, when possible, all of the items in the preceding list except item 1. Extended coverage usually requires a survey and an investigation of known and visible underground cables or utilities. The surveyor is frequently asked to certify that there are no underground utilities.

Principle 2. *Title insurance states only that if a defect is found in the title that is insured, the insurer will either defend the title in court or make good the value for the loss.*

If it is found that there is an insurable defect in an insured title, the landowner usually must inform the insurance carrier, in writing, of the claim, and the insurance company has the choice of either defending the defect in the title or paying the value of the depreciation of the property to the insured. In most title policies, questions of survey are usually not insurable. Some companies will write exceptions which will insure the survey that was conducted.

Today, the American Congress on Surveying and Mapping (ACSM) and the American Land Title Association (ALTA) have agreed on minimum standards that must be adhered to for all surveys that are to be insured under a title policy of a member organization. Many surveying organizations have adopted these standards as the minimum they will accept in their states. The major problem with these standards is that they address technical work that is being conducted, in that any work conducted will meet these standards. They do not address the requirements that the surveyor when starting the survey should be on the correct corner(s). These standards do not address corners; they only address measurements. To be able to measure any line, one must have the two proven, not resurveyed, corners, one at each end of the line.

There may be instances when a claim is made on a title insurance policy and the policyholder is not satisfied with the desire of the title insurance carrier either to pay on the claim or to defend it at trial. The choice of which selection to make is not that of the insured but that of the insurance company. Usually, after weighing the value of the claim against the cost of litigation, the least expensive course of action will be taken—but not always.

3.16 ABSTRACTORS

An abstractor compiles a chain of title on parcels of land from its origin or from a set time in the past (usually 40 years or more, set by accepted community practice or by statute) to the present time, all in accordance with the public record; unrecorded items are not included. An abstract is a collection and chronological summary of any instruments or documents of record affecting rights in the property. An abstractor is responsible for the accuracy of the records but is not responsible for the legality of each recording.

Abstracting ancient land records is fast becoming a lost art. Years ago, most young attorneys put in time in local courthouse record rooms "searching" or abstracting titles. These attorneys went on to become circuit judges and appellate judges, taking with them their knowledge of titles and boundaries. This is no longer so. In courthouse record rooms, paralegals and registered surveyors are the usual visitors. We now have a generation of judges who are knowledgeable in contract and criminal law but have very little knowledge of real property law and boundaries. In fact, today, one may find attorneys and judges who never have "run a title" to a parcel of land.

Today, it seems that abstracting is being replaced by buying title insurance and the prayers needed to assure proof of ownership and lines.

3.17 ATTORNEY'S OPINION

After the preparation of an abstract of title, a title opinion by an attorney versed in the intricacies of land law—such as probate, bankruptcy, corporation, divorce, and other areas of law having a bearing on the capacity of the parties to the transaction—is needed. An attorney's opinion is usually limited in liability to mistakes in judgment regarding items appearing on the record.

3.18 GENERAL LAND DESCRIPTIONS

The land description is a necessary and vital part of a valid land conveyance. The description is one of the required criteria that can determine whether the deed is valid, void, or voidable. For years, there has been a debate about who is "most qualified" to write a description. The debate rages on; for up until now, the question has not been answered. One will find landowners writing their own descriptions without the benefit of legal or survey assistance, attorneys writing descriptions without the benefit of adequate information, courts writing descriptions without the benefit of input or professional assistance from companion professions, and surveyors writing descriptions without full knowledge of the facts about which they write. The person most qualified to write a description is that person who has the greatest knowledge, who possesses the best vocabulary, and who has the greatest use of the English language.

3.19   WHAT IS IN A DESCRIPTION?

Principle 3.   *A land description in a deed conveys only the interest described by the wording, terminology, and survey measurements that are depicted at the time the survey was conducted and the description was written.*

When conveying real property, the law requires the conveyance to be in writing to be legally enforceable.

The deed cannot transfer any real property interest without having an adequate and legal description of the parcel's rights, including some description of having the "key" to locate the parcel and its boundaries.

In any description, one should find, as a minimum, the identification of corners, monuments, adjoiners, bearings, and distances. Each of these offers untold opportunities for the creation, the description, and the interpretation of the original description. The wording one should apply is the wording in effect when the deed was written and the original description was created, and when it is recorded or retraced. Any surveyor or attorney who works with deeds and descriptions often finds that there may be a considerable lapse of time between when the deed is written and when it is recorded and a subsequent survey for a retracement is conducted. Thus, the key time regarding the meaning of words and phrases is when the conveyance was written or delivered.

Principle 4.   *To be absolutely controlling in a resurvey or retracement, a corner and its monument must be called for in the original description.*

In some manner, every land description must be related to existing or formerly existing monumented corners. This principle has been sealed in history since before the birth of Christ. A professional burden is placed on the retracing surveyor who is employed to locate a boundary line to be able to use the documentary evidence of title, which should contain the proper relative description to retrace the desired boundaries.

In creating descriptions, people have used an array of monuments to identify corners: a skull, a hole in the ice, a wagon skein, a thigh bone, millions of IPs (iron pins or iron pipes), and even wooden stakes, including what one description called "a corner formally an oak stump." Before minimum technical standards were adopted, surveyors were left to their own imaginations as to the material that suited the needs of the project, but since minimum technical standards have been instituted, states have designated the minimum monument that a surveyor can use. Failure to call for a monument at a corner point can be a serious flaw in the description and may be considered a technical or legal violation of a statute or rule.

States apply this principle differently. States with liberal philosophies hold that if a corner monument was placed at or near the time of the original survey, it controls. Conservative states apply a strict application. To control, the corner monument must be placed and called for in the survey and the deed. A surveyor must know how the respective state applies this principle. This is not a legal question but is actually a survey question.

A recent California decision held that subdivision monuments that were set some 20 years after the original subdivision was laid out could be considered as original monuments.

3.20 MEASUREMENTS

Principle 5. *The unit of measurement indicated in the description is the unit of measurement used at the time of the survey or when the description was written and should be the reference measurement to which the surveyor should refer.*

Second to the originality of monuments that have been used is the use of measurements. Surveyors frequently relate tales about the odd measurement they encountered in a retracement that possibly defied logic and application in light of today's situations. Some odd measurements have included "2 smokes," of Spanish origin; "795 feet DBH" (distance by hollering), from North Carolina; "20 outs," from Arkansas; and "fence lengths," in New Hampshire and Vermont.

Measurements become important in several ways:

1. A measurement, accompanied by a bearing, is an important element of a boundary line, as it then becomes a course.
2. A bearing is a measurement; it is the measurement of any angle.
3. A measurement identifies objects along the lines or in the property surveyed.
4. The units of measurement used in describing the lines determine the area.
5. The surveyor should not dismiss the validity of the unit of measurement; if he or she cannot determine its value, the court will adjust to make the description valid.

Although there is no codified unit of measurement in the United States, custom has made the *foot* the accepted unit of measurement. The length of the US foot has varied slightly from time to time, usually with negligible importance to land measurements of a local nature. The *standard meter*, as defined in France, is the distance between two marks on a platinum–iridium bar kept in Sevrès, near Paris. Prior to 1959, the *standard foot*, as adopted, was 1200/3937 meter. In 1959, this was changed to 1 inch equals 2.54 centimeters exactly, or 1 foot equals 0.3048 meter. The meter was then defined as equal to 1,650,763.73 wavelengths of electrically excited krypton-86 gas.

In 1983, the definition of the meter was again redefined to equal the distance light travels in a vacuum during 1/299,729,459 of a second. Obviously, these later refinements are for the benefit of scientific measurements—land is not measured with such precision. For purposes of land measurement, the old value of the foot (1200/3937 of a meter) is still applicable.

Table 3.1 lists some of the various units of measurement that surveyors have encountered in surveys and descriptions and then have tried to retrace. The most prominent unit, starting in colonial times, was the *chain*, consisting of 100 links, with 4 poles or rods of 25 links each. This was developed in 1620 by Edmund Gunter, an English mathematician, and in all probability derived from the pole. About the same time, there was the Leybourn chain of 50 links or 33 feet. In the early days of our country, the *pole* enjoyed a lack of actual length. It was found to vary from as little as 12 feet to as much as 22 feet. Gunter developed the 66-foot chain and probably devised the fictional unit of area, the *acre*, which is equivalent by custom to 10 square chains or 160 square poles. Unfortunately, we have become "conversion crazy" and now readily accept that a chain equals 66 feet and an acre equals 43,560 square feet.

The *rod* (16½ feet) was established in the sixteenth century as the length of the left feet of the first 16 men out of church on a certain Sunday (*The Amazing Story of Measurements*, The Lufkin Rule Co.). In John Love's textbook *Geodaesia* (1796), "to lay out new lands in America, or else where," is the statement: "In some parts of England, for wood-lands, and in most parts of Ireland, for all sorts of land, they account 18 feet to a perch (rod), which is called *customary measure*." In some of the eastern towns of New England, the Scotch-Irish did use such a measure. In Virginia, New Hampshire, and other colonial states, 5% excess was allowed for variation of the chain. In some early sectionalized land surveys, the custom was to add 1 inch to the chain, for good measure. In one survey in Maine, the surveyor stated, "Add one link per chain to take up for sag." In *Geodaesia*, John Love discussed the various types of chains as Leybourn's *being 2 poles or 33 feet and the Gunter's being 66 feet in length.*

In Louisiana, 100 *French feet* are equivalent to 106 feet, 6 inches, and 7 lines US measure (written 106′6″7‴). A *line* is equivalent to 1.8 inches. In a few states, 12 lines equal an inch.

A large portion of the public domain was made up of lands acquired from Spain, Mexico, and France by cession or purchase. Before the acquisition of these areas by the United States, many land grants using foreign measurements were made to private persons, and such land grants, when duly authenticated and confirmed by court decree, were segregated from the lands subject to disposal. In the southwest and in Florida, the Spanish and Mexican unit of length, the *vara*, is equivalent to approximately 33 inches. The French crown grants prior to the Louisiana purchase were expressed in terms of the *arpent*, which is an area unit equal to approximately 0.85 acre. The *arpent frontage* unit is the length of the side of 1 square arpent or approximately 192.5 feet. Owing to a lack of exact standards, the lengths of the vara and the arpent differed in various localities, as can be noted from the average values shown in Table 3.1.

TABLE 3.1   Common Units in Deed Description

1 barleycorn	= 1/3 inch
1 line	= 1/8 inch (Louisiana)
1 chain	= 66 feet
1 chain	= 100 links
40 chains	= 1/2 mile
80 chains	= 1 mile
1 chain	= 4 rods
1 chain	= 4 poles
1 chain	= 4 perches
1 link	= 0.66 foot
1 rod	= 16 1/2 feet
4 rods	= 1 chain
1 mile	= 5280 feet
1 mile	= 80 chains
1 geographic mile	= 6076.1033 feet
1 acre	= 43,560 square feet
1 hide	English land measure between 60 and 100 acres
1 acre	= 10 square chains
1 sq rod	= 1/4 acre
1 vara av	= 33.372 inches (Florida)
1 vara av	= 33.333 inches (Texas)
1 arpent	= 0.8507 acre (Arkansas and Missouri)
1 arpent	= 0.84625 acre (Mississippi, Alabama, and Florida)
1 arpent	= 0.845 acre (Louisiana)
1 arpent	= 30 toises
1 out	= 10 placements of 1/2 chain, or 5 chains
30 toises	= 160 French feet
1065.75 feet	= 1000 French feet (Louisiana)
Side of a square arpent	= 192.50 feet (Arkansas and Missouri)
Side of a square arpent	= 191.994 feet (Mississippi, Alabama, and Florida)
1 foot	= 1200/3937 meter
1 circle	= 360°
1 step	About 3 feet by case law
1 pace	About 6 feet by case law
1°	= 60′ (called *minutes*)

For practical purposes, one could say that the 2-pole chain (33 feet) is probably the "official unit" of measure in the GLO states, with the 1-pole chain being used in the early metes and bounds states and the foot in modern surveys. When it enacted the Land Act of 1785, Congress stated that the lines shall be measured with a chain. However, it failed to identify the length of the chain. It was not until the Land Act of 1796 that Congress specified that the chain would be "two poles in length, yet returns will be kept in units of four poles."

Today, few surveyors use a metal tape or chain. Most land survey measurements are made using electronic distance equipment, which includes the generic term *total station* as well as *global positioning system (GPS) measurements*. Simply put, the total station is an instrument that encompasses both angular measuring capabilities and distance-measuring capabilities. Whatever distance indicated in the scale of the instrument is a converted distance, because all electronic distance equipment measures time frequency and converts that to metric units, which in turn are converted to feet. These final measurements are only as accurate as the indexing of the instrument and the correction for atmospheric conditions.

In some descriptions, certain recited units may indicate both a distance and an area. The retracing surveyor must be familiar with the distinction.

> In interpreting descriptions and then retracing those descriptions on the ground, the surveyor should think only in the units of measurement that created the boundaries of that parcel.

One of the principal bases on which this principle rests is the legal and survey responsibility that a surveyor faces in conducting a retracement: "finding and following the footsteps of the original surveyor." When the original creating surveyors surveyed and described the original boundaries, the survey errors, blunders, and relationships were in the units that field surveyors used. If they surveyed in 2-pole chains, all distance references, errors, and mistakes were in units of 2 poles, not 33 feet, the converted unit. Or if the unit was the 1-pole chain, a distance of 10 poles is 10 poles, not 660 feet (converted). The surveyor must look at significant numbers, in that 10 poles actually means a distance of 9.5 poles to 10.5 poles, or a spread of 66 feet. Usually, the early distances were recited to the nearest whole unit.

This reasoning also applies to the reciting of angles. Most early surveys and descriptions were read and written to fractions of degrees. Thus, an angle recited in a deed as "north 52 and one-half degrees east" (usually indicated as N 52 ½ E) does not equate to a modern measurement of "north 52 degrees and 30 minutes east."

> The units of measurement in which the original boundaries were created become the controlling units for retracement of and other relationships to these boundaries.

Today, the advent and universal acceptance of measurements by global positioning methods places the retracing surveyor in new areas of conflict in meeting this legal requirement, in that with GPS measurements, only the endpoints of the lines are determined and the resulting lines between these corners are a product of calculations.

Principle 6. *Distances cited in modern descriptions are presumed to be along a horizontal straight line. Distances in early descriptions may or may not be presumed to be along a straight line. The contrary must be proved.*

In GLO surveys, the presumption is that all measurements are horizontal along a straight line because the law required surveyors to measure in that way. In metes and bounds states, early measurements are presumed to be "slope" or "along the lay of the land." The contrary must always be proved. This presumption has not always been in effect; in a few localities, proof has been found indicating that original measurements were made along the surface. In 1960, a Kentucky court ruled that ground distance was proper where it is the custom of the locality or where it is dictated by circumstances. The court said:

Where deed to lower lot called for a depth of 60 feet "up the hill," and if a horizontal rather than surface measurement were used in determining such 60 feet, a portion of grantor's house would have been on the lower lot, and for some 13 years after the deed was executed the parties treated the boundary line as being from a tree behind one of the grantor's houses, a point approximately 60 feet by surface measurement up the hill from a street, and one of the grantees in the original deed, a predecessor in title to owners of lower lot, testified it was his understanding that the distance called for in the deed was by surface measurement; finding that surface measurement rather than horizontal measurement should be used in measuring depth of lower lot on the hill was proper.[7]

3.21 MAGNETIC DIRECTIONS

Principle 7. *One can presume that when a magnetic bearing is indicated in a description, the declination to be applied for correction to a true bearing is the declination in effect at the time the description was written or when the survey was made.*

Until the early twentieth century, the magnetic compass was the principal instrument used in conducting surveys in the United States. Today, a vast array of instruments, from now-antiquated transits and theodolites to total stations and global positioning coordinates, have been used and in some instances are still being used to conduct surveys for descriptions.

At one time, magnetic bearings were understood by all surveyors, students, judges, and attorneys, but today most modern surveyors, students, and attorneys do not comprehend magnetic surveys, as the practice is not taught as a major subject in the schools that teach surveying courses. Because of the numerous early original compass surveys that created boundaries, both GLO and metes and bounds, it is necessary to have a foundation in and understanding of magnetic surveys and of the Earth and its magnetism in order to conduct retracements.

Because the celestial poles and magnetic poles arc not coincident, the compass needle points to, or is attracted to, the magnetic poles. Depending on where one surveys, the degree of mispointing varies from approximately 23° east of the North Pole to approximately 24° west of the North Pole within the contiguous 48 states; in Alaska the variation reaches 35° east of north. The angle between true north and magnetic north is known by several names. The terms *magnetic declination, variation,* and *delineation* have all been used to identify this phenomenon. Declination does not remain stable. One can find a daily variation of as much as 15 minutes in a 24-hour period and an annual variation or change of minutes per year. Magnetic bearings are also affected by local attractions and magnetic storms. These unpredictable changes have troubled surveyors for hundreds of years. In some areas of Michigan and New York, local attractions of 45° or greater have been noted, whereas magnetic storms of 1° are common. Measured magnetic bearings have been and still are being affected by many misunderstood forces.

Many surveying texts and several US government agencies publish isogonic charts, which indicate probable declination, both nationwide and worldwide, every five years. If a surveyor wishes to know what the declination was in the past, the information is easily available, usually on a 15-quadrangle basis, beginning from approximately 1785.

Molded with the inaccuracies of magnetic declination are the mechanical attributes of the compass itself. Controversy has raged for years about what a surveyor could expect from a compass. Most compasses were usually graduated to half a degree. By interpolation, readings could be made to one-eighth of a degree. This could be reduced to an angular closure measurement of 1/300, or 1 foot in 300 feet.

If a compass was used with care and understanding, one could expect a survey closure of 1 in 300 or more. Early court decisions usually gave precedence to compass measurements over distance measurements, and several states have enacted statutes stating that when a magnetic bearing and a distance are in conflict, the magnetic bearing will be given preference or will take priority.

In working with original compass bearings, the trend today is to convert these historical magnetic bearings to modern-day true bearings. This requires the ability to determine what the magnetic declination was in the year and at that point of the survey, to determine a true bearing at that time, and then to determine the true bearing today. Making reference to a magnetic bearing of 1810 and a magnetic bearing of 2003 does not lend itself to sound modern surveying practices.

In working with historic magnetic bearings today, few of the younger retracing specialists realize that the law of retracements places little confidence in historic measurements, but judges tend to love them, because they are "here and now." In the 1820s, Isaac Briggs, deputy surveyor in the Mississippi Territory, was having instrument problems with his men. He created a true North–South line and had each of his twelve surveyors "read" the line with each individual compass. His results were amazing. Only two of them read the same bearing. The others' compasses varied as much as three-fourths of a degree. In practical application, if each tried to run a single line, on the same bearing, in 1 mile the lines would diverge as much as 66 feet.

In 1854, Justice Lumpkin of Georgia wrote in his classic opinion, *Riley* v. *Griffin*, one comment of eighteen, as follows:[8]

A. Courses and distances occupy the lowest, instead of the highest[,] grade, in the scale of evidence, as to the identification of land.

B. Courses and distances, depending for their correctness on a great variety of circumstances, are constantly liable to be incorrect; difference[s] in the instrument used, and in the care of surveyors and their assistants, lead to different results.

Over 150 years ago, a Supreme Court justice, who was familiar with surveying practice, had the insight to see beyond what none of the learned justices understood.

3.22 REFERENCE DATUMS

A *reference datum* can be defined as any position or element from which angular measurements are determined. All measurements, whether distance, angular, or elevational, must have a reference datum.

A datum can be either vertical or angular. Some reference datums that have been used are:

True (whatever that is)
Magnetic
Geodetic
Approximate
Assumed
Grid
Estimated

Some references may seem absolute, but further investigation may reveal otherwise.

Unless the reference datum is referenced to an absolute reference it should be assumed to be "ambulatory," or subject to change and not fixed.

There are special boundary lines that may be created by people or by nature. These are *contour boundaries*, which are usually identified by elevation. The usual methods to locate these are either to run levels along the contour elevational line and locate the line on the ground physically or to identify the contour line on topographic maps. Some surveyors may run a traverse line along the contour line and locate this line by measurements. This identified line then becomes the original located boundary line, regardless of where subsequent leveling may place it. This elevation can then be located on paper by platting.

In most instances when descriptions refer to a contour elevation, they usually omit the reference datum. Whenever a datum is referenced in a description, the scrivener should identify the reference datum in the Caption or another place so there will be no problem when individuals attempt to retrace the line.

Mean sea level datum is determined at a given location by the average of the hourly tide readings over an 18.6-year period. A description reading "all of that portion of Section 2, T 15 5, R 2 E, SBM, lying below elevation 2110.00 feet" would ordinarily be interpreted to be "all of that land lying below 2110.00 feet, said elevation being relative to mean sea level datum," as benchmark elevations are normally referred to that datum. If, however, the deed had read "all of that portion of Section 2, T 15 5, R 2 E, SBM, lying below elevation 2110.00 feet, said elevation being relative to the crest of Singer Dam whose crest is 2100.3 1 feet," the datum is now the crest of said dam, not mean sea level.

The vertical boundary line (contour line) in the deed was created in 1957 and was described as "all that land below the 1085 contour." The controlling 1085 contour was that location at the time the deed and description were created, not the 1085 contour as located in the year 2003. Vertical boundaries present unique problems for surveyors who create them and for those who retrace them.

In the Lambert or Mercator Grid Projection, two datums are used: (1) the origin of X-coordinates and (2) the origin of Y-coordinates. To define the geodetic datum to which the X- and Y-coordinates are referred, eight elements are needed.

Most metes and bounds descriptions have a datum that is better known as the *point of beginning* or the *true point of beginning*. Such an extension of the meaning of *datum* is not common. *Latitude* has a reference datum called the *equator. Longitude* is determined from a datum point arbitrarily selected in Greenwich, England, called the *prime meridian*. Other datum points and lines include Clark's spheroid of 1866, magnetic north, astronomic directions, and geodetic directions.

3.23 ELEMENTS OF LAND DESCRIPTIONS

To cook a perfect omelet or recipe, you must have all the ingredients; a land description must have its perfect elements as well. Descriptions of land usually are composed of the following:

1. The *caption*, which cites the general locality, the map number or reference document, city, town, county or state, and other matters of general interest.
2. The *body* of the description, which includes the calls for corners, monuments, bearings, distances, adjoiner properties, and a call for a precise area being conveyed.
3. *Qualifying clauses*, which take away something included within the body of the description.
4. *Augmenting clauses*, which may give something in addition to what was conveyed in the body, such as an augmenting easement for ingress and egress.

3.24 TYPES OF DESCRIPTIONS

Like show dogs, descriptions come in all sizes, shapes, forms, and colors. Although we can indicate the most common forms, one may always encounter a unique description that has never been observed before. Some of the most common forms of descriptions are (1) perimeter; (2) bounds; (3) strip; (4) reference, including aliquot portion; and (5) a combination of any or all of the preceding.

The names used for the various types of descriptions differ somewhat from state to state. *Perimeter* metes and bounds descriptions may be running or bounding. In a *running description*, the scrivener proceeds from a point of beginning and travels in a clockwise or counterclockwise direction around the parcel, reciting monuments, directions, distances, and sometimes abutting owners, and ends at the point of beginning. *Mete* means to measure or to assign measure, and *bounds* means the boundaries of the land or the limits and extent of the property. Within the generally accepted use of the term *metes and bounds*, it is not necessary to recite measures of a property as implied by the word *metes*, as "Beginning at an oak tree blazed on the north; thence to a boulder located on the bank of Lake Victoria, thence along the lake shore to … "

A *bounding description* is written or interpreted as if one were in the center of the parcel looking out, and recites information regarding the sides of the tract in sequence, in either a clockwise or a counterclockwise manner, as "bounded and described as follows: westerly by the land of Smith 200 feet more or less; northerly by the land of Jones 150 feet more or less; easterly by the land of White 200 feet more or less; southerly by the highway 150 feet more or less."

Bounds descriptions name adjoiners or monuments but do not have a direction of travel and often have no measurements, such as "all of that land bounded on the north by Thelma Lane; bounded on the south by Alvarado Creek; bounded on the west by the land of Thomas L. Brown; and bounded on the east by the land of Ruth Almstead." The sequence of reciting bounds is immaterial. A *metes description* is a perimeter description reciting measurements but not bounds. Often, metes descriptions are included within the meaning of metes and bounds descriptions. In court reports, the term *course* is often used to mean direction only, as "course and distance." Surveyors normally use *course* to mean both direction and distance, as in the "fourth course," which is cited as N 10° 12′ E, 300 feet.

Perimeter descriptions are those most sought by researchers and surveyors, since details of lines and corners increase the comfort level of evidence collection. In some areas, many early metes and bounds perimeter descriptions have been replaced in conveyances with less detailed ones, usually of the bounds type. Those perimeter descriptions that are more or less complete generally include directions and lengths for the lines and monument calls for the corners. Abutting owners and additional evidence in the form of natural monuments, such as water bodies, and artificial monuments, such as fences, are frequently mentioned. An expression of area may or may not be included.

Bounds descriptions generally eliminate gaps and overlaps, since abutting descriptions calling for one another ensure one common line between them. Courts have generally ruled that abutters, if identifiable, are classed as monuments and must be honored. Therefore, named abutters have to be identified. The inherent problem with this is that those calls are often not current, sometimes requiring extensive tracing of abutting parcels to ensure harmony of the calls. This research, however, can act as a safeguard to guarantee that the parcels do, in fact, abut one another.

Special care must be taken with abutting calls in that one abutter can convey part of its land to the abutter and still have the same abutting call in the description even though the acreage of both parcels has changed along with the location of their common boundary. In using *strip descriptions*, one must be very careful how they are used and written. A problem with a strip description is the difficulty the retracing surveyor may encounter in determining the point of beginning, because the initial reference may be destroyed or incapable of being located. Also, in a strip description that describes a boundary as being "50 feet perpendicular the described centerline," the description may include some areas twice and omit other areas.

Reference to prior deeds, plats, and/or maps is used by many to describe land parcels. This has advantages and disadvantages. Using a prior plat or map can cause serious problems, especially when the original monumentation has been destroyed or if adverse rights are present on the original lots. Another form of reference is the use of an *aliquot part* description. Technically, an aliquot part means a strict subdivision of the original portion. Applying this to the GLO system of description, a section is the largest portion that can be described as an aliquot. As a section is subdivided further, a parcel described as "the SE quarter" would be an aliquot part, as well as "the North half," or "the SE quarter of the SW quarter" of the section. There may be times when these aliquot portions may be or may be found to be modified.

Frequently, one will encounter a description that is a combination of several types. This is more prevalent in some areas than in others. Although seemingly adding to the descriptive information and therefore making it better, conflicts between calls often arise, making interpretation complex and more difficult. This result necessitates reference to one or more rules of construction for the resolution of conflicts. In writing descriptions, extreme care must be exercised to avoid such potential conflicts.

Description types, elements, and examples are discussed in greater detail in Chapter 5—Creation and Interpretation of Metes and Bounds and Other Nonsectionalized Descriptions.

Principle 8. *The trial team should determine the exact nature of the problem, whether it is one of boundary, title, or equity.*

At the very start of any land problem usually the client will be very concerned of the overall litigation costs and time involved. The team should jointly determine what issues they are going to present, where their best evidence is and what the evidence can and will prove. The opposing party files counterclaims, the dismissal of the plaintiff's claims does not automatically dismiss the counter claims. So the client should realize the old saying "In for a penny-In for a pound," applies.

Principle 9. *The complaint, representing the plaintiff and with input from the surveyor, should raise as many issues as possible.*

With the lack of understanding as to real property law, it is amazing what will catch the judges interest. What may seem insignificant or not important to the team, the judge may hold that issue is the important one. If it not presented and argued and a ruling made by the trial judge, it cannot be appealed.

Principle 10.   The attorneys representing the defendant, with input from the surveyor, should raise as many defenses in the answer as possible.

Here, the defendant has the opportunity to start their case. Excellent defense lawyers should understand and plead such little-understood legal principles as laches and estoppel.

Principle 11.   To have a positive line capable of being positively located, the original must create a corner at each end and the retracing surveyor must find them.

As early as 1687 John Love wrote; "A line has Length, but no Breadth or thickness, and is made by many points joined together in length ... " No person, including a judge cannot make a positive line without its endpoints (corners).

Principle 12.   Boundaries and their controlling corners are invisible until the creating surveyor places objects that must be described and referred to in the original documents.

This principle is a very basic legal principle, as to how a corner and line can exist, and not be visible. Until the creating surveyor or some knowledgeable person documents the lines and points created, these points and lines only exist legally and in words alone.

3.25 CONCLUSIONS

This chapter should set in students' minds the concept that boundaries of land interests may be varied and complex as well as interesting. The student should have an appreciation that boundaries are created by certain methods, that they are then subsequently described, by words, usually first by survey, then by legal documents, and then the retracing surveyor is given the responsibility of determining the "best" description and placing that description on the ground, using modern methods to the best of the retracing surveyor's capabilities.

NOTES

1. *Ski Roundtop Inc.* v. *Wagerman* (1989)
2. 79 Md.App. 357, 556 A.2d 1144 (1989).
3. 258 Pa.Super. 401, 392 A.2d 855 (1978).

4. *Kaplan* v. *Bernstein*, 2 N.J. Misc. R 762 (1924).

5. *Rivers* v. *Lozeau*, 539 So.2d 1147 (Fla. 1989).

6. California Civil Code, Section 1057.

7. *Justice* v. *McCoy*, 332 S.W.2d 846 (1960).

8. *Riley* v. *Griffin*, 16 Ga. 141 (1854).

CHAPTER 4

BOUNDARIES, LAW, AND RELATED PRESUMPTIONS

4.1 INTRODUCTION

The roles of law and technology are very prominent with regard to boundaries, from their initial creation to their subsequent retracement and their modification and alteration through legal doctrines. In the United States, we recognize five basic areas of the law. Each in its own way may affect boundaries and the people who create them as well as the subsequent surveyors who are asked to retrace them, even extending to the courts, which may be required to adjudicate disputes.

Five basic areas of the law can be recognized and broadly categorized, as follows: (1) constitutional law, (2) statute law, (3) common law, (4) case law, and (5) administrative law.

Although in litigation the surveyor's main function is to collect and present evidence of measurements, monuments, prior surveys, identification of possession, testimony, and, at times, opinions, it is important that the surveyor perform these functions in a professional manner and then possibly serve as a consultant to attorneys and even extend this service to the courts. To perform the functions of boundary creation and boundary retracement adequately, the surveyor should have a working understanding of the various laws and the structure of the court systems that may have an effect in this area of surveying.

Brown's Boundary Control and Legal Principles, Eighth Edition.
Donald A. Wilson, C.A. "Tony" Nettleman III, and Walter G. Robillard.
© 2024 John Wiley & Sons, Inc. Published 2024 by John Wiley & Sons, Inc.

The following principles are discussed in this chapter:

PRINCIPLE 1. Unless a landowner or person who claims an interest in land or its boundaries is named a party to an action, his or her rights cannot be affected.

PRINCIPLE 2. When a surveyor creates boundaries, the law dictates how they will be created and what elements are controlling in their retracement.

PRINCIPLE 3. A surveyor's decisions are based on the evidence available or considered at the time the decision is made. If the evidence changes or if new evidence is discovered or recovered, the surveyor's opinion may change based on this new evidence.

PRINCIPLE 4. Presumptions at law are conditional on the existence of certain facts; to avoid liability, those who rely on presumptions must eliminate contrary possibilities.

PRINCIPLE 5. The metes and bounds system of surveys is probably the oldest survey system on which land descriptions are predicated. Its principles are based on common law that traces its origin to England and probably originated in Rome and ancient Persia.

PRINCIPLE 6. The General Land Office (GLO) (PLS, Public Land Survey) system of surveying land is predicated on federal statutes, but its surveying principles may be founded in early English and Roman methods and surveying principles.

PRINCIPLE 7. Boundary disputes require advice and input from the two associated professionals: the attorney and the surveyor.

4.2 CONSTITUTIONAL LAW AND THE SURVEYOR

The controlling law of the land in the United States is constitutional law. Our entire existence as a sovereign nation is predicated on a written constitution, which every citizen is subject to and controlled by. Every property right of citizens or noncitizens is protected by both the US Constitution and state constitutions.

This area affects both the land surveyor and the lawyer in several ways. First, when identifying parties to a boundary dispute or a survey problem, all persons who have a possibility or are even remotely tangential to the problem of being affected by the line or lines should be identified and named parties to any possible action. If, in his or her research, the surveyor identifies any person(s) who may have a propriety interest in the parcel, they should be identified for possible inclusion as a party. If a person is not named as a party, any decision relative to any property interest concerning the lines will have no effect on the unidentified parties or landowners.

Furthermore, in the area of eminent domain proceedings by government agencies, there are several constitutional amendments—notably the Fifth and the Fourteenth Amendments—that prohibit any government agency or any agency with quasi-governmental authority from taking private property or property interests without due process proceedings, as prescribed by law.

A second relationship is the area of professional registration. Interestingly, the Fifth Amendment, which protects property rights, may also protect the private rights of each registered or licensed surveyor: namely, the registration that was earned and granted to the individual, which by law, is the personal property of the registrant. Under this philosophy, there is a question as to whether a licensing agency can revoke a surveyor's license without a hearing and without the licensee having a direct appeal through the courts.

4.3 JURISDICTION

Jurisdiction is the right and power of a court to adjudicate concerning the subject matter in a given case.[1] It is a term of comprehensive importance and embraces every kind of judicial action.[2] It has also been described as the authority of courts to take cognizance of and decide cases.[3] Although this is a matter for the attorney, surveyors should be familiar with the jurisdiction of courts before which they appear. Jurisdiction may be exercised over subject matter and over persons. Personal jurisdiction is the authority of the court to have its decision enforced.

There are certain areas of boundaries in which jurisdiction becomes important; for example, the US Supreme Court has exclusive jurisdiction over boundary disputes between states, and in some states certain inferior courts have jurisdiction over boundary disputes, whereas other courts in the same state may have separate jurisdiction over property rights.

4.4 FEDERAL JURISDICTION

In disputes over the boundaries between land that has always been in federal ownership (public lands) and a parcel of private land, the dispute is tried in a federal court in accordance with federal survey law but also considering state property laws as to adverse possession, and so on. Since there are only federal laws as to boundaries, it is under these laws that the boundaries were created, so it is under these laws that disputes must be tried, but there are no federal laws as to property rights or interests. Once land has passed from federal ownership to private parties within a state, the jurisdiction over boundary disputes passes to the state courts, and state law is applicable.[4]

In the event that the federal government purchases land within a state, court cases pertaining to boundary disputes are in accordance with that state's laws. For example, suppose that the federal government issues patents to all the land within a township. Now boundary disputes between private parties are under the jurisdiction of the state court. But suppose that at a later date the federal government purchases one of its formerly owned sections and a boundary dispute occurs. Although the trial over this boundary dispute may be held in either state or federal court, the applicable law will be that of the state.

Now suppose that the federal government disposes of all the land in a township except one section or even a 40-acre parcel, which was retained as an Indian reservation or for some other purpose. Because this land has never been under a state court's jurisdiction, all actions pertaining to boundary disputes are tried in federal court in accordance with federal law.

Actions between parties in different states can end in federal courts. Ms. A, a resident of state X, wishes to institute an action against Mr. B, a resident of state Y, for trespass and title problems of land in state Z. Because the action has diversity, Ms. A may initiate the action against Mr. B in a federal court or the court of state Z. Whichever court is chosen, the applicable law will be that of the state where the land is located (state Z). At times, the attorney may be wise to consider diversity jurisdiction so as to get a case tried in federal court, since the federal court's rules of evidence and discovery are more liberal than in many state courts. If a corporation is a party to an action, federal jurisdiction may apply. If the corporation is incorporated in more than one state, any state of incorporation may be selected for a trial. The surveyor's main responsibility is to define the location of the lands in litigation.

Because federal and state laws are not always identical, when resurveying land boundaries adjoining federal lands, the surveyor must determine which law is applicable, and this may require a title search. If parties appear in federal court, the federal court will apply the real property law of that particular state. As there is no federal law of real property, all lands must be decided according to the real property law of the state. Because no state court has authority to pass on the validity of or to question a federal law, it is not clear whether the jurisdiction of a state court can interpret or apply federal survey laws. This question has not yet been raised in the courts.

4.5 FEDERAL GOVERNMENT, AGENCY, OR OFFICER AS A PARTY

If a federal agency acts as a plaintiff, the agency will in all probability initiate the action in a federal court. If the surveyor's client seeks to initiate an action against a federal agency, however, he or she must first exhaust all administrative appeal processes provided for by the agency. If, after administrative appeals are completed, legal action is begun in a state court, the federal agency will probably seek to have the case dismissed for lack of jurisdiction, will claim sovereign immunity, or will ask to have the case tried in a federal court.

The initial jurisdiction for all suits "commenced by the United States, or by any agency ... is in the U.S. district courts."[5] There may be instances when private parties seek to bring a federal employee as a party to a boundary dispute between two private individuals. As a federal employee, the federal surveyor is not subject to the jurisdiction of a state court, and, acting through a US attorney, a dismissal may be granted so as to prevent the surveyor from testifying in a private action.

There are instances when private attorneys may attempt to involve federal surveyors as witnesses in private disputes. In such a situation, the federal surveyor may be protected from appearing as a witness by federal law.

4.6 SOVEREIGN IMMUNITY

Historically, the king is supreme and can do no wrong. Although the United States has no king, in its Constitution it adopted the age-old English doctrine that the king cannot be sued without his specific consent. This consent can be permitted only by Congress and not by any individual or agency. Over the years, by legislative acts, Congress has permitted people to sue the United States in tort claims or wrongdoings of agencies or federal employees against private citizens. Public Law 92-562 (28 U.S.C. Sec. 1346[a]) specifically reduced immunity in the areas of boundary problems and questions of title. All trials are in a US District Court without a jury. Section (g) states: "Nothing in this section shall be construed to permit suits against the United States based upon adverse possession." This includes adverse possession based on prior ownership. Section (b) states: "The United States shall not be disturbed in possession or control of any real property ... The United States may retain such possession and control of the property ... upon payment to the person ... of an amount which ... the district court in the same action shall determine."

4.7 UNITED STATES AS A DEFENDANT

A person who wishes to initiate an action against the United States over a title or boundary question must first point to the specific federal statute that grants jurisdiction to the particular court. The person must then strictly follow the method of serving process on a particular person or officer, including the specified number of copies of all documents. Then all actions and discovery will be conducted under the Federal Rules of Civil-Appellate-Criminal Procedure and the Federal Rules of Evidence for United States Courts and Magistrates.

In most instances, any litigation between a private person and the federal government can be initiated only as to boundary disputes and title, not as to questions of property rights, such as a boundary by agreement or adverse possession.

4.8 DISPOSING OF FEDERAL LANDS

When federal lands are concerned, Congress dictates the exact methods by which an agency may give up control of its lands, and it also dictates under what conditions land may be sold. Some agencies may sell or exchange land; others may not. Congress has enacted laws to help persons who have title and boundary problems with federal agencies. When a federal agency disposes of lands, only certain government officials have the authority to sign any documents (usually quitclaim deeds) that dispose of the subject lands.

Methods of disposal usually are exchange of public lands for private (value for value) and in a few instances by sale.

4.9   COLOR OF TITLE ACT

An act of December 22, 1928 (45 Stat. 1069 as amended; 43 U.S.C. 1068, 1068a, 1068b) provides for the sale of public domain lands by the secretary of the interior when it is shown that:

1. A tract of public land has been held in good faith and in peaceful adverse possession by the claimant, his or her ancestors, or grantors, under claim of or color of title for more than 20 years.
2. Valuable improvements have been placed on such land or some part has been reduced to cultivation.
3. Such possession began no later than January 1, 1901. Evidence must be presented that the use and possession were prior to 1901.
4. The taxes were paid on the property.
5. The area is not more than 160 acres.

4.10   PUBLIC LAW 120

An act of July 8, 1943, Public Law 78-120 (57 Stat. 388, as amended; 7 U.S.C. 2253), was specific legislation enacted by Congress in an effort to provide relief to landowners who were adjacent to lands under the administration of the secretary of agriculture. If it is determined that the lands, other than lands obtained by exchange, are wanting or deficient in title or the color of title because of a mistake, misunderstanding, or inadvertence, the lands could be reconveyed.

4.11   SMALL TRACTS ACT

The Small Tracts Act (36 C.F.R. 254, Subpart C) is a limited land adjustment act for the relief of persons who had built improvements on National Forest System land because of title, error in a survey by either a private or government surveyor, or some other error. A new metes and bounds survey of the parcel is required. Conveyance is not automatic. The claimant must produce the required proof for encroachments (36 C.F.R. 254.32), for road rights of way (36 C.F.R. 254.33), or for mineral survey fractions (36 C.F.R. 254.34). Then it must be determined if a conveyance is in the public interest (16 C.F.R. 254.36).

Although Congress has taken a humane attitude toward private persons who have occupied federal lands in error, this does not mean that reliance on an erroneous survey conducted for a cheap price will qualify a person for a reconveyance. Proof must be strictly in accordance with statute; conveyance is not automatic.

Federal laws can cause problems even between federal agencies. One law states that only the Bureau of Land Management (BLM) can dispose of public domain land, yet the Small Tracts Act permits claimants who have a legitimate claim to recover the land. If a person resided on public domain land under the management of other federal agencies, these agencies can dispose of public domain land under this act without the permission of the BLM.

4.12 RESEARCHING THE LAWS

Today, more and more states have enacted laws that cover such areas as minimum technical standards, plat acts, and registration laws that affect surveyors; thus, it is presumed and expected that every surveyor knows and understands these laws and will practice in accordance with them. Surveyors must also be knowledgeable of other laws of the respective states in such areas as adverse possession, easements, descriptions, and many others.

> Surveyors should speak with knowledge of the laws that affect them, but they should refrain from giving legal advice.

Many surveyors maintain a personal library that specializes in this area. The surveyor must always be careful and is warned not to deal in those areas that are strictly the domain of the attorney.

In the more populous areas of states, law libraries usually are available to the public, especially at the state capitol and state universities. Within these libraries, the laws that determine ownership of land and boundary locations can be found in statute books, in encyclopedias of law, in court reports, and in case digests.

Statutes are laws enacted by state legislatures, and they may be found in statute volumes such as the *Indiana Code of 1971* or *Barnes Indiana Statutes*. In some states, the laws of evidence are codified, and when they are, they can be found in statute books. As most laws pertaining to procedures used in locating land boundaries are common law derived from court opinions expressed in boundary litigations, surveyors should know how to research court findings.

Law encyclopedias consist of a large number of volumes arranged alphabetically by subject and are found under such titles as *American Jurisprudence*, second series (Am. Jur. 2d), and *Corpus Juris Secundum* (C.J.S.). *American Jurisprudence* has been published in two series, and "2d" indicates the second series. If surveyors consult these volumes under such headings as "Boundaries," "Deeds," "Waters," "Adverse Possession," and "Evidence," they will find the common laws of all states with specific references to cases. Because these volumes do not cite all the court cases of a state, other volumes for a particular state may have to be consulted.

In most states, a specific encyclopedia is available under a title such as *West's Indiana Law Encyclopedia* or *California Jurisprudence*. In these volumes, common laws of the named state and relevant court case citations pertaining to specific subjects are to be found. If a case relevant to a survey situation is not found, the laws of other states should be consulted.

Many state surveying organizations have published *digests of surveying laws*. These publications cover many areas that surveyors are interested in for their states. Many of these were prepared by committees or groups of surveyors who paid for the "privilege" of doing so. Before these publications are relied on by surveyors or attorneys, they should be examined for their depth of research and their timeliness. These publications can be a valuable asset to any surveyor's library.

4.13 COURT REPORTS

In modern times, surveyors may find themselves in court either as a party or as a witness. To keep current on legal thinking and as an aid, they should read pertinent case law in the general areas of interest. Surveyors should remember that reports are written for and in the language of the attorney, not that of the surveyor. To fully understand a rule of common law as established by the courts, specific court reports should be consulted. The process by which written court opinion is developed is as follows. In a dispute over land boundaries, the case is tried in a superior court or other competent court, wherein a written record (not a published record) is kept of the testimony and proceedings. If, after the evidence and law phases of the trial are concluded, one of the litigants disagrees with the court's findings of law (not facts), the case may be appealed to a higher court based only on the misapplication of the law to the facts of the evidence presented during the trial. If an appeal is heard by a court higher than the superior court level (either an appellate or the supreme court), the findings are published in book form and are available in law libraries.

4.14 LEGAL RESEARCH

Within the United States, there are 50 state jurisdictions plus a commonwealth. Each, in addition to the federal government, has its own court system, thus totaling more than 52 systems. Although laws regulating how land should be located or relocated are, in general, similar in all states, this is not always so; in a few instances, drastic differences may exist. In this book, laws that are generally true in all states are presented, and, in many instances, deviations from the norm are noted. Because of the variations in the details of some laws, surveyors must be able to research laws of their own state. As in many cases that are appealed and are reported, and as most of them are of no interest to surveyors, the problem is to locate those cases relevant to a particular situation.

After a person understands the general theory of boundary law, he or she should consult general jurisprudence books of that state (e.g., *California Jurisprudence, Second Series*). Under the subject of boundaries, easements, and so on, there are subheadings such as control of monuments, control of distance, control of direction, control of area, and so on. Each of these is discussed briefly, and portions of court cases are usually quoted and cited. Each citation is abbreviated, such as *Jones* v. *Brown*, 68 N.J. 123, which means that in volume 68 of the reports of the state of New Jersey on page 123, the report of the case of Jones (the plaintiff on appeal or appellant) versus Brown (the defendant on appeal or appellee) will be found. If the citation is 73 Cal. App.2d 321, it means that the case will be found in volume 73 of the second series of California Appellate reports on page 321. If the case appears to be relevant to a particular problem, it should be read and understood.

Some publishing companies divide the United States into areas (Atlantic, Northeast, Northwest, Central, Pacific, South Eastern, Southern, South Western, etc.), and for each district the laws of the states within that district are summarized and cases pertaining to states within that district are cited. A California case may be cited as 488 P.2d 213, which means that the report of the case can be found in volume 488 of the second series of the *Pacific Reporter* on page 213. At the present time, some states do not have state reports; they rely on area reports. Reporters that cover large geographical areas often issue unofficial advance reports. Official reports follow after appeal periods have passed, and then a case is finally made part of the common law. The official citation should appear first, with the unofficial citation following; thus, there may be more than one citation for a given case summary. The official report should be relied on as authoritative, and each case should be "Shepardized" for its present status.

Shepard's Citations lists later cases on the same subject, whether statute, case, or other. *Shepardizing a case* consists of looking up the case reference from the official reporter and reading the accompanying notations. Abbreviations are explained in the introductory discussion and consist of such items as "a" for affirmed, "r" for reversed, "o" for overturned, and so on. Other important notations are other courts' citations of the case and attorneys general's comments and opinions. Each state has a separate *Shepard's* citation, and there are separate citations for the various other reporters.

Each reported case is headed by a syllabus, a summary of what the case means to the reporter (usually written by a staff assistant). Because the syllabus is an opinion of the reporter and on rare occasions is in error, it is advisable to read the entire case, analyze it, and then form your own statement as to the holding of the court.

American Law Reports (*A.L.R.*), now in its fifth series, reports court opinions on selected topics. It reports on a principle of law, usually one that was significant in a recent decision, and discusses the cases relating to that principle. *A.L.R.* is a tremendous resource for studying the history and details of a legal principle or an area of the law. Many state reports and cases, including the *A.L.R.* third, fourth, fifth, and sixth reports, can be purchased on compact disks and are updated quarterly. These can be purchased and used on a computer.

4.15 JUDICIAL NOTICE

Judicial notice is a little-used tactic but should be used more by attorneys at the recommendation of surveyors. This procedure, although not evidence, helps strengthen evidence during the course of a trial or in framing a decision. The court, of its own motion, or at the request of an attorney, and without the production of evidence, will recognize the existence and truth of certain facts bearing upon the controversy, which by their nature do not lend themselves to testimony by experts. This information is usually of common knowledge. The court and the jurors already know these facts; thus, they do not need evidence to prove them.

The court may take *judicial notice* of those facts that usually are not considered by many attorneys as being evidence. Understanding judicial notice, at times, can make the attorney's trial work much easier. Usually this information is common knowledge, such as water running downhill, incorporation of towns, important well-known dates, the sun coming up every day, and the like. These things do not have to be demonstrated or proved, merely stated, and the court asked to take judicial notice of the fact. If every item of knowledge had to be proven, court trials would take forever.

Principle 1. *Unless a landowner or person who claims an interest in land or its boundaries is named a party to an action, his or her rights cannot be affected.*

In conducting surveys, it is important that all landowners along a line or owners of a parcel be identified. Constitutional law protects the unnamed or unidentified owner. The attorney may find that failure to name all landowners who share a common boundary line or corner will result in only a partial solution to the problem.

This basic rule has its foundation in constitutional law and the Fifth Amendment. To have a final adjudication on a boundary or title matter, the person who has the title to the possession estate must be made a necessary party to any litigation. This requirement is present in title matters and extends to questions of boundary lines. If two contiguous neighbors are disputing a common corner as well as the lines leading from it, any judicial determination of the validity or legality of the corner and the lines affects only those who are parties to the legal action. In the event that there are additional landowners who either meet at the common corner or who are owners along the line(s), to have any judicial determination of the finality of the corner or lines, each landowner must be identified and named as a necessary party.

Where it is the surveyor's responsibility to name the owners along a boundary line, this is a factual situation; the identification of parties who have or may have a property interest is a legal question and should be addressed by a lawyer. However, the surveyor may have intimate knowledge of who is in possession or in whom title is listed.

When a specific corner or corners are being questioned through litigation, every person who has that corner as a point must be included in the action; otherwise, only a partial solution is obtained and it is possible that the problem may arise again in the future, depending on the circumstances.

This places the surveyor expert in a position of conducting investigation to determine who has property interests in the area if the problem, and what interest they hold.

> The surveyor should become conversant and use judicial notice knowledge. Judicial notice is one form of evidence that can be useful in working on boundary issues. What will be accepted as judicial notice is left to the determination of the trial judge. Such evidence as universal knowledge: "water runs downhill," "survey field notes from governmental agencies," and laws of other states are good examples.

4.16 EVIDENCE

Principle 2. *When a surveyor creates boundaries, the law dictates how they will be created and what elements are controlling in their retracement.*

This principle is quite important to the surveyor, whether the person is a metes and bounds surveyor or a GLO surveyor. The federal GLO surveys were created under at least three early and distinct federal statutes that are still in effect today and are identified in Title 43 United States Code Annotated (USCA). They identified how the corners and lines would be created, what methods and instruments would be used, how the lines were to be marked, how retracements should be conducted, and what weight a surveyor and the courts should give to evidence of the original surveys.

Early metes and bounds surveys did not have this privilege, but their methods and instruments were identified by common reputation (law) and principles based on several centuries of use by surveyors and early English courts. Today, though, most states have enacted minimum standards for how lines should be surveyed (created), how they should be marked, and the resulting plats and descriptions that should be prepared.

Principle 3. *A surveyor's decisions are based on the evidence available or considered at the time the decision is made. If the evidence changes or if new evidence is discovered or recovered, the surveyor's opinion may change based on this new evidence.*

Court decisions and the law are usually based on the evidence presented to the trial court and the jury, and once the evidence phase of a trial is completed, the judge and jury have a right to assume that all of the evidence available that will influence the decision has been presented. The attorney and the surveyor should never leave any questions unanswered in the minds of the jury. After a period of time, if the case is not appealed, the findings become final, and any additional findings of evidence will not alter the final decision. Surveyors do not have this advantage.

As most surveyors or attorneys realize, if the available evidence is changed, the applicable law may also be changed. In agreeing to make a boundary survey, survey- ors also agree to locate a conveyance on the ground correctly, knowing full well that their work is always open to collateral attack from other surveyors. In so doing, they are charged with finding as much of the evidence as possible at the time of the survey that may have an influence on the location of the client's parcel of land and then apply- ing the correct and proper survey rules to ascertain the location of the boundaries. If they fail to do so, they may be liable in tort for their error. Sometimes, situations arise in which the first surveyor fails to find sufficient evidence necessary for a correct location, and most frequently, this is due to the failure to locate the evidence of an original monument position or retracing the description of the parcel in haste.

In the following discussions, necessary evidence is cited prior to stating which law is applicable. When a surveyor makes his or her decision, it is always to be assumed that no other evidence can be found that will influence the conclusion of law as stated by that surveyor. In a classroom, the question may be asked: What if this piece of evidence is added? The usual answer is: When you change the evidence, you often change the applicable law. Before any conclusion of law can be cited, all relevant evidence must be recovered, identified, and evaluated.

> A decision on a boundary issue is predicated on the evidence recovered and used in that decision.

4.17 PRESUMPTIONS

Principle 4. *Presumptions at law are conditional on the existence of certain facts; to avoid liability, those who rely on presumptions must eliminate contrary possibilities.*

Definition

Presumptions at law are conclusions that the law expressly directs to be deduced from certain established facts; to avoid liability, those who rely on presumptions must eliminate contrary possibilities.

The surveying of boundaries is dependent on presumptions, and many surveyors' decisions as to boundaries are also based on presumptions. A presumption is a state- ment, sometimes of fact, sometimes of law, sometimes mixed, that can be considered as being true without further proof. Presumptions are not evidence but may substitute for evidence where the latter is lacking. A presumption may not be based on another presumption.

One decision described a presumption as a conclusion reached by means of the weight of proved circumstances.[6] Most of these surveys are rebuttable presumptions that evidence may prove wrong.

According to *Black's Law Dictionary*, a presumption is a rule of law that courts and judges shall draw a particular inference from a particular fact, or from particular evidence, unless and until the truth of such inference is disproved. Examples of presumptions are:

1. A child proved to be four years old or younger cannot commit a felony.
2. A person is innocent until proven guilty. Some presumptions are conclusive, as cited in presumption 1 for the child, and others are rebuttable (assumed true until proven otherwise).
3. If one can prove a survey, it was performed according to the law.

There are several presumptions that surveyors should know or be familiar with: (1) if you have field notes and a plat of a survey, it is presumed a field survey was conducted. (2) The surveyor is presumed to know the law of boundaries. In a subdivision, in many states, the presumption is that a lot adjoining a street has ownership to the centerline of the street. This presumption is a rebuttal presumption and can be rebutted by showing that the original subdivider did not own the property to the centerline of the street. Presumptions are not evidence; they are substitutes for evidence. A presumption influences the burden of offering evidence.

A conclusive presumption is irrebuttable and absolute. No rule of law is permitted to be overcome by any proof that the fact is otherwise. Thus, everyone is presumed to know the law. Conclusive presumptions are few in number, and most authorities consider them to be substantive rules of law, not rules of evidence.

Rebuttable presumptions constitute most of the rules of law that control the location of real property. The courts state that the presumption in the order of importance of evidence in a resurvey is senior rights, monuments, bearing or distance, and area. Of course, the order of importance of each item can be rebutted by satisfactory proof to the contrary, and it is the obligation of the surveyor to seek proof that may refute a presumption.

Presumptions may be established by statute or by common law. Thus, the rule as to the age at which a child is presumed incapable of committing a felony is usually established by statute. In establishing the intent of the parties to a deed, the courts have usually ruled on the order of presumed importance of deed elements.

Inferences are not to be confused with presumptions. An inference is a deduction of fact that may logically and reasonably be drawn from another fact or group of facts.

4.18 COMMON PRESUMPTIONS

There are certain presumptions that all surveyors should know, accept, and understand. The most important of all assumptions is that when the surveyor accepts a job, the surveyor is presumed to have the capabilities to perform that job to the highest degree of technical and professional skill.

Some presumptions are applicable to all types of surveys. In the case of official surveys, it is always presumed that:

1. The government surveyor performed the work competently and according to the laws in effect at the time of the survey.
2. The work was accurate.
3. Surveys were made as stated in the field notes.
4. The lines of survey were run on the ground.
5. The corners were established as returned and were marked so as to be easily identified.
6. The field notes and plats made by government surveyors are correct.
7. Corners have been established at the places indicated in the field notes.
8. Where the government corner has been lost and the proper method in relocating was followed, the survey followed the original lines.
9. An unlocated section line is to be run according to the statute.
10. Land abutting a highway, street, or nonnavigable watercourse extends to the center of the natural object.
11. Descriptive words naming townships and sections or subdivisions thereof have reference to government plats and government surveys.
12. Where a depth is given, the distance was measured at right angles to the frontage.
13. In a proceeding in partition, a building lies wholly on one side of the dividing line.
14. An ancient survey is presumed to be correct when so accepted and treated by the parties interested.
15. GLO surveys were conducted in accordance with the law.
16. The early metes and bounds surveys were magnetic surveys.
17. GLO survey distances are horizontal.
18. GLO bearings are true bearings
19. Early metes and bounds bearings are based on the magnetic meridian.
20. Early metes and bounds distances are slope (not true in most jurisdictions, but perhaps in some). The actual fact will be determined by the court.
21. Early metes and bounds distances are slope (with the lay of the land).
22. When you see a description with courses and monuments, a survey was conducted.
23. An approved US GLO survey is unassailable by any court.

Under certain circumstances, a surveyor does not apply a presumption. For example, there is an inference that the grantee is entitled only to the land described within the limits of the boundaries in the deed (*Town of Refugio* v. *Strauch*, Com. App. 29 S.W.2d 1041 [1930]), but there is also a presumption that the grantor did not intend to retain a narrow strip along an outside line (*Mahan v. Blankensop*, 108 W. Va. 520 [1930]). Assuming that a surveyor does in fact discover a narrow vacant strip between the client and the adjoiner, he or she must prove that a former owner in the client's chain of title did in fact own the strip in question. Assuming that this condition is satisfied, should the surveyor monument the strip as being a part of the client's land? To avoid liability, the surveyor normally stakes the deed as written, prepares a report informing the client of the conditions, and advises the client to see an attorney.

In some instances a corrective deed is needed. In one instance, a surveyor did monument a narrow strip and the client built a house encroaching on the strip. When the time came to convert a construction loan to a conventional loan, the conversion was denied because the strip was not in the written deed. Since the interest rate was higher on the construction loan, the client suffered damages and the surveyor was liable. Although it is the obligation of a surveyor to inform a client of existing conditions, it is not the surveyor's function to give the client that which should be obtained by legal action. Although the surveyor should be knowledgeable of presumptions, he does not always assume the liability of executing the presumption.

For specific citations and further reading, *Corpus Juris Secundum*, Vol. 11, page 690, should be consulted. Other presumptions that exist are discussed elsewhere in the text.

4.19 SURVEY SYSTEMS PRESENT IN THE UNITED STATES

The type of survey system the surveyor and the lawyer find has a great effect on how problems are approached and decided. The methods of boundary creation, how and by whom, will have an impact on how they are retraced.

All landowners, surveyors, and attorneys should have both a working and a legal understanding of the basic systems of land boundaries in order to be responsive landowners, competent surveyors, and effective attorneys. The United States has a very ancient land survey system as well as one that supersedes the US Constitution. In the following sections, we introduce the student to the two major land boundary systems in the United States: the time-tested metes and bounds system and its companion, the GLO system.

Principle 5. *The metes and bounds system of surveys is probably the oldest survey system on which land descriptions are predicated. Its principles are based on common law that traces its origin to England and probably originated in Rome and ancient Persia.*

In subsequent chapters, the creation and retracement of boundaries are discussed and explained. Although limited rectangular surveys were discovered in early Egypt and other Middle Eastern countries, there was no wide-scale adaptation or survey of any rectangular parcels other than some towns and villages. The system of metes and bounds surveys was perhaps the oldest and most basic form of surveying and land descriptions. Many descriptions of land and land interests were identified in Mesopotamia, Egypt, and ancient Rome.

Actually, the metes and bounds descriptions often offer a myriad of possibilities for interpretation and surveying. Perhaps the terminology "metes *or* bounds" would better describe the system. Unlike the GLO system that enjoys one specific definition, there are numerous combinations that could make up a metes and bounds description. Some of the possibilities include the following:

1. Call for courses only (bearings and distances).
2. Call for courses and artificial corner monuments.
3. Call for courses and natural monuments and/or boundaries.
4. Call for 2 and 3.
5. Call for adjoining landowners only.
6. Call for 2, 3. 4, and 5.
7. Call for a single line, with measurements referenced to that line.
8. Call for area.
9. Call for all the elements listed, either in combinations or individually.

In one of the few academic studies in this area of boundaries, a 1972 report from the University of Wisconsin identified eight distinct types of metes and bounds descriptions:[7] (1) true metes and bounds, (2) metes and bounds, (3) strip, (4) (true) bounds, (5) divisional line, (6) proportional parts (of a whole parcel), (7) linear, and (8) area.

A true metes and bounds description is seldom found in old descriptions. Yet this is the type of metes and bounds description that a modern surveyor and/or attorney should strive to create today from the remnants of ancient descriptions. McEntyre explains a *true metes and bounds description* as follows: ... [A] description that contains a full caption, a call for all ties and monuments, either record or physical, that determine the boundaries, all references to adjoining lands by name and record, and full dimensional recital of the boundary courses, which close mathematically, in succession around a boundary."[8]

This brief discussion of metes and bounds descriptions is expanded in later chapters to include their interpretation, description, and retracement.

Principle 6. *The GLO (PLS) system of surveying land is predicated on federal statutes, but its surveying principles may be founded in early English and Roman methods and surveying principles.*

The second major system of land boundary descriptions in the United States is what is referred to as the GLO (, PLS system, or simply the federal survey system. Whereas the metes and bounds system is steeped in common law for its creation and retracement procedures, the GLO system is predicated on federal statutes for its creation and upon federal statutes and federal case law for its retracement procedures. Geographically as well as in terms of the number of states encompassed by this system, it is far bigger than the metes and bounds system.

A surveyor retracing boundaries in some metes and bounds states, including Tennessee, Texas, South Carolina, and western Canada, some New England states, and isolated areas in other states, may come upon descriptions for townships, ranges, and sections as the foundation for land descriptions. These should not be considered as pure GLO, but the metes and bounds surveyor may be required to apply GLO principles in retracing boundary lines. In several metes and bounds states, legislatures, state surveying minimum standards, and case law have referenced or adopted the *Manual*[9] for the metes and bounds surveyor.

GLO was a well-conceived system that was not favored by all of the Continental Congress; many of the legislators sought to retain the metes and bounds system as the primary survey system in the United States. The GLO system and its ramifications for creation and retracement are discussed in Chapter 6—Creation and Retracement of GLO Boundaries.

Principle 7. *Boundary disputes require advice and input from the two associated professionals: the attorney and the surveyor.*

The area of boundary creation and boundary retracements are the two major functions of surveyors relative to boundaries.

In applying the rules of boundary creation and boundary retracement, much has been written about retracing, recovering, and interpreting previously created boundaries. But little has been written relative to creating boundaries. The most difficult surveying is recovering the boundaries of previously created parcels and lines and then defending this work in litigation. Once again, we should defer to Mulford:[10]

Surveys are usually made in order to furnish the descriptions to se be used in legal instruments and the data necessary for legal proceedings. The surveyor does not necessarily look upon a survey in the same light as does the man of the law and he must be governed largely by the particular legal requirements of the case in hand.

While the surveyor naturally looks for the intention implied in the earliest conveyance of a piece of land and seeks to get back to the original boundaries, the lawyer may sweep aside all this exact and careful work and require a survey of the boundaries of today on the ground of the undisputed possession for a number of years.

Yet the lawyer may promptly admit that the course followed by the surveyor was the proper course for the surveyor, the responsibility of departing from the ancient record resting entirely with himself ...

In the same paragraph Mulford writes:

I have said that the surveyor must largely be governed by the legal requirements of the case in hand, but I do not mean by this that he should endeavor to make legal decisions for himself.... I think there is no doubt that the legal intricacies connected with the search and guaranteeing of complicated title are beyond the province and full appreciation of the ordinary surveyor, yet he must render intelligent help to the lawyer who is attending to the same.

4.20 CONCLUSIONS

The United States enjoys two major, distinct, and separate land survey systems. Few surveyors have the privilege of being capable of working in both systems, but that does not restrict any professional from offering advice to landowners and attorneys. However, the professional surveyor should realize that to offer advice places him or her in the same category as a professional who is totally immersed and trained in the one particular system. Each system has its strengths and its weaknesses. Neither one is perfect, but both are adequate and legal. The practicing surveyor should understand the basic principles that apply to each system. These principles cannot be learned simply by reading about them. The surveyor must have practiced in the system. The principles that the surveyor should know are unique to that particular system. Few surveyors become totally competent and well-versed in both systems. A surveyor who practices in the system in which he or she is not totally competent and well-versed in knowledge, experience, and training does a great disservice to the client.

BIBLIOGRAPHY

American Jurisprudence. San Francisco: Bancroft-Whitney Co.

American Law Reports. Rochester, NY: Lawyers Co-operative Publishing Co. and San Francisco: Bancroft-Whitney Co.

Black's Law Dictionary, 5th ed. St. Paul, MN: West Publishing Co., 1982.

California Jurisprudence. San Francisco: Bancroft-Whitney Co.

Corpus Juris Secundum. Brooklyn, NY: American Law Book Co.

Shepard's Citations. Colorado Springs, CO: Shepard's/McGraw-Hill.

West's Indiana Law Encyclopedia. St. Paul, MN: West Publishing Co.

NOTES

1. *Bidinger* v. *Fletcher*, 224 Ga. 501 (1968).
2. *Federal Land Bank* v. *Crombie*, 80 S.W.2d 39 (Ky. 1935).
3. *Fireman's Relief* v. *Brooks*, 67 P.2d 4 (Okla. 1937).
4. It is possible that in a GLO state where federal GLO survey laws either are questioned or are the basis of the litigation, jurisdiction may be in the federal courts and not in the state courts.
5. 28 U.S.C. Sec. 1345 (2002).
6. *Marquet* v. *Aetna Life Insurance Co.*, 159 S.W. 733 (Tenn. 1915).
7. Fant, Freeman, and Madson, *Report 4: Metes and Bounds Descriptions* (Madison, WI: Department of Civil Engineering, University of Minnesota, and Land Surveyors' Association, 1972).

8. John McEntyre, *Land Survey Systems* (New York: Wiley, 1978), 319.

9. Bureau of Land Management, *Manual of Instructions for the Survey of the Public Lands* (Washington, DC: U.S. Department of the Interior, 1973).

10. A. C. Mulford, *Boundaries and Landmarks* (New York: D. Van Nostrand Co., 1912). Reprinted by W. G. Robillard, Atlanta, GA.

CHAPTER 5

CREATION AND INTERPRETATION OF METES AND BOUNDS AND OTHER NONSECTIONALIZED DESCRIPTIONS

5.1 INTRODUCTION

The oldest form of boundaries identified by landowners, surveyed by surveyors, and litigated by lawyers and the courts are metes and bounds boundaries. In this chapter, we examine the legal and surveying historical aspects of this system. The modern aspects, including the surveying aspects, the retracement aspects, and the legal aspects and ramifications of those boundaries that are referred to as *metes and bounds surveys and/or descriptions* will be discussed. We describe and explain the various methods of creating nonsectionalized land parcels and their descriptions. Metes and bounds descriptions should not be lumped into the single category of metes and bounds, for the boundaries may be created by various methods that are discussed and may be predicated on an ancient rectangular system recovered in Roman Africa. These non-sectionalized methods of creating and describing boundaries are the most ancient known and have enjoyed a reputation for acceptance and familiarity.

In Chapter 1—History and Concept of Boundaries, the metes and bounds method of describing boundaries was recognized as being used before the birth of Christ. Ancient Egyptians, Romans, and many other civilizations used words to describe boundaries. Their methods, words, and applications, although ancient, showed themselves to be adaptable and retraceable in the modern world. The method is recorded on clay tablets from ancient Egypt to Babylonia. An actual kudurru or stele from these

Brown's Boundary Control and Legal Principles, Eighth Edition.
Donald A. Wilson, C.A. "Tony" Nettleman III, and Walter G. Robillard.
© 2024 John Wiley & Sons, Inc. Published 2024 by John Wiley & Sons, Inc.

early times is pictured in Figure 5.1, and its translation is provided in Figure 5.2. Notice the similarities with today's deeds and descriptions. Figure 5.3 is a reduction of a clay tablet of a deed for a land division.

Unlike the General Land Office (GLO) system or Public Land Survey System (PLSS), the metes and bounds method was created out of human knowledge and needs, rather than by law. When people eventually became agrarian, the first need was to define parcels of farmland that would ultimately become the life-sustaining element for survival. Then the obligation and burden were placed on the landowners to defend these parcels from intruders. At that time, there were no laws on which to

Figure 5.1 Babylonian boundary stones, from the reign of Marduk-Nadin-Akhè, front and rear views. (Source: L. W. King, 1912/The Trustees of the British Museum/Public Domain.)

KUDURRU (Boundary Monument) of the time of Marduk-Nadin-Akhè (reproduced by courtesy of the Trustees of the British Museum). The inscription on the stone is the wording of the deed and the names of the gods invoked to protect the land. A surveyor of three thousand years ago was named! The partial text, taken from the translation given in *Babylonian Boundary-Stones* and *Memorial Tablets* in the British Museum (edited by L. W. King), is as follows:

Twenty gur of grain-land (a gan, measured by the great cubit, being reckoned at thirty ka of seed), in the district of Al-Nirea, on the bank of the Zirzirri Canal, in Bit-Ada, Marduk-nadin-akhé, king of Babylon, during the victory in which he defeated Assyria, upon Adad-zerikisha, his servant, looked with favour, and to Marduk-il-napkhari, the son of Ina-Esagila-zeru, the minister, said "A charter for the king of Babylon!" and according to the word of the king of Babylon twenty gur of grain-land, a gan, measured by the great cubit, being reckoned at thirty ka of seed, for Adad-zerikisha, his servant, he measured and he presented it to him for ever: on the upper length, to the north, the Zirzirri Canal, adjoining Bit-Ada and the field of the Governor's house; on the lower length, to the south, the Atab-dur-Oshtar Canal, adjoining Bit-Ada; the upper width, to the East, adjoining Amel-Eulmash; the lower width, to the West, adjoining Bit-Ada. According to the word of Marduk-nadin-akhe, king of Babylon, was the deed sealed. Enlil-zer-kini, the son of Arad-Ishtar, was the surveyor of the land. The city of Dindu-E . . . , the twenty-eighth day of the month Elul in the tenth year of Marduk-nadin-akhè, king of Babylon. In the presence of Eulmash-shurki-iddina, the son of Bazi, the . . . -officer of the lands; . . .

Whensoever in later days of the brethren, sons, family, relatives, or household, of Bit-Ada, there be anyone who shall rise up and shall put forward a claim concerning that land, or shall cause one to be put forward, or shall say: "The land was not a gift!" or shall say: "The seal was not sealed," whether he be a future head of the House of Bit-Ada, or a governor of Bit-Ada, or a . . . -official of Bit-Ada, or a ruler, or an agent, or other future official of Bit-Ada who shall be appointed, and shall say: "The land was not measured," or shall say: "The seal was not sealed," or shall present this land to a god, or shall appriate it for himself, or its limit, boundary, or boundary-stone shall alter, or a curtailment or diminution in this land shall bring about, may all the gods who are upon this stone, and all whose names are mentioned, curse him with a curse that cannot be loosened! May Anu, Enlil, and Ea, the great gods, tear out his foundation and destroy it, may they tear away his offspring, may they carry off his descendants! May Marduk, the great lord, cause him to bear dropsy as a bond that cannot be broken!

May Nabu, the exalted minister, change his limit, boundary, and boundary-stone! May Adad, the ruler of heaven and earth, fill his canals with mud, and his fields may he fill with thorns, and may his feet tread down the vegetation of the pastures! May Sin, who dwells in the bright heavens, with leprosy as with a garment clothe his body! May Shamash, the judge, the ruler of men, the great one of heaven and earth, decree the refusal of his right and oppose him with violence! May Ishtar, the lady of heaven and earth, before the gods and the king of Babylon bring him evil! May Gula, the great lady, the wife of Ninib, set destructive sickness in his body so that light and dark blood he may pass like water! May Ninib, the lord of boundary-stones, remove his son, who pours the water for him! May Nergal, the lord of spears and bows, break his weapons! May Zamama, the king of battle, in the battle not grasp his hand! May Papsukal, the minister of the great gods, who goes in the service of the gods, his brothers, bar his door! May Ishkhara, the lady of victory over the lands, not hear him in the mighty battle! May the great Anu, the great Lord, cause him to take a road that is obstructed!

May all the gods who are upon this stone, and all whose names are mentioned, with a curse that cannot be loosened curse him!

Figure 5.2 The inscription on the stone is the wording of the deed and the names of the gods invoked to protect the land. A surveyor of 3000 years ago was named (Arad-Ishtar)! The partial text is taken from the translation given in *Babylonian Boundary Stones and Memorial-Tablets in the British Museum* by L. W. King. (Source: L. W. King, *Babylonian Boundary Stones and Memorial-Tablets in the British Museum*, London, 1912.)

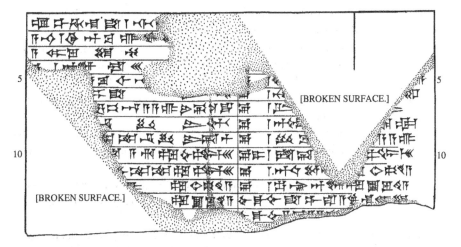

Figure 5.3 Drawing of a Babylonian stone tablet describing a division of property. (Source: L. W. King, *Babylonian Boundary Stones and Memorial-Tablets in the British Museum*, London, 1912.)

base legal systems; the landowners and early surveyors needed help in their tasks. As a result of this need, the metes and bounds system became molded in, and dependent on, the common law. There were no written laws on which to rely; it was necessary to improvise with what was available. Applying certain basic principles founded on common sense and religious doctrines, the law ultimately set a foundation for modern surveyors to follow.

The term *metes and bounds* should be considered as a collective description in that it can be a metes description, a bounds description, or a combination metes and bounds description. Whereas GLO boundaries were and are created under the law, it could also be said that metes and bounds boundaries are likewise created under a much older law, an ancient law, the common law, the unwritten law. The following principles are discussed in this chapter:

PRINCIPLE 1. A landowner is free to create any boundary desired as long as it does not infringe on the rights of an adjoiner or on a senior right.

PRINCIPLE 2. The lines and corners of the original survey control the location of a parcel. Lines are presumed to have been established in accordance with accepted rules of survey and any minimum standards in existence at the time the original lines were created.

PRINCIPLE 3. There was no standard unit of distance or angle measurement used in early metes and bounds descriptions and no standard method of describing monuments set at corners.

PRINCIPLE 4. Words and their meanings and distances recited in a description are to be interpreted as of the date of the description, not as of the time of the retracement.

PRINCIPLE 5. A course is composed of two elements: distances and direction of the line, usually called *bearings* or *azimuths*.

PRINCIPLE 6. For a corner and its monument to be controlling in a description, the corner must be called for in the written conveyance and the monument identified at the time the conveyance was made.

PRINCIPLE 7. In a metes and bounds description, a boundary (property) corner should not be placed on a curved line.

PRINCIPLE 8. A boundary line may be created by words in a legal document. That line is as binding and as legal as if it were run on the ground by a surveyor.

PRINCIPLE 9. In conducting a retracement survey of a metes and bounds boundary, the retracing surveyor must be knowledgeable of and apply the priority of calls.

PRINCIPLE 10. Although the surveyor should not practice law, in conducting original surveys or retracements he or she must follow established case law and statute law.

5.2 METHODS OF CREATING METES AND BOUNDS OR NONSECTIONALIZED DESCRIPTIONS

As pointed out earlier, the important factors in the creation of a GLO description under the law were the actual running of the lines on the ground; the preparation of the map, which in turn may substitute for the actual description, then the acceptance of the map (survey) by the proper authorities; and the patenting of those lands to that description, the map. The metes and bounds or nonsectionalized descriptions may originate either by running lines on the ground and marking these lines, and then establishing the monumenting of corners, or by creating the boundaries in writing and even through words.

5.3 METES DESCRIPTIONS

Principle 1. *A landowner is free to create any boundary desired as long as it does not infringe on the rights of an adjoiner or on a senior right.*

A metes description can basically be thought of in terms of a landowner (or surveyor) with a piece of paper (the deed and description) walking around the perimeter of a parcel of land, following the courses, looking for the objects recited in the document, and then relating those found objects to the words in the description or on the map. In a metes and/or bounds description, the courts do not look for perfection, but they do look for the ability of a subsequent surveyor or person to be able to locate the boundaries of the parcels at a later date.[1]

One of the most liberal definitions of metes and bounds was given by the Texas Courts, which usually are quite conservative when it wrote:

It is not necessary to a metes and bounds description that length of line be given when all boundaries of the involved area are fully set forth by calls for course and adjoinder. The term "metes and bounds" as used by the legislature was undoubtedly used in its generally accepted meaning as found in the dictionaries, which define metes and bounds as the boundary lines of land, with their terminal points and angles, and the boundary lines and corners of a piece of land. The term is also identified as simply a boundary line. Length of lines would not appear essential to such a description for the obvious reason that in determining the boundaries of land a call for distance is considered the most unreliable, ordinarily yielding to calls for course and adjoinder, either natural or artificial.

This decision is interesting also because it, probably without realizing it, reaffirmed the accepted "priority of calls" or "dignity of calls" recognized in Texas courts for over 100 years.

Today, a landowner is free to create any type of description of a parcel of land, as long as it meets the legal requirements. Legally, it has been held that for a conveyance to be valid, it requires the name of the grantor, the name of the grantee, a description of the property, the recitation of consideration, words of conveyance, and a delivery, and in many jurisdictions the document must be witnessed. There are no basic legal requirements as to how the description must be written. In fact, when questioned, courts will attempt to make a description valid whenever possible.

> Courts will be lenient about making a description valid, but courts will hold to the letter of the law with deeds.

The boundaries in metes descriptions are created by starting at a *point of commencement* that may or may not be on the parcel that is being described, and proceeding by a single course or courses (bearing and distance) to a *point of beginning* (POB) or *true point of beginning* (TPOB), a point on the parcel that is being described. It proceeds, either clockwise or counterclockwise (preferably clockwise) by courses, in a systematic manner encompassing a closed figure, always calling for a corner point (monumented or unmonumented) at the termination of each course and *returning to the* POB. In a metes and bounds description, the POB, when subsequently recovered, has no greater legal significance than do any other corners in the description. Usually, each course is composed of a bearing and a distance. Each course is subject to error in measurement of the angle/direction and/or the distance. In surveying terminology, the course is a vector (a bearing and a distance), which may or may not have a monument at its terminus. It must be remembered that every bearing and every distance are subject to certain errors and, thus, they lend themselves to a statement made by John Love in 1687 in *Geodaesia*:[2]

> There are but two material things (towards the measuring of a piece of land) to be done in the field; one is to measure the lines (which I have shewed you how to perform by the chain), and the other is to take the quantity of the angle included by those lines; for which there are almost as many instruments as there are surveyors.

A line in a metes description could legally exist as "thence one line north 20 degrees and 31 minutes west, for a distance of 201.11 feet, to corner 1." However, the call for no monument states one principle, while the call of "thence north 20 degrees and 31 minutes, for a distance of 201.11 feet, to corner 1, a 12-inch beech tree marked with three hacks and a blaze" (or "to a 1/2-inch crimp top pipe painted red") states another principle.

The basis of the bearings should be indicated and should not be left to speculation or assumption on the part of the surveyor or by the courts. The basis can be a magnetic bearing, a true bearing derived from a Polaris or solar observation, a reference to a geodetic triangulation station, a previous bearing from an adjacent tract, or a bearing from a previous survey (assumed, astronomic, geodetic, or grid). But the basis of the bearings must be indicated in the description. The purpose of the description is to provide sufficient information to find and retrace the boundaries of a particular parcel. In fact, the Minnesota court stated in 1942, in the case of *City of North Mankato* v. *Carlstrom*: "Descriptions are not to identify land but to furnish the means of identification."[3]

The second element of the course is the distance. Various units of distance have been used; feet, meters, poles, perches, rods, chains, varas, smokes, arrow shoots, paces, and arpents are but a few.

Many references may be parochial in that they may be used in a very limited area and for a limited period of time.

This element becomes very important to the retracement of lines and is discussed later. Some believe that a course, the bearing, and the distance are but a "finger pointer" to show a retracing surveyor where to search for the corner and its monument. This philosophy has been suggested in several early court decisions.

This philosophy was certainly stated in both a metes and bounds state and a GLO state. In the Georgia decision *Riley* v. *Griffin*, point 16, Justice Lumpkin wrote: "In *Doe* v. *Paine & Sawyer*, 4 Hawk's N. Rep. 64, the Court refers to course and distances as pointers or guides, to ascertain the natural objects of boundary." Also, in *Andrews* v. *Wheeler*, 103 P. 144, 10 Cal. App. 614 (1909), the opinion states: "It has ever been held that the marks on the ground constitute the survey; that the courses and distances are only evidence of the survey."[4]

> The *truth* of a survey is the evidence left on the ground and then subsequently recovered.

This legal philosophy makes for sound surveying when it comes to retracing a metes and bounds description and survey. The typical metes description has led to the development of what courts call the primary elements of the *priority of calls*. If there is any principle that perhaps is most often misinterpreted and possibly misapplied by surveyors, attorneys, and courts, it is the priority of calls. There are no restrictions that prohibit a landowner from conducting his or her own survey and creating his or her own description from that survey. A metes description basically describes a tract of land that is identified as falling within a closed parcel description. Its area is predicated on the courses described in the instrument of description.

Principle 2.   The lines and corners of the original survey control the location of a parcel. Lines are presumed to have been established in accordance with accepted rules of survey and any minimum standards in existence at the time the original lines were created.

It is legally accepted that if a survey line is run on the ground, that line, run and marked, regardless of where it is found, controls because it is the *best evidence* of the intent of the parties. If corners are established and monuments set and then are called for in the instrument, these monuments and lines control—regardless! Monuments have been classified by modern courts as either natural or artificial. Early case decisions refer to (or make reference to) natural boundaries, but within the last 50 years the courts have redefined the term *natural boundaries* to mean natural monuments. These are two separate and distinct terms that should not be confused.

In locating a metes description, there is no requirement that the retracing surveyor follow the description of the lines as presented in the description, in sequence, or that the survey begin at any corner identified adequately as being in the description.

Courts have generally held that a called-for monument in an instrument or survey, if found and undisturbed, controls the line, regardless of where it is found.

5.4 BOUNDS DESCRIPTIONS

A bounds or bounding description can best be described as one that is written and then interpreted as if the person were standing in or placed in the center of the parcel of land and then looks outward to the boundaries of the parcel that is being described. The description can be in either a clockwise or a counterclockwise manner. For example: "A parcel of land bounded on the north by the lands of Smith; on the east by the lands of Jones; on the south by the country road; and on the west by the lands of the church lot." There have been instances when bounds or a bounding description was prepared with a minimum of positive information and possibly in a flippant manner. This makes the retracing surveys all the more difficult because of the ambiguity of the boundary description.

Of course, the more abbreviated the named boundary, the more difficult it will be for the retracing surveyor to place the parcel with great certainty. This means that to locate that one unique parcel, the surveyor will be required to locate *all* four described elements of control. In a bounds description, it has been held that one cannot create a legal bounding description by calling for the person's own land. An example would be "a parcel of land bounded on the north by lands that are being retained by me." This would be an insufficient description in some areas but is very common and may be sufficient in New England and elsewhere.

In a bounds description, the calls for the adjoiners become the monuments. Many times they do not have a direction of travel, a bearing, or even a distance. In a bounds description, it does not matter what sequence the calls for the adjoiners follow. An example could be "bounded on the north by the creek, on the south by the road, on the west by the lands of Smith, and on the east by the lands of Jones." This is contrary to a metes description, where the direction and sequence of travel are very important elements.

In a bounds description, the parcel being surveyed is identified only after the elements of the boundaries are ascertained and located. The parcel being surveyed gets all that is left after the elements are located. In surveying a bounds description, all documents referred to in the bounds called for, as well as all monuments and courses, become part of the description just as if they had been recited in the document itself.

5.5 COMBINATION METES AND BOUNDS DESCRIPTIONS

Many of the descriptions that are prepared to describe boundaries are combination descriptions. Of all the descriptions, these are perhaps the easiest to write but the most difficult to ascertain to conduct title search and to finally to survey, because of possible conflicts. This is when the priority of calls should be considered and applied. A typical combination description is: "Commencing at the intersection of Fifth and Broad Streets, thence north 500 feet along and with the centerline of Fifth Street to the POB, a 2-inch iron pin set flush with the pavement of Fifth Street, located on the south boundary of the land of J. R. Smith; thence along and with the land of Smith, N 23 E. a distance of 721 feet to the lands of Jones, a 3-inch iron pipe on the north right-of-way line of Highway 20 . . . ."

This description gives rise to the application of the "priority of calls," in that the quality of the retracement depends on the quality of the original measurements and the quality of the original description as well as the quality of the original monuments placed at the time the parcel was created and then the quality of the created description. The first question is: Did the creating surveyor actually locate the exact intersection of the two streets? Second, did the surveyor actually locate the corner of Smith, or was the "iron pin" accepted by the surveyor without verification? Then, was the line of Smith actually run, or was the course just accepted without any verification? Did the surveyor run the description of Jones or just accept the pin as being Jones's corner? Did the surveyor actually locate the right-of-way line of the highway? These questions and more must be answered in order to conduct an adequate field retracement.

In a combination description, when two elements are called for and there is a conflict between them, the creating surveyor or scrivener must make certain that the two elements used in the description contain no conflicts when they are cited. It is presumed that a combination description has been run and verified, or it is intended that there be as few conflicts as possible. In both a metes and bounds description and a combination description, the term *course* is used. The surveyor refers to *course* as being a combination of a bearing and a distance, whereas the attorney and the courts often use the term *course and distance*, meaning "bearing and distance."

5.6 STRIP DESCRIPTIONS AND STATIONING

A strip description is one form of metes and bounds description that should enjoy a very limited use and may be subject to numerous misinterpretations and legal questioning. In many instances, this type of description has been attacked in judicial circles. This form of a description is very basic in its creation. A single line is usually described, and then the limits of the parcel are described relative to the line described. This form of description is subject to possible future problems and is often lacking in the area of monumentation.

Usually, a strip description is narrow and long, customarily being used in creating roads and utility easements.

In describing a road easement or right-of-way by a strip deed, the form such as "a right-of-way for road purposes over and across a strip of land lying 30 feet on each side of the following described centerline" is sometimes used. Generally, the *stationing system*—starting from an arbitrary point called 0 + 00 and assigning each point on the line a station that is dependent on its distance from the starting point—is employed. If a point is 1327.62 feet from the arbitrary starting point, said distance being measured along the centerline of the strip, the station is 13 + 27.62. Every 100 feet *along the centerline*, be it on a curve or an angle, is an even station; the plus number is the added distance beyond the station. The stationing system was devised for the convenience of the surveyor in note keeping and map notations. Any object along a right-of-way, such as a power pole, may be located by a simple note: "Sta. 13 + 12.60, 30 rt." When looking toward increasing station numbers, right is to your right. The "30 rt." means that the pole is located 30 feet to the right and is 90, or radial, from the station indicated. Although this stationing system is legal, it deviates from the norm and is not seen in most modern descriptions, nor is it a recommended form of description.

In a strip conveyance, where the road ends or begins on a diagonal line, the side lines must be extended or shortened to terminate on the diagonal line. In Figure 5.4, danger areas, which are not usually covered by strip descriptions, are shown. Where the road is a straight line, a recital stating "extending and shortening the side line so as to terminate at the boundary or property line" is usually sufficient to cover area *A* as shown in Figure 5.4. Where a curve is involved, *continuing* instead of extending is preferable. Parcel *C* is best described as a separate exception.

Some surveying personnel believe that using a road stationing system is a perfectly acceptable practice, but seldom do the construction personnel identify the stationing of the design elements used by the planning engineers in order to adequately locate the "strip" after the construction has been completed. The retracing surveyor must

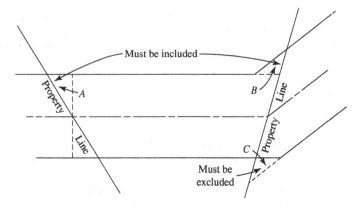

Figure 5.4 Strip conveyances should be written to include or exclude areas *A*, *B*, and *C*.

go into the original plans on which the stationing elements were originally identified. (Authors' note: The authors, in working on strip description for over 60 years, remember only one instance where the road department identified the road stationing by casting the station numbers into the concrete at the time the road was constructed. Wonder what they did when the road was resurfaced?)

5.7 DESCRIPTIONS BY REFERENCE

The simplest and most economical form for conveyance of a written land description and its boundaries is one that refers to a map or a plat. Legally, it may or may not matter whether the map is recorded. The important element is that the map be available for positive location of the parcel(s) being described. In the event that the map referred to cannot be found, serious legal problems could be encountered in that there may be uncertainty in ascertaining the true boundaries.

The description "lot 1, block 49, Fama Heights Part 2. Dekalb County, Georgia, as prepared by Mary Giggens. RLS 1307" is valid for conveyance purposes, but the boundaries are those created in the original survey of the subdivision. The entire description is predicated on the fact that a copy of the reference map is available for consulting. In the event that the map cannot be found, there is a legal possibility that the description may be considered legally void. The only way to ensure that this does not happen is to attach a copy of the map to the description (see Figure 5.5).

5.8 ALIQUOT DESCRIPTIONS

As discussed in Section 3.24, an aliquot description is a perfect description. According to federal law, the least aliquot description available (or legal) is a quarter-quarter or one-sixteenth or 40 acres, according to the GLO survey of a normal nominal

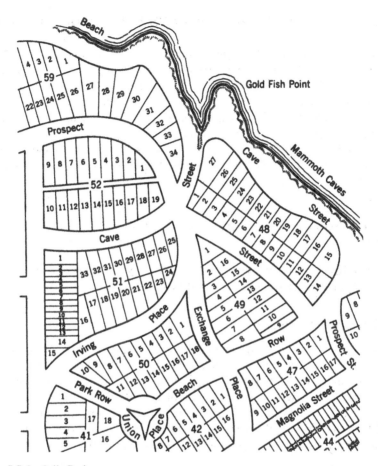

Figure 5.5 La Jolla Park.

section. Yet many people may describe a portion of a section as small as 1/64 or 1/256, or even smaller. It must be remembered that these boundaries are invisible and set by law as being a direct proportion, without a remainder.

Federal patents are issued by aliquot descriptions and not by reference to monuments. This is because federal law and state case law hold that a reference to a surveyed federal township carries with it all related surveys, monuments, instructions, and laws that were in effect at the time the survey was created.

Once the aliquot portion is described by the aliquot boundaries, any modification by adding calls for adjoiners or natural features may cancel its aliquot nature. Thus, the description "all of the NE quarter of section 2 T 26 N R 32 W, being north of Highway 6 and west of Small Creek" ceases to be an aliquot description; it becomes a combination description. An aliquot description may also be applied to a metes and bounds parcel: for example, "the east ½ of lot number 4, Happy Times subdivision, or the NE ¼ of lot 16, block 3, Reserved Acres."

5.9   OTHER MEANS OF CREATING BOUNDARIES IN DESCRIPTIONS

There are numerous ways of creating boundaries in descriptions. At times, one is unable to accurately describe the numerous ways or means by which land descriptions and the resulting boundaries are described. These means are often used to shortcut the need for a survey, for expediency, to save the cost of preparing a new description, or because the people involved do not know any better. Giving a name to these descriptions, they could possibly be called *quasi metes and bounds*. Any time an abbreviated means is used to describe a parcel of land and its resulting boundaries, there is always the possibility of future problems. These types of descriptions should be discouraged, because, to coin a phrase:

> There is no substitute for an adequate and correct description.

Some of the methods used by landowners, attorneys, and surveyors to "shortcut" descriptions and to create boundaries are described next.

Division Line Description

The division line is very easy to compile, write, and describe. This is probably the easiest form of description to describe but perhaps the most difficult to survey. In most instances, in this type of description, no survey is required, only an intent: for example, "all of Section 17 lying west of U.S. Highway 80, as it now exists" (see Figure 5.6) or "all of lot 12 lying south of Boulder Creek." Of course, a longer description may result if the surveyor actually performed a survey of the road or creek and chose to show the traverse lines or meander of the features described. This

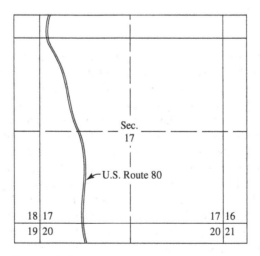

Figure 5.6 Division line description.

form of a description is just a conflict waiting for a future conflict, once U.S. 80 changes. In the future, the retracing surveyor may encounter numerous problems in trying to locate such boundary lines.

Or possibly, the "east 5 acres of my 10-acre parcel, more particularly described as the east ½". In this situation, a latent conflict may present itself on a survey when one surveyor surveys the east 5 acres and a second surveyor surveys the east ½, which legally may indicate the same parcel, but surveywise two lines may exist in the mind of the grantor between the two estates.[5]

In many older descriptions, a person may find an element in the description as "and thence along a *provisional line.*" These lines were run with the anticipation of becoming a newly created boundary between two or more parcels, created from one parent parcel. Ultimately, they may have become the final line(s).

Distance Description

The shortest form of a quasi metes and bounds description is the following: "the easterly 50 feet of lot 2." This type of creation must be the first line described. If the lot is 100 feet in length, the grantor must determine what the true intent is—is it to describe the east half or the east 50 feet? They may or may not be the same.

Any subsequent descriptions must maintain the same proportions. A junior deed would read "all except the easterly 50 feet" or "all except the easterly half," depending on how the original deed was written. A description should not be by distance and the second by fraction in the same description of the dividing boundary. They do not describe the same line; they describe two separate and distinct lines.

Proportional Conveyance Description

Proportional conveyances are used to convey land and to describe the resulting boundaries. By common law in the metes and bounds states, the presumption as to proportional conveyances is that any proportion recited is to the area contained in the whole. That is, the "north half" is the north half as to area contents within the boundaries of the parcel. In the GLO states, the opposite presumption is applied; that is, the presumption is that any proportional recital is to the distance of the respective lines. According to federal law, a section is divided into respective halves by connecting alternate quarter corners. In some metes and bounds states, statutes have reformed and clarified the common-law rule.

There may be instances when the distance description and a proportional description come into conflict: When is a half not a half? To the uninformed, such a situation can happen in the following cases:

- A person has title to a 10-acre parcel in the shape of a square. The owner sells to A the north 5 acres and retains the south 5 acres.
- He then sells the south half he retained. Two invisible division lines were created. One for the north 5 acres and one for the south half; in all probability they are not superimposed.

■ The original owner could end up with a strip/gore or a retained parcel between the two invisible lines.

If possible, it is the surveyor's responsibility to keep situations like this from occurring.

Exception Description

Land parcels have been conveyed and boundaries described by exception; for instance, "all except the easterly 50 feet of lot 2." This form of a description has a great potential for multiple interpretations and conflicts in location with its resulting litigation.

Area Description

A description such as "the south five acres of lot 7" clearly describes 5 acres, no more and no less. However, the questions remain: What is the direction, and what is the length of the dividing line? The line between the 5 acres and the remaining parcel is ambiguous. One can clearly see the advantages of this type of description: no survey is needed, and it is easy to write. If this type of description is used, the method of determining the dividing boundary must be identified.

A boundary described in this manner is fraught with possibilities for judicial interpretation and consequent problems. A description of "one acre in the south-east corner of my lot" is not the same as "one acre at the south-east corner of my lot." *The "in" description is probably void for lack of certainty, but the "at" description can be legally located, with its boundaries probably being set by the judge or the court.*

"Of" Description

Many landowners, attorneys, and others commonly use the "of" type of description, as in "the easterly 50 feet of lot 6," "the south 3 acres of lot 3," and "the east half of lot 1." Surveyors should strive to refrain from using this type of description because of the possibility of creating future problems for both landowners and other surveyors. Such descriptions are fraught with possibilities for misunderstanding and errors (see Figure 5.7).

5.10 NOMENCLATURE IN METES AND BOUNDS DESCRIPTIONS

In many instances, metes and bounds descriptions should be considered as complex and fraught with possible survey and legal conflicts. They should be studied in their entirety before one is written or before one is retraced (see Figure 5.8). The correct words must be used in their proper form and continuity in the written document.

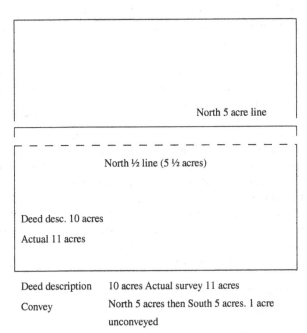

Figure 5.7 Deed description 10 acres, actual survey 11 acres; convey north 5 acres then south 5 acres. 1 acre unconveyed.

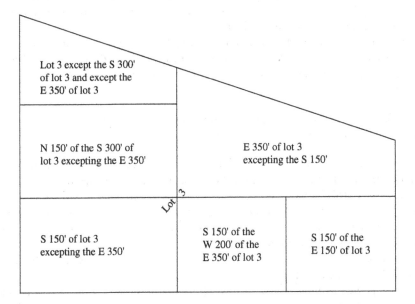

Figure 5.8 "Of" description.

Principle 3.   *There was no standard unit of distance or angle measurement used in early metes and bounds descriptions and no standard method of describing monuments set at corners.*

In most states that use metes and bounds descriptions, retracing surveyors will find mention of distances in feet, chains, poles, rods, varas, paces, and numerous other units. In many instances, references to bearings are silent as to whether they are true or magnetic. Unless the retracing surveyors can agree about the determination of the "truth," it is left of to the court to determine what was meant by the words used.

This may also apply to words within the description that describe other elements. Trees have different names today, landowners are different, and local terminology has changed. Many words that were in common use 100–200 years ago no longer have any modern meaning or the meaning has changed.

Many of the early surveys contain references to distances in feet, yards, poles, steps, and numerous other elements as well as standard units recognized by the surveyor. These discrepancies probably resulted from the liberal acceptance by the courts that "any man can survey his own land."

Principle 4.   *Words and their meanings and distances recited in a description are to be interpreted as of the date of the description, not as of the time of the retracement.*

In examining deed terms commonly used in metes and bounds descriptions, a person should have an understanding of both their historical and their modern meanings. Some of the problems one may encounter are as follows: What was the length of the chain used? If magnetic bearings were used, what was the declination at that time? Were the bearings true bearings? What is the meaning of some of the words in the description that are no longer used today?

To understand a metes and bounds boundary, the retracing surveyor has to possess this basic knowledge in order to interpret the metes and bounds description. Today's surveyor may even find that the same word has different meanings in neighboring states.

Principle 5.   *A course is composed of two elements: distances and direction of the line, usually called bearings or azimuths.*

In retracements, modern courts tend to place great reliance on historic described courses retraced by modern methods. Accepting the fact that a line may be described as "North 29° 15′ 21′ West, a distance of 31 chains," this line has two vectors: a line and a distance. Each is independent of the other, but the line is a product of both. If one element is missing, the line has no significance. It is the corner at the end of the line that controls its termination.

Each of the two elements has inherent errors that should be understood and, if possible, compensated for.

Early judges, in most instances, gave little weight to the significance of courses; in fact, one Supreme Court Justice Lumpkin, in 1855, wrote in his famous rules for retracing lines the following:

Courses and distances occupy the lowest, instead of the highest grade, in the scale of evidence, as to the identification of land.

Courses and distances, depending for their correctness on a great variety of circumstances, are constantly liable to be incorrect; difference[s] in the instrument used, and in the care of surveyors and their assistants, lead to different results.

Looking back even further into surveying history, Love wrote the following in1689:

There are but two material things (towards the measuring of a piece of Land) to be done in the Field; the one is to measure the Lines (which I have shewed you how to perform by the Chain), and the other is to take the quantity of an Angle included by those lines; for which there are almost as many Instruments as there are surveyors.

Direction of Travel

True metes and bounds descriptions and many quasi metes and bounds descriptions have a direction of travel. A bearing may be stated in either of two directions on a map or plat, but only one can be used in a written perimeter description. In Figure 5.9, starting at the POB, the direction of travel is to the southeast, making the first written bearing in the description S 45°00′ E, not N 45°00′ W. Because the relationship of one line to another is shown by the plotting of the lines in Figure 5.9, it is immaterial whether the bearing on the plat is written S 45°00′ E or N 45°00′ W. Although no fixed rule exists, in many areas the surveyor's common practice is to prepare the direction of travel in a clockwise manner.

> The retracing surveyor is likely to find some situation that will not or does not fit into a particular category. It will be at these times that he or she will have to rely on ingenuity and experience to seek a solution.

Measurements of Distance

The second element of a course, after the direction, is the distance to be measured along the line. Throughout history, numerous units have been used to describe distances. Table 3.1 shows various units that have been used over the years. The unit of measurement recited in a description is the unit that was used or accepted at the time the survey was conducted or when the description was written.

In retracing an ancient description, the retracing surveyor must make certain that the distances given in the current deed have not been converted an earlier description. Many of the older descriptions in the eastern United States cited distances in chains, whereas modern deeds now refer to the same distances in feet. This is not proper and should not be done. The retracing surveyor should attempt to retrace ancient metes and bounds boundaries in the units of measurement by which they were created. Nor should a surveyor convert an older distance without first surveying the line in the

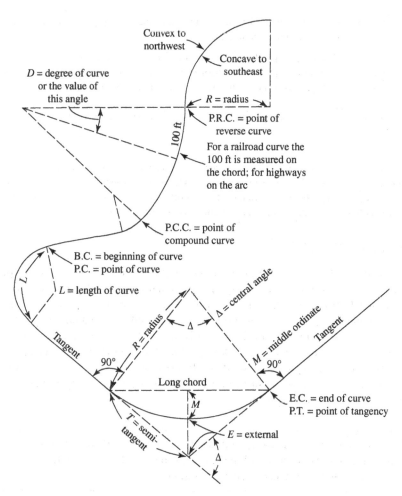

Figure 5.9 Direction of travel.

units to be used in the description. New plats should indicate the original boundary distance and the newly retraced distance, not the converted distance in today's units. If the retracing surveyor maintains references in the original units of measurements, correlations and errors can be ascertained and more easily explained.

In analyzing distances in descriptions, there are several presumptions the surveyor must accept or attempt to discredit as being *rebuttable presumptions*. One is that the distances indicated are straight-line distances between the corners. The second is that the indicated distances are horizontal in nature and not calculated.

If you, the retracing surveyor, determine that the distances have been modified that presumptions may be addressed. Today, few boundary lines are actually run directly between proven corners in a straight line. Random lines are run between the proven corners, and then the resultant unoccupied straight line is calculated and reported as if it had been actually run. Such a line may have multiple errors—not only the measurement error but calculation errors as well. Into this category we should place the results of global positioning system (GPS) calculations.

Monuments

Principle 6. *For a corner and its monument to be controlling in a description, the corner must be called for in the written conveyance and the monument identified at the time the conveyance was made.*

This principle is one of the most difficult for retracing surveyors and for the courts to first understand and then to accept. It would seem logical that if, in conducting a retracement, the surveyor should uncover a monument in close proximity of a corner, that should be the monument—*regardless*. But it does not work that way. To be controlling, that monument must have a "chain of authenticity" to the creation of the parcel—absent the chain, *no acceptance*.

> The acceptance or rejection of found evidence does allow for the logical human errors, such as a misjudgment in size, calling a 1-inch pipe a ½-inch pipe, or calling a red oak a white oak or a hickory a pecan.

The key element of a metes description is the call for a monument at each corner. When monuments are established at corner positions on boundaries and then called for in the written description of that boundary, the monuments control over bearings and distances. The monuments established must be described with certainty: no "IPs," no "pins," no "stakes." Instead, the surveyor should describe the monument as a 5/8 iron rod with surveyor cap 'Nettleman Boundary FL #6633'. The writer should describe accurately what was set for the original boundary, what was found at that point, and what was done there. If no monuments were set at the initial survey and no marks were made along the boundary lines, those lines still remain invisible, exist only in a legal sense, and are not capable of being precisely surveyed.

Lines exist because of the corners. If the corners are set and the line marked, both the corners and the line become controlling. If at any time the corners defy location, the line may still exist independently. For this reason, it is important to call for both the corners and the line—for example, "To corner 2, a 2-inch iron pipe set 2 feet in the ground and painted red. Thence along and with a marked line . . . ." In the priority of calls, the lines actually run, if called for, are superior to all other elements.

Courts have classified monuments in broad categories as natural, artificial, record, or legal. Naturally occurring monuments such as rivers, lakes, oceans, bays, sloughs, cliffs, trees, hills, and large boulders are permanent objects found on the land as they were placed by nature and are usually considered controlling over artificial monuments (human made), such as iron pipes, wooden stakes, rock mounds, stones, and wooden fences, but if the writings clearly indicate a contrary intent, especially where the lines of a survey are called for, the control might be reversed. Some human-made monuments, because of the certainty of location, visibility, stability, and permanence, are considered equal in rank to natural monuments. In this classification would fall sidewalks, street paving, curbs, wells, canals, concrete buildings, and concrete fences. A legal corner is one set by law, i.e., the 1/4 or 1/16 corner of a section, in the PLSS. In applying the proper rules of evidence to the recovery and identification of monuments, it is imperative that the surveyor or attorney understand the various interpretations of what both the trial and the appellate courts in their jurisdictions have held in relation of applying to the local *Rules of Evidence*.

The term *natural monument* is of recent origin, in that, until the 1950s, courts referred to monuments as *artificial and natural boundaries*. Whether the monuments are called natural or artificial, in retracing the boundaries the surveyor should do sufficient research to determine the holdings of the local courts. Some courts seem to be giving greater preference to the new *natural monuments* than they should. These courts seem to have overstepped the weight given some monuments that they now determine to be in this category. In one instance in a bench trial, the judge held that a small pond indicated on a GLO plat was a natural monument because the section line was shown as passing through the pond even though it was not indicated in the field notes.

Found record monuments can vary from the written record in many aspects that would give the surveyor cause for concern. Usually, the monument found should not appreciably differ from the monument called for in the writings. Differences in size, composition or materials, and markings should be minimal. The retracing surveyor should realize that the referenced or recorded measurements probably were "estimated." The authors have no personal knowledge that the early surveyors carried tape measures to measure the length of posts, or diameter tap or calipers to measure tree diameters.

A surveyor's outstretched hand was considered to be 6 inches from thumb to little finger, the width of the thumbnail was 1 inch, a step was 3 feet, and a pace was 6 feet. Today, these references do not translate to exact measurements.

Record Monuments and Adjoiners

A *record monument*, sometimes called a *legal monument*, is a monument referred to in a conveyance description and is often interpreted to mean a call for an adjoiner or abuttal property. A tree is visible and presents no major problem as to its location, provided that it is the proper species and marked as indicated, whereas a call for an adjoiner, such as a call "to the property of Daniels," requires obtaining and reading a copy of the adjoiner's conveyance to determine the adjoiner's true location. Certainly, a call for an adjoiner is a call that has size and shape (is physical in this connotation), but the limits of the size and shape cannot be seen until marked by physical objects. However, no assurance exists that the visual objects are in their correct position until the adjoiner's conveyance is researched and located on the ground. The difference between a natural or artificial monument and a record monument is as follows. To locate a natural or artificial monument, the surveyor goes on the ground with research data and searches for the monument; to locate a record monument, the surveyor must first obtain a copy of the record describing the monument, then he or she must locate the boundaries in accordance with the record.

In some reports of court cases, a call for an adjoiner has been classified as a call for a natural monument! This strange use of the word *natural* arose from reasoning that the dirt and earth that composed the adjoining property were naturally occurring. The fallacy behind this reasoning is that it overlooked the legal definition of a natural monument: a natural monument is a naturally occurring visible object. Most adjoiners (the side line of Jones) are not visible until marked, and in no way are the limits of the property naturally occurring. Once again, we look at what Justice Lumpkin had to say in *Riley* v. *Griffin:*[6]

> Any natural object, and the more prominent and permanent the object, the more controlling as a locator, when distinctly called for and satisfactorily proved, becomes a landmark not to be rejected, because the certainty which it affords, excludes the probability of mistake.

Authorities are numerous which hold that, where the call for a boundary in another deed or for the boundary of another tract is expressed as "by such a line" or "by the north line of such a tract," the line or boundary referred to is locative and fixes the boundary definitely.[7]

In the case of *McClausland* v. *York*, the Maine court stated that whether the adjoiners' deed was recorded or not was immaterial.[8]

An adjoining parcel, in order to control, must be called for, must have existed at the time of the conveyance, and must be clearly established and identified and accurately located.

Properties of Monuments

A good monument should possess the quality of being easily visible, certain of identification, stable location, permanent in character, and nondependent on measurement for its location and identification. An artificial monument possesses the qualities of a natural monument to a lesser degree. Thus, an iron object or a wooden stake placed in the ground will rust or rot with time and is less permanent than a naturally occurring large boulder. A stake is easier to move than a boulder and is therefore less stable. The visibility of record monuments is wholly dependent on natural or artificial monuments (fences, stakes, cultivation, plantings, or marks) that delineate the limits of the record monument. In the priority of calls, monuments are classified just below the lines actually run. It is interesting to note that the term *natural monument* is not found in early case decisions. There are many references to natural boundaries, but the more recent thinking is that the courts have probably misapplied the two terms.

5.11 ADJOINERS

> It is the surveyor who must determine whether the call for the adjoining boundary is controlling or descriptive.

Next in the priority of calls are adjoiners. An adjoiner is a critical element in a bounding description, but it may be only a supplemental element in a metes description. When a metes description calls for a course, an artificial monument, and an adjoiner, and no two of these calls can be reconciled, problems result. "South, 300 feet to a 6-inch concrete monument set on John Smith's line" calls for a course, an artificial monument, and an adjoiner. In the priority of calls, the ranking would be as follows: (1) the artificial monument, (2) the adjoiner, and (3) the course. To eliminate future boundary location problems, it is imperative that the creating surveyor make these elements compatible.

This element, *adjoiners*, has found its way into legal acceptance as being a controlling element, following lines actually run, and monuments, both natural and artificial. It should be noted that in order to be controlling, the adjoiner must be senior in title; however, if the adjoiner is senior in title but not called for in the survey, the surveyor should be prepared for a possible conflict.

5.12 DEED TERMS FOR CURVES

Principle 7. *In a metes and bounds description, a boundary (property) corner should not be placed on a curved line.*

When curves are included in a description, it is important that the proper terminology be used. Three elements of a curve should be recited to enable location of the curve. A description "thence with the curve of the road" may be suitable for conveyancing purposes, but it is greatly lacking for survey purposes. To locate a curve properly, a minimum of two of its parts or elements must be identified. Figures 5.10 and 5.11 depict the proper terminology.

A curve can always be described, but it cannot be surveyed. Only the elements of the curve can be surveyed.

Dubeau identifies the elements of a curve in Figure 5.12.

Curves

To define a curve in a description, at least two elements of the curve must be stated, and, in addition, (1) the relationship of the curve to the previous line, (2) the direction of the curve, and (3) the direction of travel of the curve must also be stated.[9] Figure 5.9 shows common curves and their relationship to one another. The *direction of concavity* of a curve is defined by the direction of a line drawn from the midpoint of the arc of a curve to the center point of the circle of the curve; it may be defined as

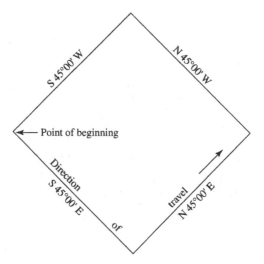

Figure 5.10 Deed terms.

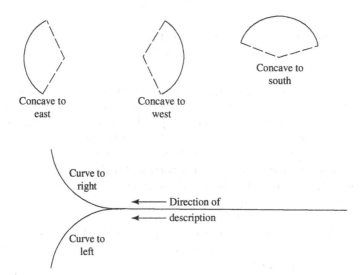

Figure 5.11 Curves.

easterly, southerly, southeasterly, and so on. The direction of travel along a curve can be stated to the right or left, as indicated in Figure 5.9, or by stating the direction as "southeasterly along the curve," or by stating a number of directions as "southeasterly, southerly, and southwesterly along the curve." A statement that a line is tangent to a curve (i.e., it is 90° to the terminal radial line of a curve), or a definition of the bearing of a line and also a definition of the bearing of the radial line at the point of contact of the curve and the line clearly indicates the relationship of the curve and the line.

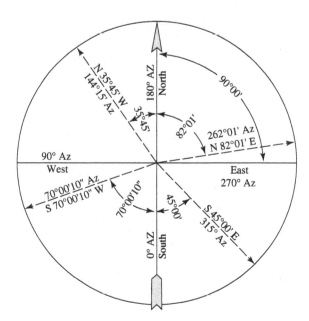

Figure 5.12 Direction of lines. Azimuth can be from north or south. The US Coast and Geodetic Survey uses south as the reference datum. Most land descriptions use north as their point of reference.

In deeds, only two curve elements are needed, but at least three are normally given: radius, central angle (called *delta*), and length of curve (see Figure 5.10). Length of curve in most deeds is a computed quantity depending on the given central angle and the radius. Rarely in metes and bounds descriptions are the middle ordinate, external, long chord, and tangent used, but on plats they are seen more frequently. The *degree of curve*, often given on railroad right-of-way plats, is defined as the central angle of a curve subtended by a *100-foot chord* on the said curve. Along highways the degree of curve is usually but not always defined as the central angle subtended by a *100-foot arc* of said curve (see Figure 5.10).

Because curves are described by their elements and the boundaries are created by connecting the elements, boundary (property) corners should not, if possible, be described or placed on curves.

5.13 LINES AND THEIR ELEMENTS

There are other elements that should be understood in boundaries and descriptions that are created, described in descriptions, and then retraced. Following are several of the elements that one should consider in order to understand lines.

A line that is described by reference to a bearing, azimuth, etc., and a distance is subject to two possible errors: distance and angle. Each is independent.

Lines

A line in a description is assumed to be the shortest horizontal distance between the points called for presumptively at the ends of the lines (corners), unless the contrary is indicated by the writings. To be absolutely correct, a straight line curves with the surface of the Earth. In the usual land boundary survey of small neighborhood lots, this correction is usually so minute that it usually is overlooked because it cannot be measured. The curvature is so slight that it is not considered in local land descriptions. A line to be identified must have a positive definition of its starting point, a direction, and length. *Free lines* are not terminated by an adjoiner or monument as "beginning at a 2-inch iron pipe; thence N 60°00′ W, 200.00 feet." If the same phrase were reworded "beginning at a 2-inch iron pipe; thence N 60°00′ W, 200.00 feet to a blazed sycamore tree," the terminus of the line is fixed by the tree; the line is not free. Many of the lines described in deeds depend on monuments and are not free lines.

A bearing quoted for a line defines it as a straight line. If a line is defined by monuments, without bearing or distance, the words "in a straight line" or "in a direct line" are sometimes added to emphasize the presumed fact that the line is straight. An ambiguity may exist if a bearing is recited, the term *straight line* inserted, and then explained further "along a natural feature," such as "thence N 10° W, a distance of 210 feet, in a straight line along an existing fence." If one element varies from the others, the relative dignity of the elements must be considered.

Compass Direction

As commonly practiced in this country, direction is defined by either a call for monuments or a bearing making reference to north or south; but azimuths, deflection angles, or coordinates have also been used to define a line. If a deed is written "commencing at a blazed sycamore tree located approximately 100 feet west of Jones's well; thence to a blazed white oak, etc.," the direction is clearly defined. It is very desirable to quote the bearing of the line for plotting purposes, but it is not essential to the legality of the conveyance. Bearings in modern descriptions are always read in degrees and minutes (plus seconds if fractions of a minute are involved) from the *north* point. Rarely are lines described from the *south* point of reference except in Hawaii, and *never* from the east or west points (see Figure 5.12). However, this may not be so in earlier descriptions. A very few early colonial boundary descriptions were described from the east or west. The direction of a line depends on which end of the line you are standing at; thus, on a northwesterly line, the direction would be southeast if you were at the northerly terminus of a line, whereas it would be northwest if you were at the southerly terminus of the same line. On a map, it is immaterial which bearing you write, as the drawing shows the relationship of one line to another, but in a written metes and bounds description, the exact direction of travel along the line being described must be stated. Most metes and bounds descriptions are usually described in a clockwise direction. This should be the modern practice.

The majority of the bearing references on old plats refer to compass bearings. These early bearings were usually magnetic and in most instances were to the nearest full degree (e.g., N 52° W or S 20° E). This line is subject to the declination *at the time of the survey* as well as the least reading indicated of the compass.

One of the first decisions a retracing surveyor must ascertain is the precision of the original survey descriptions. Since case law holds that the original survey is accurate, the retracing surveyor should "run" a closure of the original information to determine how good the original work was.

To ascertain the true direction of the same line today, the surveyor would have to correct for the amount of declination from the time of the initial survey until today's retracement. As well as being subject to this natural error, the line was also subject to the ability or exactness of the surveyor to read the compass and to the mechanical errors in the compass.[10]

Deflection Angle

Many angle observations in the field were turned by the deflection method, that of sighting on a given line, transiting the telescope, and turning the angle to another line, either to the right or to the left. Normally, the deflected angle is then converted into bearings. In some descriptions, the deed author quotes the deflected angle rather than determining a basis of bearings and computing the bearings from the angles turned. A deed reading "commencing at the southwest corner of lot 10: thence easterly 200 feet along the southerly line of said lot; thence 20°00′ to the right, 200.00 feet: thence ... " (see Figure 5.13) means that the new line leaves the lot line in a direction that is 20°00′ to the right as you are looking in the direction of deed travel. To avoid possible ambiguity, the form "thence 20°00′ to the right from the prolongation of the last course, 200.00 feet" is sometimes used. In modern deeds, the terms "to the right" and "to the left" are used so commonly that objections to the phrases are disappearing. It should be noted that if deflection angles are used, the deflection is always in the line of travel.

Interior and Exterior Angles

After a bearing of an initial course is determined, a surveyor may take a backsight and then turn an angle to the right or left. The direction turned is recorded as an interior or exterior angle. By calculations, the remaining bearings of other lines are then determined.

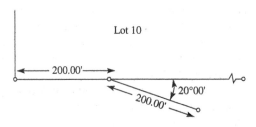

Figure 5.13 Deflection angle of 20°00′ right.

Azimuth

An azimuth, as used in surveying, is the angle measured clockwise from the meridian (usually from a north or south reference line) to the line being described. The military services commonly use north as the datum; geodetic azimuths were usually from the south. In boundary surveying, either has been adopted as the reference meridian (see Figure 5.12). If north is used, 0° azimuth is N 0° E, 135° azimuth is S 45° E, and 270° azimuth is west. In a conveyance using the azimuth system, it is imperative that the assumed zero line be clearly defined; otherwise, ambiguity results.

Local history may dictate what is used; for instance, in Hawaii azimuths are, for the most part, from a south reference meridian because many of the early surveyors were ship captains who used southern stars for orientation.

Strictly speaking, any straight line other than a true north line or the equator has a changing azimuth or bearing as one travels along the line. This is because azimuth or bearing is determined by the angle from the line described to a true north line. Because all true north lines converge toward the North Pole, the angle turned at each point along a line is being turned to a true north line, which is not parallel with any other true north line at any other point on the line. The differences in bearings at the terminus of the ends of straight lines within a local area are insignificant, and all deeds are assumed to refer to one spot (usually, the POB) for their basis of bearings.

On large geodetic maps and government topographic maps, because of the vast area surveyed, corrections for convergence are given, and the true bearing at the point considered is known. To avoid the confusion of changing bearings on any straight line, the *Lambert* or *Mercator coordinate system* has been adopted. In this system, within a specific zone, all bearings are referred to true north as defined relative to one datum line in the zone. Within any one legal description that uses azimuth or bearings, the assumption is that unless the contrary can be proved, all bearings and azimuths are referred to the same basis at the same point and that the bearings of deed lines are constant.

Compass Points

In areas near the seacoast, or where lands have been surveyed by seafaring persons, sometimes directions were expressed in compass points. The ship's compass circle is composed of 32 points, or 8 per quadrant, each point containing 11°15′. They are accumulative in a clockwise direction from due north, as illustrated by Table 5.1.

One will find many descriptions like this in New England and in Hawaii. The reason for this was many of the ship captains and navigators would compass time doing surveying and they would use the reference they were familiar with.

Parallel Lines

Parallel lines are equal distances apart, said distance being measured at right angles to the lines or on radial lines of curves. Curves, to be parallel, must be concentric (having the same center of a circle). In descriptions, parallel with a bound means parallel with

TABLE 5.1   Conversion of Compass Points to Degrees

Points	Angular Measure			Points	Angular Measure		
	Degrees	Minutes			Degrees	Minutes	
North to East				South to West			
North	0	0	0	South	16	180	0
N by E	1	11	15	S by W	17	191	15
NNE	2	22	30	SSW	18	202	30
NE by N	3	33	15	SW by S	19	213	45
NE	4	45	0	SW	20	225	0
NE by E	5	56	15	SW by W	21	236	15
ENE	6	67	30	WSW	22	247	30
E by N	7	78	45	W by S	23	258	45
East to South				West to North			
East	8	90	0	West	24	270	0
E by S	9	101	15	W by N	25	281	15
ESE	10	112	30	WNW	26	292	30
SE by E	11	123	45	NW by W	27	303	45
SE	12	135	0	NW	28	315	0
SE by S	13	146	15	NW by N	29	326	15
SSE	14	157	30	NNW	30	337	30
S by E	15	168	45	N by W	31	348	45
—	—	—	—	North	32	360	0

all parts of the bound, be they curved, angular, or straight lines. A line parallel with a creek is parallel with all the bends and angles of the creek and is an equal distance from any bend or angle unless otherwise indicated. Sometimes the width of the strip can be wider than called for (see Figure 5.14), but it is never less.

Coordinates

Any point can be defined with respect to any other fixed or assumed point by stating the distances north, east, south, or west from a datum, reference, or POB. If the southwest corner of a tract of land is assumed to be zero coordinates and the northeast corner of the land is found to be 5281.62 feet north and 5271.68 feet east of the assumed zero or datum point, the coordinates of the northeast corner would be 5281.62N, 5271.68E. Likewise, if the southeast corner of the land is 3.17 feet south

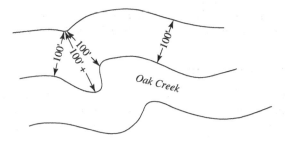

Figure 5.14 In a description, a call of "100 feet from and parallel with Oak Creek" may be more than 100 feet, as shown.

and 5279.81 feet east of the zero point, its coordinates would be 3.17S, 5279.81E or 23.17 and 15279.81. Coordinates given for any two points define completely the direction and distance between the two points and also define the direction and distance of each point from the assumed datum point. By adopting the same datum point or datum lines for all surveys within a given region or zone, as in the *Lambert system*, all surveys can be related to one another, thus eliminating many mapping problems.

Plane coordinate calculations are based on the assumption that the Earth is flat. Within a limited area, such an assumption introduces minor errors that do not exceed the limits of accuracy attained by ordinary measurements. A simple illustration will clarify the reason. An orange peel, a sphere, cannot be flattened without tearing, compressing, or folding the skin. Similarly, the surface of the Earth cannot be represented on a flat plane surface, such as a map, without distortion. A long, narrow strip of orange peel can be flattened with a minimum of distortion. If coordinate systems are limited to long, narrow strips of land, a minimum of mapping error results.

There is a movement to place coordinates as one of the elements in the priority of calls. In fact, some surveyors who are associated with academic institutions have advocated complete dependence of boundaries and descriptions on coordinates alone. This may be premature and is totally without legal foundation at this time.

Based on the fact that coordinates are a product of the instruments that collect the raw data, the instruments that calculate the information, the formulas that compute the coordinates, and the education and training of the people who do the computations, it hardly seems possible that the courts will place this element of line definition on a sound legal foundation for retracements. It also invites a basic problem of error in that a parcel can "close" but still have error (even mistakes) in position. Transposition of numbers is a common mistake with no possible check.

Recently in a retracement survey, the engineer/surveyor computed all the distances as grid distances for a description of a land parcel, yet the attorney did not know how to handle them, and to retrace the lines indicated on the plat, a reconversion to plane distances had to be done so that the lines could be retraced.

One of the major concerns with coordinate descriptions is that many professional or laypersons ultimately use description to locate the parcels. These individuals usually lack the basic knowledge of coordinates.

One of the major problems encountered may be the failure of the person who created the original coordinates to identify the reference system used.

Lambert and Mercator Grids

Many states have established long, narrow zones wherein a coordinate system is fixed by statute. If the system is narrow in a north–south direction, it is called a *Lambert grid*. If it is narrow in an east–west direction, it is called a *transverse Mercator grid*. By limiting the width of the coordinate system to 158 miles, the limits of error due to curvature distortion can be confined to 1 in 10,000, or about 1 foot in 2 miles, well within the limits of accuracy of ordinary property surveys. The state of Tennessee has one Lambert zone; New York has three transverse Mercator zones and one Lambert zone (Long Island).

Figure 5.15 illustrates a Lambert coordinate system showing squares with 100,000 feet to a side. A network of parallel straight lines, intersecting at right angles, is called a *grid*. The *central meridian*, generally located in the middle of the area of the grid, is the true north–south reference line with which all grid lines are made parallel or to which all grid lines are at right angles. As all true north lines converge toward Earth's poles, the central meridian is the *only true geodetic north line*. The central meridian forms the *Y-axis* of the grid, and the line at 90° to the assumed zero point on the *Y*-axis is called the *X-axis*. To avoid negative numbers, the central meridian is assigned a large number such as 2,000,000 feet for the *X*-coordinate (easting).

Lambert bearings have a mathematical relationship to geodetic north. The angular difference between a geodetic north line at a given point and the Lambert grid line is called the *theta* (Θ) *angle*, as shown in Figure 5.15. On the transverse Mercator grid, the angle is called the *delta* (Δ) *angle*. As a survey extends farther east or west from the central meridian, the theta angle becomes greater. A deed description based on a Lambert bearing must be clearly defined as such; otherwise, true astronomic north may be implied.

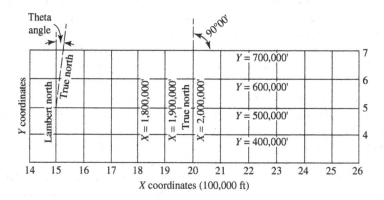

Figure 5.15 Lambert coordinate system.

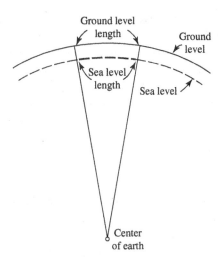

Figure 5.16 Lambert distances are reduced to sea level distances.

In the Lambert or Mercator system, horizontal distances are reduced to sea level. This correction, as shown in Figure 5.16, is necessary because of the spherical nature of the surface of the earth. When a surveyor utilizes coordinates for the purpose of a legal description, he or she should convert the datum distances to ground distances at the elevation of the survey.

The discussion of grid zones is of little importance in property or boundary surveying and boundary descriptions, but now that many states permit the use of grid distances it may become an important element when the boundary lines are quite long. Before a surveyor decides to use grid distances, the surveyor should realize that few laypeople understand this form of distance and problems may result from their use.

5.14 TAX DESCRIPTIONS AND ABBREVIATED DESCRIPTIONS

These two types of descriptions are found either within or outside a chain of title. There are times when landowners, attorneys, and surveyors will make reference to descriptions that are maintained in a tax office or some other government agency. These descriptions are usually abbreviated and unchecked, and in most situations the basic data from which they were compiled is not available. In describing boundaries or in trying to relocate boundaries, one should not rely on a tax map, a description, or an abbreviated description.

Today, there is a trend for people to describe parcels of land by a parcel identification number (PIN) or tax identification number (TIN) as a means of saving time. These numbers have no legal significance, in that they do not identify any parcel or its boundaries. Surveyors, landowners, attorneys, and even courts try to place a greater significance on these numbers than they deserve, since they do not enjoy permanency in identification as a boundary description does. These numbers can change at the whim of the government official who granted them. Referring to such numbers as this is like using a house number in a description. This has been held to be inadequate in most states, because these numbers are subject to change.

Recently many counties are initiating tax mapping projects and are showing parcels as PIN or a TIN. The courts have held these numbers do not constitute valid identification for land parcel identification. These descriptions are for administrative purposes only.

Tax Statements

The purpose of a tax statement is to identify the land being taxed. Often, the tax assessor shortens a lengthy metes and bounds description, thus avoiding paperwork yet retaining sufficient information to identify the land. Because tax statements and tax deeds are not complete deeds, they should not be used as a basis for a survey or a legal document and should not be relied on by surveyors as the primary source of a property or boundary description. Courts have permitted the introduction of tax maps to indicate who "paid the taxes," but this was for limited use only.

Abbreviated Descriptions

Lengthy metes and bounds descriptions would occupy excessive space on tax bills if it were not for abbreviations devised for that purpose. The following deed description illustrates the usefulness of this method:

Abbreviated: All that por of lot 40 of L M C acc map thrf #346 fld O of Rec of S D C Cal m p daf: beg at a pt on Nly li of lot 40 dist thon 140.84′ Ely fr N W Cor of sd lot th S 42°44′ E 241.26′ th N 89°01′ E 37.5′ th to the rt alg a tang cur whose rad is 87.03′ a dist of 76.72′ th N 47°53′ W 241.46′ to a pt on the Nly li of sd lot 40 dist thon N 84°14′30″ W 97.25′ fr the NW Cor of lot 46 of D W Tr th Wly alg the Nly li of Lot 40 to p o b.

Translation: All that portion of lot 40 of La Mesa Colony according to the map thereof number 346 filed in the office of the Recorder of San Diego County, California, more particularly described as follows:

Beginning at a point on the northerly line of lot 40, distant thereon 140.84 feet easterly from the northwest corner of said lot: thence S 42°44′ E a distance of 241.26 feet; thence N 89°01′ E a distance of 37.5 feet: thence to the right, along a tangent curve whose radius is 87.03 feet, a distance of 76.72 feet: thence N 47°53′ W a distance of 241.46 feet to a point on the northerly line of said lot 40 distant thereon N 84°14′30″ W, 97.25 feet from the northwest corner of lot 46 of the DeWitt Tract: thence westerly along the northerly line of lot 40 to the point of beginning.

An examination of the original document from which the preceding was abbreviated reveals that in shortening the deed, the assessor changed the POB and eliminated a number of courses from a distant starting point. If a surveyor were to use this tax deed to locate the land, he or she would find the land in a different position from that found by a surveyor using the true deed.

In many states, the assessor is authorized to prepare arbitrary maps and assign arbitrary numbers to each land parcel. Such maps should not be used as a basis for a land survey, only as supportive evidence.

In practice, surveyors should shy away from using these form of descriptions. The possibilities for errors and misidentification are numerous.

5.15 SUBDIVISION DESCRIPTIONS

If a map depicting parcels of land is filed with a public agency and the parcels thereon are designated by numbers or letters, the map is commonly referred to as a *subdivision map*. The precise meaning of the term *subdivision* varies from state to state, and the meaning within a given state is whatever the law defines it as being within that state (see Figure 5.17). In a number of states, the law authorizing preparation of subdivision maps is sometimes called a *platting* act rather than a *subdivision act*.

Now most states regulate how land may be divided and how it may be used. This was not so in the early part of this century, when there were no standards for subdivision creation. Even to the point that many of these old plats were never recorded. Zoning, planning, and regulating land use, although integral parts of the knowledge of all surveyors, are not within the scope of this book and are not included. An

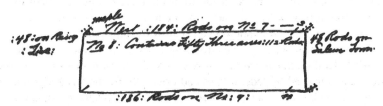

This Plan Difcribeth no : 8 : Third Divifion Jn naraganfett Town no : 5 : Containing Fifty Three acres & one hundred & Twelve Rods Lying Jn Rang : 4ᵗʰ : haveing a six Rods wide way Crofs yᵉ eaft end of The Lott said Lott Bounds as folows Begining at a ftake & heap of stones Jn Salem narraganfett town Line beeing the north weft Corner of no : 7 : & Runs from thence Eaft : 184 : Rods on No : 7 : To a maple Tree then Runs north 19° weft : 48 : Rods on Rang Five To a ftake Then Runs weft : 186 : Rods on no : 9 : To a ftake Jn faid salem Town Line Then Runs southerly on said line : 48 : Rods To The Bound Firft mentioned Laid out Jn yᵉ year : 1788 : by order of the Committee : Laid Down by a scale of : 40 : Rods to an Jnch :

pʳ Stephen Hofmer Junʳ : Surveyor :

Figure 5.17 Original proprietors' records, Bedford, New Hampshire. In 1739, ranges and townships were used. As of that date, "s" resembled the present day "f." Also, "j" was used instead of "i," and "J" was sometimes the numeral 1.

understanding of the map itself and its contents are essential for retracement surveys and is included.

From the standpoint of the retracement surveyor, the important features of a map are as follows:

1. What monuments were set or found?
2. What are the record sizes and locations of the lots and blocks?
3. What restrictions, easements, or conditions have been imposed on the land since the filing of the map?

From the standpoint of the scrivener, subdivision maps offer the simplest means of describing land, as they present the maximum of information and the minimum of words. "Lot 40 of La Mesa Colony, according to Map 346 as filed in the Office of the Recorder, San Diego County, Calif." or "Sec. 16, T15N, R20E, Principal Meridian" or "lot 2 according to the partition map filed in Superior Court Case 17632" constitute complete descriptions of land that can be identified from all other parcels of land. The simplicity of the title wording does not mean that a lot and block description of a section of land is easier to survey than a parcel of land described by a metes and bounds conveyance. Certainty and ease of location are totally unrelated to the length of the deed describing the land.

Most subdivision laws require that before filing a map, survey markers (monuments) must be established on the land. Many older maps made before the passage of such laws were "office maps" made from the record without benefit of survey. In defining the word *subdivision*, survey is not an essential consideration.

The partitioning of the public domain into townships and sections created true subdivisions made by a set of rules differing from those employed by the various states. All federal sectionalized land was surveyed and monumented in accordance with the rules imposed by federal laws and originally by the GLO. Except for federally owned land, subdivisions of parcels of land are regulated by state laws. State laws have varied from time to time, and the title interest of a person owning a lot within a subdivision must be viewed with respect to the laws in force at the time the map was made or approved.

Early subdivisions executed by private interests were poorly regulated by law, and any sheet of paper presented as a subdivision map to the record's office was usually filed upon payment of the filing fee. Occasionally, important data, such as street widths, lot sizes, and what was being subdivided, were omitted. The map of La Jolla Park was compiled with undimensioned lots and undimensioned curved streets laid out with a varying radius French curve. On old maps, the surveyor or engineer rarely made a statement describing what monuments were to be found or set. How could a later surveyor, ignorant of important facts such as what markers were originally set, retrace a subdivision? Most modern subdivisions are regulated by rigid laws, which have corrected many of the conditions mentioned. Because of numerous problems arising from poorly made maps, many court decrees have been rendered outlining or reconfirming common law. These are discussed in Chapters

12—Locating Simultaneously Created Boundaries and 13—Locating Combination Descriptions and Conveyances.

Many subdivision maps may seem complete when examined by the landowner, the surveyor and the attorney but one does not know what the jury or the judge will hold. Usually, a jury and a judge will only look at the lot in contention of a subdivision and will not "go beyond the lot lines."

Consider a recorded subdivision in which the boundaries of two lots were questioned, the question being whether the lot terminated at an O on the map or extended 10 feet to the riparian boundary of a waterway. The holding created a 10-foot strip along that one lot and potentially for 200 similar lots in the same subdivision. That one lot was taxed as having a water boundary, but upon application to have the status changed to nonriparian, the tax commissioner stated that the judge did not have the authority to do so. *What do you think?*

5.16 PARCELS CREATED BY PROTRACTION

Parcels of land or lots drawn on a subdivision map but not surveyed or monumented on the ground by an original survey are said to be created by *protraction*. This is particularly so in the GLO states, but it could apply to the metes and bounds states. If a surveyor divides a parcel of land into blocks 200 by 300 feet by setting monuments at each block corner and then draws 12 lots in each block of the map, the lots are said to be protracted. In the sectionalized land system, sections are created by prior survey and monumentation; parts of sections are created by protraction (SW ¼ of NE ¼ of Section 10). The words *protracted lots* or *protracted parcels* imply "created on paper without the benefit of an original survey." In some areas, entire townships were created by this protraction. If several parcels or lots are protracted on the same map, all are created simultaneously at the moment of approval of the map and no parcel or lot has senior rights over an adjoiner. In New York, the date of sale determines prior rights; this is an exception to the general rule. In considering protracted lots, the surveyor should realize that lots created by plat or protraction, without a survey on the ground, are as valid as those created with a survey. This statement also applies to roads and easements that are depicted on plats. If *one* lot is sold in reliance on the plat, all lots and roads become legally sealed.

5.17 FEATURES OF PLATTING ACTS

A subdivision act is a statute law that usually includes regulations of land use, how land must be improved, zoning, and so on. It is usually broader in meaning than a platting act, which regulates only the size and shape of parcels, the monumentation of parcels, and other items pertaining to the description of the land. Platting acts also regulate how original surveys for subdivisions shall be made; after an original survey is made, the courts interpret how a resurvey shall be made. Because this book pertains to retracements and resurveys, original surveys of subdivisions are largely omitted.

The essential parts of platting acts are the following:

1. Guarantee that title is marketable (taxes and assessments all paid and all encumbrances cleared)
2. All known easements disclosed
3. Adequate monumentation clearly described on the plat
4. Complete map description of all parcels
5. Certificate of survey
6. Dedication of easements

Resurvey procedures are dependent on the subdivision laws existing as of the date of the subdivision. In some areas, only subdivision boundary monuments are required; in others, all lots must be monumented. Although it is important for the local surveyor to know the history and former laws of his or her area of activity, compilation of all local conditions could not be included within this book because of the variation between states and the lesser political subdivisions within the states.

5.18   WRITING LAND DESCRIPTIONS

Principle 8.   *A boundary line may be created by words in a legal document. That line is as binding and as legal as if it were run on the ground by a surveyor.*

For centuries, there has been the question: Who will write the description? The answer is still pending. No one can give a positive answer.

To be competent at writing land descriptions, a person must know the legal meaning of words and phrases and know how the courts have interpreted the intent of the entire contents. The information presented thus far has covered the general contents of land descriptions; the information following should be studied before an attempt is made to write descriptions. Questions arise as to who is most qualified to prepare a land description. A description is the product of an adequate survey together with the proper use of words and proper rules of punctuation. A description represents the person who prepared it. An adequate survey accompanied by improper English and poor punctuation is as unacceptable as a description in which all words and sentences are proper but are based on an inadequate survey.

In writing a description, the person should be familiar with modern techniques and methods as well as historical methods. Although it is more important for the retracing surveyor to have a firm understanding of the early surveys, it is also important that the creating scrivener of the description have a familiarity with and understanding of early survey methods and customs.

When one understands the *priority of calls*, that person has acquired unique tools to create solid, legal, understandable descriptions that should withstand future tests of litigation challenges. The elements identified can be an aid to any scrivener creating a land description and should be applied whenever possible. See Section 5.20.

Scriveners may create many types of descriptions for deeds. Assuming each individual right may have a boundary description, separate and distinct from every other right, then we may have mineral deeds, timber deeds, easement deeds, water deeds, and claims of adverse possession deeds, and each of these may have a boundary separate and distinct from every other boundary.

5.19 EARLY SURVEYS

Today, some individuals, including courts, wonder why many gross errors exist in the early surveys. Most errors were due to poor equipment, a hostile environment, the abundance of land, and a lack of skilled surveyors. In early surveys of Texas and the prairies, it was probably more important to be an Indian fighter than a skilled surveyor. To illustrate some of the problems of early surveyors, two examples have been selected.

The first selection is from J. Stuart Boyles in *One League to Each Wind* (Texas Surveyors' Association, 1966, p. 303):

Besides being "shot at for laying and driving over fences," fighting rattlesnakes off driftwood from the 1900 Galveston storm, and pulling bogged down cows out of ponds, he worked in hot summertime when there was an "on shore wind," wore boots, gloves, rubber coat, flannel shirt, and mosquito bar on hat. Mosquitoes were like sand storms. Many cattle smothered to death with their nostrils and mouth choked-up.

In the same book (p. 77), there is the following commentary by W. S. Mabry (dated 1873):

We walked from ten to twenty miles a day, giving us ravenous appetites. After eating hearty meals during the day, we would sit around the campfire at night and toast buffalo humps and marrow bones, eat a big part, go to bed on the ground in the open, and awake the next morning feeling fine.

The wild turkey were fat, and a favorite dish was to slice the breast and drop the slices in a frying pan of hot grease making delicious turkey breast steaks.

During our nine months between the Pecos River and the Canadian river we had no fights with the Indians. On one or two occasions a few shots were exchanged, and one night they ran into our camp making hideous noises to stampede our horses. As every man went to bed with his Winchester by his side, every man was up, firing at the noise and keeping up such a fusilade that the Indians disappeared. The next morning when we rounded up our stock we were one horse short, but found one Indian pony, with saddle and bridle, which we could not explain unless its rider was either killed or wounded and carried off by the other Indians.

When our entire party was together, we could have stood off quite a force of Indians. But while north of Red River the party was often divided into three surveying parties and would be separated for about a week at a time. Each of these parties had to have along extra men to carry the guns of the compassman and the chainmen.

While in the upper Panhandle every day or so we would see bunches of Indians: on seeing us they would ride around us.

For an 1847 Wisconsin survey (Reports to United States General Land Office reprinted in Lowell O. Stewart, *Public Land Surveys* [Ames, IA: Collegiate Press, Inc., 1935], p. 84), Harry A. Wiltse wrote the following:

The aggregate amount of swamp traversed by the two lines was about one hundred and seventy-five miles, a considerable portion of which might be termed windfall (fallen trees, etc.)

During four consecutive weeks there was not a dry garment in the party, day or night.

Consider a situation like the above connected with the dreadful swamps through which we waded, and the great extent of windfalls over which we clumb and clambered; the deep and rapid creeks and rivers that we crossed, all at the highest stage of water; that we were constantly surrounded and as constantly excoriated by swarms or rather clouds of mosquitoes, and still more troublesome insects; and consider further that we were all the while confined to a line, and consequently had no choice of ground ... and you can form some idea of our suffering condition.

Our principal suffering, however, grew out of exhaustion of our provisions, coarse as they were.... Worn out by fatigue and hardship, and nearly destitute of clothes, they had now to make a forced march of three days for the lake in search of provisions, of which, during that three days, they had had not a mouthful.

I contracted to execute this work at ten dollars per mile ... but would not again, after a lifetime of experience in the field and a great fondness for camp life, enter upon the same, or a similar survey, at any price whatever.

The next excerpt is from the *Empire State Surveyor* (July 1967):

Actually, the Dutch system of surveys and measurements, differing only in their linear standards from the existing English System, came to a close with the termination of the official career of Corteljou in 1671.

What the Dutch Standards of measurement actually were at that time are not precisely known. The New York State Bureau of Weights and Measures (now administered by the State Department of Agriculture and Markets) has no record relative to these measures or any standards that have been identified as belonging to this period.

Boundaries, however, as then marked must still be adhered to, and the limits of the patents granted to the Van Rensselaers, the Coeymans, the Schuylers, Courtlands and others were located and marked in accordance with these standards, whose length we are now able to infer from a comparison of the recorded distances in the ancient deeds with existing landmarks which are now rarely found.

The importance of a background knowledge of the Standards of Measurement during this period is appreciated when it is remembered that even the foot measure of Europe, at that time varied in every kingdom and principality.

Many of the early grants were made during the Dutch ascendancy at which time the mile (old measure) of the European Continent was far greater than the English mile. The English foot, however, was greater than the old standard continental average foot measure by from one-half an inch to nine-tenths of an inch.

It is apparent that caution must be used in restoring the ancient boundaries and particular care must be exercised where the lines of the old colonial grants are being restored.

The problems, unfortunately do not end here, as it is most probable that the surveyors of the colonial period sold or transmitted their surveying equipment to their successors, so that where the Dutch measure ceased and the English standards began is a frequently unanswerable question.

It would indeed be difficult for a surveyor from California to try to survey in New York. In any state, it would be assumed that the road widths would comply with the law in force as of the date of dedication. In New York, the legislative body prescribed road widths as a maximum or minimum as follows:

June 19, 1703, old roads a minimum of 4 rods and all new roads a minimum of 6 rods (99 feet).

1704. Richman County could have lesser roads with a width of 3 rods.

1801. Public roads laid out by the Commissioners shalt not be more than 4 rods wide and all private roads not more than 3 rods wide.

1811. In the counties of Westchester, Rockland, Duchess, Orange, Ulster, and Sullivan, roads laid out by the Commissioners may not be more than 4 rods wide nor less than 2 rods wide.

Other laws followed. Space limitations in this book prevent inclusion of many factual items like those just listed. Each state and local area has its own unusual conditions, which are known by those who practice in the area. This book focuses on general principles found in most localities; precise details of a given area, especially items of local importance, are generally omitted.

One of the functions of any boundary retracer is to have a "working knowledge" of the history where he or she will be working.

5.20 PRIORITY OF CALLS IN METES AND BOUNDS SURVEYS

Principle 9. *In conducting a retracement survey of a metes and bounds boundary, the retracing surveyor must be knowledgeable of and apply the priority of calls.*

Principle 10. *Although the surveyor should not practice law, in conducting original surveys or retracements, he or she must follow established case law and statute law.*

By their very nature surveyors want to solve problems. This desire leads many to practice and make decisions far beyond their knowledge and capabilities. The practice of law is reserved exclusively for attorneys. But in order to conduct their work, which is so intimately tied to law and its precepts and principles, surveyors who choose to become qualified retracement specialists should make a sincere effort to become and remain current on legal decisions and what is expected of them.

In this chapter, the history and creation of metes and bounds surveys were discussed. An examination of numerous state and federal statutes, case law from numerous states, and common-law principles leads to the following list of the priority of calls for the retracement of metes and bounds boundaries (and GLO boundaries as well). This list reflects the law in the majority of states, although a particular state may place one element in a somewhat different position on the list.

1. The lines actually run on the ground by the creating surveyor(s). This includes corners established, monuments set, and lines marked on the ground at the time of the creating survey.
2. Calls for monuments set at the time of the survey or that enjoy a clear chain of history back to the original monuments:
   a. Natural monuments set at the time of the original survey.
   b. Artificial monuments set and referred to in the description.
3. Calls for adjoining parcels that are senior in title, called for and identified in the survey, and called for in the description.
4. Calls for courses:
   a. In most states, except in rare or unusual circumstances and some federal surveys, bearings control.
   b. In federal surveys, distance controls for corners, but distance can control in states only in unusual cases and under odd circumstances.
5. Control of area:
   a. Area may control in some instances.
   b. Area may be evidentiary in some instances.

5.21   APPLYING PRIORITY CALLS

Surveyors really need to have a clear understanding of these elements. These are rules of construction, not rules of survey or evidence. These are to apply only in cases of conflict between elements within a land description. Some believe they resolve discrepancies between record and physical evidence, or between items of physical evidence, or where there is a lack of physical evidence. There may be other rules that provide the same result, but for entirely different reasons. In applying the priority of calls, it is not a selection of "pick the one I like most" but a rule of survey of evidence: evidence that was created at the time of the survey, evidence that was described in the instruments, evidence that was searched for in the retracement, and evidence evaluated as accepted to locate the boundary line.

Lines Actually Run

The foundation for this rule is probably the federal Land Act of February 11, 1805, which stated that the original lines "are without (survey) error." This one act places the responsibility on the retracing surveyor to first see what the creating surveyors were to do and then determine what they did.

The first rule in retracing a boundary is to "find the lines actually run on the ground by the creating surveyor(s)." This includes corners established, monuments set, and lines marked on the ground at the time of the creating survey. This rule is paramount in both metes and bounds and GLO surveys.

All lines marked and surveyed, when positively identified, identify the true intended boundaries of the parcel and as such will control over less certain elements in the description. These lines constitute the true intended boundaries and will yield to less certain elements.[11] The surveyor is directed to "follow the footsteps of the creating surveyor" if they can be ascertained with certainty on the ground. Once found, the conflicts in calls or any ambiguities in field notes or plats are considered as subordinate. The actual survey and all of the marks on the ground, when ascertained with certainty, are conclusive.[12]

If a field survey is conducted and monuments are set, if you have the notes identifying the corners, and if you go on the ground and find them, the original survey controls. *But* if no lines are called for or if you go onto the ground and find nothing but called-for monuments identified in the notes, either natural or artificial, these control over the marked lines. Then you go to the next category.

Monuments Set

Calls for natural or permanent objects or monuments, definitely located, will generally control over other and conflicting calls unless a different intent is determined. These are the next elements in the order of control only if they can be located on the ground.[13] This rule is flexible and open to construction in that it must be proven that the natural monument was intended. If natural monuments are not called for or cannot be identified, the surveyor has to rely on artificial monuments called for, if it can be determined that the artificial monument is the more certain of the two.[14]

One major legal problem that courts encounter is what to do when monuments are set after the original survey was run. Some courts will look at the time lapse and make their decision as to the intervening time frame as to determine if these should be considered as original monuments or not.

Adjoining Parcels

An adjoining parcel may be considered as a natural boundary in the light that natural or artificial monuments are neither called for nor unascertainable. If the adjoining line is senior in time, well defined, and called for as an established line, the surveyor may accept the calls for the adjoining line. One must be careful to make certain that the line called for is senior and is the line and not its extension. To be absolute in location, the line must be marked to a degree of certainty that there can be no mistake about its validity.

If no lines are run or found, no monuments set or called for or found, and no adjoining parcels called for or identified, then of necessity we must resort to courses or bearings and distances. The surveyor can then place the lines on the ground, taking into account changes in magnetic declination, if necessary, from the date of the original survey. Whether bearing controls or distance controls depends on the state in which the lines are being retraced. In most states, bearings will control over distances.

If it is determined that the surveyor must resort to courses, each line must be analyzed independently and in conjunction with all other lines.

The one basic requirement that courts seem to overlook is that the adjoiner must be senior in title and/or survey; preferably both.

Area

Area is the least controlling of all the elements. In many instances, people place great reliance on area—for example: "I bought 6 acres and my surveyor surveyed only 4 acres; where are my 2 acres?" It is difficult to convince a person who bought land that in a metes and bounds description they get *all* of the area that is contained within the lines of the description. Area depends for its determination on other calculations and adjustments; there is no absolute area ascertainment.

5.22   CONCLUSIONS

As the "Old Guard" surveyors are gradually disappearing, they are taking with them years of knowledge and volumes of information, without which many of the old surveys are a "foreign language," subject to translations from varied sources. The only translation that legally counts is "What does the court say?"

Unlike the GLO system of boundary creation, for which system and retracements are described by written statute law, the metes and bounds system of boundary creation places a great responsibility on the surveyor or person who is to write the description and is then given the responsibility to retrace the results of the creation. To make certain that the various elements recognized in common and case law do not conflict, they must be made to harmonize at the time the description is written and the boundaries created. Cost must not be a controlling factor. In the event it is determined that the description is wanting in technical attributes, the courts will be forced to intercede and apply age-old legal principles. Recently, during the trial of a boundary problem that resulted from a poorly written description, three surveyors testified as to their location of two boundaries of a parcel. One stated that the description could not be placed on the ground; thus, the deed was void. The second placed the deed on the ground literally, using no common sense or judgment, and the third attempted to apply proper interpretation of the legal survey principles. Upon completion of the testimony, the judge stated, "It is apparent that no two persons will be happy with my decision." The trial ended in a hung jury because the 12 jurors could not agree on the meaning of the description, and a new trial was ordered.

In applying the basic principles of both writing descriptions of boundaries and retracing the boundaries, the key elements or priority of calls is an absolute that must be understood and applied in an everyday retracement. If a surveyor fails to understand this requirement and uses his or her own priority, if ever questioned, in all probability the courts will hold that the retracing surveyor was negligent, as a matter of law.[15]

This chapter also places a great responsibility on the readers and those individuals who either continue as professional surveyors or who join the ranks of a few registered surveyors who go into the legal profession to become trained in the proper methods of correctly retracing those historic lines of years past as well as those created yesterday.

NOTES

1. *Lefler* v. *City of Dallas*, 177 S.W.2d 231 (Tx. 1943).
2. John Love, *Geodaesia*. Reprint by W. G. Robillard, Atlanta, GA, 2000, p. 56.
3. N.W.2d 130, 212 Minn. 32.
4. *Riley* v. *Griffin*, 16 Ga. 141 (1854).
5. *Bryant* v. *Blevins*, 16 Ky. 91 (1809).
6. *Riley* v. *Griffin*, 161 Ga. 141 (1854).
7. *Fagan* v. *Walters*, 197 P 635, 115 Wash. 454 (1925).
8. 174 A. 383, 133 Me. 115 (1934).
9. In "Surveying and Land Information Systems," Francois Dubeau states that only two elements are needed. The authors of this book basically agree, but feel that the third element is needed as a check. The two elements, if used, will not allow the computation of the remaining curve elements, or the curve itself. Dubeau's article is an excellent explanation of curves and should be referenced by all.
10. The authors have used staff compasses for more than 50 years in creating and retracing old boundary lines in many states, both GLO and metes and bounds, and have supervised numerous crews doing the same. In one instance, a true meridian was established, and 15 surveyors each read the bearing of the line with their own compasses and then compared the results. There were variances of up to 1 ½ degrees of the bearings reading the same line on the various compasses. Only 3 of the 15 compasses read the same bearing.
11. *U.S.* v. *Doyle*, 468 F.2d 633 (Colo. 1957).
12. *Stuart* v. *Coldwell Banker & Co.*, 552 S.W.2d 904 (Tex. Civ. App. 1977).
13. *Fordson Coal Co.* v. *Spurlock*, 19 F.2d 820 (Ky. 1927).
14. *Land* v. *Dunn*, 241 S.W. 580 (Tex. Civ. App. 1922).
15. *Spainhour* v. *Huffman Associates*, 377 S.E.2d 615 (Va. 1989).

CHAPTER 6

CREATION AND RETRACEMENT OF GENERAL LAND OFFICE (GLO) BOUNDARIES

6.1 INTRODUCTION

Of the two major classes of boundaries, metes and bounds and General Land Office (GLO; otherwise known as the Public Land Survey System or PLSS), GLO retracements probably provide a greater number of specific guidelines and rules for the retracing surveyor. Yet, because of its history, the GLO system is, in all probability, least understood by surveyors in private practice. The primary reference for surveying the public lands is the *BLM Manual* which was designed and created for in-house use of the Bureau of Land Management (BLM). This manual was intended to be guidance for people employed by the *BLM* of lands in public ownership. The principal reason being so few of the surveyors in private practice have ever had the pleasure of conducting original surveys, from laying out baselines, correction parallels, scribing bearing trees, taking Polaris observations to determine local declination, "enjoying" camp life, to preparing field notes and plats on the top of an old sheet of plywood with a Coleman lantern. Anytime there is a problem between a US agency and a parcel of land in private ownership, the case is litigated in the federal courts, using the federal rules, and applying federal law to survey issues and state law relative to real property rights.

This chapter is dedicated solely to the GLO system of boundary creation, which is unique to the United States. Little did the founding fathers realize on July 4, 1776,

Brown's Boundary Control and Legal Principles, Eighth Edition.
Donald A. Wilson, C.A. "Tony" Nettleman III, and Walter G. Robillard.
© 2024 John Wiley & Sons, Inc. Published 2024 by John Wiley & Sons, Inc.

that the Declaration of Independence would result in one of the most ambitious and extensive boundary creations known to the world. The basic system was created by statute in 1785, modified in 1796, and finally solidified on February 11, 1805, in one of the most important pieces of land legislation ever enacted by Congress. GLO boundaries are still being created under the same federal statutes. The federal survey system has had its admitted weaknesses, primarily in the technical aspects, but so far as the legal aspects are concerned, the federal laws that created the system, beginning in 1785, are still recognized and are consulted and interpreted by all courts, from the various state courts to the Supreme Court of the United States. The original concept for a federally created system of land descriptions and land disposal was good, legal, and responsive to the needs of the people and the newly developing country.

In the New World, the original settlers were content with what they had always known: the metes and bounds system. The confusion of land titles, boundary locations, and the inability to have boundary surveys performed quickly and at a reasonable cost all impeded development to the west. British forts on the Appalachian Mountains acted as a barrier to western expansion. British troops kept the people hemmed in along the eastern seaboard. Upon the signing of the Treaty of Peace with Great Britain, one of the first acts that Congress adopted was the rectangular land survey system. This system, with the stroke of a pen, cast aside a system of land survey and description that was over 1000 years old and was universally accepted and adopted, and then that single stroke of a pen terminated and discarded the old system and created an entirely new system.

Historically, little is known about who first suggested or introduced the system. An early writer wrote: "The memory of the founder of this system, which has proved such a benefit to our country, is, indeed, worthy of a monument; yet if only asked what his name was, there are very few persons who could answer correctly."[1] The system worked because of the willingness of the various colonies to cede to the newly formed federal authority all of their western lands, which were granted to them by the English kings, that had not already been committed for disposal. Political and land boundaries are a creation of the state, in this case the federal government. Needless to say, the development of a land cadastre, land descriptions, and boundary determination, historically and legally, go hand in hand.

The initial legislation for the survey of public lands was proposed by a committee headed by Thomas Jefferson. In an attempt to depart from the old metes and bounds system, Jefferson proposed in the Land Act of 1784,[2] "The territory shall be divided into hundreds of ten geographical miles, each containing 6086 feet, and 4/10 of a foot.... These hundreds shall be subdivided into lots of one hundred squares each, or 850 acres and 4/10 of an acre...." To say the least, the new country was not ready for the metric system, and the proposal was summarily defeated because they lacked a survey chain for the standard unit of measurement, the meter. Jefferson tried hard to maintain the historic metes and bounds system. But the failure of having a metric standard was the critical factor. The legislators still had their roots in the historic metes and bounds system that had served the early colonists so well. The Land Act of 1784 was in turn replaced by the Ordinance of May 20, 1785.

Whereas the Land Act of 1785 stated that "all lines will be measured with a chain," it failed to identify the length of the chain. This was not a limiting factor, because the original surveyors used the chain that they customarily used, namely, the *Rathborne* chain of "two perches containing 33 feet." This oversight was corrected in 1796 when the Land Act stated that "all surveys will be conducted with a two pole chain."

Before boundaries can be redefined or relocated, they must first be created. In this chapter, we describe and explain various methods of creating boundaries. There are no pure systems of boundaries in the United States. Generally, the country is divided into two basic categories: GLO and metes and bounds. The GLO or PLSS states are those whose lands were subdivided by the federal government according to the various federal laws, beginning with the Land Act of 1785. The metes and bounds states are recognized as those states that rely on the ancient method of calling for courses, monuments, adjoiners, and several other elements. In the United States, there is no pure system; however, within GLO states a surveyor can find metes and bounds surveys, and in metes and bounds states one can find rectangular systems.

There are areas that have a similar system. Maine, South Carolina, Texas, and Tennessee all have descriptions that refer to sections, townships, and ranges. Western Canada refers to a PLSS. None of these encompassed the magnitude of the PLSS or GLO system.

GLO boundaries were created and are still being created under these same laws. The following principles are discussed in this chapter:

PRINCIPLE 1. The approved original survey does not ascertain boundaries; it creates them. An original monument is one that was created, identified, called, and recovered on an approved original survey.

PRINCIPLE 2. The lines and corners of an original survey control the location of a parcel. The lines are presumed to have been established in accordance with law, the special instructions, and the *BLM Manual* in force as of the date of the original survey fieldwork and the date on the signature of the plat.

PRINCIPLE 3. Special instructions or manuals instructed the original surveyors on how to survey the land; courts interpret the laws and how patented parcels of land shall be resurveyed or retraced.

PRINCIPLE 4. In the public land states, to practice in the modern technical world, the surveyor must have an intimate knowledge of the historic legislation that created the Public Land Survey System.

PRINCIPLE 5. There can be only one original survey. Any other survey of an original survey is a retracement and cannot alter the original corners or rights or section lines.

PRINCIPLE 6. Unless created by state regulations, the *BLM Manual* is not mandatory for private surveyors but can be used as guidelines for private registered surveyors.

PRINCIPLE 7. The edition of the *BLM Manual* that applies to a retracement is the *Manual* that was current at the time the retracement was performed.

PRINCIPLE 8. A surveyor who conducts retracements of the GLO/PLSS boundaries should be acquainted with historic and current court decisions.

6.2 ORIGINAL SURVEYS AND CORRECTIVE SURVEYS

Principle 1.   The approved original survey does not ascertain boundaries; it creates them. An original monument is one that was created, identified, called, and recovered on an approved original survey.[3]

The courts have repeatedly held that an original survey of public lands does not ascertain boundaries but creates them. Until the townships, sections, and ranges have been run on the ground, the monuments established, and the plat and notes approved, these parcels simply do not exist in the eyes of the law. A wise surveyor will not use a federal government section or quarter corner until satisfied that the location was approved as part of an official survey. Confusion has resulted when set monuments that had not been approved were used by local surveyors.

Prior to the issuance of a patent to public lands, the government can make an approved *corrective survey*. If more than one approved survey for a parcel of land exists, the most recent survey controls. A corrective approved survey made prior to the issuance of a patent is also an original survey.

The statutory authority to correct erroneous surveys was sanctioned in *Kittridge v. Landry*.[4] It has been held repeatedly that as long as no land has been patented, a corrective survey can be made and substituted for the first survey.[5] On the other hand, after the public lands within a section have been disposed of by the federal government, any rights that were acquired cannot be affected by a corrective survey.[6] The application of this principle generated and coined the term *bona fide rights*.

It is important that the person who attempts to "find the footsteps" of the original surveyors has an understanding of what "footsteps" were created and what is left to recover. These "footsteps" are defined not only by actions but also by the laws under which these actions were authorized.[7]

This principle is best explained looking at a blank piece of paper that we will call the parcel to be subdivided. As owner of the paper, you can choose how it is to be subdivided—by means of words or by identifiable lines that you create. You can *estimate* the paper is 5 inches square and then grant out the north ½. When divided into two pieces, each half should get 2 ½ inches of paper. Then you can either measure 2 ½ inches or you can fold the paper, creasing it at the middle.

6.3 LAW, MANUALS, AND SPECIAL INSTRUCTIONS

Principle 2.   The lines and corners of an original survey control the location of a parcel. The lines are presumed to have been established in accordance with law, the special instructions, and the BLM Manual in force as of the date of the original survey fieldwork and the date on the signature of the plat.

Today, the Bureau of Land Management (BLM) is the only agency of the federal government with the statutory authority and responsibility for surveying public lands. Initially, the responsibility was with the Treasury Department, then it was transferred to the GLO, which later became the BLM. For the purpose of uniformity in surveys, starting in approximately 1855, from time to time the BLM issued and still issues special instructions or manuals for use by BLM surveyors. Because of the presumption stated in Principle 2, knowledge of the various instructions or manuals is necessary for the retracing surveyor. Manuals, unless adopted by state statute or state courts, are advisory to land surveyors for that specific state.

The depiction of the PLSS/GLO system has been attempted by other entities, including the US Geological Survey. This agency attempts to depict the location of the land net on its topographic quadrangle maps. Few surveyors or courts understand how the locations of the townships, ranges, and sections were depicted. With a skeleton of positive corner locations, if any, the entire townships will be drawn, without correlating the original field notes to any positive features depicted on the official maps.

For the first sectionalized land survey, that of the Seven Ranges in Ohio, no written instructions were found or known to have been issued. The first known written instructions, *General Instructions to Deputy Surveyors*, were issued by Jared Mansfield in August 1804 and applied to the Northwest Territory. Later, Tiffin's instructions modified those of Mansfield. In other districts, such as Mississippi and Florida, written instructions were also issued; however, details of the survey procedure were not identical to those of Mansfield or Tiffin.

In 1851, a book titled *The Manual of Surveying Instructions for the Survey of the Public Lands of the United States* (now known as the *Oregon Manual*) was the first attempt to provide uniform instructions to all surveyors. Because it was printed without all the illustrations, however, it was not a complete manual. In 1855, except for instructions about correction lines, a reprint of the *Oregon Manual* with all the illustrations was issued (now known as the *1855 Manual*). Each subsequent reprinting (hereafter referred to as the *Manual*) contained instructions on how to make original surveys, and each *Manual*, distinguished by its date of publication, was in force until the next *Manual* was published. Some of the later editions of the *Manual* included instructions on how federal surveyors resurvey and subdivide sections.

Although each *Manual* explained how the federal deputy surveyors were supposed to make original surveys, the procedures were not always followed because of special instructions or for other reasons, some of which were personal to the field surveyors, which may have resulted from not following the instructions. Because field notes of the original surveyor have superior standing to *Manual* instructions, deviations from *Manual* instructions as found in field notes are accepted.

6.4 EFFECT OF MANUALS ON RESURVEYS

Principle 3. *Special instructions or manuals instructed the original surveyors on how to survey the land; courts interpret the laws and how patented parcels of land shall be resurveyed or retraced.*

The making of original surveys by the United States is and was strictly controlled by the legislative body; how to interpret and locate an existing conveyance is controlled by the judicial body and its courts. Although the original procedure used by surveyors has an influence on the court's thinking, not all of the instructions given in the *Manual* in force at the time of original survey are applicable during a resurvey or retracement. For example, in some editions of the *Manual*, the original surveyor's permitted tolerance of error in setting a corner was about 3 feet in 1/2 mile. In a court case requiring the reestablishment of a lost regular quarter corner, the corner must be set equidistant between section corners, not halfway plus or minus 3 feet. Most modern surveyors do not retrace using the same instruments, such as a compass and a 2-pole chain called for in early editions of the *Manual*. The federal statute today requires original surveys to be made using a 2-pole (33-foot) Gunter's chain. Obviously, this law has not been enforced or strictly followed by either government or private surveyors because many of the original instruments are not now permitted in that most surveyors use total stations and GPS.

There are a few modern surveyors who employ the same type of equipment to retrace historical descriptions and boundaries. But many states prohibit this type of equipment to be used to conduct modern surveys for clients.

In this chapter, we will explain how the early surveyors prepared their work so that the modern-day retracement surveyor will know what to look for by research and where to look for the evidence of the original survey, or, as the courts state, the *footsteps* left by the original surveyors.

6.5 HISTORY OF THE PUBLIC LAND SURVEY SYSTEM

Before Congress enacted the original laws creating the Public Land Survey System (PLSS), it met in 1784 and drafted proposed legislation, which, because of a minor technicality, was never enacted. The basic concepts that went into the system were the following:

1. The official US mile would be a geographic mile.
2. The area unit acre would be reformed.
3. Townships would be 10 geographical miles square.
4. Each section would be 1 mile in length.
5. No land would be sold prior to survey.
6. Any preemptive rights or valid prior grants would be honored.
7. A system of surveys predicated on legal statutes would be enacted to reduce or eliminate any future problems or questions of boundary.
8. Valid land claims would be identified and adjudicated.
9. Title to Native American land would be considered and then eliminated or purchased.

Armed with this philosophy, the Continental Congress initiated the largest land survey in the history of the world. A system that has been in existence continually for over 200 years is still in effect today.

Surveyors and attorneys should be knowledgeable about the historical facts of original surveys. No single volume can give a complete history of original surveys. Perhaps the best information available at present is *A History of the Rectangular Survey System*, authored by C. Albert White and published by the US Department of

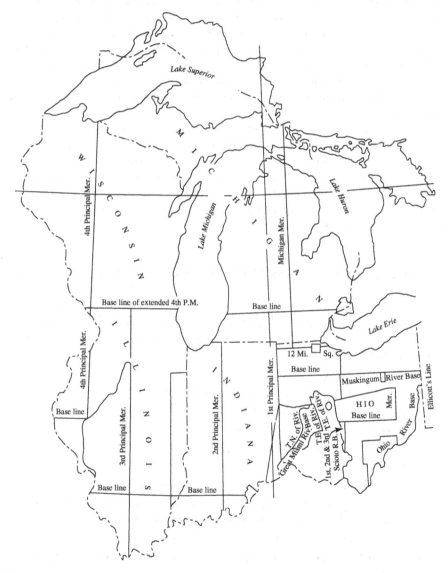

Figure 6.1 The Northwest Territory.

the Interior, Bureau of Land Management. Included in the volume are many reprints of special instructions issued to deputy surveyors.

This particular volume is a must for surveyors who practice in the PLSS states, and it should be available for consultation.

6.6 TESTING GROUND: THE SEVEN RANGES

In 1785, the Continental Congress, under the Articles of Confederation, passed the first land ordinance designed to presurvey the area west of Pennsylvania, north of the Ohio River, and south of Lake Erie (see Figures 6.1 and 6.2). Although not all of the area designated was surveyed, the portion that was surveyed became known as the *Seven Ranges*.

Square townships, 6 miles on a side, on cardinal directions were laid out with 36 lots numbered, as shown in Figure 6.4. The interior lots were protracted. Corners along township lines were set at mile intervals as shown; no interior section or quarter corners were set. (*Note*: At a later date, some of the townships were surveyed into sections by government surveyors.) As surveyed on the ground, many township corners were not common corners, as shown in Figure 6.3. A summary of the original procedure is best presented in the following outline and in Figures 6.3 and 6.4.

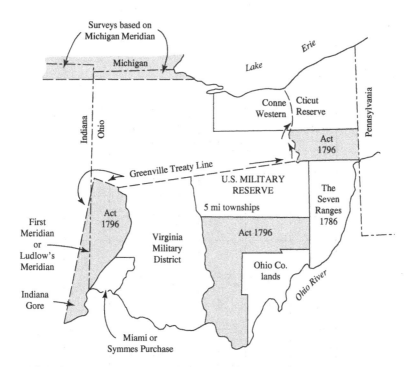

Figure 6.2 Ohio surveys.

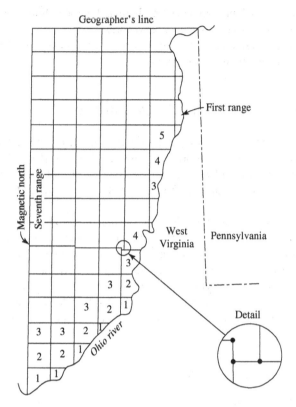

Figure 6.3 The Seven Ranges.

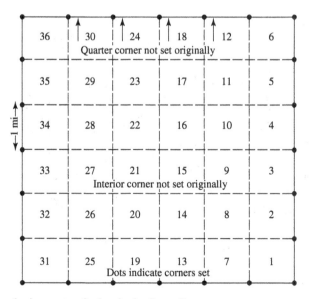

Figure 6.4 Numbering system for lots in the Seven Ranges.

Outline of the Provision of the Ordinance of May 20, 1785, the Northwest Ordinance

To a student, a layperson, or one who is not familiar with US history, the title of this act may seem a misnomer. One must realize that in 1785, all American land interests stopped at the Mississippi River as a result of the peace treaty with the British. After the French and Indian War in 1765, the French were awarded all lands west of the Mississippi to the Pacific Ocean, except some British claims in the Pacific Northwest, and the British agreed that all lands east of the Mississippi River would be under their control. As a result, it would come to pass that in less than 20 years the greatest land purchase in history from the French would triple the size of the new nation and add new opportunities toward solidifying the PLSS and the resulting surveys. Thus, the United States could only claim what it was that the British had to relinquish until 1803 and the Louisiana Purchase. Until that time, the western boundary of the new nation was the Mississippi River.

The Land Act of 1785 provided for the following:

A. Structure of lots (corresponding to sections in later acts).
  1. Townships to be 6 miles square (no allowance for convergence).
  2. Ranges to be numbered westward (only seven surveyed).
  3. Townships to be numbered in sequence from the Ohio River northward to Lake Erie (none went that far).
  4. Lines to run north–south and east–west on the true meridian (changed to magnetic meridian in 1786).
  5. Fractional townships to have lots bear the same numbers as if the township had been entire.
  6. Lots (sections) were to be 36 in number and 1 mile square (see Figure 6.4).
B. Survey
  1. The point of beginning of the survey was "on the River Ohio at a point that shall be found [to be] due north from the western termination of a line which has been run as the southern boundary of the State of Pennsylvania."
  2. Lines measured with a chain (no mention of length).
  3. Lines run on the true meridian (changed to magnetic meridian in 1786).
  4. Surveyors to note all mines, salt springs, salt licks, mill seats, watercourses, mountains, and the quality of the land.
  5. Only township lines run. (An act of May 1, 1802, required that all land sold in the Seven Ranges prior to May 10, 1800, be surveyed in sections or fractional sections in the manner most consistent with the supposed boundaries at the time of the sale.)
  6. Corners on township lines marked at mile intervals.
  7. Geographer (Thomas Hutchins) appointed in charge who "shall occasionally form such regulations for the conduct of surveyors as deemed necessary."

     8. No written instructions issued for surveyors (personal guidance on the ground).

C. Plats

     1. To be submitted to Board of Treasury.

     2. To show 36 lots protracted (not surveyed).

D. Sale of land

     1. Every other township in its entirety sold (not many were sold).

     2. Every other township by lots (sections in later surveys).

E. Reservations

     1. Lot 16 reserved for schools (still true).

     2. One-seventh of the land reserved for the Continental Army.

     3. Lots 8, 11, 26, and 29 reserved for future sales (on the theory that prices would go up).

     4. Land patents also given to Christian Indians.

By today's standards, the work was not precise. But it would (in 1805) be deemed accurate, by law. However, it is difficult to criticize, considering the number of times the process was interrupted by Indians and the fact that the lines were apparently run by the circumferentor's magnetic needle, which leads to the questions: How could anyone prove today where the surveyor of 1786 set his corners? and "Should the right to land as possessed be jeopardized merely because of loss of original monument material.?" Within the Seven Ranges as well as within any of the older states, limitation titles—titles that can be proved only by prolonged occupancy—set to rest an otherwise endless argument regarding how the protracted lots should be surveyed.

The two pertinent paragraphs pertaining to survey procedures were as follows:

> The geographer and surveyors shall pay the utmost attention to the variation of the magnetic needle, and shall run and note all lines by the true meridian, certifying with every plat what was the variation at the times of running the lines thereon noted.

> The lines shall be measured with a chain; [no length indicated until 1796] shall be plainly marked by chaps on the trees and exactly described on a plat, whereon shall be noted by the Surveyor, at the proper distances, all mines, salt springs, salt licks and mill seats that shall come to his knowledge, and all water courses, mountains, and other remarkable and permanent things over or near which such lines shall pass and also the quality of the lands.

In later use, the word *chap* became known as *blaze* or *hacks*. It is particularly interesting that *chaps* and notes of mines, salt springs, and the like were incorporated in all later acts. A *line tree*, or a tree standing directly on line between corners, was marked by notches (sometimes hacks). All early surveyors except Ludlow made one notch on each side of the tree; Ludlow made two on each side. Ever since that time, the practice of requiring two notches on each side has been approved.

In 1786, the requirement of lines run "by the true meridian" was deleted. Since almost all of the Seven Ranges were run after 1786, most lines were run with a compass bearing. Later acts of Congress reinstated "by the true meridian," but not until the Seven Ranges were completed.

It was not until 1788 (the year that the Seven Ranges were completed) that the Continental Congress required the following: "That each surveyor, upon making any survey, shall project and lay the same down in a general map, to be kept and preserved, and shall make a record of each survey in a book to be kept for that purpose … etc. Each surveyor … shall take an oath … etc." Before this act, Thomas Hutchins and others were already turning in notes and plats even though notes and plats were not specifically required (though implied) by law. Plats as filed failed to note declination of the compass, did not disclose inaccuracies of direction and distance, and often ignored the existence of double or open corners usually existing where townships joined.

6.7 ACT OF MAY 18, 1796—CLARIFICATION OF 1785

As Congress completed treaties with the various tribes of Native Americans and land became available for survey, for laws prohibiting surveys without the "Indian title being extinguished," it was found that the Land Act of 1785 was deficient in several areas. First, no length of the chain or measurement instrument was noted, and it provided for a departure, in certain cases, from 6-mile townships. After 1788, no further public sectionalized land surveys were made until 1796, when the system was revived.

Although the first US Congress met in March 1789, the first sectionalized land act was not passed until 1796. In this act, provision was made for the survey of the US Military District, Connecticut Western Reserve (see Figure 6.2), and much of the remaining Ohio land (plus some land in the present state of Indiana now referred to as the *Indiana Gore*). Under the Ordinance of 1785, a geographer, Thomas Hutchins, was in charge; under the Act of 1796, a surveyor general was provided for. Rufus Putnam, the first surveyor general, was appointed by George Washington primarily because of his military background rather than his technical qualifications. He was not an expert on "how to run the true meridian," and this feature of the law was not strictly enforced.

By 1799, the areas designated in the Act of 1796 (see Figure 6.2) were almost completed, and the second phase of the survey of the public domain was nearly finished. The Act of 1796 made some changes in the Ordinance of 1785; in general, it added refinements. The provisions of the act are summarized by the following outline and by Figures 6.5 and 6.6.

Summary of the Act of 1796

    A. Structure of sections (term *lot* dropped).

        1. Townships were to be 6 miles on a side (no allowance for convergence).

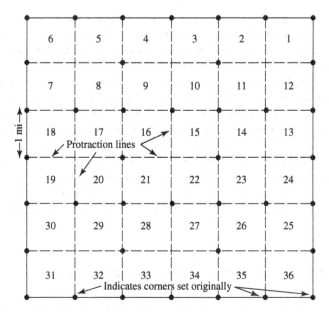

Figure 6.5 Act of May 18, 1796. For every other township interior lines were run every 2 miles. Alternate townships had only exterior lines.

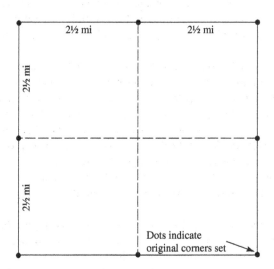

Figure 6.6 Act of May 18, 1796, created 5-mile townships.

B. Survey.
   1. All township lines were to be marked every mile.
   2. Every other township was to be surveyed by lines run every 2 miles in each direction (see Figure 6.5).

3. Section corners on lines run were to be monumented.

4. Four witness trees were to be marked (scribed for section number) at each corner (also township number to be marked over the section number).

5. All lines were to be plainly marked on trees (blazed).

6. Lines were to be run with a chain containing 2 perches (33 feet) subdivided into 50 links each. (This law is the same as it exists today! It is not obeyed.)

7. Field notes were to be made in field books.

8. Note salt licks, as in 1785 ordinance.

C. Plats (same as required in act of 1788).

D. Sale. Alternate townships sold entire; remainder by sections; salt springs reserved; sections 15, 16, 21, and 22 reserved for future sales.

E. Five-mile townships (see Figure 6.6). This act provided for the lands of the United Brethren Society to be divided into townships of 5 miles to a side, each corner to be set at 2½-mile intervals (center quarter corner not set). Lots were protracted in various units of 100 acres or more as provided. This act was used for the following lands (see Figures 6.2 and 6.6); alternate townships had only the exterior lines run:

1. US Military District.

2. Connecticut Western Reserve.

3. Society of United Brethren (religious—not many acres involved).

F. Section 3 of the Act of 1796 provided "that a salt spring lying upon a creek which empties into the Scioto River, on the east side, together with as many contiguous sections as shall be equal to one township, and every other salt spring which may be discovered, together with the section of one mile square which includes it, shall be reserved for future disposal." Obviously, salt was at a premium. Gold, silver, lead, and copper were also reserved but none was found.

Until 1805, there were several acts that were minor in nature. Some are described next.

6.8 ACTS OF 1800

In 1800, two acts were passed, one on March 1 and a second on May 10. These two acts are of little importance today but are mentioned to provide continuity.

Act of March 1, 1800

The Act of March 1 pertained to the 5-mile townships (US Military Tract; see Figures 6.5 and 6.6). It did establish the important principle that corners regularly set by the original surveyor are to be held as the true corners even though later surveys may show the measurements to be in error.

That the respective points of intersection of the lines actually run as the boundaries of the seven townships surveyed by virtue of this act ... accordingly as the said lines have been marked and ascertained at the time when the same were run, notwithstanding the same are not in conformity to the act aforesaid, or shall not appear to correspond with the plat of the survey which has been returned by the surveyor general, shall be considered, and they are hereby declared to be, the corners of the said townships; etc.

All irregular federal surveys, including the 5-mile townships, are described in Chapter 7—Federal and State Nonsectionalized Land Surveys.

Act of May 10, 1800

The Act of May 10 provided for placing errors and convergence in the north and west tiers of sections; it also provided for running all the sections and placing quarter corners on north and south section lines only (see Figure 6.7). The rate for surveys was $2 per mile for the Seven Ranges, $3 per mile after that. In this act, $4 per mile was provided for lines in the vicinity of Vincennes. For the first time, the sale of lands in half sections was authorized (see Figure 6.7).

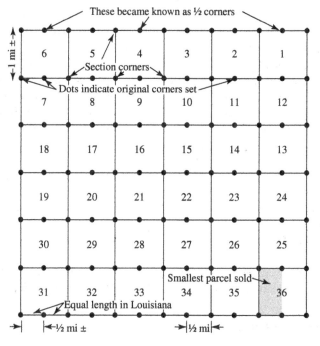

Figure 6.7 Act of May 10, 1800. Errors and convergence are to be placed in the north tier of sections and in the west tier of half sections. All section lines were to be run and half corners set (they were not known as quarter corners). In early Louisiana surveys, the error of closure was divided equally in the last mile.

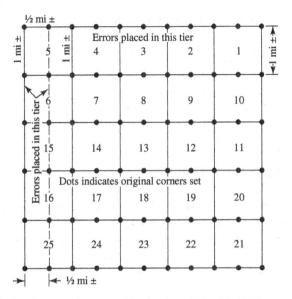

Figure 6.8 Ohio 5-mile townships created by the Act of May 10, 1800.

Structure of Sections

1. Quarter corners were to be set on the north and south sides of sections, and sections to be divided into half sections.
2. Excess or deficiency were placed in the northern and western tier of sections.
3. Other features were the same as in previous acts.

> And in all cases where the interior section lines of the townships, thus to be divided into sections or half sections, shall not exceed six miles the excess or deficiency shall be especially noted, and added to or deducted from the western and northern range of sections or half sections in such townships, according as the error may be in running the lines from east to west, or from south to north; the sections and half sections bounded on the northern and western lines of such townships shall be sold as containing only the quantity expressed in the returns and plat, respectively, and all others as containing the complete legal quantity.

In this act, special provision was made to subdivide the 5-mile townships "west of the Muskegon [River]" into sections as shown in Figure 6.8. These townships exist only in Ohio and represent a redivision of 5-mile townships created by the Act of 1796 for the US Military Reserve (see Figure 6.2).

> That the surveyor general shall cause the townships west of the Muskingum, which, by the above mentioned act, are directed to be sold in quarter townships, to be subdivided into half sections of three hundred and twenty acres each, as nearly as may be, by -running parallel lines through the same from east to west, and from south to north, at

the distance of one mile from each other, and marking corners, at the distance of each half mile on the lines running from east to west, and at the distance of each mile on those running from south to north, and making the marks, notes, and descriptions, prescribed to surveyors by the above mentioned act.

The ranges in this district are designated by progressive numbers from the eastern boundary westward; the townships are numbered from the southern boundary toward the north.

6.9 1803—THE SYSTEM EXPLODES

Until 1803, lands available for survey and subsequent distribution were limited to the Mississippi River on the west and the Canadian border on the north. The boundaries of the United States were fixed. The system exploded; this one signature would ultimately provide survey work for thousands of surveyors and their successors, many fathers and sons, and at times entire families were public surveyors, for more than 200 years.

Jared Mansfield was appointed surveyor general in place of Putnam by President Jefferson. Mansfield, a professor of mathematics at the US Military Academy, wrote a book on mathematics, and well understood the convergence of the meridians, knew how to determine the meridian, and was better technically qualified. Putnam believed that his removal was solely due to his being a Federalist (politics existed even then). Mansfield was the last to make significant changes to basic principles. Most changes after his time were refinements in techniques.

Before Mansfield, townships were generally numbered from the Ohio River northward or from the south boundary of the tract northward (an exception is the Symmes Tract). The first ranges numbered eastward and westward from a principal meridian, as was done from Ludlow's Meridian (see Figure 6.2). Mansfield laid out the first baseline in Indiana that had townships numbered northward and southward from the baseline. He also initiated the first correction line (standard parallel), located near the city of Indianapolis.

In 1803, the Louisiana Territory was purchased from France by the United States, and shortly thereafter government surveys began. Although Spain owned Louisiana for the greater part of the preceding time interval, most of the land was acquired by France in accordance with French measurements and customs. When Louisiana was admitted as a state in 1812, Napoleonic codes were adopted. Today, Louisiana is somewhat different from English common-law states.

Valid grants and valid possessions made prior to 1803 were surveyed and excluded from the public domain after the Act of 1806. These surveys were irregular, as described in Chapter 7—Federal and State Nonsectionalized Land Surveys. Surveys of the public domain of Louisiana began in earnest after the Act of 1811 under the jurisdiction of two separate principal deputy surveyors; thus, there developed survey differences from the Northwest Territory.

6.10 ACT OF MARCH 26, 1804

The year 1804 was important in that the federal government, by law, made smaller parcels of public domain sections available to more people. Lands could be sold either in entire sections, in half sections, or in *quarter sections*: "and it shall be the duty of the said surveyor general … to ascertain by astronomical observations the positions of such places as may be deemed necessary for the correctness of the surveys and to be most important points of geography of the country." As smaller and smaller parcels of land were able to be conveyed, the names that were given to respective corners always changed in accordance with the laws.

Except in the initial legislation, when present-day sections were called *lots*, the corner names, except for section corners, have been modified from time to time. In the Act of 1800, the north and south corners, at points where the modern-day north and south quarter corners are located, were called *half corners*, because these corners, when connected, would cut the section in half. This was the least amount the government could sell (see Figure 6.9). It was not until April 20, 1820, that Congress enacted legislation that permitted the land offices to sell the land in half quarter sections, or eighth sections. This act created the sixteenth corners and permitted the land offices to sell smaller parcels. Following this, subsequent legislation permitted the government to sell one-sixteenth of a section, or 40 acres. Today we use the terminology "a 40" and "1/16 corners" to describe these points.

6.11 ACT OF FEBRUARY 11, 1805

The second most significant legislation of the public land laws was the Act of February 11, 1805, and it is probably the most quoted of all the acts in terms of identifying the surveying principles that have made the GLO system so unique. This act has been referenced in more state and federal case decisions relative to the survey, resurvey, and retracement of the public lands than any other federal land law. The initial legislation and its interpretation are must-reads for all practicing surveyors and attorneys.

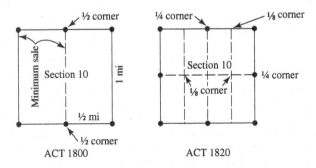

Figure 6.9 Corner names depend on the date of subdivision.

In order to conduct a modern-day retracement, or as the modern courts instruct, *follow the footsteps*, the modern surveyor and lawyer should understand what the creating surveyors were told to do.

This act contains the last important statute law change to the original ordinance of 1785 and is applicable in Indiana, Illinois, and later states. Among other things, it provided for the completion of surveys already made by running alternate mile lines under the Act of 1796. It also provided the method of dividing sections into quarter sections by straight lines connecting opposite corners. It provided for subdivision into sections of those townships that had been subdivided by "running through the townships parallel lines, each way, at the end of every two miles, and by marking a corner on each of the said lines at the end of every mile" (Act of 1796). This was to be done "by running straight lines from the mile corners, thus marked to the opposite corresponding corners, and by marking on each of the said lines intermediate corners, as nearly as possible equidistant from the corners of the sections on the same."

The Act of 1805 provided for a partial subdivision into quarter sections by causing "the boundaries of all the half sections, which had been purchased previous to the first day of July last ... to be surveyed and marked by running straight lines from the half mile corners heretofore marked to the opposite corresponding corners" ... (This applied to lands surveyed under the 1796 act.) For new lands to be divided, the act provided (these statutes are as they exist today):

Sec. 2396, R.S. (Readopted in 1874). The boundaries and contents of the several sections, half sections, and quarter sections of the public lands shall be ascertained in conformity with the following principles:

First. All the corners marked in the surveys returned by the surveyor general shall be established as the proper corners of sections, or subdivisions, of sections, which they were intended to designate, and the corners of half and quarter sections, not marked on the surveys, shall be placed as nearly as possible equidistant from two corners which stand on the same line. (*Note:* This is in conflict with the second part that follows. The center quarter corner of a section is set by the rule of straight lines connecting opposite corresponding corners, *not* by this rule.)

Second. The boundary lines, actually run and marked in the surveys returned by the surveyor general, shall be established as the proper boundary lines of the sections or subdivisions for which they were intended, and the length of such lines as returned shall be held and considered as the true length thereof. And the boundary lines which have not been actually run and marked shall be ascertained by running straight lines from the established corners to the opposite corresponding corners (*Note:* Conflicting with the preceding, this principle is used to set the center of a section); but in those portions of the fractional townships where no such opposite corresponding corners have been or can be fixed, the boundary lines shall be ascertained by running from the established corners due north and south or east and west lines, as the case may be, to the water course, Indian boundary line, or other external boundary of such fractional township. (*Note:* East and West or North and South has been interpreted by the courts to mean East and West or North and South in the *average direction* as run by the original surveyor. *It is not due east or west* as determined by an astronomic observation or as determined by a magnetic observation.)

Third. Each section or subdivision of section, the contents whereof have been returned by the surveyor general, shall be held and considered as containing the exact quantity expressed in such return, and the half-sections and quarter-sections, the contents whereof shall not have been thus returned, shall be held and considered as containing the one-half or the one-fourth part, respectively, of the returned contents of the section of which they may make part. (U.S.C., Title 43, Sec. 752.) (*Note:* Sections do measure differently from the record. The legal idea is that the lengths of lines remain the same—you merely use a longer or shorter chain to get the same distance as originally reported. After determining the right length of chain that will fit between proven corners, missing corners can then be replaced. After determining the right length of the chain to get the same distance as originally reported, the lost corners can then be repositioned. When determining the right length of the chain that will fit the original measurements missing corners are set in between at the record distances. Proportionate measurement should produce the same result and should be used in preference to "finding" the right correlation of the chain for every line.)

Of all the land acts enacted since 1785, this is perhaps the second most important to the surveyor, in that it sets out the basic principles for future retracements of public lands. It establishes the basic principle that surveys are "without error" in that original corners are where surveyors find them and that the original measurements recorded in the field are the "true" measurements, error and all. The interpretation of the meaning of the "true measurement provision" has required court opinion, as stated previously.

Today, of all the land acts that seem to give surveyors survey and legal problems, the provision for setting the center 1/4 of the section seems to be the most difficult for the boundary surveyor to accept. The provision for running straight lines from each of the four quarter corners 1/4 north–south and east and west, and the intersection of these two lines is deemed to be the center 1/4 corner. Although it was never monumented, this corner exists legally. The law does provide for a smaller subdivision to a 1/4 corner, or a 1/16 corner. Under this law and other laws, 40 acres is the smallest aliquot parcel identified for sale.

One month after the passage of this Act, on March 13, 1805, Secretary of Treasury Albert Gallatin wrote to Isaac Briggs, Surveyor General for Lands South of the State of Tennessee, explaining the Act of 1805 and what was expected. This is a very concise and important letter in that it definitely sets in mind the expectations that were to be anticipated. The letter reads as follows:

Sir,

I have the honour to enclose an Act concerning the mode of surveying the public lands of the United States, which, although principally intended to palliate the errors made in the surveys north of the Ohio, contains certain general principles, in relation to the mode of establishing corners and running interior lines, which apply to all of the public lands.

Permit me earnestly to repeat my request that you would take immediate measures for running the township lines & for executing generally all of the surveys within the tracts lying in the Mississippi Territory to which the Indian Title has been extinguished. The Legislature has fixed the price at four dollars per mile; that price will not be enhanced; and although very great correctness cannot be attained for that price in that part of the Country, it is our duty to carry the law into effect, and all that can be expected is that the surveys will be as correct as can be done at that rate. You will also perceive from the enclosed act that the principal object which Congress has in view is that the corners and boundaries of the sections & subdivisions of sections should be definitively fixed; and that the ascertainment of the precise contents of each is not considered as equally important. Indeed it is not so material either for the United States or for the individuals, as the purchasers should actually hold a few acres more or less than their surveys may call for, as it is that they should know with precision, and so as to avoid any litigation, which are the certain boundaries of their tract. It is true that you will not be able to complete your work in a scientifick manner which was desirable, & that it will not be possessed of that merit, in a geographical point of view, which your abilities enable you to give it. But those are only secondary though very desirable objects; and it is of primary importance that the land should be surveyed and divided, as well as it can be done, so as at least to connect the whole work, to ascertain the claims affirmed by the Commissioners, and enable [the] Government to dispose of the vacant lands. I hope, therefore, considering the time which during your absence has been already, that you will not fail to take the necessary measures for carrying, without further delay, the law into effect.

(Signed) Albert Gallatin

In reading and applying this philosophy, Gallatin was telling Briggs: Get going and don't waste time worrying about how well the work was being done. We know that Briggs must have passed the words of his boss on to his deputies.

6.12 LAND SURVEYS AFTER 1805

By the year 1805, the PLSS as we know it today was in place and operating. Surveys were being made, notes prepared, plats drawn and approved, and lands were being conveyed according to these plats, and people were going into possession to the lines. Most changes in the rectangular survey system after 1805 were made by the surveyor general within the framework of the laws stated previously.

After 1803, Mansfield laid out the Indiana baseline and the second principal meridian. He devised a system for taking care of the convergence of the meridians; he also started our present range and township numbering system: townships are numbered north or south from the baseline, and ranges are numbered east or west from the principal meridian. These features were not enacted into law; they were rules and regulations of the department.

Tiffin followed Mansfield in 1814. He issued the first known written instruction for the Northwest Territory (not Louisiana or Florida) in 1816. He used guide meridians and standard parallels in 1824. The Act of April 24, 1820, provided for dividing land into half-quarter sections (80 acres). The act of April 5, 1832, provided for selling land into quarter–quarter sections (40 acres). In these acts, the setting of corners was limited to quarter corners (not sixteenth corners). Division into smaller parts was by protraction. Double corners on township lines were permitted until 1843, and double corners are used occasionally today.

The solar compass was invented in 1836 and was specified for limited use in the 1846 instructions. The *Oregon Manual* of 1851 applied to California until 1855 and for a short time thereafter. Technically, the 1851 *Oregon Manual* was the first *Manual* published; however, because it did not have all the diagrams in it, it was not considered a complete *Manual*. In 1855, the first complete *Instructions to the Surveyors General of Public Lands* was issued, and it soon became a part of contracts by instructions. In 1862, a statute law (2399, R.S.) required the 1855 *Manual* to be a part of all contracts.

6.13 SURVEY INSTRUCTIONS

Laws enacted to accomplish a specific purpose, such as the survey of the public lands, cannot be specific enough so as to leave no questions by those who are given the task of performing the work. The original Land Act of 1785 and subsequent acts of 1796 and 1805 still left some questions unanswered. Today, instructions or special instructions are an everyday part of a modern survey.

In 1815, Edward Tiffin issued special instructions that have since been referred to as the "Tiffin Instructions." Similar instructions were issued by John Coffee, principal deputy surveyor of the Mississippi Territory, in 1817. In 1819, Thomas Freeman, Surveyor General of the Public Lands South of the State of Tennessee, wrote the excellent *Freeman Instructions*, which gave insight and guidance for the South and Florida.

It is noted that there are no known surveys that mention Tiffin's Instructions. Furthermore, research has not recovered any handwritten preliminary copies; but a complete copy of the Tiffin's Instructions was recovered, and, upon research, proved to be in Mansfield's handwriting.

Tiffin's Instructions

The first known written instructions to deputy surveyors in the Northwest Territory were prepared by Tiffin in 1816. As these instructions explain how double corners came into being, parts are reproduced here. Instructions for keeping the 2-pole chain level, and so on, are omitted. Although none of these instructions have been recovered in Tiffin's own handwriting, a set of these handwritten instructions were discovered, and after an analysis of the handwriting, they appear to have been written by Mansfield. (Research has never been able to verify whether these instructions were ever followed.)

1. When the township lines are completed, you must begin the survey of sections at the southeast corner of the township and move on in continued progression from east to west and from south to north in order that the excess or defect of

the township as to complete sections may fall on the west and north sides of the township, according to the provisions of the Act of the 10th of May 1800.

2. Each side of a section must be made 1 mile in measure by the chain, and quarter section corners are to be established at every half mile, except, when in the closing of a section if the measure of the closing side should vary from 80 chains or one mile, you are in that case to place the quarter section corners equidistant, or at an average distance from the corners of the section, but in running out the sectional lines on the west or north side of the township you will establish your quarter section posts or corners at the distance half a mile from the last corner and leave the remaining excess or defect on the west or north tier of quarter sections, which balance or remainder you will carefully measure and put down in your field notes in order to calculate the remaining or fractional quarter section on the north and west side of the township; also in running to the western boundary, unless your sectional lines fall in with the posts established there for the corners of sections in the adjacent townships, you must set posts and mark bearing trees at the points of intersection of your line with the town boundaries, and take the distances of your corners from the corners of the sections of the adjacent townships, and note that and the side on which it varies in chains, or links or both.

3. The sections must be made to close by running a random line from one corner to another except on the north and west ranges of sections, and the true line between them is to be established by means of offsets.

General Instructions for Deputies

1. You will provide a good compass of Rittenhouse's construction, having a nonius-division and movable sights, and a two pole chain of 50 links; the chain must be adjusted by the standard chain in the office of the Surveyor General, and it will be of importance that both it and the compass be frequently examined in the field in order to determine any errors and irregularities which may arise from the use of them.

   (*Note:* The term *Rittenhouse* may have been a reference to a design rather than to a manufacturer.)

4. All township or sectional lines which you may survey are to be marked in the manner hitherto practised in the surveys of the United States land, viz: all those trees which your line cuts must have two notches made on each side of the tree where the line cuts; but no spot or blaze is to be made on them, and all or most of the trees on each side of the line, and near it, must be marked with two spots or blazes diagonally or quartering towards the line.

5. The posts must be erected at the distance of every mile, and half mile from where the town or sectional line commenced (except a tree may be so situated as to supply the place of a post) which post must be at least three inches diameter and rise not less than three feet. All mile posts must have as many notches cut

on two sides of them as there are miles distant from where the town or sectional line commenced, but the town corner posts, or trees shall be notched with six notches on each side, and the half mile sectional posts are to be without any marks; the places of the posts are to be perpetuated in the following manner, viz: at each post the courses shall be taken and the distances measured to two or more adjacent trees in opposite directions, as nearly as may be, which trees, called bearing trees, shall be blazed on the side next the post and one notch made with an *axe* on the blaze, and there shall be cut with a marking iron on a bearing tree, or some other tree within and near each corner of a section, the number of the section, and over it the letter *T* with the number of the township, and above this the letter *R* with the number of the range, but for quarter-section corners, you are to put no numbers on the trees; they are to be distinguished by this mark, 1/4 *S.*

6. You will be careful to note in your field book all the courses and distances you shall have run, the names and estimated diameters of all corners or bearing trees, and those trees which fall in your line called station or line trees notched as aforesaid, together with the courses and distances of the bearing trees from their respective corners, with the letters and numbers marked on them as aforesaid; also all rivers, creeks, springs and smaller streams of water, with their width, and the course they run in crossing the lines of survey, and whether navigable, rapid or mountainous; the kinds of timber and undergrowth with which the land may be covered, all swamps, ponds, stone quarries, coal beds, peat or turf grounds, uncommon natural or artificial productions, such as mounds, precipices, eaves, etc., all rapids, cascades or falls or water; mineral, ores, fossils, etc.; the quality of the soil and the true situation of all mines, salt licks, salt springs and mill seats, which may come to your knowledge are particularly to be regarded and noticed in your note books.

7. In all measurements the level or horizontal length is to be taken, not that which arises from measuring over the surface of the ground when it happens to be uneven and hilly; for this purpose the chainmen ascending or descending hills must alternately let down one end of the chain to the ground and raise the other to a level as nearly as may be, from the end of which a plumb should be let fall to ascertain the spot where to set the tally rod or stick; and where the land is very steep, it will be necessary to shorten the chain by doubling the links together, so as to obtain the true horizontal measure.

8. Though the line be measured by a chain of two perches, you are notwithstanding to keep your reckoning in chains of four perches for of one hundred links each, and all entries in your field books, and all your plans and calculations must be made according to the decimal measure of a chain.

9. Your courses and distances must be placed in the margin of your field books on the left, for which purpose it should be large, and your remarks made on the right in the manner following:

North		
Chains	**Links**	Between Sections 35 and 36, Town 4. Range 6.
20	30	A white oak 20-inch diameter.
37	40	A stream 30 links wide. S.E.
40	—	Set half mile post, from which a B oak 18-inch diameter bears 5. 50 E. 40 links, and a sugar tree 15-inch diameter bears N. 10 W. 34 links.
East		
Chains	**Links**	Between Nos. 25 and 36. Town 4. Range 6 on a random.
16	40	A brook 30 links wide, course S 20 W.
40	00	Set temporary quarter section post. This half mile over broken land. Timber oak, ash, etc.
64	30	A stream 25 links wide, course SE.
79	90	Intersected N. and S. line 20 links south of section corner. Over hilly land, soil rich and good for farming. Timber oak, hickory, poplar, ash, etc.
West		
Chains	**Links**	Between Sections 25 and 36 Town, 4. Range 6 on true line.
39	95	Moved temporary post to the average distance for 1/4 section corner, from which a black jack 10-inch diameter bears N. 25 W. 20 links.
55	00	A white oak 11 inch in diameter.
79	90	Section corner.

In this manner you must enter all courses and distances in your field book; the date must follow the close of each day['']s work, which field book, written with a fair mind, of each township separately, or a true and fair copy, together with the original you will return to the office of the surveyor general.

10. The plat of each township and fractional part of a township must be neatly and accurately protracted on durable paper, by a scale of 2 inches to a mile, or 40 chains on an inch, and must be in such measure and proportions in every line and part as actually was determined by measurement in the field. A compass having the true and magnetic meridian, and the scale by which the lines are laid down, are to be placed on the SE corner of the plat. Figure 6.10 illustrates double corners possible according to Tiffin's instructions. Only when the section corner happened to fall in with the existing corner on the township line was there a single corner, as is shown in Figure 6.10 for the corner common to 31-32-6-5 at the bottom. Triple corners (see Figure 6.11) came into being by some of the earlier instructions wherein the surveyor set a new corner whenever

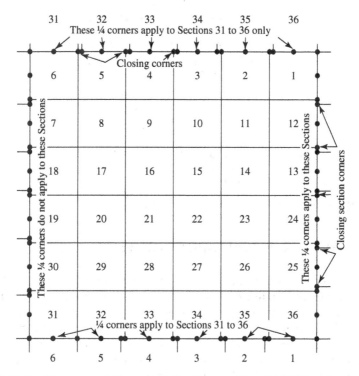

Figure 6.10 Double corners along township lines in accordance with Tiffin's instructions.

he did not fall in with the existing corners on the township line on *all* sides of the township. When running easterly from the northwest corner of Section 36 toward the township line, where the line run did not fall in with the existing corner, a new corner was set. When running westerly from the northeast corner of Section 31 toward the same township line, where the line run did not fall in with an existing corner, another corner would be set. There could be (1) that when the township line was run, (2) the post set when running easterly to the township line, and (3) the post set when running westerly to the township line.

It is not known whether any of the deputy surveyors ever followed these instructions, but, historically, they are important to know and understand since they set the basic foundation of knowledge and expectancy that was supposed to be provided by the deputy surveyors. The two areas that are probably the most litigated sections are the center-section definition and the fact that the area recited is the "official area," which should not be questioned and is without error. Every student, attorney, and surveyor should consider this act as a major milestone toward the development and survey of a great national land system.

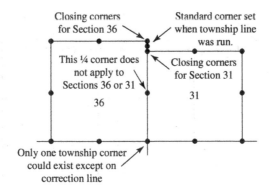

Figure 6.11 Triple corners.

6.14 STATE INSTRUCTIONS AND STATUTES

Instructions issued to explain the enacted laws must be flexible enough to get the job done but specific enough to adhere to the letter of the law. Each township that was surveyed was separate and unique. At the beginning of the surveys, there were no standardized instructions. If any were needed, in all probability handwritten instructions were given by the surveyors general to surveyors on an individual basis. A good example are those that were written to support and aid the surveyors who were given the responsibility of surveying the public lands of Louisiana.

Act of March 3, 1811 (Louisiana)

Congress provided for the survey of lands claimed by persons in the Territories of Orleans and Louisiana and provided for the subdivision of the public domain as follows (only parts pertaining to surveys are quoted):

> Sec. 2. *And be it further enacted*, That the two principal deputy surveyors of the Territory of Orleans shall, and they are hereby authorized, in surveying and dividing such of the public lands in the said Territory, which are or may be authorized to be surveyed and divided, as are adjacent to any river, lake, creek, bayou, or watercourse, to vary the mode heretofore prescribed by law, so far as relates to the contents of the tracts, and *to the angles and boundary lines*, and to lay out the same into tracts as far as practicable, of fifty-eight poles in front and four hundred and sixty-five poles in depth, of such shape, and bounded by such lines as the nature of the country will render practicable, and most convenient: *Provided* however, That such deviations from the ordinary mode of surveying shall be made with the approbation of, and in conformity with the general instructions which may be given to that effect, by the surveyor of the public lands south of the State of Tennessee.

> Sec. 4. *And be it further enacted*, That the powers vested in the President of the United States by the eleventh section of the act, entitled "An act supplementary to an act, entitled An act for ascertaining and adjusting the titles and claims to land within the Territory

of Orleans, and the District of Louisiana," passed on the twenty-first day of April, one thousand eight hundred and six, in relation to the public lands, lying in the western district of the Territory of Orleans, and all the other provisions made by the same section, for the sale of said lands, and for obtaining patents for the same, shall be, and the same are hereby, in every respect, extended to the public lands, lying in the eastern district of the Territory of Orleans.

Sec. 6. *And be it further enacted*, That every person who, either by virtue of a French or Spanish grant recognized by the laws of the United States, or under a claim confirmed by the commissioners appointed for the purpose to ascertaining the rights of persons claiming lands in the Territory of Orleans, owns a tract of land bordering on any river, creek, bayou, or water-course, in the said Territory, and not exceeding in depth forty arpents, French measure, shall be entitled to a preference in becoming the purchaser of any vacant tract of land adjacent to and back of his own tract, not exceeding forty arpents, French measure, in depth, nor in quantity of land that which is contained in his own tract; at the same price, and on the same terms and conditions, as are, or fit may be, provided by law for the other public lands in the said Territory. And the principal deputy surveyor of each district respectively shall be and he is hereby authorized, under the superintendence of the surveyor of the public lands south of the State of Tennessee, to cause to be surveyed the tracts claimed by virtue of this section; and in all cases where by reason of bends in the river, lake, creek, bayou, or water-course, bordering on the tract, and of adjacent claims of a similar nature, each claimant cannot obtain a tract equal in quantity to the adjacent tract already owned by him, to divide the vacant land applicable to that object between the several claimants, in such manner as to him may appear most equitable: *Provided however*, That the right of pre-emption granted by this section shall not extend so far in depth, as to include lands fit for cultivation, bordering on another river, creek, bayou or water-course. And every person entitled to the benefit of this section shall, within three years after the date of this act, deliver to the register of the proper land office, a notice in writing, stating the situation and extent of the tract of land he wishes to purchase, and shall also make the payment and payments for the same, at the time and times, which are, or may be, prescribed by law for the disposal of the other public lands in the said Territory; the time of his delivering the notice aforesaid being considered as the date of the purchase. And if any such person shall fail to deliver such notice within the said period of three years, or to make such payment or payments at the time above mentioned, his right of pre-emption shall cease and become void; and the land may thereafter be purchased by any other person in the same manner, and on the same terms, as are or may be provided by law for the sale of other public lands in the said Territory.

Sec. 8. *And be it further enacted*, That the surveyor-general shall cause such of the public lands in the Territory of Louisiana, as the President of the United States shall direct, to be surveyed and divided in the same manner and under the same regulations and limitation as to expenses as is provided by law in relation to the lands of the United States, northwest of the river Ohio and above the mouth of Kentucky River.

Sec. 12. *And be it further enacted*, that all navigable rivers and waters in the Territories of Orleans and Louisiana shall be and forever remain public highways.

The lands of Louisiana were not surveyed exactly in accordance with the procedure used in the Northwest Territory. Double and triple corners (see Section 6.13) were avoided and the northern and western sections of townships differed.

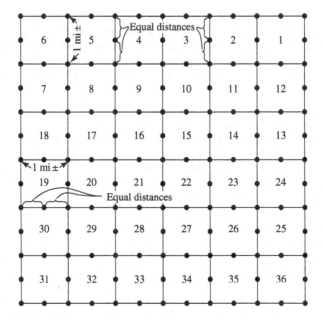

Figure 6.12 In early Louisiana, survey errors due to convergence and measurement were divided equally in the last half mile.

Although the Act of May 10, 1800, provided for placing errors of measurement and convergence in the last half mile of the northern and western tiers of sections, this was not done in parts of Louisiana and Florida. In Louisiana, the errors were placed equally in each half mile of the last mile, as shown in Figure 6.12.

In Florida, errors of closure and convergence were sometimes placed on the south and east sides of a township, as is noted in the following letter of instruction of 1845: in Florida, a letter from Butler to his surveyors contained a note that each surveyor should "add a link of one inch to each half chain." There is no explanation for this recommendation, except that he felt this would compensate for sag of the chain.

Office of the Surveyor General

St. Augustine, February 10th, 1845

Sir:

Having entered into bond with approved security and your chain and compass adjusted by the standard instruments of this office, you are expected to repair with all possible dispatch to the work assigned you. You will run all the exterior lines of the Townships which are indicated on the accompanying diagram of District E. As it has been reported impracticable to connect with the surveyed lands East of the marshes of the St. Johns, you will *begin on the North and West sides of the District and throw the errors on the South and East in conformity with the survey heretofore made.* After running those exterior lines you will proceed to subdivide into sections five Townships, selecting for this purpose that portion which in your judgment would seem most desirable for location and settlement and an affidavit will be required of you that the Townships sub-divided were occupied by settlers or were the best lands contained in your contract. When you report any part of your contract impracticable an affidavit will be required of you to that effect and a full statement of the causes that render it so. (Emphasis added.)

When published, the *Manual* of 1855 placed all survey districts on the same basis. For a short time after 1855, irregular procedures did occur in some areas.

Louisiana Instructions and Statutes

Statutes pertaining to Louisiana have been quoted here. Each surveyor general had the authority to put statutes into effect by issuing rules and regulations to his deputies. The following circular and letter explain some special instructions applicable in Louisiana.

In a general circular dated September 23, 1831, signed by Elijah Hayword, are found the following special notes pertaining to Louisiana:

AS TO MARKING

20th. The greatest possible caution is to be observed in marking the corners of townships etc., in a plain, distinct and permanent manner.

Where a tree is not found immediately at the corner, a corner is to be established by planting a post on which is to be marked the number of the township, over which is to be marked the number of the Range, and underneath the number of the Section.

The bearing and distance, also the names and respective diameter of the nearest trees from such corner are to be carefully taken and noted in the field book. The nearest of such trees (where there are more than one) is to be marked to correspond with the *marked corner*. The mark should be in a *regular chop, squared off*, to be made into such tree, so as to be always distinguishable from a mere *blaze*. The letters B. T. to denote the fact of its being a *"bearing tree"* should be distinctly cut into the wood some distance below the other marks. All these particulars are to be most intelligibly and minutely noted in the field book. The posts used in forming the corners of townships must always be larger (for the sake of distinction) than those which denote the Sectional and quarter Sectional corners, and should be neatly *squared off* at the top to correspond with the cardinal points.

The marks on the posts and bearing trees should be deeply burnt into the wood with marking irons. The posts must always be made of the most durable wood that can be had, and should be set in the earth to the depth of two feet & very securely rammed in with earth and stone. It is highly important in reference to their durability that the portion of each post below the surface should be charred and the whole of it rubbed over with tar, except the portion which bears the Surveyor's marks. The Sectional posts are to indicate by a number of notches on each of the four corners directed to the cardinal points, the number of miles that it stands from the outlines of the township—the side of the post will be numbered to correspond with the number of the Section it faces—Each half mile post on a Section line, and quarter section post on a township line should be marked to indicate that it is a quarter sectional "(1,4,S)" post, and the nearest adjoining tree on *each side* of such post, must be similarly marked—The Surveyor to note in his field book the kind of tree, its diameter, bearing & distance from the true corner.

Posts denoting the same kind or character of boundary, should be of uniform construction, and there should always be a striking difference between posts denoting different kinds of boundary.

To create *additional* and *increased* facility in the discovery of boundary lines by the purchasers of public lands and to prevent errors of entry, you are to require your Deputies to fasten to the Sectional and quarter Sectional posts, near the ground, but so as to be plainly seen, a finger board on which is to be *distinctly* marked with *black oil Paint*, the appropriate number of the tract. This board to point diagonally across the tract and to be marked similarly to this: NW 1/4 S. 1 T. 1 N. R. 1 E.

In prairie countries where hearing trees cannot be had, *mounds*, to be covered with sod, are agreeably to contract, to be erected. Such mounds should be of uniform size and conform precisely to instructions to be given by you. As mounds are subject to be worn away by the action of the weather and other causes, I would recommend that a stone be planted in the centre of the mound and that a few handfuls of charcoal be enclosed therein. I would further recommend that at each corner of a square which will enclose the mound and conform to the cardinal point, there be planted a chestnut, hickory nut, walnut or acorn.

A stake to be set up in the centre of the mound to which is to be fastened a finger board on which is to be designated in black oil paint, the appropriate numbers.

All the Particulars relative to the construction of a mound are to be minutely indicated in the field book.

The perpetuation of the corners of the public surveys is a subject at *primary importance*. Every possible care and precaution to secure correct and durable corners must be observed by your Agents whose fidelity you should test by every means in your power.

The following letter, dated 1844, was put into effect soon after.

Surveyor General's Office

Baton Rouge, (La.)

For

Honorable Thomas N. Blake

Comm. Gen Land Office

Washington

D.C.

Sir:

I have the honor to enclose herewith a printed copy of the revised instructions which I have found it necessary to prepare for the information and guidance of my Deputies in the field, for your sanction.

The only material alteration I have required in marking and establishing corners in the field occurs in the directions for prairie surveys, which I have made for the following reasons, viz:

1. That mounds in prairie country are destroyed entirely in a few weeks, owing to the herds of cattle seeking them to protect themselves from the fly and other insects—with their horns and hoofs they soon obliterate all the marks placed at a township corner in connection with a mound—and in two or three years it cannot be distinguished from the surrounding plain or prairie.

2. In the method proposed by me of sinking a pit—the following advantages are attained—If sought by cattle at all, the stake prevents the use of their horns—if the stake is loosened so that they can be free to act upon it, the result is greater depth obtained for the pit, making it more recognized as a land mark—the water which collects in the bottom of the pit, causes the growth of a strong reed grass, which attains from eight to ten feet heights over the grass of the surrounding country, with a much deeper green color—so that a corner can be discovered on foot at two or three miles distance making at all seasons of the year a permanent and easily ascertained land mark.

To this change of the establishment of mounds, I think your approval is necessary, and if you deem it proper. I would be glad to have it at as early a day as practicable—in order that I may circulate the instructions—Should it not be approved of, I can restore the old plan, by a marginal note.

I have the honor to be with much respect

Your obedient servant

F. D. Newcomb

Surveyor General,

Louisiana

On the margin of a printed general circular to Deputy Surveyors (uncertain date) were handwritten the instructions contained in the preceding letter. In this printed general circular was also this sentence: "It is intended that all errors should fall upon the last mile, which may either exceed or fall short of the required distance, but the half mile post should be set equidistant between the corners."

Half-Mile Posts: Alabama and Florida

In the early years of the surveys, it was recognized that the survey system had to be employed in as economical a manner as possible. It is interesting to note that modern-day surveyors who conduct retracements in these two states have known and recognized half-mile posts as being present, but it was not until recent years that they recognized the legal significance of these surveyed points. Surveyors practicing in these two states find it an absolute necessity to refer to the original notes to determine how the original surveyors treated these points in running the original lines.

In many of the early Florida and Alabama surveys, the custom was to set half-mile posts at record measurements of 40 chains on all lines, including random lines. In most instances the ½ mile posts were monumented and only a reference by distance was made to the ¼ corner. The description of the patents were by aliquot parts of the section and the un-monumented ¼ corner. The usual practice was to accept the monument points as the true aliquot corners and not the unmonumented

¼ corner. In 1815, in Alabama, the validity of these ½ posts was legally tested in the decision of *Walters* v. *Commons*.[8] The holding of the trial court and the appeal court was tested. Holding the Land Act of 1805 was controlling. Both courts rejected the holding of the ¼ post being used for subdividing the section. The holding of court that even though there was no monument at the ¼ corner, it was the proper cornet to survey to and from.

The following are notes of Clements, T10S, R29E, Fla., 1834.

Section No. 11

Beginning at NE corner section, south, enter timber, pine, and bushes.

40.00	set ½ mile post

S 75° W 104 pine
N 38° W 100 pine

-Both of these trees are on the west side of the line as none could be found east.

40.05	Qr. section post level and timber as before
80.10	to SE corner section

-Beginning at SE corner section, east, land level, scrub pine, pine and bushes.

40.00	set ½ mile post

S 8° E 33 pine
N 34° W 60 pine

level and timber as before

80.00	to SE corner of section set post

In this case, the quarter-corner post was set in one instance, and in the other the half-mile post was the true quarter corner. In other cases, such as the following for T5, R12, S&E, Boyd, 1825, the half-mile post does not represent the quarter corner (never set). The ¼ S notation was added after the notes were completed (often at a later date).

5.00	Out of pond, water runs SE
29.00	Xpond
40.00	½ mile post. Pine N 71° W 33. Pine N 59° E 46.
50.00	To a pond
79.80	To the intersection. Flat ponds, water from 1 to 3 feet deep.
―――	
39.90	¼ S.

South 34 and 35

40.00	½ mile post, Cypress N 3° W 150. Cypress N 15° E 142
47.00	Out of pond
79.80	Intersect 50 lks west. corrected hack
―――	
39.90	¼ S

In some of the notes, the quarter-section post notation (added after the notes were completed) appears in proper sequence. The following notes for T5S, R8E, Fla., Washington, 1825, include both proper sequence and an addition at the end.

South Boundary of Section 1
Beginning at the SW corner 1 run N 89° 109 E

1.00	to Cyprus pond
8.00	to flat 3rd open piney woods
30.00	crossed swamp 10.00 ch wide
39.67	quarter section post
40.00	½ mile post whence bears

pine N 45° W 21
pine S 54° E 22

79.35	the SE corner of the section
	leave 3rd open pine and palimento

Section No. 11 T5 R8
Beginning at the SW corner 1 run east

40.00	½ mile post, whence bears

Pine N 31° W 60 lks
Pine S 88° E 34

79.56	the SE corner of the section
39.78	quarter S. Post

In some cases, the surplus or deficiency was placed in the south part of a section instead of the north as required presently. In the following notes by Norris, T5S, R27E, Fla., 1848, the surplus of Section 4 is contrary to the location expected.

West boundary of Sec 4 (note: going south)

4.00	leave hammock
16.00	to pond
26.00	X do (meaning "same") to Pine andPale.
40.00	set ¼ sec.

Pine N 27° W 50
do S 30° E 30

42.50	to pond
47.00	X do to pine and Palmetto
80.62	to Cor of Sec.

Any retracing surveyor must consider that half-mile posts in these two states can be treated in either of two ways: (1) the half-mile post is treated as a quarter corner, or (2) the half-mile post is used as a point for line determination and placement of the quarter corner by proportioning or by a direct measurement from the half-mile post.

The validity of half-mile posts was questioned very early in Alabama as to what weight a retracing surveyor should give to these monumented points over the unmonumented ¼ corner.[9] The Supreme Court of Alabama held that the Land Act of February 11, 1805, controls, in that the half-mile post has no legal significance. In fact, they even recommended that the surveyor remove them and resort to the ¼ as being equidistant between the two section corners.

6.15 INSTRUMENTS USED

The instruments used in public land surveys were identified by federal statutes. Initially, the only reference was to the compass, of Rittenhouse Construction, possibly because declination (magnetic variation could be set off) but as to distance measurement it was in the Land Act of 1785, which stated that "all lines would be measured by the chain." No mention was made of what length the chain should be. At that time, most surveyors in the metes and bounds states used a chain of 2 poles, or 33 feet. It is certain that this is the unit used in the commencement of the surveys. In 1796, the chain was identified, by law, as being a "chain of two poles, or 50 links." It did state that all measurements would be kept in "measurements of four poles." Even though the law is still valid today, surveyors now use electronic distance measurements and global positioning.

The law stated that all lines would be run "with a compass of Rittenhouse construction." Initially, the federal law required magnetic declination to be compensated for. Then a few years later, Congress repealed that portion of the act, only to add it again when it realized the error of its actions.

The magnetic compass caused problems with keeping the lines true as in reading the bearings, in that the least division was probably ½ degree. Local attraction, daily variations, and many other problems made lines unreliable, as did the ability of each surveyor.

In 1839, William Austin Burt, a US deputy surveyor in Michigan, invented the solar compass, which relied on the sun for determination of the true bearings. A significant improvement over the regular magnetic compass, the solar compass permitted surveyors to run lines to an accuracy of minutes of true bearing, as long as the sun was shining.

Until about 1902, the public land surveys were made with the magnetic compass; after that date, surveys were made with instruments provided with the accessories necessary to determine the true meridian without reference to the magnetic needle (*Manual* of 1902; see Figure 6.13). The magnetic compass is now recognized as an instrument of low precision, which is very unreliable under certain circumstances. Before the discontinuance of the compass, surveys were extended into the iron ore belt of Michigan. Many local areas are known to deflect the needle as much as 10° to 20° with resulting distorted sections and surveys. This same problem existed in many of the metes and bounds states, but there is no known record that the solar compass was ever used in those states.

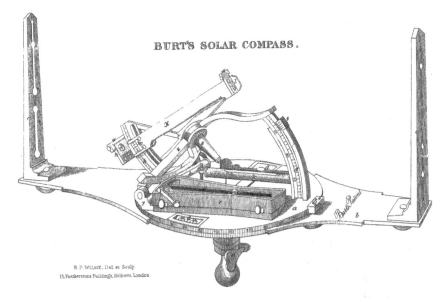

BURT'S SOLAR COMPASS.

B.P. WILME, Del. et. Sculp.
15, Featherstone Buildings, Holborn, London.

Figure 6.13 Solar compass used in early surveys. The sighting devices were crude compared with today's telescopes. (Source: William Austin Burt, *A Key to the Solar Compass, and Surveyor's Companion*, 1853.)

6.16 FIELD NOTES

By law, land is patented from an approved GLO plat. However, the plat is a product of the field notes that were kept by the surveyor. In the notes, the surveyors indicated the lines run, the manner in which they were run, and information that was helpful to the people who prepared the GLO plats. One will also find other information regarding topography, terrain, and happenings that the surveyors felt would be helpful or informational. A surveyor should read any and all notes available before any subsequent surveys or retracements are conducted or undertaken. Decisions have held that the field notes along with any other documents are as much a part of the original survey as was the running of the lines on the ground. To complete a retracement, the retracing surveyor must refer to and consider these original notes.

6.17 NOMENCLATURE FOR SECTIONS

The nomenclature shown in Figure 6.14 is used to designate separate parcels of sectionalized land. Reading descriptions of this type is simplified by following the description backward, as, for example, "the E 1/2 of the SW 1/4 of the SE 1/4" would be visualized as the SE quarter first, then the SW quarter of the SE quarter next, and then the east half of the last visualized parcel. The smallest legal size recognized

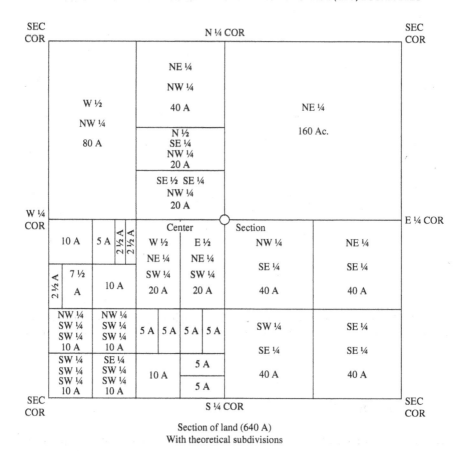

Figure 6.14 Nomenclature for portions of sections.

by federal statute for homestead purposes was one-quarter of one-quarter; however, smaller parcels are commonly designated to identify portions of homesteaded holdings.

6.18 MEANDERING

In the GLO surveys, from the earliest times meander lines were run to determine the location and area of rivers and lakes. Charges were made for dry land acreage only; meander lines were run to determine the area to be charged. Meander lines generally were run inaccurately and sometimes fraudulently. In the wintertime, when surveys were run across ice, entire lakes were omitted, and at times, gross errors were made. In most instances, meander lines do not control boundaries, but they do control area determination. Recovered meander corners may be used to control the direction of a line but not to control a property or section boundary. They may also be used to proportion lost corners.

6.19 RESURVEYS AND RETRACEMENTS

Land owned by the government and once subdivided may be resubdivided by the government at any time in accordance with the manner prescribed by law. In the event of a resubdivision of a township, two plats exist, one showing the original survey and the second showing the resurvey or retracement. Where an entryman patented a parcel of land by the original township plat, no resurvey or retracement is supposed to be executed so as to impair the bona fide rights or claims of the entryman. Usually, the entryman's land is surveyed from the original markers and given a tract number, whereas the balance of the township, owned by the government, is resurveyed in accordance with the present rules of rectangular subdivisions. Unfortunately, the government has not protected the bona fide rights of the patentee in every case. Frequently, the tract designations do not correspond to the location shown on the original survey.

There are times when individuals, both landowners and surveyors, refer to specific corners in a section by reference to the fractional breakdown; that is, "the NE corner of the SW ¼ of Section 18." In reality, this is an improper reference. Specific corners should be referred to in accordance with their positioning within a section. Figure 6.14 shows the correct reference to corners within a section.

Where the midpoint of the section may be either the NE corner of the SE 1/4, or the NE corner of the SW 1/4, or the SE corner of the NE corner or the SW corner of the NW 1/4, its true location is the center 1/4 (C 1/4).

6.20 DEFECTIVE BOUNDARIES ENCOUNTERED IN RESURVEYS

On resurveying the boundaries of an older survey to initiate a new survey, defective conditions of the older survey are not incorporated into the new surveys. Defective conditions may be caused by alignment, measurement, or both. When a new township is to be surveyed and it is found that the southerly line is defective in measurement only, new corners set at half-mile intervals will apply to the sections to the north and the old corners with irregular measurements will apply to the sections to the south. Double sets of corners may thus occur. Double sets of corners similarly may occur on the easterly, northerly, or westerly boundaries. When the easterly or southerly boundary of a township is defective in its alignment, a new sectional guide meridian or sectional correction line is run, as shown in Figure 6.15.

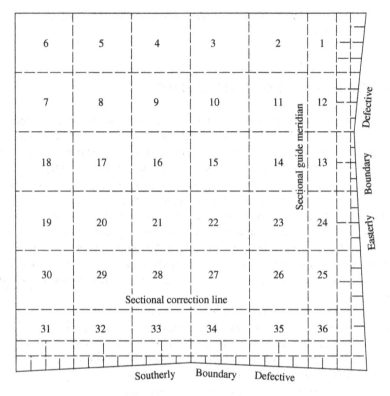

Figure 6.15 Sectional correction lines.

6.21 SECTIONALIZED SURVEYS AND INNOVATIONS

From its inception, the sectionalized system of surveys has had many innovations introduced, not only by the government but also by the surveyors themselves. From the early days, when surveyors used their own methods, to modern times, when the BLM succeeded the GLO, innovation helped get the job done. More recently, global positioning is being used to help position new surveys and retracements. In Alaska, thousands of square miles were surveyed using helicopter methods, protracting, and various untried methods. All were developed to get the job done as quickly and as efficiently as possible. These new methods permit the surveyors to use coordinates as part of the survey, resurvey, and retracement processes.

6.22 IRREGULAR ORIGINAL GOVERNMENT SUBDIVISIONS

The laws enacted to provide for the survey of the PLSS were unique in that they provided for presidential authority, through the secretary of the treasury and later through the secretary of the interior, to vary the original surveys when conditions warranted it. This was done when surveyors encountered such problems as Indian land and prior grants from France, Spain, England, Mexico, and other countries. This can be seen in Ohio during the early surveys.

6.23 TOWNSHIPS OTHER THAN REGULAR

In the course of researching and conducting surveys, a surveyor in a GLO state may encounter townships and the resulting sections that are at variance with the 80-chain sides and cardinal directions. The various laws reflect the anticipation that full 6-mile-square townships and the sections contained within them will be other than as prescribed by law. Fractional townships and sections were anticipated when a township was surveyed and its lines would not make 6 miles square, and as such there would be sections within the township that would not be 80-chain sides in cardinal directions.

Under the law, to have a fractional township, it must border lines that terminate at a meandered water body, impassible object, state line, reservation, Indian reservation, or grant boundary (see Figure 6.16). However, there are townships that bear absolutely no resemblance to what the law anticipated. In these instances, one may find as many as 47 sections in a township, township lines that are broken, and lots against these broken lines. In many instances, the surveyors at the time the lines were created were left to their own devices as to how the lines should be run. In many instances, the usual rules of retracement cannot and will not apply. Such sections and townships would be considered anomalous sections and townships, in that they do not meet the letter of the law, and usually there are no set rules for retracing the lines. Such a township or section usually will not have the legal acreage prescribed by law (see Figure 6.17).

In fractional sections and townships there are specific rules as to how the lines should be run when there are no opposite corners. Before a surveyor attempts to retrace a fractional section or township, he or she should become totally familiar with the specific methodology described in the *BLM Manual of Instructions for the Survey of the Public Lands* as well as the booklet entitled *The Restoration of Lost or Obliterated Corners*.

6.24 LOCATING GLO RECORDS IN STATE ARCHIVES

Original GLO records such as land patents, survey plats, field notes, control document index records, tract books, and the land catalogs are maintained by the BLM, a federal agency. No matter whether you are practicing in Florida or Nevada, these records may be searched for and accessed online at GLORecords.BLM.Gov.

On the other hand, GLO records created after a state took over stewardship of its own cadaster may be found in a variety of databases being administered by a variety of state or local agencies. Sometimes these records are online and easily accessible—Florida's Land Boundary Information System (LABINS) is a good

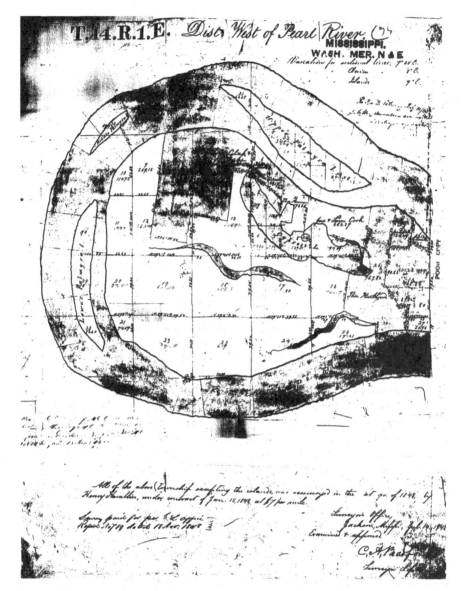

Figure 6.16 Fractional township/Mississippi.

example. On the other hand, finding field notes of retracement surveys in Wyoming requires the surveyor to travel to the county courthouse and search for hard copies of any field notes which are usually not indexed.

In some states, the survey plats may be stored in one office while the field notes are stored somewhere else. While in other states, GLO records before the state took over maintenance of the cadaster are stored somewhere different than after the cadastral

Figure 6.17 Anomalous township/Mississippi.

takeover took place. Therefore, it is extremely important that surveyors be familiar with how their home state and county index GLO records.

6.25 SUMMARY OF THE GLO SYSTEM

Principle 4. *In the public land states, to practice in the modern technical world, the surveyor must have an intimate knowledge of the historic legislation that created the Public Land Survey System.*

The PLSS of the United States is the greatest legally created land survey system in the world. The surveyor must understand that a basic requisite of this system is that no land would be patented to any person until it had been surveyed and a plat prepared and approved. The system and the laws relative to this system provided for the sanctity of the work performed by the original survey crews in the field to the point that a law was enacted, whereas the original measurements were deemed to be "without error." The laws also identified how surveyors would run the lines and set the monuments at the corners and what methods would be used for accessories. The laws provided specifically for notes, plats, and the scale to which they would be drafted. All of this was accomplished with as few instructions as possible and as much latitude as could be provided. Based on these laws, instructions and their subsequent interpretations by the courts have helped to make the system work. The modern surveyor must know what the basic federal land laws created and what principles were identified.

Some of the principles that should be understood by surveyors are:

1. The creating surveyors created the boundaries of the public lands.
2. The original surveys, by law, are without error.
3. The accessories to the original corners set by original surveyors at the time of the original surveys, when recovered and proven, have equal dignity with the corner itself.
4. The federal government can resurvey any of its lands that have not been patented as long as bona fide rights are not affected.

Principle 5.   *There can be only one original survey. Any other survey of an original survey is a retracement and cannot alter the original corners or rights or section lines.*

This concept is difficult to understand, until one realizes that land, before it is surveyed, has no legal or physical boundaries. The survey creates the boundaries.

The agency or individual who created these boundaries has full control over them. Until property rights are transferred according to that particular survey, the creator may alter the boundaries as he or she wishes. *However*, once a property right is created, then all lines are legally sealed and cannot be altered.

Principle 6.   *Unless created by state regulations, the* BLM Manual *is not mandatory for private surveyors but can be used as guidelines for private registered surveyors.*

The surveying community eagerly waited for the 2009 edition of the *BLM Manual*. Private surveyors use this manual as their principal source of guidelines for working in the GLO areas of the country. They do not realize that this *Manual*, without legislation or direct acceptance from their respective states, *does not apply to private surveyors*. It was assigned as *in-house* guidance for BLM surveyors in the conduct of their duties. The bulk of this *Manual* is directed to the creation of GLO boundaries, with a chapter on retracing these boundaries.

Principle 7.   *The edition of the* BLM Manual *that applies to a retracement is the* Manual *that was current at the time the retracement was performed.*

One of the biggest problems surveyors have is keeping current with standards. If a surveyor chooses to refer to the *Manual* the edition that controls is the edition that was in acceptance at the time the fieldwork was performed. In an unreported decision in which the *Nature Conservancy* was a party, the plaintiff's surveyor attempted to use the current *Manual* to support his work. After an extensive trial, the judge ruled that since the original plats were approved in 1845, before any *Manual* existed, the current *Manual* was inappropriate to apply and common-law principles of retracement were appropriate.

Principle 8.   *A surveyor who conducts retracements of the GLO/PLSS boundaries should be acquainted with historic and current court decisions.*

Realizing that many of our actions are guided by case law, the following decisions should be understood in order to provide a foundation for retracing GLO boundaries.

> THIS IS NOT A LEGAL TEXTBOOK, NOR WAS IT DESIGNED AS SUCH. IT SHOULD BE USED FOR INFORMATIONAL PURPOSES ONLY. Use these decisions for guidance only.

Few private surveyors ever conducted or conduct original public land surveys today. Essentially, they are engaged in retracing the boundaries of lines that were originally created decades ago.

In order to give the surveyor and/or the student guidance in this turbulent, complex legal area, the following court decisions are recommended for reading:

1. What constitutes an original survey and its sanctity. *Cragin* v. *Powell*, 128 U.S. 691 (La. 1888)
2. The distinction between an original survey and a retracement. *Rivers* v. *Lozeau*, 539 So. 2d 1147 (Fla. Dist. Ct. App. 1989)
3. The distinction between an obliterated corner and a lost corner. *U.S.* v. *Doyle*, 468 F. 2d 633 (10th cir., 1972)
4. An extensive discussion as to what was considered an original survey. Failed to consider what a property corner was. The trial judge went beyond his authority and changed an original GLO survey. *Dykes* v. *Arnold*, 129 p3d 257 (Or. 2006)

Before any surveyor undertakes a retracement of any PLS description, he or she should have read these decisions and should use the resulting decisions as guidelines.

The PLSS is a unique system that was predicated on statute law. With considerable forethought, with one very basic failing, no provisions were made to maintain the system after the lands had been patented to private parties. Whether the student will practice in the GLO states or in the metes and bounds states, each student should have an understanding of how the system evolved and its strengths and weaknesses that the surveyor of today must work with and understand.

NOTES

1. Sketches of Sac and Fox Indians, and the Early Settlements of Wapello County, *Annals of Iowa*, vols. I–V, pp. 480–536, 1836–1837.
2. *Journal of the U.S. Continental Congress*, vol. XXVII, p. 46, May 28, 1784.
3. *Cox* v. *Hart*, 260 U.S. 247 (Calif. 1922).
4. *Kittridge* v. *Landry*, 2 Rob. 85 (La. 1842).
5. *Kelsey* v. *Lake Childs Co.*, 112 So. 887 (1927).
6. *State of New Mexico* v. *State of Colorado*, 45 S.Ct. 202, 267 (1925), 267 U.S. 30, 41 (1925).
7. *Vaught* v. *McClymond*, 155 P.2d 612 (Mont. 1945).
8. Ibid.
9. *Walters* v. *Commons*, 2 Port. 38 (Ala. 1835).

CHAPTER 7

FEDERAL AND STATE NONSECTIONALIZED LAND SURVEYS

7.1 INTRODUCTION

This chapter is one of the most difficult to keep current because much of this information is not reported where it is readily available to those who are residents of the local area. Basically, this chapter is for informational purposes to the general surveying and legal communities.

Surveyors, attorneys, and the courts should realize that many surveys have been sanctioned by governmental agencies other than the federal government. One can find sections of land referred to in many states other than the General Land Office (GLO) states. Texas, New York, Maine, and South Carolina as, well as Tennessee and Kentucky, have small areas of rectangular surveys somehow created using the federal land system nomenclature. In this chapter, we show that some of those nonsectionalized surveys have their foundations in the Public Land Survey System. Yet the surveyor will also find that some survey systems are unique in their own right and are much localized in nature.

Brown's Boundary Control and Legal Principles, Eighth Edition.
Donald A. Wilson, C.A. "Tony" Nettleman III, and Walter G. Robillard.
© 2024 John Wiley & Sons, Inc. Published 2024 by John Wiley & Sons, Inc.

The following principles are discussed in this chapter:

PRINCIPLE 1. Public domain lands may be exchanged for lands in private ownership. Once the exchange is effective and deeds exchanged between
the private individual and the federal agency, those private lands that
were exchanged for public domain lands attain the dignity of public
domain lands and must be treated as such.

PRINCIPLE 2. To retrace the boundaries of foreign land grants, the decree and survey
ordered by the court are conclusive. The surveyor does not look for
facts behind the decree.

PRINCIPLE 3. Land grants usually have senior standing or rights over public
domain sectionalized land surveys, and any encroachments of
sections on a land grant usually are resolved in favor of the senior
land grant.

PRINCIPLE 4. In many areas of the country, local land survey and description systems can be found, and surveyors practicing in those geographic areas
or areas of specialty must, by necessity, have a firm knowledge of the
unique system or systems that are present in the respective state or
area of practice, the county, and even the local community relative to
the boundary creation or retracement at hand.

PRINCIPLE 5. In many of the areas in which private and quasi-government surveys
are present, the ability and possibility of finding original evidence
may be very difficult, if not virtually impossible, and any surveyor
practicing in these areas must be especially cautious.

Many eastern lands were originally divided and granted according to a uniform
system (see Figures 7.1a, 7.1b). Because there was no uniform set of instructions
governing these land divisions, each was unique; resurveys and retracements require
knowledge of the original layouts, such as where original survey markers were set
and how lines were marked. Sometimes, plans of the original layouts are available,
but when they are unknown or have been lost, it is often necessary to reconstruct
the layout using land records in conjunction with field evidence. Whenever the
original land plans are available, they should be consulted when the lines are
retraced.

Although the majority of federal surveys were made for the purpose of dividing
land into townships and sections, surveys sometimes deviated from that pattern. In
addition, federal and state governments sometimes sold large tracts of land to be
surveyed as agreed upon or in accordance with the buyer's wishes. In this chapter,
a number of the original surveys of such parcels of land are discussed. In many
instances, the federal government required and permitted departures from instructions issued for sectionalized lands, although we do not always know why. Where
there were departures, the rules for resurveying of sectionalized lands seldom or only
partially applied. Perhaps the largest volume of federal government nonsectionalized land surveys was for the purpose of locating and describing lode mining claims

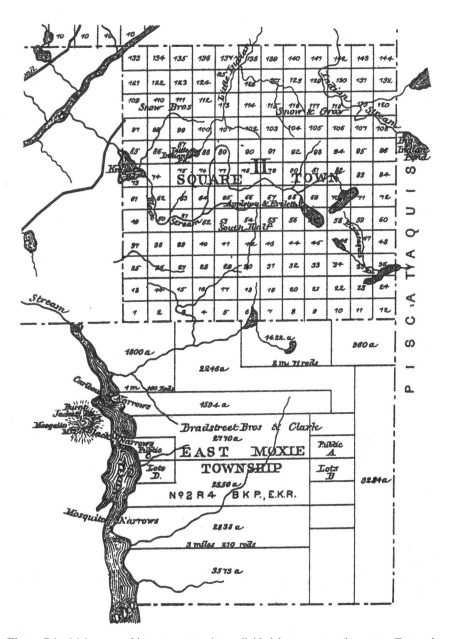

Figure 7.1a Maine townships were sometimes divided into squares (in square Town, the squares are ½ mile on a side) and were sometimes irregular (as shown for East Moxie). Resurvey procedures are determined by the state.

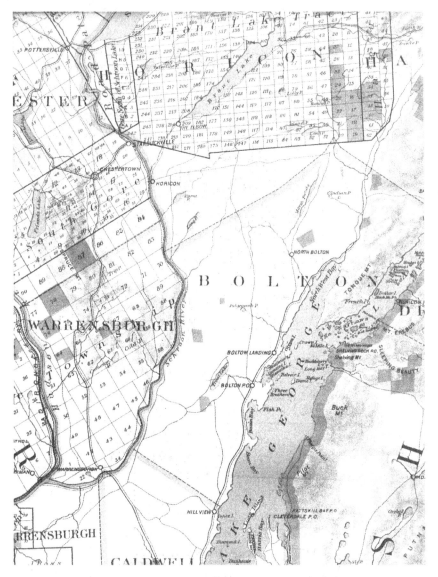

Figure 7.1b Combination of rectangular parcels and metes and bounds parcels—Adirondack Mountain region of New York, 1800s.

(minerals found in veins). Placer claims were usually located as an aliquot portion of a section of land, although not always. The more important departures from the sectionalized land system are discussed to emphasize that surveyors should always be prepared to "expect the unexpected."

7.2 EARLY NEW ENGLAND AND OTHER COLONIAL-ERA SURVEYS

Early surveyors would design and lay out the division of a "wild" or unsurveyed township according to (1) the shape of the township, (2) the number and requirements of the proprietors (original owners or grantees), and (3) instructions from the European sovereign who was claiming and granting that particular parcel. In many instances, the surveyor(s) would run each line and mark every corner.

In many references, certain terms are used to describe these situations. Some of the more commonly used terms are:

- *Lottery.* Lots were chosen by potential owners according to a lottery system, from which the term *drawing lots* could possibly have originated.
- *Land divisions.* The first division, often home lots, was comprised of small lots and was given for homesteads. Additional lots were larger and were used for farming or as a source of timber. There were sometimes six or more divisions, with a grantee receiving at least one lot in most, if not all, divisions.
- *Pitches.* Common land was usually left after division (1) to compensate those who did not get their fair share in either quality or quantity, and (2) for later divisions among subsequent proprietors or settlers. A person entitled to a share or a certain amount of land could "pitch" for it, meaning that he could lay out his lot within the ungranted lands anywhere he chose as long as it did not infringe on anyone else's claim. He was then to record his survey/claim by making a return in the "pitch book" kept by the proprietor's clerk.
- *Reservations.* In many towns, reservations were made for specific purposes, such as schools, churches, ministry, first settled minister, and glebe lands (Church of England). Many of these original grants were later combined with "common lands" and title was vested in the town (public), where it remains today unless previously disposed of. Many town divisions left strips of land between ranges or rows of lots for the layout of future roads. These are generally referred to as *rangeways.* Sometimes, portions of a rangeway were exchanged for land taken for highway purposes through a lot in lieu of paying damages; in other instances, the fee ownership remains in the town.
- *Special grants.* Grants were made for special purposes, such as to individuals for military service, to schools, for ferry services, and the like. These were sometimes on a rectangular basis but frequently were not. The important thing to realize is that they were original grants and were often surveyed at the time of the grant.

7.3 OHIO COMPANY OF ASSOCIATES

Contemporaneous with and just after the survey of the Seven Ranges, the concept of selling large blocks of land for development by private parties came into being, and one block of land west of the Seven Ranges was acquired on October 27, 1787, by the

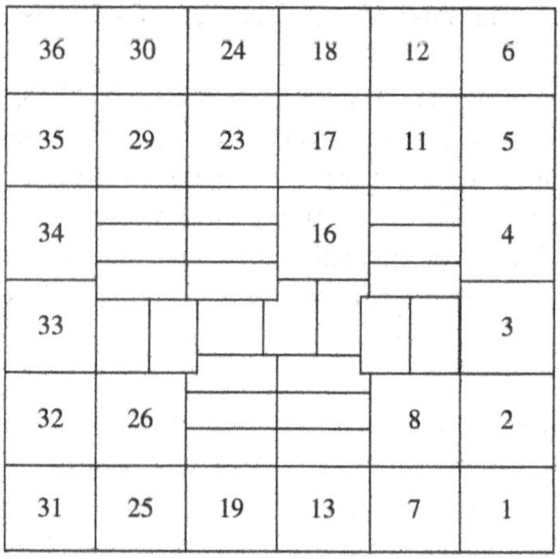

36	30	24	18	12	6
35	29	23	17	11	5
34			16		4
33					3
32	26			8	2
31	25	19	13	7	1

Figure 7.2 Land divisions in the Ohio Company Purchase. Regular lots (now called *sections*) were 640 acres, and fractional lots contained 262 acres. Lots 8, 11, and 26 were reserved for the US government, Lot 16 for schools, and Lot 29 for religious purposes. Corners set were to be the same as for the Seven Ranges.

Ohio Company of Associates. Interestingly, four of the directors intimately involved with the Ohio Company were Rufus Putman, two former Revolutionary War officers, and a minister. According to the purchase agreement, the land was to be surveyed in accordance with the Land Act applicable to the Seven Ranges. The lines were run on a compass bearing (about 4° east of true north), lines were blazed, corners were monumented with wooden posts or corner trees, and level distances were recorded. No field notes have been found, nor was the company required to file plats. Reservations were made for schools (Section 16) for Christian Indians, and two other sections were reserved for Congress.

Because of the right of selection of lands by subscribers or stockholders, some of the townships were divided into 262-acre parcels, as illustrated for a township in Figure 7.2. As a condition of the sale, the US government reserved Sections 8, 11, and 26 for itself; Section 16 for schools; and Section 29 for religious purposes.

7.4 DONATION TRACT

Originally, the Ohio Land Company had intended to donate 100 acres to any male who was willing to settle on the land and help to protect it from the Indians. However, the plan became unworkable, owing to costs and other problems, and the company appealed to Congress, who authorized the president to set aside 100,000 acres for free donation to male settlers. The area was not subdivided into sections and was located just north of the Ohio Land Company land. The exterior boundaries that identified this tract were the "Ludlow line" on the north, the "Seven Ranges plus an extension of that line so as to include 100,000 acres" on the east, and by lines that defined "The

First Purchase" on the south. The Ohio Land Company was dissolved in 1849. This is quite a unique description, in that the area, 100,000 acres, was the controlling element of the final line after the other three had been surveyed.

7.5 SYMMES PURCHASE

The area between the Little Miami and Great Miami rivers was purchased and developed by Symmes and is referred to as the *Symmes Purchase* or the *Miami Purchase*. This purchase of over 300,000 acres was ultimately sold by Congress for less than 66 cents per acre. The private surveys conducted by Israel Ludlow are the only place in the public land system descriptions where the ranges run north–south and township lines were run east–west on the magnetic meridian without cross ties to adjoining township lines. In effect, the corners set were at mile intervals on north–south lines but not on east–west lines. The settlers were to pay for the survey of all east–west lines. Symmes "jumped the gun" and sold land in areas where he did not get title, and these sales were eventually perfected when Congress sold the lands, for a second time, to these preemptors. Needless to say, gross distortions resulted. Eventually, the Ohio Supreme Court ruled that the original corners held regardless of distortion. No field notes are known to exist, and most of the original plats were destroyed at the time the Symmes house burned. Needless to say, this purchase was so badly managed and so poorly surveyed that Congress had second thoughts about making other large land grants. This purchase probably had more to do with the need for improving the quality of field surveying than any other single purchase. The government surveyors were required to tie into these poorly executed surveys in order to complete the survey of the lands between the Miami rivers. In retracing original lines in this area, the basic rule that would later be adopted by Congress was that the lines actually run, regardless of how poorly, controls.

7.6 VIRGINIA MILITARY DISTRICT

During the Revolutionary War, Virginia gave land warrants to soldiers and officers who served in its militia. In 1784, Virginia gave up its claim to vast areas of western land but reserved an area in Ohio to satisfy land warrant obligations (see Figure 7.2) if there were not sufficient land in Kentucky to fulfill the needs. This area was located between the Scioto and Little Miami rivers. Under the Virginia system, settlers were permitted to select any vacant parcel that was unclaimed, utilizing the original metes and bounds method of location. No thought was given to tying these surveys together, and today when a retracing surveyor attempts to survey in this area, "gaps, gores, overlaps, and vacancies" can be found. A survey was then ordered encompassing the number of acres to which a veteran was entitled. Because no restrictions were placed on the shape of the parcel, in the normal expected selection, parcels were laid out with the intention of including the best tillable lands. Under this system, the

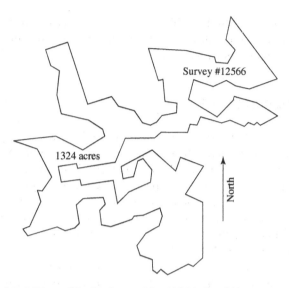

Figure 7.3 Virginia Military District Survey No. 12566. See Ohio State Auditors Records, Virginia Military District Surveys, vol. D, p. 146.

valleys and fertile flatlands were the first to be located, and less desirable lands were claimed later.

The method of surveys was indiscriminate. Usually, no effort was made to tie a new survey to existing land parcels or prior surveys; thus, at times it became impossible to determine the extent of gaps, gores, or overlaps. Although the land offices and surveyors were under the supervision of the Commonwealth of Virginia, the US Board of Treasury and later the GLO issued the patents to lands. Descriptions and resulting patents within the Virginia Military District may be long, complicated, and confusing. As an example of an old parchment patent made to fulfill three warrants purchased by William McDonald from various assignees (*Ohio Land Subdivisions*, vol. 111, C. E. Sherman, Ohio State University, Columbus, Ohio, p. 23), see Figure 7.3.

Congress put the Virginia Military District to rest once and for all on February 18, 1871 (16 Stat. 116), when it granted all unclaimed and unsurveyed lands in the boundaries of the Virginia Military District to the state of Ohio, without benefit of survey. At that time, it was estimated that 76,000 acres waited to be claimed by Virginia colonials who served in the war. In the resurvey of the Virginia Military District lands, it is important to remember that titles were issued in sequence with senior and junior deed considerations. Also, boundary and adverse right claims are adjudicated in accordance with Ohio court rules rather than those of Virginia or the federal government.

7.7 US MILITARY TRACT

For the benefit of veterans of the Revolutionary War, a small area of Ohio public lands, now known as the US Military Tract, was set aside by the Act of May 18, 1796, and

June 1, 1796. According to the act, 50 quarter townships (2.5 miles by 2.5 miles, or 4000 acres) should be divided "by the secretary of the treasury, upon the respective plats thereof, as returned by the surveyor general, into as many lots of 100 acres each as shall be equal, as nearly as may be, to the quantity such quarter township or fractions is stated to contain," and there were to be 160 perches (40 chains) by 100 perches (25 chains) "reserved for satisfying warrants granted to individuals for their military services." Today, some of these original land grants to the colonial veterans may still be outstanding.

7.8 CONNECTICUT WESTERN RESERVE AND FIRELANDS

Connecticut gave up its claims to western lands but reserved a tract of land in the Ohio area for future conveyancing. In 1796, the land was divided into 5-mile-square townships running north–south and east–west. Townships were divided into irregular lots of various acreage and size, generally of rectangular shape. Although these lands were surveyed into 5-mile townships, several transfers and agreements between the Continental Congress and the state of Connecticut transpired; these records are on file in Hartford, Connecticut, and the deeds are on file in Ohio.

Adjoining the western boundary of the reserve were the Fire Lands. During the Revolutionary War, the British burned several Connecticut towns (New Haven, Greenwich, Norwalk, Fairfield, and New London), and the Connecticut legislature set aside 500,000 acres for compensation for the victims. These lands were surveyed into 5-mile squares and further subdivided into four quarter townships, each containing 4000 acres.

7.9 MORAVIAN TRACTS

Moravian tracts are an example of the federal government's policy of granting small tracts of land for specific purposes or to specific individuals. As early as 1785, Congress recognized the work of early Moravian missionaries with the Indians, and in 1796, Congress directed that an area of reserve lands existing around each of three Indian settlements in the Military District be granted to the church for town lots. Such lands were to be held by the church in trust for the Indians, and the lines of the tracts and lots were to fit occupation, improvements or buildings, and fields. The survey was done in 1824 by Joseph Wampler, who took into consideration the preemption of settlers who held land by title or possession, causing parcels to be irregular in shape.

7.10 FLORIDA KEYS SURVEY

As the Florida surveys were extended from the baseline, the commissioner of the GLO realized that the rectangular system could not be extended to that portion of Florida known as the "Keys" and its adjacent islands. Hence, Congress enacted the act of June

28, 1848, nine Stat. 242, which permitted the commissioner to survey "in such a mode and manner" as he saw fit. In 1848, the commissioner issued "Special Instructions" to the US Coast Survey (USCS) to conduct surveys for the public domain in the Florida Keys. The initial plan was to tie the USCS triangulation net into existing township surveys in south Florida and then extend these into the Keys. Where section lines fell on an island, a meander corner would be set. If a section corner or quarter corner fell on land, it would be established and set, and all islands were to be meandered. In addition, plats were to be made and field notes written, all in accordance with the existing law for sectionalized lands. But it did not work that way. The Keys were mapped, and control was established, but very few corners, either section or meander, were established. The USCS was more interested in control than in land subdivisions; it did not seem to understand the distinction between a control map and a subdivision plat for land conveyancing.

7.11 DONATION LAND CLAIMS

Donation land claims were authorized in Florida, New Mexico, and Oregon in 1842. The surveys were intended to describe regular plots, but ended as irregular parcels with cardinal lines, starting with DLC No. 37. They were conducted by Bureau of Land Management (BLM) (GLO) or Forest Service surveyors acting under special instructions. Because the surveys often followed lines of possession, they constituted what would be called a metes and bounds survey. Corners were monumented, lines were run, plats were prepared, and a patent was issued after approval by the GLO. The resurvey of these parcels usually followed the rules for resurvey of sequentially created parcels.

7.12 EXCHANGE SURVEYS AND THEIR STATUS

Public law permits the US Forest Service and other federal agencies to exchange federal lands under their administration for lands in private ownership. The objective is to consolidate the public lands to facilitate management objectives. Initially, exchanges were on the basis of equal area for equal area, but in recent years it has been equal value for equal value. Early exchange surveys were made by Forest Service surveyors under special instructions from the BLM, but more recent surveys have been conducted by the Bureau. It was intended that surveys would be by aliquot parts; however, experience and actual surveys have produced irregular parcels that may be referenced to the GLO system but in effect are irregularly shaped parcels.

Principle 1. *Public domain lands may be exchanged for lands in private ownership. Once the exchange is effective and deeds exchanged between the private individual and the federal agency, those private lands that were exchanged for public domain lands attain the dignity of public domain lands and must be treated as such.*

7.13   PRIOR LAND GRANTS FROM FOREIGN GOVERNMENTS

In some states, particularly Louisiana, Florida, New Mexico, Arizona, and California, the federal government approved numerous land grants that had been granted by sovereign powers that controlled the land before it became public domain. France, Spain, and Mexico had granted extensive parcels to individuals while the lands were still under their control. Once the various treaties were executed, provisions were made to protect the bona fide rights of the purported owners of the lands. At various times, a special court of claims was established to investigate each claim. Not all were verified by the court, but once a claim had been approved, the individual grant was surveyed, and a verifying patent was issued to the original owner. Contrary to the law, many of these parcels were called sections, but some are also referred to as lots. In working in these areas, one would have to look at the survey to see what was done.

Principle 2.   *To retrace the boundaries of foreign land grants, the decree and survey ordered by the court are conclusive. The surveyor does not look for facts behind the decree.*

The majority of Spanish or Mexican land grants were located in California, New Mexico, and Arizona. The majority of the Florida land grants were from Spain. Others originated from English grants. In most instances, the original grant was surveyed by metes and bounds, monumented, and platted as a condition of the issuance of a patent by the United States.

Principle 3.   *Land grants usually have senior standing or rights over public domain sectionalized land surveys, and any encroachments of sections on a land grant usually are resolved in favor of the senior land grant.*

7.14   FRENCH GRANTS IN THE LOUISIANA PURCHASE

In many areas settled by the French (Michigan, New Hampshire, Maine, Ohio, and states along the Mississippi), there are many land grants from France made prior to the Louisiana Purchase of 1803. The Act of March 3, 1811, provided for their surveying. As French grant rights came into being prior to development of the sectionalized land system, they have senior standing relative to sections of land. The majority of grants were made in the state of Louisiana, and details of these grants will be described.

Along the Mississippi River and its connecting bayous, grants were usually 1 arpent river frontage by 40 arpents deep. Actually, an arpent is a measure of area. The arpent frontage unit became the length of the side of a square arpent. After the Louisiana Purchase, the federal government permitted many grantees to enlarge their original arpent grants by the purchase of additional lands, not to exceed a total of 80 arpents deep (16,697 feet or 3.16 miles in depth). Also, those in possession were permitted to claim the land they were occupying (see Figure 7.4).

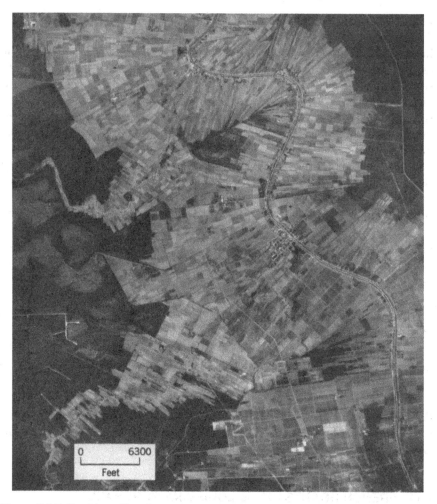

Figure 7.4 In Louisiana, French arpent grants along rivers were common. Small squares are cultivation areas; the main radial lines extending all the way back are original grant boundaries. Most cities and houses are built near the river where the natural levee gives suitable foundations. The back area is swampy. (Source: Courtesy of Edgar Tobin Aerial Surveys, San Antonio, Texas.)

Today, in many parishes of Louisiana, the original grants are recorded in French, along with resurvey data. An excellent historical account of these enlargements is on file in the State Land Office in Baton Rouge. Among the records is this letter from Thos. Freeman to the deputy surveyors:

Surveyors Office Washington M. T.

June. 1811

Instructions to Deputy Surveyors for surveying the public lands adjoining navigable streams, Lakes, Bayous, etc. in the Orleans Territory under the 2^d Sec^t. of the Act of the 3^d of March, 1811.

The Surveyor should first take an accurate survey of the margin of the water course so far as surveys of tracts are to be extended thereon. Then lay down his survey on the large scale; and draw thereon right lines in the direction of the general course of the water course, and on these lines lay off the lengths of fronts of tracts 58 Rods or 14.50 (Ch. Lks.) chains and through these points draw right lines at right angles to the line first drawn which shall extend back from the margin of the water course 465 Rods or 116.25 (Ch. Lks.) chains and close his tract by drawing a back line at right angles to his side lines, or parallel to his first line, when the water course happens to be straight or nearly so, the back line of one tract may be extended so as to become the back line to several adjoining tracts.

The side lines of the same tract will frequently be of different lengths and insect the dimensions & contents of these tracts, and will frequently differ from each other, but that cannot be avoided. The law in this case can very rarely be accurately complied with in consequence of the very great irregularities of the water courses. Any unavailable and unimportant deviations from the Law should be in favor of the neatness and convenience of the survey. On large curves or bends of the water course the side lines of tracts should be drawn converging, or diverging, as the case may require. Converging on the concave side to prevent the tracts from interfering with each other and diverging, on the convex side to avoid the inconvenience of small angular vacancies which would remain between the tracts if the side lines were drawn parallel to each other; In these instances the back lines may be drawn at right angles to one of the side lines and at 465 Rods from the margin of the water course, or front, which will necessarily make all the tracts having diverging side lines something larger than required. The tracts having converging sides would contain less, but the fronts of these should be extended so that the lines may include the quantity required or nearly so. It appears to be the object of Government to attach the fronts on the water courses all the lands from thence within the distance prescribed 465 Rods.

In short we may suppose any lake, bayou, water course, etc. to be circumscribed by lines drawn at 465 rods from the general course of its margin, and that the space included between these lines and the water course is to be divided into tracts of 58 Rods front and 465 Rods in depth. It should be the first object of the artist whose duty it becomes to designate those tracts to take an accurate survey of the water course, to lay that survey down on a large scale, and to divide the space as above into tracts as nearly conformable to the law as possible; To draw his lines on his Map both side and back lines noting their respective courses and distances; Thus prepared the surveyor can with great ease and accuracy transfer these lines of tracts from his plan to the ground and complete his survey agreeably to the intention of the Law and wish of the Government.

The first and principal object of the surveyor should be to have his lines accurately run, distinctly marked, and the contents of his survey correctly ascertained. It is much more desirable both to the Government, and purchaser, that the lines of a tract of land should be plainly designated amid its contents correctly determined, than that its dimensions should be precisely a given number of chains and links. Should one tract occupy a larger or better front on a water course than another, it will be more valuable, and consequently sell for more than the other.

The surveys contemplated by the 5th section of the Act above mentioned will be so very few in number if any for 2 years to come, that it appears almost unnecessary to say anything on that subject at this time, should application be made by any of the owners of front tracts, to have a back tract surveyed adjoining him, and only on application of proprietors of front tracts are these surveys to be made. The law is very plain on that subject. The front tract should not exceed 40 arpents in depth to Entitle its owner to an adjoining back tract: The back tract is not to exceed 40 arpents in depth, nor to contain a quantity greater than the front tract, neither shall the back tract in any instance extend so far in depth as to include lands fit for cultivation on another water course, etc.

The only difficulty that can arise in marking these surveys is when, by reasons of bends in the River, lake, etc., bordering on and in rear of front and adjoining tracts; each claimant cannot obtain a tract equal in quantity to his front tract; in that case the vacant land, in the rear is to be divided between the claimants in the direct ratio of the quantity contained in their respective front tracts.

Should the parties not consent to abide by the decision you may deem proper to make, you will send me a correct statement of the case, and I will make a division of these lands which shall be carried into Effect.

The Law does not point out any mode of marking or numbering the tracts to be surveyed under the 2^d Sect. of the Act, so that they may be distinctly known from each other, this appears to be a defect which if not remedied will be productive of great trouble and inconvenience both to the Register & purchaser.

The following manner of marking those tracts is recommended. Let the tracts be numerically numbered from some well known point or land mark, such as a Bluff, the junction of some Bayou or water course, or the intersection of some of the meridians or parallels already run. Thus lot No. 1, 2, 3, etc. as it may be above or below the land mark (naming it) and on the right or left of the water course as it may be situated. The surveyor should set a strong squared picket in the side of each track near the margin of the water course noting its distance there from. These pickets should be set firmly in the ground and numbered on each side with a marking Iron the number of the adjacent tract; the course and distance from this picket should be taken to a tree if convenient on each tract, which tree is to be numbered with No. of the tract on which it stands; a picket should in like manner be set in the ground at the termination of the side lines and the bearing and distance taken to two trees which shall be marked and No.^{d} as above. The whole is to be carefully noted in his field Book.

When the Surveyor shall find it impracticable from the interposition of lakes or impenetrable swamps etc. to extend his side lines their full extent required and to complete his tract by running the back lines he should set a picket at the termination of his side line which picket is to be marked with the No. of the adjoining lots and courses and distances taken from it to two trees which shall also be marked and numbered as in the first instance.

Should the depth required (465 Rods) extend so near another water course as to interfere with, or include lands fit for cultivation on it. Fronts of tracts should be laid off on both water courses, and the lands between these water courses should be equally divided between the fronts on each. Provided the distance between the fronts or water courses be two miles or nearly so.

These surveys will frequently lie between private claims or tracts already surveyed, in that case the front or tracts may be numbered from one of those former surveys to the other; and should a fraction remain it should be annexed to the adjoining tract without running the side line between them.

I must here repeat a request I made when I first wrote to you on this subject. To report to the office as early as possible the number of tracts that can and ought to be surveyed either under the 2^d Section of the Act above mentioned or in the usual way into Townships and Sections and what No. of the latter description are already prepared for sale. Taking it as a positive order that no new surveys of either description shall be made this year but such as are immediately saleable, or will be certain to be sold in a reasonable time.

The enclosed diagram exhibits the mode recommended for laying off tracts on water courses etc. under the 2^d Sect. of the Act above mentioned. The Red Lines represent the general course of the River etc. on which the fronts of the tracts are laid down. The courses and distances of the side and back lines can very readily be ascertained on the Map; and from thence with care and accuracy be transferred to the ground.

Any difficulty that may arise to you in the execution of this or any other part of your duty as connected with this office you will from time to time communicate to me and I shall with pleasure give you my advice and assistance thereon.

I am respectfully

Your Obt. Servant

Thos. Freeman

Gideon Fritz Esgn.

PD Surveyors

7.15 MISSISSIPPI TOWNSHIPS

In an area bordered by the Mississippi River and the Homochitto River in the state of Mississippi lay a number of abnormal or anomalous townships created because of prior (preemptive) occupancy and prior rights. Land was designated for survey by the GLO after the extinguishment of Indian rights and the questionable claims of former residents of Georgia in the Yazoo land frauds. A few French grants were also present. On July 25, 1803, in a letter of instructions to Isaac Briggs (the surveyor of lands south of Tennessee), Albert Gallatin wrote in part:

Sir,

Although by my letter of the 8th April, ult. you were generally instructed to divide the whole of the two tracts to which the Indian title has been extinguished into Ranges & Townships, yet, as that mode may present some difficulties which might induce you to prefer another, I have thought it necessary to repeat particularly that a deviation from that plan would be attended with great confusion in the several offices connected with yours and particularly in this Department. But, although the whole country shall without exception be divided into regular Ranges & Townships. The lands for which Certificates shall have been granted by the Commissioners must nevertheless be surveyed in conformity to such Certificates. It will only result that a person will often have a part of his tract in one and the remainder in another Township & that the tract will be returned by you as consisting of two or more Sections lying contiguous but in different Townships. But as a single patent will nevertheless issue for the whole tract, neither inconvenience nor additional expense will be experienced by the party. The outlines of all of the Townships must, however, be surveyed at the expense of the United States, though running amongst lands for the expense of surveying which Individuals must pay, and ...

The misapplication of the word *section* to identify prior rights was most unfortunate; it is not identical in meaning to *section* as used in regular township plats. One such township had two section 47s and ultimately had a total of 53 sections. Many individuals, both surveyors and attorneys, cannot understand why some Mississippi townships contain more than 100 sections, have two sections with the same number, or have the placement of one section inside another. In later surveys the word *tract* was used to identify land located in accordance with prior rights. In retracing lines within this area, the surveyor must cast aside the accepted rules of GLO survey, for they simply do not apply.

7.16   SOLDIER'S ADDITIONAL HOMESTEAD

During the Civil War, Congress enacted the Homestead Act, which provided free land to citizens, soldiers, and other veterans. A second act in 1872 provided additional land to the same veterans, and the act of March 3, 1909, extended these provisions into Alaska. In most instances, homesteading was done in surveyed sectionalized lands.

7.17   INDIAN ALLOTMENT SURVEYS

Indian allotments were surveyed under various instructions, laws, and provisions. Although they were intended to be rectangular subdivisions of sections, the rectangular form did not align itself with the cardinal points or directions. Perhaps the consistency of these allotments is their inconsistency. Before a surveyor attempts to retrace any of these allotments, he or she should understand their origin; some were created simultaneously and some were created in sequence.

7.18   NATIONAL FOREST HOMESTEAD ENTRY

President Theodore Roosevelt withdrew millions of acres of land from the public domain to form the nucleus of the National Forest System, and later other presidents followed his example. Once these lands were placed into the National Forest System, they were no longer available for homesteading. The act of June 11, 1906, provided for homesteads within the National Forest boundaries of lands that were more suitable for homestead than for forestry purposes. Most of these surveys were nonsectionalized within established townships, with only exterior lines run.

7.19 TENNESSEE TOWNSHIPS

Perhaps the least known federal townships are located in Tennessee near Cleveland and close to the Georgia border. Although originally surveyed by US deputy surveyors, titles to these lands came from Tennessee. Tennessee was the western area of North Carolina; thus, laws passed in North Carolina applied to Tennessee as well. In 1783, North Carolina enacted legislation to provide western lands (now in Tennessee) for the veterans of the Revolutionary War. As part of the act of April 18, 1783, the Cherokee Indians were assured that their lands in what is now Tennessee "are hereby reserved unto the Cherokee Indians and their Nation forever, anything herein to the contrary notwithstanding ... and be it further enacted that no person shall enter and survey any land within the bounds set apart for the said Cherokee Indians, under the penalty of fifty pounds specie.... All such entries and grants, if any should be made, shall be utterly void." By an act of November 2, 1789, the state of North Carolina parted with both the legal title and sovereignty of the lands described, with one exception: sovereignty of the Indian territory. On April 2, 1790, President George Washington accepted the Tennessee lands, and in 1796 Tennessee became a state. Through these acts and treaties, the United States gained title to the Indian lands.

In 1836, Tennessee enacted "an Act to provide for the survey of lands ceded to the United States by the Cherokee nation of Indians within the State of Tennessee by the Treaty of 23rd day of May, 1836." The act provided for a name (the Ocoee District) and a Surveyor General who "(1st) shall establish a basis line, extending from a point on the bank of the Hiwassee River, opposite the termination, on said river, of the main street in the town of Calhoun, South twenty degrees West, to the South boundary line (to Georgia), and (2nd) he shall run out the whole Ocoee District on each side of the basis line into ranges of six miles in width, by means of lines run parallel to the basis line, which ranges shall be distinguished by progressive numbers, East and West of said basis line, and (3rd) he shall then divide the said ranges by lines running at right angles with said basis line into townships six miles square, and (4th) in running the said basis line he shall establish on it corners for townships at distances of every six miles, and corners for Sections at the end of every mile, and quarter section corners at the end of every half mile, and (7th section) the whole of said townships shall be subdivided into sections, containing as near as practicable six hundred and forty acres each," and (8th to 13th sections) provide for numbering sections, marking corners, fractional townships, marking and measuring lines, topographic notes, and school sections.

Although this was a state act, the federal government sent deputy surveyors into Tennessee to survey their public lands. The surveys were conducted, monuments set, and plats prepared. The federal government, for reasons unknown, abandoned these lands upon completion of the survey and apparently turned over all of the notes and plats to the state. The state then granted patents to individuals according to the "government survey." Today, township plats exist of the surveys but no field notes have ever been found. Yet references to sections, townships, and ranges are still used in preparing descriptions for deeds.

Today, none of the plats or field notes provided for in the act exist in any public office, either in Tennessee or in the National Archives. According to a title attorney's report, the state had the authority to issue original titles; the book of original entries disappeared from the Land Office before 1911. Since some of the records were destroyed, stolen, or mutilated during the Civil War, and the general plan of the district of Ocoee was removed to the office of the Land Commissioner of the State of Tennessee at Nashville, little is of public record for Tennessee lands.

The retracement surveyor practicing in the old Ocoee District is plagued by lack of information. Original grants, surveys, maps, and plats are not in existence. For generations, land has been conveyed according to "aliquot portions of sections," and everyone knew where the possession lines were. In recent years, with the subdivision of the original sections into smaller and smaller parcels, surveyors have resorted to metes and bounds descriptions to supplement the record. Even when an aliquot portion of a section is surveyed, it is reduced to a metes and bounds description.

7.20 FLORIDA: FORBES COMPANY PURCHASE SURVEYS

Immediately south and west of Tallahassee, Florida's capital, there are nearly 2 million acres of land acquired by the Forbes Company through negotiations with Spain. Although this land was subdivided into parcels called townships and sections, the survey was not performed with the sanction or under the authority of the federal government. Whatever retracement rules Florida courts approve for resurveys are applicable. The *Manual for the Survey of the Public Domain* has no legal standing in retracing boundaries in this area.

Three men—Patton, Leslie, and Forbes—all left Georgia in 1776 as Tories sympathetic to the British cause and settled in Spanish Florida as traders. Initially, they sought a grant of land from the United States in south Alabama, but this was denied. Later, a grant was sought from Spain as compensation for extensive debts owed to the Forbes Company by the Creek and Seminole Indians. The Spanish king and 22 Seminole chiefs agreed to "cede, concede, give, sell, and transfer the aforesaid district of land" bounded east by the Apalachicola and Wakulla rivers, by the Gulf of Mexico, including the offshore islands to the south, and an undefined line to the north. In 1811, a second extensive land grant was made.

Asa Hartsfield, a surveyor from South Carolina, was engaged to locate the first metes and bounds survey from the original description. After he located the boundary, he then surveyed a portion on the Wakulla River into "sections 60 chains on a side and in a northeasterly direction." Prior to the War of 1812, the British were permitted to build a fort on the land and in 1816 the fort was destroyed by an American expedition.

On May 29, 1819, while the United States and Spain were negotiating the sale of Florida to the United States, the Forbes Company sold the land at $13 per acre to a trusted US citizen who would be acceptable to the United States. Since the treaty between Spain and the United States did not go into effect until 1821, the company sought to have the grant "legitimized" by Congress. In 1824, while Congress was still debating, a second survey was conducted on the Little River with the "sections" 40 chains wide east and west and 80 chains wide north and south. Congress refused to "legitimize" and forced a decision through the courts. In 1834, the Supreme Court verified the claim, except for an area near St. Marks (a Colonial fort), which was ordered to be surveyed by the surveyor general of Florida into "the United States Land Office's range and township system."

Once the grant was approved, the Forbes Company established the Apalachicola Land Company to sell land based on the early surveyor's field notes, including (1) the Hartsfield survey of 1802 of 31,065 acres with its sections 60 chains square and lines running N 17½W, (2) the exterior boundary survey of Daniel Blue in 1811, (3) an 1821 survey by Samuel Brown near Sweetwater Creek, and (4) an 1822 magnetic north survey by Brown and McBride on the Little River with sections of 800 acres.

In 1838, W. R. Hopkins was engaged to survey the remainder of the land into the US township and range system, using similar monumentation, marking bearing trees, noting natural features, and so on. The work was finished by Melver and Williams in 1856. The surveys included a resurvey of some of the initial townships, and in some instances two plats are on record. A close examination would make a surveyor believe that the plats and survey were an "official" GLO survey, which they are not. The surveys are no more than a private subdivision prepared by a private individual; the federal government never had an interest in them, and as such, the rules of retracement do not apply within these borders.

Copies of most of the original plats are available from the official records of the state of Florida; most are very abbreviated and contain little information. No individual or agency has an entire collection of the original field notes. In 1964 or 1965, in an old warehouse, a surveyor discovered a box of musty, leather-bound survey books containing a portion of the original field surveys. Today, most of these lands are within the boundaries of the Apalachicola National Forest and lands owned by the St. Joe Paper Co. Not all sectionalized land surveys originated with the federal government!

7.21 GEORGIA LOT SYSTEM

Although Georgia was one of the original 13 states, and it is considered to be one of the metes and bounds states, over half of the state was surveyed with rectangular parcels prevailing. Very early in its statehood, the "headright" system of land granting was abandoned in favor of a rectangular survey prior to disposal. The state was divided into districts, and the size of parcels in each district was based on the carrying capacity of the soil in relation to the needs of a farm family. Sandy areas required larger parcels. In all, there were six acts creating parcels of various sizes and with some differences in survey requirements. In general, the surveyors were required to survey with a half chain (2 poles), mark lines, and mark corner trees, all at a rate of $2.75 a mile under the supervision of the surveyor general. All surveyors, deputies, chain carriers, axmen, and others were required to take an oath to use the utmost skill at all times. The land was then distributed by placing the description in a barrel and a person pulling by random the number of "lots" to which he was entitled.

Original notes and plats are on record in the Surveyor General's Office in Atlanta. As with many of the early states, few present-day surveyors in Georgia have recovered any of the original survey evidence. No one was charged with the responsibility of perpetuating the system; thus, most of the evidence was lost with time. From 1924 to 1927, a federal agency conducted extensive retracements of hundreds of original lots first surveyed in the 1830s. New monuments were set and new bearing trees were noted, all based on the best evidence available at the time. A surveyor practicing within the lot area of Georgia must not only know surveying, but because of the manner in which each district was surveyed, he must also know the history of county development. Figures 7.5–7.8 depict the survey history of a "typical" Georgia land lot and its resulting patent. Figure 7.5 is a reproduction of one page of a survey of a 20-chain lot of 40 acres, also known as a *gold lot*. Figure 7.6 is a single page from the field book of a survey of a 40-chain (160-acre) lot, Figure 7.7 is a portion of the final plat, and Figure 7.8 is a reproduction of the patent of a 160-acre land lot in the district shown in Figure 7.6.

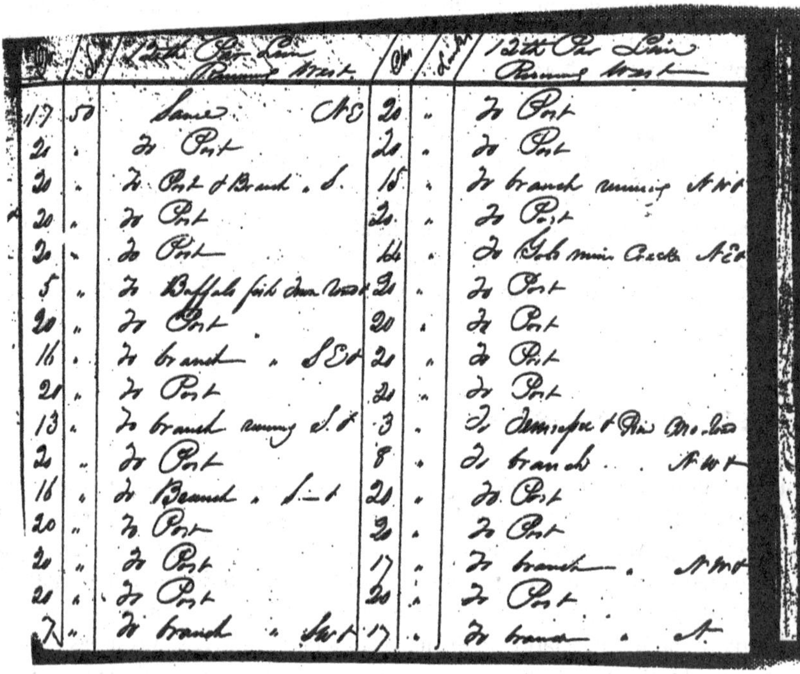

Figure 7.5 Reproduction of a page of a field book of a district with 40-acre land lots. Note that all distances are given to the nearest full chain, there is no identification of the species of corner post, and there are no calls for line trees.

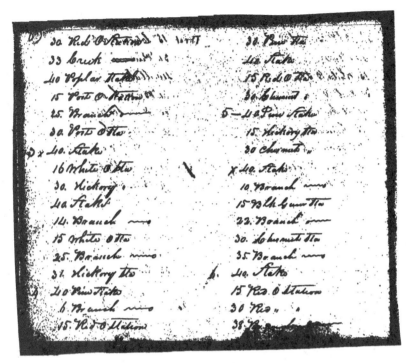

Figure 7.6 Page from the field notes for a 160-acre land lot. Note calls for 3 × = the third parallel line; 30 Red O Station (or Sta) means a station tree on line at 30 chains from the previous corner. Usually, these station trees were scribed with the lot number. 40 Pine Stake 5 a pine state was set at the corner.

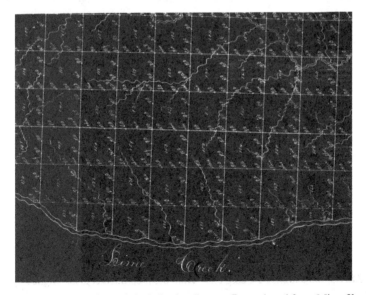

Figure 7.7 Small portion of an original district. Source: Reproduced from Microfilm.

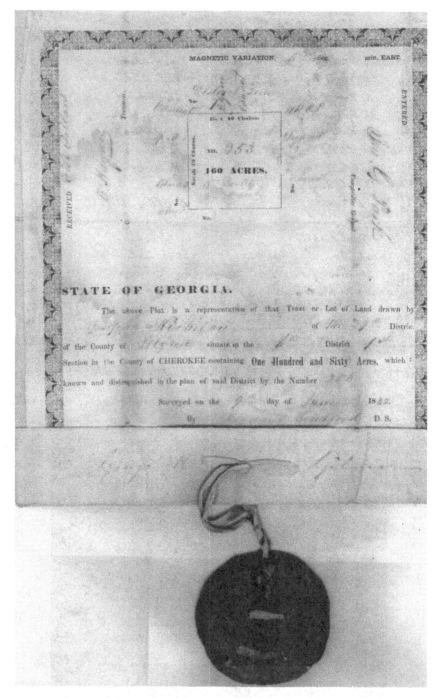

Figure 7.8 Reproduction of a Georgia lot. Note the preprinted courses. Signed by Governor Gilmer.

The retracing surveyor must understand the abbreviations used by the surveyor and/or draftsperson who created the plat and the surveyor who created the original boundaries:

P.O. = Post Oak line tree or Post Oak station

B.J. = Black Jack line tree or station tree

B.J.P. = Black Jack corner post (This is probably the worst wood a surveyor could use as a corner post.)

P.O.P. = Post Oak corner post (It does *not* mean *poplar corner*, as some attorneys and surveyors believe. If the notes were correlated to the plat, this would become quite evident.)

Today, the retracing surveyor should consult both the plat and the field notes of both the original survey and any subsequent surveys.

In many states that have similar situations, the retracing surveyor must assemble a sufficient amount of information as to the creation of the boundaries that are being retraced to be able to make sound decisions based on sufficient evidence. Like all other surveys that were conducted under private or state authority, there was little or no standardization of the field notes. Each deputy surveyor was permitted to keep the notes for his particular surveys in just about any manner that he wished. Figure 7.6 depicts pages of field notes from two different surveyors. Note the absence of identification of lot numbers and similar details. In retracing these original notes and lines, the modern-day surveyor must, in many instances, sit down and plot each individual line, and a correlation must be made to the map of the district. This one effort may take hours of reading the notes, trying to decipher them, and then fitting each line together.

7.22   LAND TENURE SYSTEMS OF TEXAS

The state of Texas has one of the most interesting and diverse land tenure systems in the United States. When determining property ownership within Texas, it is important to ask two questions: first, at what date was the parcel originally created? And second, in what geographic area of Texas does the parcel lie?

The first question concerning the date of the parcel's inception is important because Texas has been ruled by six sovereigns over the past 200 years. While the French originally "found" Texas, the state has also been ruled by Spain, Mexico, the Republic of Texas, and the Confederacy, in addition to the United States government. Depending on what sovereign oversaw the creation of this parcel, there may be special rights for the owners or special rules under which the retracing surveyor should conduct himself. As an example, if a waterfront property was created under the auspices of Spain, the freshwater boundary determination process may be different than simply locating the "gradient boundary."

The second question concerning the geographic area of Texas in which the parcel exists is also a very important determinant to make before beginning the boundary retracement process. For example, much of East Texas simply employs the metes and bounds method of describing property, very similar to the Eastern United States. That is for good reason: Eastern Texas was settled before the Tejanos moved west. On the other hand, in Western Texas, land is often divided into "blocks and sections." This blocks and sections system mimics the Public Land Survey System in principle. But there can be as many as 100 sections in a single block—something that would make a GLO surveyor blush.

In conclusion, any out-of-state surveyor practicing within Texas should spend ample amounts of time learning about the unique land tenure system administered within the Lone Star state and also work with a local surveyor to familiarize himself with the Texas and county-specific land knowledge before practicing on his own.

SURVEYS IN THE NONCONTINENTAL UNITED STATES

7.23 GENERAL COMMENTS

The United States of America has a state, Hawaii, a commonwealth, Puerto Rico, and several territories, including Guam and American Samoa, that are islands and are not attached to the mainland. Each of these islands has a unique land tenure and land survey history separate and independent from the mainland. Special requirements are expected of those who seek to practice in each of these areas. An applicant for the land surveyor examination in Hawaii is expected to translate a land description from the native Hawaiian language, as found in deeds and surveys. In Puerto Rico, reading and understanding Spanish deeds and surveys are an absolute necessity.

7.24 HAWAIIAN LAND LAWS

At the time of the death of King Kamehameha I in 1819, the king owned all lands. No written Hawaiian language existed, no missionaries were present, and no English language was used. By the time the Land Commission was established and subsequently dissolved in 1855, English was the second language of the government and business, and the Supreme Court of Hawaii had rendered its first decision eight years earlier.[1] Thus, Hawaii became "modernized" in 36 years. A statement in *Keelikolani* v. *Robinson* should be remembered: "Hawaii's land laws are unique in that they are based on ancient tradition, custom, practice and usage."[2]

Hawaiian Land Titles

Kamehameha I owned all of the land in the kingdom, whether under the sea or above it.[3] Although Kamehameha I owned all lands, his people required the fruits of this land to live. Kamehameha did divide certain parcels with his warrior chiefs and a very few select others. The chiefs in turn subdivided their individual parcels to lesser chiefs and to others in their own society for service. Then each of these persons in turn subdivided his parcel until the least common denominator was obtained. As far as can be ascertained, all the lands were included in some form of description; the basic major subdivisions were by natural features, but at times the divisions were only known to and pointed out by the elders of villages. Unique descriptions existed, such as the treeline of a particular species, a grass type, or the habitat of a bird.

The basic land holding was the *ahupuaa*, "so called because the boundary was marked by a heap (*ahu*) of stones surmounted by an image of a pig (*puaa*), or because other tribute was laid on the altar as a tax to the chief."[4] The *ahupuaa* was further divided into *ilis*. Whereas the *ahupuaa* could be from 100 to 200,000 acres, the *ili* could be any area. An *ili kupono* consisted of several unconnected parcels of land that might be located randomly throughout the *ahupuaa*. The individual parcels that constituted the *ili* were further called *lele* or "jumps."[5] All of these specific parcels were described in one way or another, either by calls for specific bounds or by personal knowledge.

Hawaiian Law and Kamaaina

Hawaiian law is unique and is predicated on English common law of boundaries, with a Hawaiian flavor. Hawaiian courts have considered such elements as expert testimony on boundaries (*kamaaina*), survey evidence, the meanings of words and phrases, adverse possession, riparian rights, and the seaward boundary of measurements. Hawaiian courts have long recognized the necessity of using "experts," called *kamaaina*, who have been acquainted with the property in general from childhood. In the holding of the case of "In Re Boundaries of Pulehunui," the courts relied on the testimony of ancients to fix the boundaries of a 16,000-acre parcel on Maui that was known by name only.[6] Since 1879, *kamaaina* testimony has been relied upon in many instances and in many situations. In 1980, the Supreme Court, in *Haiku Plantation* v. *Lano*,[7] used *kamaaina* testimony to determine whether an area was used for ingress and egress or was used by the natives for a parking area.

The Great Mahele of 1848

Under the king and the subsequent chiefs of Hawaii, the idea of fee simple ownership of land was unknown. In 1847, the Privy Council appointed a committee to formulate rules to identify clear property titles. The suggestion was to provide for a *mahele* or land division, consisting of a series of quitclaim documents starting at the top with the king and extending downward through the chiefs (*konohiki*) to the individual landowners. The releases were recorded in the *Mahele Book*, with the king and the specific chiefs releasing each particular right or interest. Although the process extended over a period of 39 days, the courts in *Harris* v. *Carter*[8] added: " ... although these maheles were executed day after day until the work was completed, it was because it was too great a task to be all completed in one day, and they might well have been dated on the same day. It was one act." In other words, the *mahele* was a simultaneous act; deeds were not created in sequence, and proportional rights exist. Today, the *Mahele Book* is held in the same reverence in Hawaii as is the *Doomsday* (or *Domesday*) *Book* in England. The original is preserved in the offices of the Division of Archives, State of Hawaii, in Honolulu.

Hawaiian Land Court

Hawaii operates a land court, which was established by constitution. Upon application and after meeting the requirements for survey, platting, and research, Hawaii will insure title to and register a property.

Hawaiian Surveys

Because of the large number of land divisions that were being made in a short period of time during the *mahele*, no parcel was surveyed. In 1852, the Land Commission permitted the chiefs and *konohikis* to obtain title by the name of the parcel only. It was not until 1862 that it became forbidden to obtain title by patent without a survey or the boundaries being defined,[9] and then the survey cost was to be at the expense of the tenants.

The original *ahupuaa* grants may or may not have been made without benefit of survey, and if a survey was made, the grant contained *"aka koe nae na kuleana of kanaka maloko"* or "excepting, however, the kuleanas of the natives."[10] Surveys conducted at the time of the *mahele* were imprecise because of the instruments and methods used. With the destruction of monuments, resurveys are difficult to perform.

The surveys of Hawaii have generated unique terms reflecting island conditions. On each island is usually a high peak. The term *mauka* has come to mean "north," although it more frequently meant "as toward the mountain." *Makai* means "toward the sea."

Because of the method used to identify lands, usually without benefit of survey, the court findings differ from those of other states. In *Ookala Sugar Co.* v. *Wilson*,[11] the court stated: "It is also settled that although as a rule natural monuments control artificial monuments and the latter control courses and they in turn control distances and these control areas, yet any of these may control any others where that appears to be the intentions as gathered from the entire grant and its application to the land, for the object is to ascertain the intentions, and monuments and courses and distances for instance as a rule control area because they are as a rule more certain and there is a less likelihood of a mistake being made as to them, but where this reason fails the rule itself fails." In the *Application of Sing Chong*[12] is stated: " ... when the application of the stated course and distance of the fourth side of a parcel fails to close the boundary, then the specified area may be relied upon to determine the appropriate course and distance of the fourth side."

Water Boundaries

Over the years, the location of the seaward boundary of land parcels has varied considerably. At one time, surveyors placed the seaward boundary at the *limu*, or seaweed line.[13] In the decision rendered *In re Ashford*,[14] the court redefined the seaward boundary for Hawaii. The location by one surveyor was at mean high water, which was represented by a contour placed at the mean high-tide plane of the sea as located from US Coast and Geodetic Survey information. The state argued that the *ma ke kai* (along the sea) was along the vegetation line or the line of debris left by the wash of the waves some 30 feet *mauka* (away from the sea).

The court observed: "It is not solely a question for a modern-day surveyor to determine boundaries in a manner completely oblivious to the knowledge and the intentions of the king and old-time *kamaainas* who knew the history and the names of the various lands and monuments thereof." Then, with tongue in cheek, it stated: "Cases from other jurisdictions cannot be used in determining the intentions of the King in 1899. We do not find that the data or information published and contained in the publications of the U.S. Coast and Geodetic Survey were relied upon by the kamaainas for the purpose of locating seaward boundaries ... All of the matters contained in such publications were unknown to ancient Hawaiians and foreign to the determination of boundaries in Hawaii. Property rights are determined by the law in existence at the time such rights were vested." The vegetation and debris line now control.

In *State* v. *Zimring*,[15] the question of a lava addition to ocean parcels was addressed. One individual, whose frontage was cut off by the lava, claimed the lava, based on the doctrine of riparian rights. Without much law on its side, the court held that the land belonged to the state. In an effort to rectify the problem, the legislature enacted a law that would grant any lava additions to the frontage owners. The bill was vetoed and did not receive a sufficient majority to override the veto.

Adverse Possession

Hawaii recognizes adverse possession as an unwritten means of creating new title in landowners (unlike other title registration systems). Whereas on the mainland the collection and presentation of evidence relative to adverse possession provides a preponderance of the evidence, in *Redfern* v. *Kuhia*[16] Hawaii raises that requirement to a higher burden of proof, stating: " ... the law presumes that possession by the adverse possessor is in subordination to that of the rightful owner. Thus, title by adverse possession must be established by clear and positive proof." The elements required in court are actual, hostile, open, notorious, and exclusive. As on the mainland, the surveyor in adverse possession cases is merely a collector of evidence for the attorney; he or she does not rule on the law.

7.25 PUERTO RICAN LAND SURVEYS

Puerto Rico has land boundary laws pertaining to titles, registration, and reservations that are not applicable on the US mainland. Although most private land titles were derived from Spain, some were derived from original Crown lands that were ceded to the United States by the King of Spain at the termination of the Spanish–American War at the Treaty of Paris on December 10, 1898.

Old Spanish Surveys

The Division of Public Lands and Archives has on file original copies of the field notes and parchment paper plats of many of the Spanish surveys conducted prior to 1898. These documents, written in fine Spanish script with waterproof ink, date back to the early 1830s. In analyzing them, evidence indicates that the early bearings were according to the magnetic meridian and distances were slope distances.

At the time of the early surveys, most parcels of land were quite large; subsequently, they were subdivided and re-subdivided into smaller and smaller tracts or parcels. Like many early surveys of the mainland, the works contain errors, but these original data references must be examined and are absolutely indispensable when attempting to resurvey or build a modern land configuration map.

Modern Land Subdivisions In Puerto Rico

Rural lands in Puerto Rico are subdivided into irregularly shaped parcels or tracts of various sizes, seldom square or rectangular. Most parcels are described by metes and bounds, with a limited amount of basic information for the retracement surveyor. *Hitos* (corners) and *lineas* (lines) may be described by reputation and in all probability by natural features, including streams, trails, fences, or ridges, along with a brief statement as to what type of tree or material the *hito* is composed of. Today, most distances are in meters with bearings or azimuths in degrees or grads (azimuths with angles between 360° and 400° usually indicates the use of grads). Areas of individual parcels are usually in hectares and cuerdas, either one or the other or both. Many government agencies purchase land by the acre but also indicate the cuerda area (1 cuerda = 0.9712 acre).

Prior to the early 1920s, lands were known by name rather than by a land parcel number. Individual ownerships (called *fincas* or *granjas*) were and still are known by a name given by an early owner, often to reflect a nearby natural feature. The actual land description often contains information as to the present owner, the new owner, the ward or barrio, the municipality, the area in cuerdas or hectares, names of adjoiners, and the original owner.

Natural Watercourses

Watercourses in Puerto Rico are identified by the poorly defined terms *rios* (rivers) and *quebradas* (brooks). If the waterway is defined as a river, the public owns the bed; if defined as a brook, the public does not own the bed. If the survey is for a river, the boundary line is run on the bank on each side (high-water line) and the people own the bed. If a river runs through a parcel, the surveyor locates the boundary of each bank and subtracts the area of the river from the total holdings. On the mainland, disputes arise over whether a waterway is navigable or nonnavigable; here the dispute is over whether a watercourse is a rio or a quebrada.

Land Parcel Research

In rural areas, land parcel research is difficult, and at times it is necessary to reconstruct the lines of division of the parent property. In the area of Toro Negro, dozens of parcels are identified as belonging to "Figueroa." Added to this difficulty is the custom of dividing the same parcel among numerous heirs, and the heirs taking the mother's name followed by the father's name, or the use of *de* by some married women attached to their husband's name. All of these customs, applied in a nonuniform manner, add interest, intrigue, and quandary to legal research. Titles of tracts that can be clearly traced from very early Spanish times to the present are known as *Dominion Titles*.

Survey Authority

Modern-day *agrimensores* (land surveyors) in Puerto Rico may receive a degree in land surveying at the university and may become qualified after an examination in Spanish. The Commonwealth of Puerto Rico has elevated the profession of land surveyor to one of special importance through its educational and licensing requirements.

Preparation of Deeds and Descriptions

To prepare a modern deed, a complete traverse is required, with a plat showing numbered corner identifications, boundary measurements, watercourses, roads, any topographic features forming a boundary, and adjoiner names. Unlike on the mainland, in many areas all abutting landowners must be notified in advance of the survey and are often present during the survey. In addition, any landowner may file a written objection to the survey.

In Puerto Rico, only a licensed public *notario*, who must also be an *abogado* (attorney), can prepare deeds and titles to land parcels. After preparation of a document, he or she completes registration, transmits a certified copy to the client, and retains the original in his or her official records. Only a notario may swear to documents.

Registration of Land Titles

Registration of property is maintained in many places on the island, and deed transfers are recorded in books of registry using prepared descriptions. A geographic record of parcels does not exist; grantee–grantor indexes are used. Surveyors may spend considerable time researching an adequate description from which to survey. Registries are well run and in most cases can be equated with the average county clerk or register of deeds office on the mainland.

7.26 FEDERAL MINERAL SURVEYS: GENERAL COMMENTS

Minerals were one consideration of Congress when on May 20, 1785, it reserved "... one-third part of all gold, silver, lead, and copper mines, to be sold, or otherwise dispensed of as Congress shall hereafter direct." In the early years of mineral development, land holdings were based on the Spanish concept of possession. In the absence of statutory mining laws and any form of law enforcement, each mining camp enacted its own laws, with excellent results. Enforcement was by ropes, guns, and clubs.

Once a claim was abandoned by a failure to work it actively, it could be re-staked. As territories became states, legislation was enacted to protect the rights of individuals.

Under federal mining law, a mining claim is a potential right to fee title to the land and any minerals contained in the area described. In addition, for lode claims, the claimant has a right to follow the claimed mineral vein through any public domain land that was open to mineral claims as of the date of discovery. Mineral laws recognize two types of mineral claims: lode and placer. A *lode* claim is for a mineral deposit, such us gold, silver, or mercury, located in a vein of rock in place. The word *placer* includes all forms of deposits except veins of lodes or rock in place, and generally means minerals in a loose state, such as gold in gravel located at or near the surface. A placer may be hundreds of feet below the surface, such as a petroleum or gold deposit found in a former riverbed that is now covered with lava. Certain types of placer deposits—petroleum, phosphate, potash, sodium, borax, oil shale, and sulfur lands—may now be obtained only by lease arrangements. A placer claim is limited to the surface area described and its vertical extension downward. A lode claim may follow its vein through any public domain land open for location as of the date of discovery.

7.27 WATER AND MINERAL RIGHT LAWS

In the operation of placer claims, large amounts of water were often needed. In the West, according to Spanish law, water could be appropriated by usage, and, in general, water rights evolved from usage. In some mining districts, local laws were enacted by miners to regulate water usage. On July 26, 1866, Congress passed "An Act Granting Right of Way to Ditch and Canal Owners over Public Land, and for other purposes" (14 Stat. 251). The act states that "the mineral lands of the public domain, both surveyed and unsurveyed, are free and open to exploration and occupation." Section 3 of the same act provides: "It shall be the duty of the surveyor-general, upon application of the party (mining claimant), to survey the premises and make a plat thereof." In southern California, runoff waters of practically all watersheds have been appropriated and claimed by communities for domestic and irrigation water supplies, and many creeks or rivers have been dammed.

One of the very specialized areas of boundary creation and surveys is the area of mineral lodes or mineral rights. Minerals, one of the bundle of rights, has a unique place in boundaries in that, in most instances, different people own the surface rights and the mineral rights. The mineral interests may or may not have the same boundaries as do the surface rights. In some instances, the surveyor may be required to identify boundaries that are thousands of feet below the surface boundaries.

7.28 LAND OPEN TO APPROPRIATION OF MINERALS

Lands indicated as being held by the United States for disposal under the land laws are open to mineral location. Lands specifically withdrawn, such as national parks, national monuments, military reservations, and Indian lands, are not subject to location. Minerals found within a national forest are open to location, provided that the discovery is such that it would justify an ordinary prudent person's expenditure of time and effort in developing a paying mine. Without the existence of commercial value, mineral claims within a national forest are not valid locations. Because each state owns the bed of meandered lakes, the bed of navigable streams, and land between high- and low-water marks, these lands are not open to location under federal mining laws.

States whose land never belonged to the United States, and states specifically exempted from federal mining laws, are not open to location under federal laws. These states are Virginia, North Carolina, South Carolina, Pennsylvania, Rhode Island, New York, New Hampshire, New Jersey, Massachusetts, Maryland, Georgia, Delaware, Connecticut, Maine, Vermont, Kentucky, Alabama, Kansas, Illinois, Indiana, Michigan, Minnesota, Missouri, Ohio, Oklahoma, Wisconsin, Texas, and Hawaii.

7.29 VEINS, LODES, OR LEDGES

The terms *veins, lodes,* and *ledges,* although probably having different shades of meaning, are used synonymously by the Mining Act and the courts. A vein is a continuous body of ore or mineral-bearing substance found on or within the Earth, having depth, width, and length, and being bounded on each side by country rock. It need not be straight or uniform, but it must be identifiable by vision or assay. The term *lode* often includes more than one vein. Tilted sedimental beds containing ores deposited by streams, and not by mineralizing agencies, come within the meaning of lode, vein, or ledge.

A vein or lode is continuous if it can be traced through the country rock. Ore found in disconnected pockets or cavities does not constitute a vein, but occasional pockets, connected by ore, constitute a vein. A vein may curve, vary in dip or strike, or divide into branches, which may or may not reunite. A *broad lode* or *zone* is one having the same genetic origin and lying within clearly defined boundaries. The broad zone may be fractured or faulted. If a vein is traceable for a great distance and has definite walls, it is treated as a separate vein even though many veins are grouped close together. An extensive area, such as a limestone deposit, cannot be treated as a vein, nor can a broad metalliferous zone.

A vein has an apex, strike, dip, and walls. The apex is the summit or top of a vein or lode at or below the surface. The strike or course of a vein is the direction of the vein across the land; the dip is the downward direction. A bearing is normally given to define the direction of the strike. Dip is measured by the downward angle formed with the horizon. The same vein may have variable dip or strike throughout its extent. The wall of a vein or lode is the line of demarcation between the vein and the country rock. The *hanging wall* and the *foot wall* (the wall walked on) may be of different or the same material.

7.30 EXTRALATERAL AND INTRALIMITAL RIGHTS

An *extralateral right* is the right to follow the downward course of a vein, with certain limitations, to the end of its depth (see Figures 7.9–7.12). The limitations are: (1) the apex of the vein must be within the boundaries of the claim; (2) only the dip or downward course may be followed; (3) the direction followed must be within the limits of the end lines of the claim extended vertically and outward; and (4) the vein must

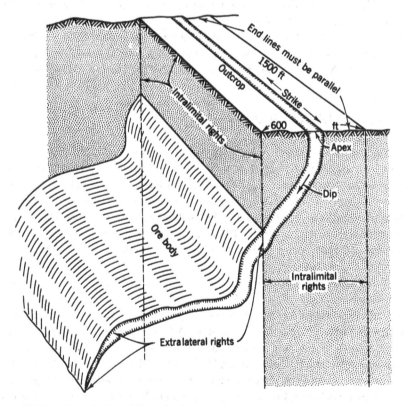

Figure 7.9 Extralateral rights follow the downward course of a vein.

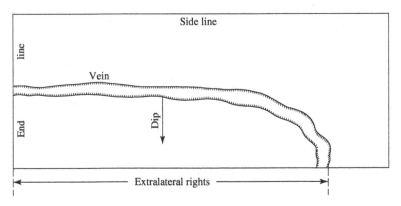

Figure 7.10 Only the downward course may be followed.

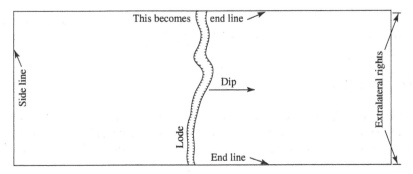

Figure 7.11 Direction must be within the limits of the end lines of the claim.

travel through land open to location. The vein may be followed, whether irregular, curved, waving, faulted, broken, or brecciated, as long as the vein is continuous with the same mineralizing agent. Extralateral rights are not lost where the vein is found to run crosswise instead of lengthwise to the claim; the normal side lines become the end lines, and the end lines become the side lines (see Figure 7.12).

Intralimital rights to a claim extend downward within the limits of the claim to the center of the Earth. All surface rights and everything within the limits of the owner's intralimital rights, except the extralateral rights attached to other veins apexing in another's claim, belong to the owner of the claim. The owner of another vein has the right to follow the vein through the claim of another, but such rights are limited to that extent. There are no rights to explore in another's land, nor are there rights to approach the vein from any other location than the vein itself. Country rock adjacent to the vein and within the limits of another's claim may be cut where the vein is too narrow or too crooked to mine. All the ore within the intersection of intersecting veins belongs to the senior claimant, but a right of way through the intersection is ensured. The issuing of a patent to a homesteader does not invalidate the extralateral rights

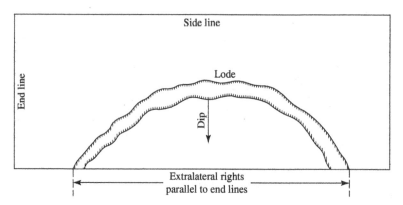

Figure 7.12 Extralateral rights are not lost where the vein runs crosswise to the claim.

of a prior vein or lode claimant. The date of discovery of minerals determines the extralateral rights of a claimant, not the date of patent.

7.31 MILL SITES

In addition to lode and placer claims, mill sites of 5 acres or less may be claimed by either a vein or a lode proprietor or by a mill owner for the purpose of processing minerals. Mill site locations are limited to mineral lands that do not contain valuable minerals.

7.32 TUNNEL LOCATIONS

Tunnel sites are acquired in accordance with local rules and customs but may not exceed 3000 feet, as imposed by federal law. When a lode is discovered within a tunnel, the owner is called upon to make a surface location of the vein or lode as required by law. Discontinuing operations for six months constitutes abandonment of all rights to the veins along the line of the tunnel. At times, it is technically difficult to superimpose the surface boundaries in the location as the subsurface.

7.33 SIZE OF CLAIMS

Lode claims are limited to not more than 300 feet in width on each side of the center-line of a vein and not more than 1500 feet in length. End lines must be parallel with one another. The number of claims that a person may file is not limited. Although lode claims may overlap one another for the purpose of determining extralateral rights, the overlapped portion normally belongs to the senior locator. Local legislation can be passed to limit the size of a claim to less than 600 by 1500 feet, but such limitations cannot be less than 25 feet on each side of the middle of the vein.

Placer claims are limited in size to 20 acres, or 160 acres for an association of not fewer than eight people. Each placer claim, whether upon surveyed or unsurveyed lands, is required to conform as nearly as possible to the sectionalized system of surveys.

7.34 DISCOVERY

Discovery of minerals occurs when there is reasonable evidence that there is a vein or lode containing valuable minerals or that there are valuable minerals in a placer deposit. Discovery cannot be based on conjecture or the mere possibility of locating minerals. If a reasonable prospect of success in developing a paying mine exists and sufficient evidence exists to justify a reasonably prudent person in making expenditures in money and labor, discovery is justified. Minerals in paying quantities at the time of location are not necessary, but there must be reasonable hope and assurance that there will be such minerals. The amount and manner of the discovery work are not stated in the federal law. Local laws often require posting a notice of discovery and sinking a shaft of a minimum depth or length.

7.35 LOCATIONS

In a broad sense, there are four types of locations: lode or vein, placer, tunnel, and mill site. There is a limitation on the size of a location, but there is no limit on the number of locations. Discovery of minerals usually precedes location, but if the order is reversed and there are no intervening rights, it is immaterial that the statutory order is reversed, since location becomes effective upon discovery. The claim must be marked upon the ground as an indispensable condition of location and its boundaries identified. This does not mean that each corner of a claim must be monumented, only that the location must be marked so that the boundaries may readily be traced. Thus, if an unnavigable stream subject to location is marked by a notice on the bank stating the boundaries and limits of the claim, the claim is sufficiently marked.

Specific requirements by revised statutes, Sec. 2324 (30 U.S.C. Sec. 28), are that "the location must be distinctly marked upon the ground so that its boundaries can be readily traced" and "all records of mining claims made after May 10, 1872, shall contain the name or names of the locators, the date of the location, and such a description of the claim or claims located, by reference to some natural object or permanent monument, as will identify the claim."

There may be a conflict between the mineral survey boundaries and any GLO boundaries if they have been previously created.

Many state statutes allow a period of time after discovery before it becomes mandatory to make a location. This gives the claimant time to explore and determine the direction of the strike of the vein. During such time, the position of the claim may be laid in any direction from the discovery point, along the vein, a distance not exceeding 1500 feet. The monumenting of the boundaries of a placer claim location is not vital since the boundaries are usually described by the public survey system. Because lode claims are irregular in shape and size, monumenting of the boundaries is an essential part of final location. If there are any local laws in addition to the federal laws concerning monumenting, they must be adhered to. The caveats that a locator must adhere to are that he or she (1) must not stake a claim on lands that have been withdrawn from mineral entry by the BLM, (2) must not locate or stake a claim on a prior existing claim that has been patented to a third party, and (3) must not stake a discovery point on a vein location already staked by and in the possession of a prior locator.

7.36 POSSESSION

Exclusive possession follows location. Title of the locator, prior to discovery, is good against all except the paramount owner, the US government. Title of the locator followed by discovery in sufficient quantities to be profitable is good against all, including the United States. Location without discovery does not confer a right to the minerals that might exist within a claim, but it does confer a right of possession as long as the claimant occupies the land to the exclusion of others and in good faith pursues the work of trying to discover minerals.

If it is determined that a description is ambiguous, continuous ancient possession, couples with this ambiguous title, can be used to provide the *key* to cure the ambiguity. The more ancient possession, coupled with the ability to trace the ambiguous description, may cure the problem of location. This information of possession attached to the historical history of property may solve the problem.

A good example presented itself when the surveyor was asked to locate a parcel of land that was originally described by four adjoining parcels. The most recent landowner purchased the parcel based on the original description. The surveyor, understanding the historical process of the original method of conveyancing, *livery of seisin*, suggested to the attorney that the possession was the *key* to locating the boundaries of the original parcel. The degree of proof was of the lowest standard, that is convincing or credible, and could carry the burden of proof. This nexus of history, possession, and a 200-year-old description all came together to locate the parcel.

7.37 ANNUAL EXPENDITURES

The Mining Act provides that a minimum of $100 of labor or improvements for the purpose of development shall be expended annually upon each claim until a patent is issued. The amount required can be increased by local rule. If claims are contiguous, but not just touching on one corner, all of the annual expenditure of $100 per claim can be spent on any one of the claims.

7.38 REQUIREMENTS FOR PATENT

Application for patent to a lode mining claim must be accompanied by a request and deposit for survey along with a certified copy of the location notice. No survey is required for a placer claim described by the sectionalized system. An order for a survey will be issued, and the survey will be administered by the government office cadastral engineer. Prior to patent, at least $500 must be expended upon each mineral claim.[17] Only labor and improvements having a direct relationship to the development of the mine, such as shafts, or tunnels, are included within the $500. Buildings, machinery, roadways, trails, and the like are excluded except where it is clearly shown that they are associated with actual excavations. Mills and equipment to transport ores are not included. If the application for patent embraces several locations held in common and adjacent to one another, the expenditures spent on one claim can be prorated among all the claims.

7.39 UNITED STATES MINERAL SURVEYORS

Federal mineral surveyors are appointed by the government after the applicant proves by extensive tests that he or she is proficient in the use of surveying instruments and has proper knowledge of mining surveys and the pertinent laws. The particular mineral surveyor selected for the patent survey and the fee paid to the surveyor are arranged for by the applicant. Mineral surveyors are in private practice.

7.40 SURVEY OF THE CLAIM

All mineral surveys must be made by the mineral surveyor in person. No observations or measurements previously made by himself or by others may be accepted as true. Historically, transits with a solar attachment or other means of correctly determining the meridian and a steel tape graduated in feet have been used to make mineral surveys. Bearings are referred to the true meridian, and distances are returned as the shortest horizontal measurement in feet.

The position of the claim is fixed by determining the bearing and distance of the claim to the nearest section or quarter-section corner. If no public corner exists within 2 miles, a location monument is established on some prominent point having good visibility from every direction. Monuments set by the original locator are used as the true location of the claim. However, the size of the claim cannot exceed 1500 feet in length or 600 feet in width, nor can the end lines be other than straight and parallel. An error of closure in the survey must not exceed the statutory limit. In addition to determining the size of a claim and setting permanent boundary corners, the surveyor records topography, location of shafts and tunnels, location of the vein where it crosses the end lines, conflicts with adjoining claims, and location of springs, salt licks, and mill sites.

Legal lode mineral claims have the right to follow a vein through any public land open for location as of the date of discovery. If public domain land is patented after the date of discovery and location of a nearby lode mining claim, it has a cloud on its title. If the mineral vein claimed lies under such patented land, the mineral can be mined without the surface owner's permission. In many states, a subdivider must own the mineral right to land before the land can be subdivided. A mineral claim, until patented, is a partial right, because title to the land is retained by the government.

7.41 CONCLUSIONS

Principle 4.   *In many areas of the country, local land survey and description systems can be found, and surveyors practicing in those geographic areas or areas of specialty must, by necessity, have a firm knowledge of the unique system or systems that are present in the respective state or area of practice, the county, and even the local community relative to the boundary creation or retracement at hand.*

The average surveyor may never have the opportunity to conduct surveys or retracements in some of the discussed areas, so these are included to expand the knowledge one should have.

After reading this chapter, the student can see the importance of being versatile and knowledgeable about the basic land boundary systems in his or her geographic area of work. The original public domain was vast, and the early years were the years of testing as to how to proceed and what was needed to define the boundaries adequately. As we have seen, although there were laws in effect, the federal government was under no obligation to follow its own laws. Thus, a myriad of systems were in place, which places on the retracing surveyor today a professional responsibility to understand what was done.

Of course, there was no responsibility for the respective states to follow the methods and laws of the federal government, and they didn't. They acted independently and established new systems or accepted or expanded existing systems. We have much more ancient systems in Hawaii, Puerto Rico, and perhaps other areas of the country than were in place when the GLO system was being initiated.

In this chapter, we have included a number of survey situations that may be parochial in nature, yet any surveyor who is registered is presumed to be capable of performing in these specialized areas. There is no way in which a textbook can instill the necessity of having knowledge and experience in these areas before an original survey or retracement survey is undertaken.

Principle 5.   *In many of the areas in which private and quasi-government surveys are present, the ability and possibility of finding original evidence may be very difficult, if not virtually impossible, and any surveyor practicing in these areas must be especially cautious.*

Perhaps this comment is overly broad in that such a statement may lead the reader to believe that there is no hope to adequately retrace the original lines of such nonsectionalized boundaries. This is not so. However, the retracing surveyor should look at each individual retracement as a separate and distinct opportunity for record research, a field examination searching for evidence indicated in the original survey, and then an evaluation of this evidence to formulate an opinion of where the original lines that are described are actually located. Of course, this is only the surveyor's opinion, and it will always be subject to collateral attack by other surveyors. There are certain absolutes that many of the nonsectionalized surveys are lacking in original notes or plats because they either were not created, were lost, or are woefully lacking in information.

The best suggestion one can leave with the surveyor who undertakes a retracement of a federal or state nonsectionalized boundary is: Search, search, and search the records and the field. If it is there, find it. If not, be able to explain why. Never leave any stone unturned, either in the office or in the field.

RECOMMENDED READING

White, C.A. (ed.) (1983). *A History of the Rectangular Survey System*. Washington, DC: U.S. Department of the Interior, Bureau of Land Management.

Burke, T.A. (1987). *Ohio Lands: A Short History*. Columbus, OH: Auditor, State of Ohio.

NOTES

1. *Wood* v. *Smart* (Haw, 1847).
2. *Keelikolani* v. *Robinson*, 2 H 514 (1862).
3. *Hawaii* v. *Liliuokalani*, 14 H 88 (1902).
4. *Hawaiian-English Dictionary*, 2nd ed. (Honolulu: University of Hawaii Press, Honolulu, 1986).
5. *Horner* v. *Kumuliili*, 10 H 174 (1895).
6. *In re Boundaries of Pulehunui*, 4 H 239 (1879).
7. *Haiku Plantation* v. *Lano*, 618 P.2d 312 (1980).
8. *Harris* v. *Carter*, 6 H 195 (1877).
9. *Hawaii* v. *Midkiff*, 49 H 456 (1966).
10. *Re Kakaako*, 30 H 666 (1928).
11. *Ookala Sugar Co.* v. *Wilson*, 13 H 127 (1900).
12. Application of Sing Chong, 1 H Ct. App. 236 (1980).
13. *County of Hawaii* v. *Sotomura*, 517 P.2d 57 (1973).
14. *In re Ashford*, 50 H 314 (1968).
15. *State* v. *Zimring*, 58 H 106 (1977).
16. *Redfern* v. *Kuhia*, 53 H 378 (1972).
17. R.S. Sec. 2325, 30 U.S.C. Sec. 29.

CHAPTER 8

LOCATING EASEMENTS AND REVERSIONS

8.1 INTRODUCTION

In Chapter 1—History and Concept of Boundaries, it was shown that land is quite unique in that it is composed of multiple interests that may be likened to a "bundle of straws or rights" that are being held together by what we refer to as title. Each of these rights or interests has its own boundaries, which may overlap other rights or interests sometimes held by other people.

In the scheme of boundary issues, the aspects of easement boundaries play an important part in the surveyor's workday.

One of these individual rights or interests is that of easements for numerous purposes. As depicted earlier, an easement is what English law has defined as an incorporeal *hereditament*, or nonpossessory estate, or only a right to use an entire parcel of land or a portion of the parent parcel for a specific purpose or purposes.

Easements, in general, are unique, in that by their very nature they are an estate in land that does not and will not permit possession of a person's property. It does give the easement holder the right to use the property up to and including the boundaries described in the respective documents.

For the most part, people look upon an easement only as the right of ingress or egress for road purposes, or, in lay terms, the right to *come and go* across a parcel of land. This may be the major purpose of easements, but this nonpossessory right has been used in a multitude of other ways. Many agencies, utilities, and individuals obtain nonpossessory rights for many reasons and purposes.

Brown's Boundary Control and Legal Principles, Eighth Edition.
Donald A. Wilson, C.A. "Tony" Nettleman III, and Walter G. Robillard.
© 2024 John Wiley & Sons, Inc. Published 2024 by John Wiley & Sons, Inc.

Easements were seldom mentioned in either early English common law or American law. Early Roman history indicates that the Romans had organized roads from the early fourth century BCE. Records indicating ownership or possessors along the Via Appia indicated that the Roman legions obtained rights of passage over private lands for movement of their armies. These same roads or highways[1] were also used by travelers going from Rome to Egypt and the Holy Land and to other portions of the Roman Empire. Maps have been recovered depicting a system of planned roads throughout the Roman Empire, including England. Today in Great Britain, one can examine the British Ordinance topographic quadrangle maps and visit abandoned Roman roads still visible and indicated on both ancient Roman maps and modern maps.

The Roman law relative to servitudes had a strong influence on the English law in this area of easements. In Roman law, servitudes were considered as a division of *jura in re aliena*, or real rights that one landowner may have in the property of another. The Romans considered these servitudes to be of two classes: personal and praedial. The *personal* were considered to be individual rights lasting, at most, the lifetime of the landowner; the *praedial* were rights that permitted one to use the adjoining land of another in certain ways.

Easements and their reversions are unique in the area of boundaries. When comparing litigation over boundaries of land parcels with litigation over easement problems, recent research indicates a 10:1 relationship, in that there is 10 times more litigation over easements than over land boundary problems. Personal research and observations of nearly 2000 appeal cases in the area of easements indicate that nearly 50% are reversed on appeal. A valid easement or right-of-way has boundaries that may be permanent or may be measured in a time frame. Easement boundaries are created in special ways not inconsistent with regular property boundaries, and their retracement may also cause the retracing surveyor serious problems. Along with the survey problems associated with easement boundary locations and retracements, a surveyor may be requested to assist in the determination of reversionary rights.

A *reversion* or *reversionary right* is a right defined by law that permits an adjacent landowner to either gain or regain certain property rights of land located within the vacated easement. In a vacated easement, old boundary lines are revived once the reversion takes place. A surveyor working with easements will find this area consuming a large amount of litigation in the courts, and it may become an area of conflict in the everyday surveying of a parcel of land and its boundaries.

The essential qualities of all true easements are as follows:

1. They are incorporeal and may or may not be inherited, depending on the nature.
2. They are imposed on corporeal property *(in rem)* and not against the person *(in personam)*.
3. They confer no rights to profits from the property.
4. They are imposed for the benefit of corporeal property, normally a second parcel, usually contiguous.
5. There must be two tenements, the *dominant*, to which the right attaches, and the *servient*, upon which the burden attaches.
6. Usually, a true easement is nonpossessive in that the holder of the easement may use it but not "possess" it.

The following principles are discussed in this chapter:

PRINCIPLE 1. A description of a land parcel may contain more than one boundary description within the boundaries of the parent parcel. The law recognizes that rights granted may be stacked one on the other.

PRINCIPLE 2. An easement is an incorporeal, nonpossessory right or hereditament of a party to use the property of another person for a specific purpose or purposes.

PRINCIPLE 3. The boundaries of an easement may be described in many ways. There is no single, absolute way to describe an easement.

PRINCIPLE 4. In an easement, a grantor can grant only those rights that the landowner has, and no more.

PRINCIPLE 5. Whenever an easement is granted for a specific purpose, the fee title to the land encumbered by the easement remains with the grantors of the easement, their heirs, or assigns.

PRINCIPLE 6. Between private parties there is a presumption that fee simple title is intended to be granted in a deed to real property unless it appears in the writings that a lesser estate was intended. Unless there is a statement to the contrary in the conveyance or a statute otherwise, the presumption is that the public acquires an easement only in highways.

PRINCIPLE 7. All easements have three basic descriptions that may have to be created or retraced.

PRINCIPLE 8. A surveyor may have to fully understand easement descriptions in order to locate and describe a new easement and to relocate existing easements.

PRINCIPLE 9. If there is a question of whether an easement is an incorporeal hereditament or a fee conveyance, this may be a question of fact and law, but it does not affect the location of the boundaries. That is a survey question.

PRINCIPLE 10. A metes and bounds description with a road as a boundary must be written to positively exclude the road; otherwise, in those cases where the grantor owns the bed of the road, it will be presumed that the conveyance is intended to convey title to the center of the road, subject only to the public easement.

PRINCIPLE 11. Ownership lines in a street are determined by the original ownership lines as they existed before the easement or dedication of the road. A private conveyance of land abutting on a road, the fee to which belongs to the adjoining proprietors, is interpreted, if possible, to pass fee title to the centerline thereof; otherwise, the contrary must be shown.

PRINCIPLE 12. Unless a deed or map that created the easement or dedication indicates otherwise, reversion rights extend from the street termini of the property lines to the centerline or thread of the street in a direction that is at right angles to the centerline of the street.

PRINCIPLE 13. Reversion rights of a lot in the middle of a block extend radially to the centerline of the street.

PRINCIPLE 14. Reversion rights of a lot adjoining a subdivision boundary extend along the boundary line of the subdivision and cannot extend beyond the boundary line.

PRINCIPLE 15. Reversion rights at an angle point in a road extend on the bisection of the angle.

PRINCIPLE 16. By law and by definition, an easement is a nonpossessory right to use the property of another person.

PRINCIPLE 17. The creating words become important when there is a question as to what was created, what was vacated, and what was reverted to the landowners.

PRINCIPLE 18. Easements boundaries may be either surface, subsurface, or aerial. The location may restrict the use by the servient titleholder.

8.2 RIGHTS GRANTED

Principle 1. *A description of a land parcel may contain more than one boundary description within the boundaries of the parent parcel. The law recognizes that rights granted may be stacked on another.*

A major problem lawyers, surveyors, and courts must address for solutions are situations where landowners may grant multiple easements over the same parcel and including the same property, with multiple descriptions. In conducting research for a retracement, the retracing surveyor may find easements for numerous purposes: surface easements for roads and power lines; subsurface for utility sewers and water lines; aerial for communication, wind, and sunlight. Some of these may be adequately described, and some may be for cross-country blanket pipelines. Each will carry with it a description of its boundaries; all may be conflicting with each other.

Principle 2. *An easement is an incorporeal, nonpossessory right or hereditament of a party to use the property of another person for a specific purpose or purposes.*

There are numerous definitions of easements. Easements have been described by courts in several ways:

A right of one owner of a parcel of land, by reason of such ownership, to use the land of another for a special purpose, not inconsistent with the general in the owner.[2]

Or

A privilege, service or convenience which one neighbor has of another, by prescription, grant, or necessary implication and without profit; has a way over his land, a gate-way, watercourse, and the like.[3]

Or

A liberty, privilege, or advantage without profit, which the owner of one parcel of land may have in the lands of another.[4]

Or

An interest in land created by grant or agreement, express or implied, which confers right upon holder thereof to some profit, benefit, dominion, enjoyment or lawful use out of or over estate of another; thus, holder of easement falls within scope of generic term "owner."[5]

A more academic definition is

a privilege which the owner of one adjacent tenement hath of another, existing in respect of their several tenements, by which that owner against whose tenement the privilege exists is obliged to suffer or not do something on or in regard to his own land for the advantage of him in whose land the privilege exists.[6]

To look at these definitions and to put them into ordinary words, an easement can be defined as follows: A true easement is a nonpossessory right of one landowner. (dominant parcel) to use the land of an adjacent landowner (servient parcel) for a specific purpose or purposes. All easements, regardless of what purpose or purposes the use is for, are just one of those straws that go to make up the "bundle of rights" identified in English common law. The owner of the easement has only a right to use the property, not to possess the property.[7]

Principle 3.   *The boundaries of an easement may be described in many ways. There is no single, absolute way to describe an easement.*

A surveyor who examined documents that purported to create easements would see a multitude of descriptions. In many instances, the landowners must resort to litigation to determine such questions as: Is an easement present? or Is it an easement? or What is the extent of the easement? If the parties find that ambiguities exist in an easement, such as the location or the extent or any other question, the courts will try to make the easement effective and will fill in the missing terms. That is not what the parties should seek. The description of the easement and its extent should be sufficient so that a court of law should not have to "fill in any blanks."

Parties will try to create easements by calling for easements described as "for road purposes" or an easement 40 feet wide across the SW ¼ of the SE ¼ of Section 10, or some other description. These *blanket easements* encumber the entire parcel until it is placed on the ground, and the location of the pipe or transmission line then fixes the location in place. With such a description, the dominant easement holder is permitted to place the utility at any location within the entire 40-acre parcel.

The most common form of easements are *centerline easements*, in which the parcel that is the subject of the easement is described by a single line, described by bearings and distances, and the easement is described as "an easement 25 in width," "25 feet on each side," or "25 feet on either side," perpendicular to the line described. In this situation, the extent of the easement may involve serious questions of legal interpretation as to any gaps or overlaps, for example, and possibly as to whether it is a 25-foot or 50-foot easement.

Perhaps the most positive description of an easement is that of a *perimeter easement*, in which the entire easement is described from a point of beginning to a point of commencement and with courses enclosing the entire parcel that will be subject to the easement. The description takes on all of the legal and technical aspects of any boundary line. Some people try to create an easement simply by referring to "an easement across the road on my property." The various forms of descriptions are discussed later.

Principle 4. *In an easement, a grantor can grant only those rights that the landowner has, and no more.*

This principle takes us back to the *bundle of rights*. That is, if the freehold does not have that one particular stick in the bundle of rights, the titleholder cannot grant or convey it. For instance, if the person has possession of the property under a life estate, the life estate holder cannot encumber the property beyond his or her tenure, that is, the person's lifetime. A life estate holder *can* encumber the entire parcel by granting an easement, but the easement terminates upon the death of the holder of the life estate. The person seeking the easement must obtain the easement from the life estate holder as well as the remainderman to have an easement that extends beyond the life of the party who granted it. If a party has less than 100% interest in a parcel, that person can grant an easement, but the other party or parties in interest can bring an action to have the easement declared void or seek compensation for their interest.

In many states, the minimum technical standards require surveyors to conduct sufficient research to be able to locate all recorded easements with certainty. This usually requires the surveyor to conduct research to a sufficient degree to be able to locate any and all lines. This requirement does not place any burden on the creating or retracing surveyor to determine the legality of the easement.

As might be expected, easement boundaries should be identified with certainty and with specific dimensions. In reading these documents, information may be revealed depicting the amount or type of interest the grantor had. Research will often indicate that the grantor had fee title by a warranty or quitclaim deed. It may also be found that the grantor was occupying the land under a life estate. In the latter situation, any easement granted can be valid only as long as the life estate is valid.

Principle 5. *Whenever an easement is granted for a specific purpose, the fee title to the land encumbered by the easement remains with the grantors of the easement, their heirs, or assigns.*

Although this principle is easy to apply, there are a number of land conveyances made after the original that may fail to make reference to the original easement. Questions may arise as to who has ownership of reversionary rights, fee rights, and various other rights. If the surveyor discovers such a document during the course of investigating the records, it is imperative that the client be notified. In the event that the holder of a life estate grants an easement, the grantee may find that a remainderman has a legal right to the easement rights, and they may have to be repurchased. In this situation, an easement is either in gross or appurtenant. In an easement appurtenant, the easement attaches to the land encumbered and transfers when the land is transferred. If it is an easement in gross, the easement is personal (i.e., to an individual), and the presumption is that the easement terminates, but if the easement is for commercial purposes, the easement transfers or may be assigned.

8.3 FEE TITLE OR EASEMENT RIGHT

Principle 6. *Between private parties there is a presumption that fee simple title is intended to be granted in a deed to real property unless it appears in the writings that a lesser estate was intended. Unless there is a statement to the contrary in the conveyance or a statute otherwise, the presumption is that the public acquires an easement only in highways.*

Between private parties, when the word *easement* is omitted from a description that was intended to be an easement, the courts may rule that a fee title passed. A description of "a 30-foot strip of land for road purposes" may pass fee title but may be subject to a road easement. The wording "reserving the west 25 feet for road purposes" may or may not create an easement. States are divided on this issue. The wording may create a reservation of a 25-foot fee title, but subject to an easement for a road. When an easement description is written, if in the future there is a possibility of confusion as to what the parties intended, the scriveners of the document could possibly eliminate any problems by using the habendum clause of the description to its fullest intended use. This clause is usually found at the end of the boundary description to clear up or explain any possible ambiguities that may exist in the present or in the future.

When a government agency takes an easement for highway purposes, an easement right is presumed.[8] However, some statutes may provide for the taking of a fee for land appropriated for street purposes. Some states provide that all dedicated streets created by plat result in a fee simple conveyance, and the public body retains fee title to the streets on vacation of the streets.[9]

Although surveyors should not give clients legal advice, they must know what their respective state laws are; for example, all interstate highways are fee conveyances, and in California, the state requires fee title to its freeways, and some states have passed maintenance statutes. The surveyor may find that precolonial law becomes important. In applying Dutch law to ownership in early New York grants, the surveyor will find that the beds of streets vested from the Dutch are owned in fee by the state of New York. When the Dutch vested power in the British, this ceased and did not apply.[10]

8.4 THREE EASEMENT DESCRIPTIONS AND THREE BOUNDARIES

Principle 7. *All easements have three basic descriptions that may have to be created or retraced.*

With any easement, three separate and distinct parcel descriptions and their boundaries are important to the identification and location of the easement. These three descriptions are as follows:

1. Description of the dominant parcel when one exists
2. Description of the servient parcel(s)
3. Description of the boundaries of the easement proper

As a matter of law, all states in the Union hold that if a description in a deed is so ambiguous as to preclude its being able to be located by a competent surveyor, that deed may be void for lack of certainty of description. There may occur a particular situation when a surveyor fails to prepare an adequate description for the easement or that one of the parcels is so ambiguous that it cannot be located with certainty; then the easement may be void.

8.5 OWNERSHIP OF THE BED OF EASEMENTS

Who is entitled to the land under an abandoned road, power line, or railroad? If the easement was a fee conveyance, there is no reversion, and the grantee of the original easement has ownership; but if the conveyance was considered an easement with reversionary rights, the original land taken for the easement reverts to the original grantor, his heirs, or assigns. Usually, reversionary rights are returned to the original owners and to the original boundaries that were in place at the time the easement was created.

In an actual situation, the state of Florida obtained an easement by prescription. Subsequently, to perfect the prescriptive easement, the state purchased, in fee, two 10-foot strips of land on each side of the prescriptive easement. Then oil was discovered. The original landowners sought the oil royalties as their right of ownership of the fee underlying the road. The court held that since the state owned two 10-foot strips of land, in fee, on each side of the road, upon abandonment, the prescriptive road would go to the adjacent fee owners, now the state, and as such the state was entitled to the mineral interests under the road.

The New Hampshire Supreme Court stated in the case of *Duchesnaye* v. *Silva* that in order to exclude the bed of a street (or highway), the declaration must be *clear and unequivocal.*[11]

8.6 SURVEYOR'S RESPONSIBILITY AS TO EASEMENTS

Principle 8. *A surveyor may have to fully understand easement descriptions in order to locate and describe a new easement and to relocate existing easements.*

It is accepted, by fact and law, that in order to have an effective easement that is created by grant, there must be a valid easement description. If it is determined that the easement description is defective, in that it does not adequately describe the easement lines, the easement may be void for lack of certainty of description. But it is strictly a question of law, to be determined by the court, whether the description is void. The surveyor can only determine whether the easement can be surveyed. Whether the deed contains a valid description is a question of fact, and the surveyor must determine whether it can or cannot be surveyed. The court is the only authority to determine whether the description is valid or void. This responsibility of the description is placed on the surveyor in two instances: when the surveyor conducts a survey to create and identify the boundaries of a new easement and when the surveyor is asked to locate the boundaries of an easement that is described in a prior document (under extreme conditions, the surveyor may be asked to survey possession lines for prescriptive easements).

8.7 REQUIREMENTS FOR LOCATING EASEMENTS

One of the responsibilities of both creating and retracing surveyors is the area of easements. As pointed out earlier, the multiple land interests have one common element; that is, they all have exterior boundaries that may complement other boundaries or conflict with them.

Since easements are an interest in land, like all other interests in land, there are certain minimum requirements or standards for locating them. The American Right of Way Association has a manual of practice and a set of standards, and several states have identified these standards as accepted in case law. These requirements are tantamount to performing to a standard of boundary line surveying. A court has ascertained that a border of a right-of-way is a boundary line like any other.[12] The respondent argued that requiring a land user to determine exactly where the right-of-way lines are located places too great a burden on the land user. The court responded by stating that a landowner or occupier is under an obligation to know the boundaries of the property. Many of these requirements are not mandatory for the surveyor to either create them or retrace the boundaries. As such, the landowner may be completely oblivious to the product that was received.

Since an easement boundary is like any other boundary, the surveyor's duty in a retracement is to locate the boundary where the original surveyor located it.

The Wisconsin court illustrated this very well, citing two cases. In the latter case, the court held that the corner of a city block involved in the determination of boundaries is where the original surveyor of the plat fixed it by the setting of a stake or other monument, regardless of where present surveys may fix the centerline of a street, and regardless of inaccuracies in his measurements judged by such a centerline. It cited an earlier case, wherein it was stated that "the east line of the street was where the original surveyor placed it, not where it should be according to resurveys or subsequent surveys; that subsequent surveys are worse than useless; and that they only serve to confuse, unless they agree with the original survey."[13]

In the California case of *People* v. *Covell*, the court dealt with a highway survey of 1872 done by the county surveyor and approved by the board of supervisors. A later survey was done by a deputy county surveyor, who concluded that one of the abutting parcels encroached anywhere from 6 to 10 feet along the line of the highway. The court stated, in a previous case, "[I]t was sought by a later survey to move the lines established by a prior survey. This court held that whether accurate or inaccurate, the original survey granting and establishing certain rights, fixed the rights not only of the government, but of the landowners, and that the government, after establishing such a line and granting and conveying certain rights, possessed no power thereafter to change the course of that line."[14]

Principle 9. *If there is a question of whether an easement is an incorporeal hereditament or a fee conveyance, this may be a question of fact and law, but it does not affect the location of the boundaries. That is a survey question.*

Easements are a question of fact and law. The meaning of and the legality of the words describing the easement are usually a question of law that is addressed by the court. The surveyor has no input into the legality of this area, but if there is a question of the surveying and descriptive aspects of the easement, which equates to a description and/or a boundary issue, which may be determined by the surveyor's interpretation of the words and then the placement of these descriptive words on the ground as a survey, then the surveyor becomes intimately important and the surveyor's opinion becomes important.

The court will determine the legality of the conveying document that created the easement, but when it comes to the locatability of and the location of the lines described in the description of the easement, this is strictly a survey or measurement question or a problem to be addressed by a surveyor.

8.8 CENTERLINE PRESUMPTION

Principle 10. *A metes and bounds description with a road as a boundary must be written to positively exclude the road; otherwise, in those cases where the grantor owns the bed of the road, it will be presumed that the conveyance intended to convey title to the center of the road, subject only to the public easement.*

When a surveyor writes a metes and bounds description that includes a road as a boundary, care should be taken to say what is intended. The interpretation of these descriptions creates a fruitful area for litigation. An owner of land bounded by a road or street is presumed to own the center of the way, but the contrary may be shown.

Such phrases as "along said road" do not exclude the road. "Along the east side of said road," "the side line of said road," "excepting the road," and "excluding the road" are definite statements showing clear intent to exclude the road. In one situation when a landowner reserved a parcel "for road purposes," it was held that the words did not create an easement. Unless the deed clearly states that the road is excluded, or clearly indicates so by its language, the conveyance is to the centerline of the road, provided that the grantor owns the bed of the road. Upon a reversion, questions may arise: What is the easement boundary line? Is it the paved surface or the ditch line? Reference to a stake in the side line of a road, along with a distance that extends only to the side line of the road, does not exclude conveyance to the centerline because stakes are normally set on the side line whether or not the conveyance is to the centerline. For the grant to extend to the centerline, the grantor must have title that extends to the centerline.

Problems may occur when a person who has sold all the land adjoining a road is not interested in the maintenance or improvement of the road and should not have title interest in the road.

When improvements are made on a road and assessments are made, the costs are normally paid by the adjoining owners; however, if title to the road is vested in a third person, questions of the legality of the assessments arise. To circumvent such problems, most states have statutes in accordance with the principle stated here. Also, the "strip and gore" doctrine may apply in determining who has title to the unusable strip. This area is quite complicated, and many individuals who work in it do not have a full understanding of the law. The surveyor is put on notice that when working in this area, he/she should "tread lightly."

Descriptions cite running to the centerline or going along the side line. Some states, including several in the Midwest, state that "going along the side line" *excludes* the bed of the street. Most states do not hold that rule, stating that the presumption is to go to the centerline, for a variety of very good reasons, not the least of which is access upon vacation of the easement. Even a call for the side line is insufficient to exclude the fee under the street. Most hold the rule as articulated by New Hampshire.

Using a centerline as the definitive measurement from which to measure the width of an easement can be a serious problem for the retracing surveyor. There are no laws that require that the agency locate the road in the center of the easement area. The only requirement is that the facility should not be located off the easement proper. In a situation where a surveyor used the center of a 50-foot cleared pipeline easement to locate the easement proper, it was later determined that the actual pipe was offset some 20 feet from the centerline.

8.9 CONVEYANCES WITH PRIVATE WAY BOUNDARIES

If a surveyor is asked to interpret a description with the boundary or boundaries of the lands described as being along a public or private way (road), the surveyor can presume that the intention of the grantor of the deed was to convey a fee title to the center of the way if the grantor's title extended that far.[15] This does not apply if the words in the conveyance show a different intention. Like all other presumptions, the reverse is followed in some states, in that the presumption is when land is bounded by a private way, the fee is limited to the side line.[16]

There is a class of easements created by federal statute—namely, Section Line Easements. These are found in a few GLO states, Montana, Alaska, and Oklahoma being three of them. These were created as easements, 50 or 66 feet wide on all section lines. This type of easement cannot be identified until the section lines are located, and thus the section lines become the controlling element of the easement description and location.

8.10 USE OF EASEMENTS

Because surveyors are frequently the first professionals to read the descriptions of easements, each surveyor should understand that an easement provides for limited use of the parcel. As the dominant tenant can use the servient tenant's land for the specific purposes identified in the easement, any uses inconsistent with the uses identified may terminate the easement. Both the creating surveyor and the retracing surveyor should understand what an easement will be used for and what uses are permitted. If, upon retracement, the surveyor discovers that the present use is inconsistent with the use permitted, that should be reported to the client or noted on the plat that is prepared.

In general, easements are strictly limited in use to the specific purpose(s) noted in the conveyance. An easement for road purposes does not necessarily carry with it the right to place utility lines, nor may it include parking. When preparing a description of the boundaries of an easement, the specific purposes(s) to which the proposed easement is to be put should be stated. State law may, by implication, permit one who grants an easement for a road to the public to extend the right to place utility lines either above or below the ground surface.

8.11 REVIVAL OF PUBLIC EASEMENTS

Once a public easement is extinguished, the boundaries of that easement legally cease to exist. If it is desirable to revive these boundaries, this must be done only through new acts of dedication or eminent domain. When considering public easements, the surveyor must be cautious, because a public easement may be extinguished, but private easement rights may still be valid, having been perfected prior to or subsequent to the public easement.

8.12 CREATION OF EASEMENT BOUNDARIES

Many easement boundaries are created on paper—that is, without benefit of a survey or monumentation—or they may be created by law, which includes prescription, implication, or necessity. Many of these legally created easements have resulting boundaries that the courts create. This may include easements taken by eminent domain. This practice is dangerous in that a utility may be constructed off the easement described. The grantee may find it necessary to purchase additional easement rights. The importance of monumentation of boundaries of the easement cannot be overemphasized for legal and survey purposes. This is especially true in the condemnation of any easements, in that the land is condemned to the monuments set and called for in the description and not to the actual survey. The professional surveyor should treat the boundaries of easements like any other boundaries: adequate monumentation based on a survey, with the corners, monuments, and lines called for in the description.

In conducting a proper survey for an easement, the surveyor should not attempt to shortcut identification of the parent, the servient parcel across which the easement is to be placed, or the dominant parcel, the parcel that will benefit from the easement, or the description of the easement proper. It is important that surface boundaries of both dominant and servient parcels be identified, and that the easement itself be properly surveyed, monumented, and identified to a degree of certainty. In retracing an easement, the surveyor may find it necessary to survey the boundaries of all three descriptions.

Whenever a question arises as to who can use the easement and what uses will be permitted on the easement, one must look at the creating words. Basically, this is a legal question and not a survey question.

In an actual example now in the courts, referring to Figures 8.1 and 8.1a, the facts are as follows:

1. The original subdivision was created in 1957.
2. Lots were created and monuments set.

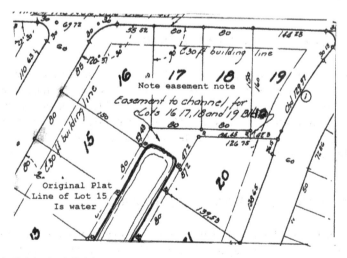

Figure 8.1 Original subdivision plat.

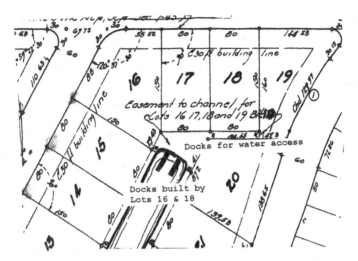

Figure 8.1a Plat showing location of docks.

3. Easements were indicated on the plat.
4. The easement in question borders Lots 15, 16, 17, 18, and 19 as indicated on the original plat.
5. The plat shows Lot 15 to be riparian.
6. The same plat indicates that the easement was created to benefit Lots 15, 16, 17, 18, and 19.
7. The easement states "Easement to channel for Lots 15, 16, 17, 18, 19."

The Problem

1. The owner of Lot 15 built a dock on his riparian land.
2. The owners of Lots 16 and 18, thinking the easement gave them the right to possess the easement proper, destroyed 15's dock and built two new docks for their lots.
3. 15 called the police to stop them. He was put under a restraining order and was told 16 and 18 could build their docks on the channel and 15's had to go.

The Questions to Be Answered in Court

1. What are the boundaries of the easement?
2. What are the boundaries of 15, 16, 17, 18, and 19?
3. What uses can 16, 17, 18, and 19 use in the easement. Can they build a dock?
4. What are 15's boundaries?
5. Can 15 get compensated for the destruction of his dock?

For the court's decision, see the conclusion statements.

After a two-day trial, the judge took only one day to render his decision. His decision was as follows: The boundary of the lots stopped at the iron pins that were indicated as being located 10 feet from the proposed canal. The lots did not have riparian boundaries at the canal, and, as such, the owners of the lots could not place docks on the canal. The area of 10 feet between the canal and the court-adjudicated boundary markers was still owned by the original subdivision owners (a family trust that is being administered by a bank). The judge's prophetic comment that "someone still owns the strip" did not fully realize the impact of his decision. He stated that the owners of the easement rights, having a water boundary, could place docks in the canal.

Today, the situation is far from resolved. The former riparian owners have now petitioned the tax assessor for a reduction of their tax bills. The riparian lots carried an additional assessment charge, having water boundaries. The tax assessor refused, stating that the court order did not affect their determination of the boundaries for tax assessment purposes. The affected landowners have now filed suit against the tax assessor, with the same judge hearing the case.

8.13 DIVIDING PRIVATE STREET OWNERSHIP

Principle 11. *Ownership lines in a street are determined by the original ownership lines as they existed before the easement or dedication of the road. A private conveyance of land abutting on a road, the fee to which belongs to the adjoining proprietors, is interpreted, if possible, to pass fee title to the centerline thereof; otherwise, the contrary must be shown.*

It is recognized that many easements are described by metes and bounds descriptions. These descriptions often define or delineate the limits of private ownership. Although falling within the easement boundaries, the private ownership lines existed prior to the creation of the easement (see Figure 8.2). When it is found that the bounds within a street are given or identified, the surveyor should not assume the center of the street as the boundary. The presumption may not exist. Deeds may be found that expressly exclude the street, but unless they do, the inference is that the street is included. Before a surveyor can positively state how the division should be made, an investigation of the original lines must be made, and if it is determined that the owners on each side did not contribute equally to the easement, the presumption fails.

Surveyors often rely on subdivision maps to depict the extent of easements, but in most states, these recorded maps will not show the extent of the private boundaries. Many presumptions are accepted with subdivision maps. Seldom does a deed for a lot mention the streets. It is usually assumed that the streets are included, to the centerline. Once all the lots in a subdivision are sold, the developer has no further interest in maintaining streets and utilities relative to maintenance, taxes, and so on, and the lot owners find that these items are assessed against their lots. This issue

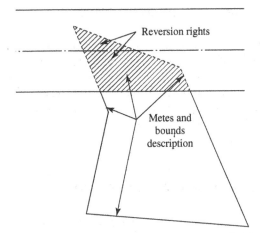

Figure 8.2 Reversion rights are determined by property lines that existed before dedication of the street.

was raised in Maine when a developer indicated that the lots extended only to the respective side lines, not to the centerline of the streets.[17]

Surveyors will find that landowners will generally try any means in an attempt to keep from paying taxes and assessments. Texas landowners refused to pay their street assessments, using the theory that their metes and bounds descriptions followed along the street but did not include the street itself. The court found that deeds to a city lot that fronted on a street carried fee to the center of the street, unless there were clauses expressly declaring a contrary intention.[18]

8.14 WORDS USED IN CENTERLINE CONVEYANCES

The assumption that an owner owns to the centerline of a street can be overcome by definitive statements to the contrary. Such phrases as "along the side line of the street," "exclusive of the street," and "excepting the street" will prevent title in the bed of a street from passing. "Bounded by the highway," "fronting on the highway," "to and along the highway," "with and by the highway," and "in line of highway" can be construed to mean "along the centerline of the highway" and are usually taken in that light in the absence of other words of exclusion.

Descriptions of property on a highway and without words expressly including the highway, but with words granting an easement in the highway, are construed to indicate an intent not to convey the bed of the street. The fact that it was necessary to convey an easement indicates that the fee to the bed did not pass. As part of the initial research, surveyors should read the words for their exact meaning. Surveyors will often be called on to interpret the meanings in descriptions. A recent Florida decision stated as one of the findings that the only person, by virtue of education, training, and experience, capable of interpreting a description is a professional surveyor.[19]

8.15 APPORTIONING REVERSION RIGHTS

In instances where original boundaries were never identified, the complex problems of apportioning limits of landownership arising from the common-law presumption that fee interest is conveyed to the centerline of the street lack substantial documentation. This is particularly true in subdivisions where roads have curved and irregular street patterns. Several principles have been identified based on accepted practices, common sense, and an understanding of property rights. These are described in the sections that follow.

8.16 GENERAL PRINCIPLE OF REVERSION

Principle 12.   *Unless a deed or map that created the easement or dedication indicates otherwise, reversion rights extend from the street termini of the property lines to the centerline or thread of the street in a direction that is at right angles to the centerline of the street.*

Figure 8.3 illustrates possible difficulties in applying this principle rigidly. This is depicted in lot 1.

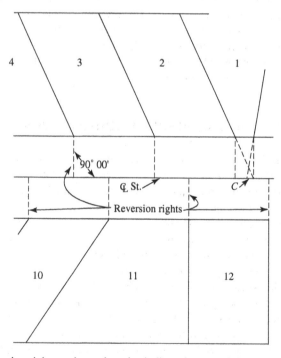

Figure 8.3 Reversion rights, unless otherwise indicated, are at right angles to the centerline of the street.

When easement boundaries are created in writing or by a map, one should expect that the corners are monumented and that the lines were run on the ground prior to the creation of the description. However, this is not always true. Many easements are created at law, in that such doctrines as estoppel, acquiescence, agreement, implication, necessity, or adverse possession are used to create the easement, but then the boundaries become a matter of interpretation by the courts and in most instances do not contain definite lines or measurements.

In locating easement lines that are created by words, the words are to be followed, yet in attempting to locate those easement boundaries that are created under the law, the surveyor must have a complete understanding of the legal requirements and limitations for each state.

8.17 REVERSION RIGHTS OF A LOT ON A CURVED STREET

Principle 13. *Reversion rights of a lot in the middle of a block extend radially to the centerline of the street.*

Radial lines are at right angles to a tangent of a curve. If Principle 12 is interpreted broadly, this principle can be considered a special application of the former principle.

Figure 8.4 illustrates the more common applications of the principle.

Lots adjoining a street intersection receive the lion's share of the land vacated, with just cause. Where a street is paved next to a corner, as in Figure 8.5, the

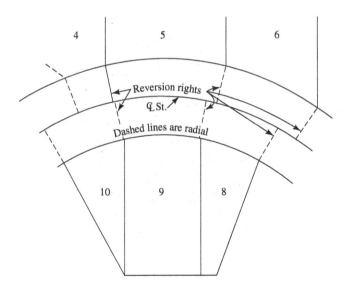

Figure 8.4 Reversion rights, unless otherwise indicated, are on radial lines.

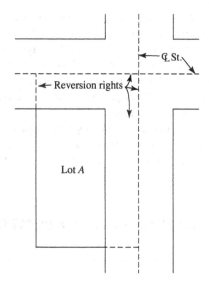

Figure 8.5 The corner lot pays for the assessment of the intersection.

corner lot must pay for the assessment of the intersection and is entitled to all the land that it pays assessments on. Where an intersection is curved, as shown in Figure 8.6, the reversion rights also extend to the centerline as determined by a curved line.

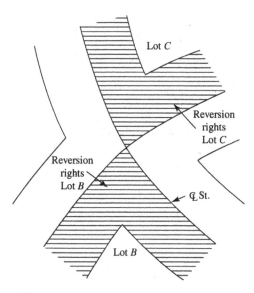

Figure 8.6 Reversion rights extend to the centerline in a curved intersection.

8.18 LOTS ADJOINING TWO SUBDIVISION BOUNDARIES

Principle 14. *Reversion rights of a lot adjoining a subdivision boundary extend along the boundary line of the subdivision and cannot extend beyond the boundary line.*

At times, a surveyor may have to identify a reversion boundary that is common to two separate and independent subdivisions. First, the senior and controlling line is the original boundary between the two separate subdivisions. This cannot be altered or interpreted in any other manner. Then the boundary lines of each separate lot within the respective subdivisions should be reidentified in accordance with these exterior lines.

In Figure 8.7, the surveyor will see that Lot 1 of Map 70 and Lot A of Map 180 represent two lots in two independent subdivisions. In relocating the common boundary line between the two subdivisions, the retracing surveyor may have to understand and apply junior–senior rights between the two subdivisions. After that is done, the separate lines of each lot are considered. After the subdivision line is located, the independent lots are located in accordance with accepted principles.

Under certain circumstances, lots adjoining a subdivision may have reversion rights extending beyond the centerline of the street. This is shown in Figure 8.8, where the street reverts to Lot 7 and the remaining lots accept different amounts.[20]

The subdivision that lies on the north side of the street had no original title interest in the street adjacent to Lot 7. In a situation such as this, when an easement is vacated or terminated, the land must revert to the original grantor or to his or her heirs or assigns.

8.19 LOTS AT AN ANGLE POINT IN A ROAD

Principle 15. *Reversion rights at an angle point in a road extend on the bisection of the angle.*

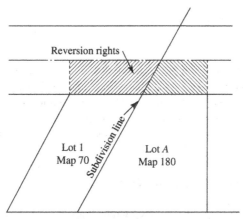

Figure 8.7 Relocating the common boundary line between two subdivisions.

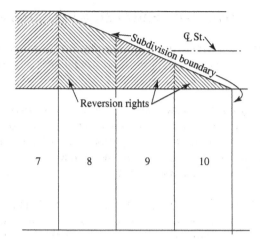

Figure 8.8 Reversion rights may sometimes extend beyond the centerline of the street.

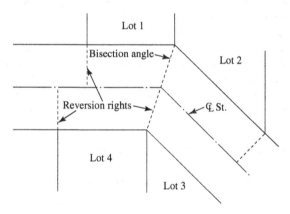

Figure 8.9 Reversion rights at angles in the street.

The majority of these principles discussed in this chapter apply a presumption of equity in reversions. Principle 15 applies simplicity. In Figure 8.9, where an angle point is present, the basic application is to bisect the angle. Notice in Figure 8.9 that Lot 4 received a greater percentage of the land than any other lot, including Lot 1, even though Lot 1 and Lot 4 had equal frontage.

8.20 INDETERMINATE SITUATIONS

There may be instances in which the surveyor searches for an answer to a reversion but finds no ready solution. In Figure 8.10, the reversion of area X would normally

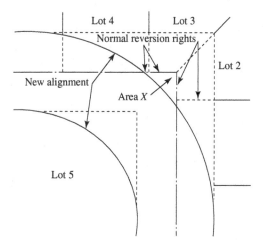

Figure 8.10 Indeterminate reversion situation.

become attached to Lot 5. A Kentucky decision held that in considering an area such as X, equity should give it to the abutting owners who were entitled to road access. As an alternative solution, if it were determined that area X would be given to Lots 1, 2, 3, and 4, how would you divide them? This may be very frustrating!

8.21 EXCEPTIONS TO THE RULES OF APPORTIONMENT

Like all rules, there are always exceptions. In this section, we discuss some of those exceptions, simply to show that there are no complete certainties in this area. In Kansas[21] and Oklahoma,[22] the courts apply the rule of apportionment along the same principle as accretions in rivers and riparian principles. Not only is this recognized in case law, but statutes also address the problem.[23] The Oklahoma statute states:

> Whenever any street, avenue, alley or lane shall be vacated, the same shall revert to the owners of the real estate thereto adjacent to each side, in proportion to the frontage of such real estate, except in cases where such street avenue, alley, or lane shall have been taken and appropriated to public use in a different proportion, in which case it shall revert to adjacent lots of real estate on proportion as it was taken from them.

Applying this statute in the states of Kansas and Oklahoma, the rule of apportionment is similar to that of accretions to rivers. This statute, in numerous cases, would be applied in accordance with the rules given in the preceding sections. In the case of a curve (Figure 8.4), the frontage of all lots remains in proportion to the new frontage at the centerline of the street, provided that radial lines are used as shown. In Figure 8.11, b is to B as c is to C, an exact proportion. But troubles ensue if the principle is adhered to strictly, at the termination of curves and at irregular boundary lines of the subdivision, as shown in Figure 8.8.

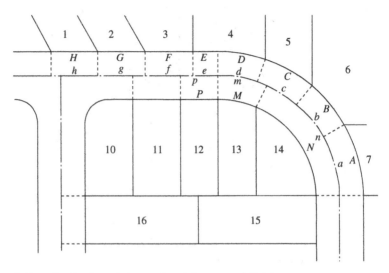

Figure 8.11 Indicating intended reversion rights on a subdivision map.

The proportion g is to G as f is to F as e is to E is exact. Also, d is to D as c is to C as b is to B as a is to A. But d is to D as e is to E is not a proportion. Difficulty in applying the foregoing statute is encountered at every intersection. How will Lot 10 receive its share of the vacated street? The simple method of obviating such difficulties is to indicate on the subdivision map being filed the intended reversion rights of the adjoining lots. If the dashed lines shown in Figure 8.11 were included as part of the original map, no doubt could exist.

In the state of Washington, a street in the tidelands was vacated and the court commented: "We think these lands must be allotted the same portion of the vacated streets that they would be in other platted lands, the lateral lot lines to be extended straight from their former property line to the center line of the vacated street."[24]

8.22 DESCRIBING VACATED STREETS AND EASEMENTS

When a street within a subdivision is conveyed, it is presumed that title in the deed is to the centerline of the street. What, then, is the answer when the street is vacated? Apparently, after a street is vacated, the vacated portion of the street must be described with certainty specifically and positively before it can be conveyed. A recommended description would be: "All of Lot 1, together with the vacated portion of Grape Street adjoining Lot 1."

In a Minnesota decision, the court wrote:

A conveyance of land bounded on an existing highway vests in the grantee the fee to the center of the highway, solely because the court presumes that such is the intention of the parties to the instrument. If a deed of conveyance to land bordering on the highway carries title to the center of the highway, solely because of this presumption, then there would seem to be no escape from the logical conclusion that when the facts which give rise to the presumption cease to exist, the presumption also falls to the ground.[25]

When the street ceases to exist, the reason for the presumption ceases, and the portion in the vacated street must be included in a description in order to convey it.

Principle 16. *By law and by definition, an easement is a nonpossessory right to use the property of another person.*

Much litigation has resulted because of poorly written easement descriptions. Writers often use conflicting words, not knowing what havoc they may create. One serious conflict occurs with the words each and either. The erroneous application can make a 50-foot-wide easement a 100-foot wide.

Such statements as "an easement deed" or "reserved for road purposes" cause the courts to wrestle with the legal consequences of what the creating individuals really meant when they wrote the ambiguous words. Since an easement grants the easement holder only the right to use the property and not to possess it, it has been held that an easement for ingress and regress, without additional rights, does not permit the parking of a vehicle or the driver to walk on the land. Although many of these problems are legal problems and usually are not addressed by the surveyor, the surveyor should be sufficiently knowledgeable as to provide the client with help should such a situation occur.

Principle 17. *The creating words become important when there is a question as to what was created, what was vacated, and what was reverted to the landowners.*

Litigation is costly! Errors and problems in easements are abundant. Unscientific research indicates that for every boundary problem, there are 10 easement problems that are addressed by the courts. Of these easement problems that are appealed, over one-half of the appellate courts reverse the trial court's decision.

Although the surveyor should not practice law, he or she can be a member of the team. In most instances, easements are considered "after the fact." In most instances, surveyors should become involved in the writing of descriptions, in what is described, in what is retained, and what is to be granted back. Every surveyor who writes property descriptions should become conversant with what words mean, how courts think, and what people want when they write a description for the boundaries of parcels of land.

Principle 18. *Easement boundaries may be either surface, subsurface, or aerial. The location may restrict the use by the servient titleholder.*

Since easements are incorporeal hereditaments—that is, for want of a better description, unable to be seen like the elements of real property—their identification places an additional burden on those professionals who attempt to locate them.

Unlike trees or minerals, easement rights are not visible to the retracing surveyor. Because of the nature of easements, creating or retracing surveyors should keep a watchful eye for visible evidence of them that is not on the surface of the lot being surveyed. Easements can be described and located either on the ground surface, below the surface, or even above the surface of a lot. A long lineal depression across a lot, power poles on each side of a lot with wires overhead, or a badly overgrown rutted road winding its way across a parcel may all be indications to the surveyor of prior rights, which may in turn indicate easement rights. The surveyor's responsibility is to indicate the evidence of the possible easement and not to offer legal decisions as to the easement's validity.

8.23 LITIGATING EASEMENTS

When a client decides to litigate, he or she, through the attorneys, must decide under what theory or theories to bring the action. The two usual avenues of litigating an easement dispute are law and equity. These are completely different and separate from each other. Two phrases that all attorneys learn early in their careers are:

Equity follows the law
And
When there is no possibility at law, equity may apply.

Equity denotes fairness, impartial justice by natural reasoning that is tried in a court of equity. Many states hold that easement questions are equitable questions.[26]

The final decision as to the area in which the case should be tried is one for the attorney, not the surveyor, to determine.

8.24 CONCLUSIONS

Real property issues are not the typical strong suit of the average judge or practicing attorney. However, the practicing surveyor is placed in a position of expected knowledge in this area. Since many states permit the surveyor to write land and boundary descriptions, he or she should have a fundamental knowledge of the primary requirements for creating and describing not only land boundaries but boundaries of any real property interest, regardless of where the boundaries may be described—on the surface, above the surface, or underground—so as to limit or eliminate any possible problems the client/landowner may encounter now or in the future.

NOTES

1. O. W. A. Dilkie, *Greek and Roman Maps*, (Baltimore: Johns Hopkins University Press, 1985), 122. The modern term *highway* probably originated from the Roman practice of

making the center portion of the road elevated (or higher) by adding bricks or gravel. At times, the road would be elevated by as much as 2 feet in wet areas.

2. *Holloman* v. *Board of Education*, 147 S.E. 882 (Ga. 1928).

3. *Harrison* v. *Boring*, 44 Tex. 267 (Tex. 1855).

4. *Magnolia Petroleum Co.* v. *Caswell*, 295 S.W. 653 (Tex. 1927).

5. *Copertino* v. *Ward*, 473 N.Y.S.2d 494 (1884).

6. *Termes de la Ley, Terms of the Common Laws and Statutes Expounded and Explained*, John Raswell, 1527.

7. *Note:* If the easement is for ingress and egress (using the property), it cannot be used for parking (possessing the property).

8. *American Law Review*, 2-25, 18-1018, and 11-551; 114 Ky. 1634 (1868, 1889, 1976).

9. *Lake City* v. *Fulkerson*, 112 Iowa 569, 98 N.W. 376 (1904).

10. *Appleton* v. *City of New York*, 218 N.Y. 150, 114 N.E. 73 (1916).

11. *Duchesnaye* v. *Silva*, 118 N.H. 728, 394 A.2d 59 (1978).

12. *National Bank of Detroit, Pers. Rep., et al.* v. *Erie County Road Commission, Huron Township, et al.*, 587 N.E.2d 819 (Ohio 1992).

13. *Johnson* v. *Westrick*, 200 Wis. 405 (1930) citing City of Racine v. Emerson, 85 Wis. 80, 55 N.W. 177 (1893).

14. 62 P.2d 602; 17 Cal.App.2d 627 (1936).

15. *McCorkle* v. *City of Charleston*, 105 W.Va. 395 (1928).

16. *Bangor Housing Prop.* v. *Brown*, 33 Maine 309 (1851); *Gerbig* v. *Zumpena*, 177 N.Y.S.2d 969 (1958).

17. *Sutherland* v. *Jackson*, 32 Me. 80 (1850).

18. *Texas Betulhic Co.* v. *Warqweck*, 293 S.W. 160 (1927).

19. *Rivers* v. *Lozeau*, 529 So.2d 1147 (Fla. App. 1989).

20. *Oberhelman* v. *Allen*, 7 Ohio App. 251 (1915).

21. *Showalter* v. *So. Kansas Railway Co.*, 49 Kan. 421, 32 P. 42 (1892).

22. *Blackwell Railway Co.* v. *Gut*, 18 Okla. 516, 90 P. 889 (1907).

23. Oklahoma Statute Law (2002).

24. *State* v. *Supreme Ct.*, 102 Wash. 331 (1918).

25. *White* v. *Jefferson*, 110 Minn. 276 (1910).

26. *Miller* v. *Kenniston*, 30 A. 114 (Me.).86 Me. 550, 30 A. 114 (Me., 1894).

CHAPTER 9

RIPARIAN AND LITTORAL BOUNDARIES

9.1 INTRODUCTION

Water boundaries are possibly the most ancient and widely used of humankind's boundaries. Today, these aspects of surveying and law are probably the least understood by the courts and the practicing surveyors. Surveyors in a few states are concerned with *tidal boundaries as well as nontidal boundaries, but have to worry about title boundaries*. Throughout recorded history, societies have recognized water boundaries as the limits of their territories. As an early example of this, very detailed laws regarding water boundaries were included in the Roman Civil Code known as the Institutes of Justinian, prepared under the direction of Emperor Justinian in CE 529. That document serves as the root of today's water boundary procedures. The following translations[1] of pertinent excerpts from that code illustrate the detail and the similarity of that code to modern water law. As may be seen, that code addresses many of the same issues pertinent in today's society, including issues such as public ownership of navigable waters, the location of the boundary between publicly owned waters and privately owned littoral or riparian land, public access to navigable waters, riparian rights to newly formed lands, and ownership of newly formed islands.

The US Constitution enters into the legal and historical when it enacted the *Interstate Commerce Clause* and its resultant application to anything that applied to interstate commerce, and declaring all navigable rivers navigable.

Brown's Boundary Control and Legal Principles, Eighth Edition.
Donald A. Wilson, C.A. "Tony" Nettleman III, and Walter G. Robillard.
© 2024 John Wiley & Sons, Inc. Published 2024 by John Wiley & Sons, Inc.

Section 1. By the law of nature these things are common to mankind—the air, running water, the sea, and consequently the shores of the sea. No one, therefore, is forbidden to approach the sea-shore, provided that he respects habitations, monuments, and buildings, which are not, like the sea, subject only to the law of nations .... The sea-shore, that is the shore as far as the waves go at furthest, was considered to belong to all men.

Section 2. All rivers and ports are public; hence the right of fishing in a port or in rivers, is common to all men.

Section 3. The sea-shore extends to the limit reached by the greatest winter flood.

Section 4. The public use of the banks of a river is part of the law of nations, just as is that of the river itself. All persons, therefore, are as much at liberty to bring their vessels to the bank, to fasten ropes to the trees growing there, and to place any part of their cargo there, as to navigate the river itself. But the banks of a river are the property of those whose land they adjoin; and consequently the trees growing on them are also the property of the same persons.

Section 5. The public use of the sea-shore, too, is part of the law of nations, as is that of the sea itself; and therefore any person is at liberty to place on it a cottage, to which he may retreat, or to dry his nets there, and haul them from the sea; for the shores may be said to be the property of no man, but are subject to the same law as the sea itself, and the ground or sand beneath it.

Section 20. Moreover, the alluvial soil added by a river to your land becomes yours by the law of nations. Alluvion is an imperceptible increase; and that is added by alluvion, which is added so gradually that no one can perceive how much is added at any one moment of time.

Section 21. But if the violence of a river should bear away a portion of your land and unite it to that of your neighbour, it undoubtedly still continues yours. If, however, it remains for a long time united to your neighbour's land, and the trees, which it swept away with it, take root in his ground, those trees from that time become part of your neighbour's estate.

Section 22. When an island is formed in the sea, which rarely happens, it is the property of the first occupant; for before the occupation it belongs to no one. But when an island is formed in a river, which frequently happens, then it occupies the middle of the river, it belongs respectively to those who possess the lands near the banks on each side of the river, in portion to the extent along the banks of each man's estate. But if the island is nearer to one side than the other, it belongs to those persons only who possess lands contiguous to the bank on that side. But if a river divides itself at a certain point, and lower down unites again, thus giving to any one's land the form of an island, the land still continues to belong to the person to whom it belonged before.

Section 23. If a river, entirely forsaking its natural channel, begins to flow in another direction, the old bed of the river belongs to those who possess the lands adjoining its banks, in proportion to the extent along the banks of their respective estates. The new bed follows the condition of the river, that is, it becomes public. And, if after some time the river returns to its former channel, the new bed again becomes the property of those who possess the lands along its banks.

Section 24. The case is quite different if any one's land is completely inundated; for the inundation does not alter the nature of the land, and therefore, if the water recedes, the land remains indisputably the property of the same owner.

Book II, Title I

As may be seen from these excerpts from the Institutes of Justinian, water boundaries were complex even seven centuries ago. Today, water boundaries continue to be possibly the most complex of man's boundaries. The complexity of these boundaries is due to the fact that both the level of the water and the shoreline itself are constantly changing. This complexity resulted in the development of unique principles, such as those illustrated in the excerpts from the Institutes of Justinian, to guide the determination of water boundaries. The currently prevailing versions of those principles are the subject of this chapter. The current principles gleaned from historical and current case law include the following:

PRINCIPLE 1. The high seas are considered as a global commons accessible to all. A marginal territorial sea along the coastline is subject to exclusive claim by each coastal nation and/or state.

PRINCIPLE 2. Although the territorial sea of the United States currently extends 12 nautical miles seaward of the mean lower low-water line, each coastal state has ownership of the area out to 3 geographical miles, except for Florida and Texas, whose boundaries extend out to 3 marine leagues (9 miles) in the Gulf of Mexico.

PRINCIPLE 3. With statehood, all coastal states received ownership of all natural submerged lands affected by the tides (tidelands) within their boundaries, regardless of their navigability. Some states have relinquished a claim to those tidelands that are not navigable-in-fact.

PRINCIPLE 4. In most of the US coastal states, the landward boundary between the state-owned submerged lands in tidal waters and bordering uplands is the mean high-water line.

PRINCIPLE 5. Exceptions to the use of the mean high-water line as the sovereign/upland coastal boundary include several northeastern states that recognize the mean low-water line as that boundary. In addition, exceptions exist in some areas where land title has roots in a grant from a sovereign power where civil law, as opposed to Anglo-American common law, prevailed. By the civil law, the shore extends as far "as the highest waves reach in winter" as opposed to the mean high-water line of the common law.

PRINCIPLE 6. The submerged lands under natural navigable nontidal waters are owned by the public in most of the United States, although some states do not claim such ownership.

PRINCIPLE 7. The boundary of publicly owned nontidal submerged lands is the ordinary high-water mark for many states, although a number of states claim only to the ordinary low-water mark.

PRINCIPLE 8. A meander line of the Public Land Survey generally does not constitute a boundary. Rather, it represents the approximate location of the shoreline at a certain point in time.

PRINCIPLE 9. Submerged lands under non–publicly owned waters are customarily owned by the adjoiners, with the ownership extending to the center of the stream.

PRINCIPLE 10. Swamp and overflowed lands are not part of the public trust submerged lands. In the public land states, such lands were granted to the states for subsequent improvement and conveyance to private owners.

PRINCIPLE 11. The United States has authority to regulate activities on all navigable waters of the country.

PRINCIPLE 12. Riparian and littoral properties are often subject to public regulation, which may create regulatory limit lines, which, although different from property lines, may also affect the utility of the property.

PRINCIPLE 13. The general rule is that water boundaries change with gradual and natural changes to the shoreline, while those boundaries do not change with sudden or human-made changes. Land that is gradually and naturally built up by accretion or exposed by reliction accrues to the riparian owner. Land that is gradually and naturally eroded away or permanently inundated by the water accrues to the owner of the bed.

PRINCIPLE 14. Regarding the apportionment of riparian and littoral rights:

- The objective of apportionment is to distribute either newly formed land or rights in water in proportion to the length of water frontage. Therefore, the length of the affected tract's water frontage, not its total area or upland lateral lot lines, generally controls the apportionment of riparian rights.

- For the division of newly formed land, the point of departure for a division line is the intersection of the upland lateral boundary and the shoreline at the time of the subdivision of the area. For the division of water area, the point of departure for a division line is the intersection of the upland lateral boundary and the current shoreline.

- For the division of newly formed land, the newly formed shoreline should be proportioned relative to the respective lengths of the old and new shoreline.

- In streams, the general rule for the division of water area is to project the division line in a direction perpendicular to the thread of the stream.

- In large bays, oceans, or wide rivers where a thread is not readily discernible, the general rule for the division of water area is to run lines perpendicular to various other baselines such as the shoreline if it is relatively straight, the bulkhead or pier line, or the channel line.

- In lakes, the general rule for the division of water area for round lakes is for side lines to be drawn to a center point, creating pie slice–shaped areas. For long lakes, it is generally held that a centerline should be constructed and lines projected out perpendicular to the centerline.

- In partially enclosed areas, such as coves, the general rule for the division of water area is to proportion the closing line across the mouth of the area based on the respective lengths of the shoreline.

PRINCIPLE 15. Emergent islands are generally considered to belong to the owner of the bed of the water body. In public land states, omitted lands (such as islands existing at the time of, but not surveyed in, the public land survey) remain in the ownership of the federal government until specifically conveyed. Islands emerging after the conveyance of the water body to the states with statehood, belong to the owner of the bed of the water body.

9.2 OWNERSHIP OF THE SEAS[2]

Principle 1. *The high seas are considered as global commons accessible to all. A marginal territorial sea along the coastline is subject to exclusive claim by each coastal nation and/or state.*

As illustrated by the excerpts from the Roman Institutes of Justinian cited in the introduction to this chapter, the high seas have long been considered as a global commons, open and available to all. Yet, as navigation of the seas became more common, coastal nations began reserving the portion of the seas bordering their nations as their exclusive territory. This concept of a marginal or territorial sea evolved as a matter of self-defense on the part of the coastal nations. That zone was, for practical purposes, the zone that the nation could comfortably defend for its exclusive use from hostile shipping. By the eighteenth century, this was more or less standardized by the so-called cannon shot rule. By that informal rule, the boundary of the marginal sea was the distance from shore that a cannon shot could typically reach. That distance was generally considered to be 3 miles.

More recently, as various nations began exploiting the riches of the sea and sea floor, widely ranging claims have been made, which vary considerably from the "cannon shot" rule. This issue has been frequently debated at international conferences on the law of the sea. In 1982, a United Nations conference adopted an international treaty that has somewhat standardized maritime boundaries. That treaty gave coastal nations a territorial sea extending out to 12 miles from shore, with exclusive economic zones of 200 miles or to a maximum of 350 miles if the country's continental shelf extends out that far. Bays and partially enclosed coastal indentions are considered to be exclusive territory of the surrounding nation under this treaty. As with earlier law-of-the-sea treaties, the latest treaty also stresses that the open seas are common to all and not controllable by any individual nation. Although not a signatory to the United Nations Treaty of the Sea, the United States has adopted similar claims to its adjacent waters.

9.3 OWNERSHIP OF THE US TERRITORIAL SEA

Principle 2. *Although the territorial sea of the United States currently extends 12 nautical miles seaward of the mean lower low-water line, each coastal state has ownership of the area out to 3 geographical miles, except for Florida and Texas, whose boundaries extend out to 3 marine leagues (9 miles) in the Gulf of Mexico.*

Federal Claims

Prior to the 1930s, the United States claimed a 3-mile-wide territorial sea and made no attempt to exert jurisdiction beyond that limit. During the 1930s, the jurisdiction of the United States for law enforcement was extended to the limits of a 12-mile "contiguous zone" in connection with enforcement of prohibition laws (rumrunners had been able to easily escape interception with the 3-mile territorial jurisdiction. That jurisdictional extension was of an extraterritorial nature and did not extend the territorial sea ownership.

In 1945, with the development of offshore oil drilling, the claims of the United States were further extended by a presidential proclamation stating, " ... the Government of the United States regards the natural resources of the subsoil and seabed of the continental shelf beneath the high seas but contiguous to the coasts of the United States as appertaining to the United States, subject to its jurisdiction and control."[3] The continental shelf is a gently sloping plain of land along the coasts of most islands and continents, and it varies in width, from a few miles to hundreds of miles. The continental shelf is considered to end where the continental slope begins to drop more steeply to the ocean floor.

In 1977, the United States claimed exclusive fishing management jurisdiction over the 200-mile limit, regardless of whether this line went beyond the continental shelf. In 1982, the previously mentioned United Nations Law of the Sea Conference adopted a treaty establishing a 12-mile territorial sea, and, in 1988, the United States adopted an equivalent-width territorial sea. This occurred when President Ronald Reagan extended the US territorial sea " ... to the limits permitted by international law ... " with a presidential proclamation. That document proclaimed as follows:

> The territorial sea of the United States henceforth extends to 12 nautical miles from the baseline of the United States determined in accordance with international law.[4]

Therefore, the current policy of the United States is to claim ownership of a territorial sea of 12 nautical miles from its coastline, to claim exclusive jurisdiction over the natural resources to the extent of the continental shelf, and to claim exclusive fishing rights to 200 miles. The coastline, from which the width of the territorial sea is measured, is considered to be the low-water line, as depicted on a nation's nautical charts. For the United States, that line is the mean lower low-water line of either the mainland or any island (any land mass, including rocks, with surfaces above the mean high-water line) lying within the nation's territory.

State Claims

Until the 1930s, it was generally accepted that the public of each of the coastal states, rather than the federal government, owned the marginal sea off its coast as well as the interior navigable waters. However, with the discovery of oil in the submerged lands off the California coast, disputes arose as to the ownership of such lands. The resulting conflict led to a series of court cases that held that the federal government, rather than the states, was the owner of the marginal sea. Following those decisions, the US Congress passed the Submerged Lands Act (43 U.S.C., s1301-1 1970) in 1953, which conveyed a band of submerged lands bordering the coastal states to the public of those states. That act defines the seaward limit of the public trust waters of each of the coastal states as follows:

> Section 4—The seaward boundary of each original coastal State is hereby approved and confirmed as a line three geographical miles distant from its coastline or, in the case of the Great Lakes, to the international boundary. Any State admitted subsequent to the formation of the Union which has not already done so may extend its seaward boundaries to a line three geographical miles distant from its coastline or to the International boundaries of the United States in the Great Lakes or any other body of water traversed by such boundaries.... Nothing in this section is to be construed as questioning or in any manner prejudicing the existence of any State's seaward boundary beyond three geographical miles if it was so provided by its constitution or laws prior to or at the time such State became a member of the Union, or if it has been heretofore approved by Congress. (Note: A geographical mile is the length of 1 minute of arc on the equator or 6087.09 feet on the Clark Spheroid of 1866. A marine league is three geographical miles.[5])

The question of whether any of the states were entitled, under the act, to submerged lands greater than 3 geographical miles from the coastline was decided by the Supreme Court in 1960.[6] That decision held that Texas and Florida were entitled to submerged lands extending 3 leagues into the Gulf of Mexico due to the extent of their boundaries at the time of admission into the Union. The same decision held that Louisiana, Mississippi, and Alabama were entitled to a marginal sea of only 3 geographical miles.[7]

As a result, the public trust lands of each of the coastal states are now considered to be the navigable interior waters of their state together with the portion of the territorial sea out to 3 miles from the mean lower low-water line (even though the current policy of the federal government is to claim ownership out to 12 miles), except for the states of Florida and Texas, with lands extending out to 3 leagues into the Gulf of Mexico.

9.4 OWNERSHIP OF INTERIOR TIDAL WATERS OF THE UNITED STATES

Principle 3.   With statehood, all coastal states received ownership of all natural submerged lands affected by the tides (tidelands) within their boundaries, regardless of their navigability. Some states have relinquished a claim to those tidelands which are not navigable-in-fact.

All natural tidal waters within their boundaries, unless validly conveyed, are considered to have been conveyed to the public of each of the US coastal states. This is the case whether or not such waters are navigable. Therefore, unless a state has relinquished its claim, such waters are considered to be owned by the public of each state. This principle was confirmed in a case in the State of Mississippi,[8] in which both the Chancery Court and the Mississippi Supreme Court ruled that as long as the tide ebbed and flowed in a natural water body, it was publicly owned. Salient excerpts from that colorful court opinion are as follows:

> The early federal cases refer to the trust as including all lands within the ebb and flow of the tide.
>
> … it is our view that as a matter of federal law, the United States granted to this State in 1817 all lands subject to the ebb and flow of the tide and up to the mean high water level, without regard to navigability.
>
> Yet so long as by unbroken water course—when the level of the waters is at mean high water mark—one may hoist a sail upon a toothpick and without interruption navigate from the navigable channel/area to land, always afloat, the waters traversed and the lands beneath them are within the inland boundaries we consider the United States set for the properties granted the state in trust.

The case was appealed to the US Supreme Court,[9] which concurred with the state court in ruling that all coastal states received all lands over which tidal waters flow, and that Mississippi still does. The opinion noted that this ruling "will not upset titles in all coastal states" since it "does nothing to change ownership rights in states which previously relinquished a public trust claim to tidelands such as those at issue here." Thus, it is presumed that in states with well-established case law stating that only navigable tidal waters were publicly owned, navigability-in-fact would be the salient criterion for public ownership. However, that issue has yet to be judicially tested.

9.5 LANDWARD BOUNDARY OF TIDAL WATERS

Principle 4.   In most of the US coastal states, the landward boundary between the state-owned submerged lands in tidal waters and bordering uplands is the mean high-water line.

Principle 5. Exceptions to the use of the mean high-water line as the sovereign/upland coastal boundary include several northeastern states that recognize the mean low-water line as that boundary. In addition, exceptions exist in some areas where land title has roots in a grant from a sovereign power where civil law, as opposed to Anglo-American common law, prevailed. By the civil law, the shore extends as far "as the highest waves reach in winter" as opposed to the mean high-water line of the common law.

Under English common law, submerged lands underlying tidal waters belonged to the Crown. In the United States, the public of each state is sovereign and thus holds title to the beds under navigable waters. That concept was first stated in the US Supreme Court case of *Martin* v. *Waddell*[10] in 1842. According to the Court in that case,

> When the revolution took place the people of each state became themselves sovereign and in that character hold the absolute right to all their navigable waters in the soils under them for their own common use....

General Anglo-American Common Law

In most of the United States, the shoreward boundary of such public trust waters is defined by Anglo-American common law. Possibly the earliest definition of the boundary was provided by Thomas Digges, an engineer, surveyor, and lawyer during the reign of Queen Elizabeth I, in a book entitled *Proofs of the Queen's Interest in Land Left by the Sea and the Salt Shores Thereof*. Digges suggested that the queen's ownership of the waters included the foreshore, which is the area between high and low tides. In the following century, Lord Matthew Hale, a jurist who was to become the British chief justice, built on that concept in his treatise *De Jure Maris*,[11] written in about 1666. In that writing, Hale posited that the foreshore, which is overflowed by "ordinary tides or neap tides, which happen between the full and change of the moon," belonged to the Crown. In 1854, this definition was clarified in English common law by the case of *Attorney General* v. *Chambers*.[12] The *Chambers* case ruled that the ordinary high-water mark was to be found by "the average of the medium tides in each quarter of a lunar evolution during the year (which line) gives the limit, in the absence of all usage, to the rights of the Crown on the seashore."

In the United States, the boundary was further clarified in 1935 with the US Supreme Court's landmark decision in *Borax Consolidated* v. *City of Los Angeles*,[13] as follows:

> In view of the definition of the mean high tide, as given by the United States Coast and Geodetic Survey that "mean high water at any place is the average height of all the high waters at that place over a considerable period of time", and the further observation that "from theoretical considerations of an astronomic character" there should be "a periodic variation in the rise of water above sea level having a period of 18.6 years", the Court of Appeals directed that in order to ascertain the mean high tide line with requisite certainty

in fixing the boundary of valuable tidelands, such as those here in question appear to be, "an average of 18.6 years should be determined as near as possible". We find no error in that instruction.

As may be seen, the *Borax* decision applied modern technical knowledge to set forth a specific technique for precisely locating the boundary in question by taking into account all of the periods of the astronomical forces affecting the tides to obtain an average height. Thus, the mean high-water line represents an attempt to define the boundary between publicly owned submerged lands and uplands subject to private ownership as the average upper reach of the daily tides. The result is a line that is exceeded by the high tide on approximately one-half of the tidal cycles.

Case law in most of the coastal states has generally followed the English common law and its updated definition as put forth in the *Borax* decision. Some states have codified their statutory law on this subject. As an example, in Florida, the Coastal Mapping Act of 1974 (Chapter 177, Part II, Florida Statutes) declares that "mean high water line along the shores of lands immediately bordering on navigable waters is recognized and declared to be the boundary between the foreshore owned by the State in its sovereign capacity and upland subject to private ownership." That statute further defines the mean high-water line using the *Borax* concept.

Exceptions Based on Massachusetts Colonial Ordinance

Several northeastern states are exceptions to those using the mean high-water line as a coastal boundary. Those states recognize the mean low-water line as the sovereign/upland boundary. The basis for that ownership is generally considered to be an early Massachusetts colonial ordinance, which was enacted in 1641 and revised in 1647. Thus, it is possibly the earliest reference to water boundaries in the New World. The ordinance provided as follows:

> ... in all creeks, coves and other places, about and upon salt water where the Sea ebs and flows, the Proprietor of the land adjoining shall have proprietie to the low water mark where the Sea doth not ebb above a hundred rods, and not more wheresoever it ebs farther.[14]

The language of the ordinance has generally been interpreted as defining the water-ward boundary of lands bordering on tidal waters as the mean low-water line unless that line is 100 rods distant from the mean high-water line. If there is more than 100 rods' distance between the two lines, then the boundary is a line 100 rods seaward of the mean high-water line. Therefore, ownership of riparian or littoral lands under this legal system includes the tidal flats often found in such areas. Much of the ordinance remains in effect in the Commonwealth of Massachusetts and the state of Maine, which was separated from Massachusetts in 1820, as part of the common law. Other states rely on parts of the ordinance as part of their statute law or common law.

Exceptions Based on Civil Law

Other exceptions to the use of the mean high-water line as a coastal boundary exist where civil law, as opposed to Anglo-American common law, controls. This is generally in areas where the land title has its roots in a grant from a sovereign power where civil law, such as the Roman Institutes of Justinian, prevailed. Possibly because the Roman civil law code was developed in an area of the Mediterranean with little daily tidal range, such law does not define the coastal boundary in terms of daily tide but, rather, in terms of seasonal water level changes. As an example, a translation of a portion of the Institutes of Justinian follows:

> The sea-shore, that is, the shore as far as the waves go at furthest, was considered to belong to all men.... The sea shore extends as far as the greatest winter floods runs up.[15]

That difference between the common law and civil law boundary definitions was clearly stated in the previously discussed *Borax* decision as follows:

> By the civil law, the shore extends as far as the highest waves reach in winter. But by the common law, the shore "is confined to the flux and reflux of the sea at ordinary tides."

As an example of the use of the civil law definition, Louisiana has adopted the line of the highest winter tide as its coastal boundary. Interestingly, the highest winter tide in the northern Gulf of Mexico is usually *less* than the highest tide in other parts of the year because of the prevailing annual constituent of the tides in the Gulf. The State of Hawaii also appears to follow the common-law approach with a coastal boundary defined as "the upper reach of the wash of the waves."[16] Another example is the State of Texas, which has recognized the civil law definition in areas of the state with origins of land title in Spanish or Mexican land grants. The Commonwealth of Puerto Rico also follows the civil law. For the latter two jurisdictions, it has been held that the limit of ownership is controlled by old Spanish law contained in *Las Siete Partidas*,[17] written in the thirteenth century, where the boundary is defined as follows:

> ... e todo aquel lugar es llamado ribera de la mar quanto se cubre el agua della, quanto mas crece en todo el ano, quier en tiempo del invierno o verano ... (Partida 3, Title 28, Law 4)

That definition appears to use the highest reach of the water whether in winter or summer ("*invierno o verano*"). The Supreme Court of Texas has interpreted that definition as being the mean higher high-water line in one case,[18] although, in coastal Texas, the mean higher high-water line is almost identical in height to the mean high-water line due to the prevailing diurnal tide in that area. In Puerto Rico, the boundary is usually determined by use of vegetative and geological features reflecting higher water levels. Tests of that approach indicate that such a line is generally higher than the highest astronomic tide and therefore reflects a line created by higher water levels such as those associated with storm events.[19]

9.6 OWNERSHIP OF NONTIDAL NAVIGABLE WATERS

Principle 6. *The submerged lands under natural navigable nontidal waters are owned by the public in most of the United States, although some states do not claim such ownership.*

The public trust doctrine also extends to navigable nontidal waters in most of the United States. This is actually a long-established concept addressed in the Roman Civil Code of Emperor Justinian I, which expressly declared all rivers as well as the seas to be commons areas. However, early English common law considered only tidal waters to be sovereign, possibly due to the fact that the English island has few inland waters with a capacity for public navigation. The English common-law position is illustrated by the following:

> That rivers not navigable (that is, freshwaters rivers of what kind so ever) do, of common right belong to the owners of the soil adjacent. But that rivers, where the tide ebbs and flows, belong to the State or public.[20]

In the United States, case law has generally differed from English common law on this issue. For example, the Massachusetts Colonial Ordinance of 1641 held that great ponds, in excess of 10 acres, could not be conveyed to private ownership. In 1876, this issue was addressed by the US Supreme Court with the case of *Barney* v. *Keokuk*,[21] as follows:

> The confusion of navigable with tide water, found in the monuments of the common law, long prevailed in this country, notwithstanding the broad differences existing between the extent and topography of the British island and that of the American continent .... And since this court ... has declared that the Great Lakes and other navigable waters of the country, are, in the strictest sense, entitled to the denomination of navigable waters, and amenable to the admiralty jurisdiction, there seems to be no sound reason for adhering to the old rule as to the proprietorship of the beds and shores of such waters. It properly belongs to the States by their inherent sovereignty ...

State common law has generally followed this lead. As a result, many of the states, although there are exceptions, claim title to the beds of navigable waters regardless of whether the waters are tidally affected. Navigability of nontidal waters for title purposes is generally a question of navigability-in-fact, with many states adopting an early test suggested by federal case law,[22] as follows:

> and they are navigable in fact when they are used or susceptible of being used in their ordinary condition, as highways for commerce, over which trade and travel are or may be conducted in the customary modes of trade and travel over water.

9.7 LANDWARD BOUNDARIES OF NONTIDAL WATERS

Principle 7.   *The boundary of publicly owned nontidal submerged lands is the ordinary high-water mark for many states, although a number of states claim only to the ordinary low-water mark.*

There was no precedent in English common law regarding boundaries of nontidal waters due to the fact that early English common law did not consider nontidal waters to be privately owned. Further, with the lack of a predictable rise and fall of water level such as that observed in tidal waters, definitions used in tidal waters could not be used directly. Therefore, American case law adopted the physical fact test to determine the equivalent of mean high water in nontidal waters. To distinguish that boundary from a mathematically derived tidal boundary (mean high water), that boundary is generally called the *ordinary high-water mark*. Prior to the *Borax* decision, the term *ordinary high-water mark* was also frequently used to describe what is now commonly called the *mean high-water line* in tidal waters. (It should be noted that a number of states use the *ordinary low-water mark* for the boundary of nontidal waters.)

The leading definition in federal case law, *Howard* v. *Ingersol*,[23] gives the following instructions for determining the ordinary high-water mark:

> This line is to be found by examining the bed and banks and ascertaining where the presence and action of waters are so common and usual and so long continued in all ordinary years, as to mark upon the soil of the bed a character distinct from that of the banks, in respect to vegetation, as well as in respect to the nature of the soil itself.

As may be seen, that opinion suggests the use of physical evidence of long-standing water to determine the boundary, and case law in most of the states has adopted this approach. Further, surveying practice has followed in that direction as suggested by the following guidelines from the *Manual of Surveying Instruction*:[24]

> The most reliable indicator of mean (ordinary) high water elevation is the evidence made by the water's action at its various stages, which are generally well marked in the soil. In timbered localities, a very certain indication of the locus of the various important water levels is found in the belting of the native forest species.

Considering both the legal and technical guidance, evidence that should be evaluated for determining the appropriate location of the ordinary high-water mark includes the following:[25]

1. *Geomorphological features*. These include features indicative of the natural limits of water bodies such as escarpments and natural levees.
2. *Vegetation*. This category includes evidence such as the lower limit of terrestrial vegetation and areas where vegetation has been wrested away by wave action or stream currents.

3. *Change in the character of the soil.* This includes evidence such as differences in organic content due to leaching and the landward limit of stratified beach deposits where waterward transport of eroded shoreline has resulted in a natural sorting of material.

4. *Hydrological records.* Although earlier case law has typically rejected the use of statistical averaging of water level data, some recent cases have suggested that hydrological data are potentially a valuable class of evidence. Further, tests of one proposed method for analyzing hydrographic data show agreement with traditional physical evidence.[26]

Some states have adopted the ordinary low-water mark as the boundary of navigable nontidal waters. Such a boundary is difficult to locate due to soil and vegetative evidence generally not existing at lower water stands. Typically, the best evidence of the ordinary low-water line includes escarpments associated with low-water stands and long-term hydrological records.

9.8 SIGNIFICANCE OF PUBLIC LAND SURVEY MEANDER LINES[27]

Principle 8. *A meander line of the Public Land Survey generally does not constitute a boundary. Rather, it represents the approximate location of the shoreline at a certain point in time.*

The meandering of navigable water bodies has been an integral part of the survey of the public lands since its inception.[28] As an illustration of this, the first survey conducted under the rectangular survey system (Township 1, Range 1, Seven Ranges, Ohio, by Deputy Surveyor Absolom Martin) included meandering of the Ohio River. Typical of early instructions for the Public Land Survey, the 1842 instructions to deputy surveyors in Florida addressed meandering as follows:

> You will accurately meander, by course and distance, all navigable rivers which may bound or pass through your district; all navigable bayous flowing from or into such rivers; all takes and deep ponds of sufficient magnitude ...

The first manual to provide detailed instructions for the location of the meander lines was the 1919 *Manual of Instruction.* That manual specifically called for the meander line to be run at "mean high water elevation" and provided further detail describing the line using currently accepted definitions of the ordinary high-water mark. In tidal waters, the instructions obviously will not yield as precise and repeatable a line as a more sophisticated mean high-water line determined by modern techniques. Nevertheless, in most cases, the line found by such instructions would be a good approximation of the mean high-water line. Therefore, it appears that the meander line was at least intended to be, and should be in most cases, a reasonable approximation of the ordinary high-water line or mean high-water line at the time of the survey.

Despite the fact that meander lines were intended to approximate the mean high-water line in tidal waters or the ordinary high-water line in nontidal waters, the location of a meander line generally may not be used as a boundary. This is due to the fact that the body of water itself is the natural monument and the current boundary when called in a description, not the representation provided by the meander line. Rather, the meander line represents only an approximation of the location of the boundary at a point in time. This concept is reflected in a number of court opinions typified by the following:

> It is an established and accepted principle, subject only to the exception hereinafter noted, that the meander line of an official government survey does not constitute a boundary, rather the body of water whose shoreline is meandered is the true boundary ... [29]

Nevertheless, there may be instances where the current water boundary cannot be located and the meander line may be accepted as the boundary by default, or where the discrepancy between the meander line and the shoreline is large enough to indicate intentional omission of certain lands or fraud. These instructions are illustrated by the following exceptions from court opinions:

> Under the circumstances of this case [Testimony had indicated that the true mean high-water line could not be located with any certainty], we hold that the meander line constituted the boundary line between the swamp and overflowed lands and the sovereignty lands ... [30]

> However, a meander line may constitute a boundary where so intended or where the discrepancies between the meander line and the ordinary high-water line leave an excess of unsurveyed land so great as to clearly and palpably indicate fraud or mistake.[31]

In addition, the meander line has value as a representation of the location of the boundary at a certain point in time in the process of apportionment of newly formed land created by accretion and reliction. Therefore, although certain limitations must be recognized, the location of a meander line can have legal significance for boundary determination.

9.9 OWNERSHIP OF NON–PUBLICLY OWNED SUBMERGED LANDS

Principle 9. *Submerged lands under non–publicly owned waters are customarily owned by the adjoiners, with the ownership extending to the center of the stream.*

Title to the submerged lands underlying non–public trust waters generally is in private ownership. As a result, boundary questions in such waters usually involve the boundary between two private owners.

Perhaps the most elementary example of this type of boundary is that involving a stream as the boundary between two parcels of land. In such a case, when the deeds of either side call "to the stream," the center of the stream would be the boundary. There is general agreement on the part of most courts on this issue. Yet there are two complicating issues relating to stream boundaries in which there appears to be a divergence among court opinions. These issues deal with the questions of what constitutes the "center" of the stream and how the lateral boundaries between the upland boundaries and the center of the stream should be run.

Regarding the center of the stream, one approach defines the boundary as the "thread of the stream," which is defined as the line lying equidistant between the banks. For example, one source,[32] supported by a number of court cases, states as follows:

> The term "thread of the stream" means the geographic center of the stream at ordinary or medium stage of the water, disregarding slight and exceptional irregularities in the banks. It is fixed without regard to the main channel of the stream.... If the stream is made a boundary in a private conveyance, then the thread of the stream will be the stream boundary.

The second approach regarding the center of the stream defines the boundary as the *thalweg*, or deepest part of the channel. As an example of this approach, *Stubblefield v. Osborn*[33] states as follows:

> Upon principle, therefore, it would appear that the thread of a nonnavigable river is the line of water at its lowest stage. The thread or center of a channel, as the term is above employed, must be the line which would give to the landowners on either side access to the water, whatever its stage might be and particularly at its lowest stage.

There also seems to be more than one approach to how lateral boundaries are extended to the center of the stream. At least one source[34] indicates that straight-line extension of the upland boundary would be the correct approach for unmeandered (not meandered in a public land survey) streams. Other sources suggest that such partition lines should be drawn perpendicular to the "line of navigation," especially if the stream is sinuous, which would create inequitable boundaries with the straight-line approach. A third approach found in the *Manual of Instructions*[35] directs apportioning the length of a median line according to the length of a meander line of the shoreline of the adjacent riparian lots.

When nonnavigable lakes or ponds form the boundary line between two or more parcels, similar approaches are used as those for the proportion of riparian rights (see Section 9.14). For substantially round lakes, the general rule is that a center point is selected at the geographical center (or, arguably, the deepest point in the lake) of the lake, and partition lines are then run from the ends of the upland boundaries to the center point forming "pie slices." For long or irregular lakes, the lines typically run perpendicular to a centerline.

9.10 SWAMP AND OVERFLOWED LANDS

Principle 10. *Swamp and overflowed lands are not part of the public trust submerged lands. In the public land states, such lands were granted to the states for subsequent improvement and conveyance to private owners.*

Uplands that were unfit for agriculture in their natural state due to inundation were given a special classification in the US Public Land Survey. They were not considered part of the navigable public trust submerged lands. However, since they were unfit for agriculture in their natural state, they were not considered suitable for conveyance to settlers without improvement. As a result, lands classified as such were granted to the individual states for improvement and resale based on a series of acts of Congress beginning in 1848. In some public land states, a large amount of land was considered to fall in this classification and be conveyed to the states. For example, in Florida, over 20 million acres were conveyed to the state as swamp and overflowed lands out of a total area of the state of 35 million acres. The state, in turn, sold much of that amount to private entities for the purpose of reclamation.[36]

For the purposes of classification, *swamplands* are those that "require drainage to make them fit for cultivation. *Overflowed lands* are those that are subject to such periodic or frequent overflows as to require levees or embankments to keep out the water and render them suitable for cultivation."[37] Swamp and overflowed lands are not tidelands, nor are they part of the beds of lakes or nontidal streams. Rather, they are part of the floodplain lying above the public trust submerged lands. This classification is emphasized in the instructions to public land surveyors,[38] where it is stressed that meander lines, which approximate the upland boundary of public trust waters, should not be established at the segregation line between the upland and the swamp and overflowed lands but at the actual margin of the bed of the water bodies.

Once the United States divested itself of these "unusable" lands to the limited number of states, the states were free to dispose of them in any manner they desired. Some were drained, some were sold to individuals and corporations, and some were retained by the states. In a recent case in Florida, the state obtained swamp land that bordered a fractional section on the Atlantic Ocean. The original surveyors who conducted the original township breakdown created a fractional Lot 5. The "water line" was surveyed and described in the field notes. The dividing line was listed in the field notes by courses and called a "delineation" line between land and swamp. The State of Florida subsequently conveyed that portion of the fractional section that lay seaward from the fractional lot to an individual who in turn conveyed it to the railroad, who subsequently conveyed it to the Nature Conservancy who donated it to a federal wildlife refuge. After nearly 100 years, the fractional lot was sold by the original patentee "according to the government survey." The new owner reading the notes and with recommendations from his surveyor claimed water access and started to build a walkway. This claim was based on the presumption that the "delineation line" was a meander line, and thus the fractional lot extended to the ocean. The "delineation line" was visible to observation as a minutely raised ridge.

After a two-day trial, the court held that the surveyor had to take the surveyor's notes literally. He did not call the line a meander line but a "demarcation" line; thus, it was a boundary that was created. This intention was manifested when the federal government conveyed the land as a separate parcel and the fractional lot as a separate parcel. In this instance, the "paper trail" from the original notes to today's deed became critical in explaining what a "demarcation line" was intended to be.

9.11 NAVIGATIONAL SERVITUDE

Principle 11. *The United States has authority to regulate activities on all navigable waters of the country.*

The commerce clause of the US Constitution confers on the United States a unique power of regulation and control over all navigable waters of the United States. For that purpose, they are the public property of the nation. This power extends to the entire stream and the streambed below the ordinary high-water mark. It allows the government to take property without compensation from private property owners. The proper exercise of this power is not an invasion of any private property rights in the stream or the lands underlying it, for the damage sustained does not result from taking property from riparian owners within the meaning of the Fifth Amendment but from the lawful exercise of a power to which riparian interests have always been subject. Thus, without being obligated constitutionally to pay compensation, the United States may change the course of a navigable stream or otherwise impair or destroy a riparian owner's access to navigable waters. It does not matter that the market value of the riparian owner's land is substantially diminished.

9.12 PUBLIC REGULATION OF RIPARIAN AND LITTORAL LANDS

Principle 12. *Riparian and littoral properties is often subject to public regulation, which may create regulatory limit lines, which, although different from property lines, may also affect the utility of the property.*

Riparian and littoral properties have traditionally been subject to certain restrictions in their use. The antiquity of this concept is evidenced even in the Institutes of Justinian, which, as indicated in the introduction to this chapter, required that privately owned riparian land remain open to the public, as follows:

The public use of the banks of a river is part of the law of nations, just as is that of the river itself. All persons, therefore, are as much at liberty to bring their vessels to the bank, to fasten ropes to the trees growing there, and to place any part of their cargo there, as to navigate the river itself. But the banks of the river are the property of those whose land they adjoin, and consequently the trees growing on them are also the property of the same person.[39]

With increased population density in littoral and riparian areas since the time of the Roman Empire, there has been a commensurate requirement for increased mediation by governmental policy and increased restriction on land use in such areas. Examples of such regulation include coastal construction control, regulation of dredge and fill in riparian and littoral wetlands, and regulation of development in floodplains.

Most coastal states have *coastal construction control lines* regulating or prohibiting construction in defined littoral areas. Typically, such lines are established by state agencies based on the height of the shoreline topography compared with the height of specific return frequency storm waves. Usually, the lines are monumented and have recorded metes and bounds descriptions.

Wetlands are natural areas where the groundwater table is normally at or close to the surface. Due to their value for plant and animal habitat, water purification, groundwater recharge, floodwater storage and flood peak reduction, and carbon storage, many states as well as the federal government regulate dredge and fill and construction in riparian wetlands. Typically, jurisdiction lines delimiting regulated wetlands are established only on an as-needed basis. Such wetlands are usually defined by the predominance of certain indicator species of vegetation, the presence of hydric soils, and/or the frequent presence of groundwater near the surface.[40]

Floodplains are natural overflowed areas adjoining surface waters. While some floodplains are also wetlands, many are not, even though they are subject to flooding during storm events. Floodplains are often defined based on the probability of being flooded in any given year. The primary concerns regarding land use in floodplains are the loss of life and property from flooding that can result when houses are constructed in these areas, together with concern that alteration of the natural topography of the floodplain could cause worse flooding. Activity in such areas is often regulated on the local, state, and federal levels. Federal floodplain regulation is administered through the National Flood Insurance Program administered by the Federal Emergency Management Agency. Generally, regulations regarding floodplains allow agriculture and similar low-impact uses as long as no topographical alterations are made. Construction is usually not allowed in the floodway zones closest to the surface water but is generally allowed in the outer portions of the floodplain with the provision that occupied areas and such facilities as septic tanks be elevated above maximum flood elevations.

In the foregoing examples of regulation of riparian and littoral lands, the land generally remains in the ownership of the upland owner, yet the use of the land may be severely restricted. In such cases, the regulation line often takes on a greater significance to the use of the land than that of the actual water boundary. As a result, it often becomes necessary to locate pertinent regulation lines in addition to or in lieu of the water boundary itself. Therefore, research for the location of riparian and littoral boundaries typically includes investigation into pertinent land use regulations.

9.13 SHORELINE CHANGES AND WATER BOUNDARIES

Principle 13. *The general rule is that water boundaries change with gradual and natural changes to the shoreline, while those boundaries do not change with sudden* *or human-made changes. Land that is gradually and naturally built up by accretion or exposed by reliction accrues to the riparian owner. Land that is gradually and naturally eroded away or permanently inundated by the water accrues to the owner of the bed.*

Water boundaries generally change with and follow changes in the position of the shoreline. Shoreline changes may be categorized as those relating to changes in water level, including withdrawal of the water (*reliction*) and rises in water level, and changes in land form, including deposition of material along the shoreline by water action (*accretion*) and the wearing away of the land by water and wind action (*erosion*).

The general rule is that the upland owner gains title to new upland created by reliction and accretion and loses title to lands submerged by rises in water level or lost by erosion. The general rule would presumably also apply to new upland created by the rising of land due to glacial rebound. That phenomenon is common in areas near melting glaciers. An exception to the general rule may occur when the shoreline changes are not gradual or imperceptible, or where such changes are artificially induced. Where such changes occur suddenly, such as during storms, this is termed *avulsion*, and it is generally held that title does not change with such shoreline changes. Likewise, it is usually held that shoreline changes resulting from human-made actions, such those associated with dredging or construction of groins, do not change title if the upland owner or a predecessor in title caused the changes.

There are exceptions to these general rules. For example, in regard to the ownership of accreted lands, it has been held in Louisiana that accretions along the Gulf of Mexico (but not along streams) belong to the state. A similar exception exists in the state of Washington.

Regarding erosion and inundation, land gradually eroded away or permanently inundated by water becomes part of the bed of the water body and generally belongs to the owner of the bed. In sovereign-owned water bodies, the ownership changes from the riparian owner to the public as owners of the bed. However, if a riparian owner already owns the bed or owns to the middle of the water course, ownership remains the same.

Unlike gradual erosion, land is not lost by avulsion. Therefore, land detached from the riparian land by the sudden process of avulsion belongs to the owner to whose land it was attached. Similarly, land created by sudden action does not accrue to the ownership to which it is attached. A river that changes its course by cutting a new channel and forming an island does not change the ownership of the new island. Similarly, ownership of a parcel of ground detached from the land of one owner and attached to the land of another by the sudden change of a river course is not lost by the original owner. In some states, a statute of limitation determines how soon the former owner must claim the detached land. For example, in some states, when a riparian owner on a publicly owned water body loses land due to avulsion, the riparian owner has the right to fill the newly submerged lands if accomplished within a certain period of time after the loss.

Human-caused shoreline changes are generally referred to as *artificial changes*. Such changes may alter a shoreline directly, as by installation of fill or a jetty, or may indirectly cause artificial accretion that changes the shoreline. Generally, direct artificial shoreline changes do not change title. As a general rule, the owner of the bed of submerged land becomes the owner of fill placed on it. But land created at the water's edge as the result of filling or dredging by a third person has been held or recognized to be the property of the adjacent upland owner as with natural accretion. It has been generally held that shoreline changes resulting from human-made actions, such as those associated with dredging or groins, do not change title if the upland owner or a predecessor in title to the upland owner caused the changes, but may change title if the shoreline changes were caused by action beyond the control of the upland owner. As an illustration of this, a Florida case[41] held that artificial accretion caused by the upland owner remains the property of the sovereign, while another case in that state[42] held that artificial accretion caused by third parties will accrue to the upland owner.

When erosion or permanent inundation causes all of a riparian tract to be submerged, the new land touched by the water becomes riparian. If the original riparian land reappears due to accretion or reliction, judicial guidance appears to vary as to whether the reappeared land belongs to the original owner or to the newly riparian owner. As an example, in *Yearsley* v. *Gipple*,[43] the court held that if lands become riparian by the eroding away of adjoining lands, the owner is entitled to the right of the riparian owner to accretions, even though they extend beyond the original boundary line of the land. Conversely, in *Allard* v. *Curren*[44], land in Section 30 was eroded away by the Missouri River until land in Section 31 became riparian. By accretion, land in Section 30 was restored. The restored land in Section 30 was ruled not to be the property of the owner of Section 31. A consideration in such decisions appears to be whether the submergence was long continued so as to preclude identity upon reliction and whether erosion transported soil beyond the limits of ownership.

Based on all of this, it may be seen that the ambulatory nature of riparian boundaries requires the use of additional procedures to those required for the more conventional land boundaries. The first of these is that it is necessary to associate a *time* with the determination of a water boundary. The location today may not be the same as the location of that boundary a few years later. A second requirement is the need to determine *the nature and causes of any changes* that have taken place, considering the foregoing discussions on avulsion and human-caused changes. A third requirement is the necessity at times to determine the *location of historic shorelines* when shoreline changes have occurred that have not resulted in a change of title.[45]

9.14   APPORTIONMENT OF RIPARIAN AND LITTORAL RIGHTS

Principle 14.   *Regarding the apportionment of riparian and littoral rights:*

- *The objective of apportionment is to distribute either newly formed land or rights in water in proportion to the length of water frontage. Therefore, the length of the affected tract's water frontage, not its total area or upland lateral lot lines, generally controls the apportionment of riparian rights.*

- *For the division of newly formed land, the point of departure for a division line is the intersection of the upland lateral boundary and the shoreline at the time of the subdivision of the area. For the division of water area, the point of departure for a division line is the intersection of the upland lateral boundary and the current shoreline.*

- *For the division of newly formed land, the newly formed shoreline should be proportioned relative to the respective lengths of the old and new shorelines.*

- *In streams, the general rule for the division of water area is to project the division line in a direction perpendicular to the thread of the stream.*

- *In large bays, oceans, or wide rivers where a thread is not readily discernible, the general rule for the division of water area is to run lines perpendicular to various other baselines such as the shoreline if it is relatively straight, the bulkhead or pier line, or the channel line.*

- *In lakes, the general rule for the division of water area for round lakes is for side lines to be drawn to a center point, creating pie slice–shaped areas. For long lakes, it is generally held that a centerline should be constructed and lines projected out perpendicular to the centerline.*

- *In partially enclosed areas, such as coves, the general rule for the division of water area is to proportion the closing line across the mouth of the area based on the respective lengths of the shoreline.*

Riparian and littoral rights are those rights associated with ownership of land fronting on water. Actually, riparian rights (derived from the Latin *ripa*, "a riverbank") apply only to lands bordering rivers and streams, while littoral rights (derived from the Latin *litus*, "the seashore") apply to oceanfront lands. Due to the more general use of the word, *riparian* will be used in this writing for rights associated with all types of water bodies. These rights include the right to newly formed upland created by accretion or reliction. In addition, these rights also may include a wide range of privileges, subject to state law, associated with the use and enjoyment of waterfront property, including reasonable use of water for domestic, irrigation, and livestock use, construction of docks and mills, access to navigable channels, and scenic views across the water.

While the method for determining limits of the riparian rights area of a tract may vary significantly with the situation, the objective is to apportion either the newly formed land or the water area over which riparian rights are available in proportion to the relative length of each riparian tract. As a result, the length of the affected tract's water frontage, not its total area or the direction of the upland lateral lot lines, generally controls the apportionment of riparian rights.

Where the limits of rights to newly formed land are at issue, it is generally held that the starting points for lines dividing the new land among the affected riparian owners are at the intersection of the lateral upland lot line and the shoreline at the time of the subdivision of the area. For the division of water areas for other riparian

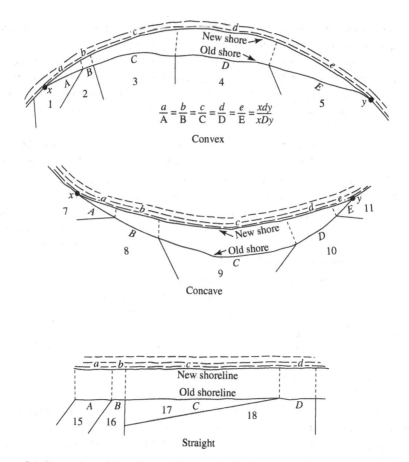

Figure 9.1 Proportionate shoreline method for distributing newly formed land.

rights, the starting points for the division lines are at the intersection of the lateral upland lot lines and the current shoreline, such as the mean high-water line or the ordinary high-water line.

For the division of newly formed land, it has been generally held that the newly formed shoreline should be proportioned relative to the respective lengths of the old shoreline (Figure 9.1). This is usually accomplished by measurement of the newly formed shoreline and use of the original shoreline from the survey subdividing the area.

For the division of water area, there are typically not two shorelines to proportion, as with the division of newly formed land. Therefore, other means must be used to provide comparable results. The method used for this purpose varies with the nature of the water body. It has been generally held that the proper procedure for this does *not* involve extension of the upland boundary without a change of direction. Therefore, the direction of the upland boundaries is generally not controlling for division lines for riparian rights.

In rivers that are significantly narrow for determining a *thread* or center of stream, the procedure suggested in most cases is to project the line in a direction perpendicular to the thread of the stream. The "thread" of a stream is the line equidistant from the two banks of the stream. Typical cases suggesting this approach include *Campau Realty Co.* v. *Detroit.*[46] Where the geometric center of the stream is considerably removed from the thalweg, or deepest part (channel) of the stream, there may be justification for using the thalweg as the center. Location of the thalweg requires the use of hydrographic soundings.

In bays or oceans, as well as in wide rivers where a thread is not readily discernible, case law generally suggests using lines run perpendicular to various other baselines such as the shoreline if it is relatively straight, the bulkhead or pier line, or the channel line. Among numerous other cases, this approach has been specifically held in the Florida case of *Hayes* v. *Bowman,*[47] which dealt with a wide water body, where the court held that lateral lines run normal to the channel line would be the proper division lines (Figure 9.2).

One use of this approach is in the State of Mississippi, where riparian rights are limited to 750 yards waterward of the mean low-water line by statute. Therefore, in that state, for rivers and other water bodies greater than 1500 yards in width, the lateral lines for the riparian rights areas are either run perpendicular to a line offset 750 yards from the mean low-water line or the area can be proportioned based on the respective lengths of the shoreline and the offset line.[48]

Where the shoreline in question is curving or irregular such as in coves, the general direction from case law is to give each riparian tract a proportionate share based on

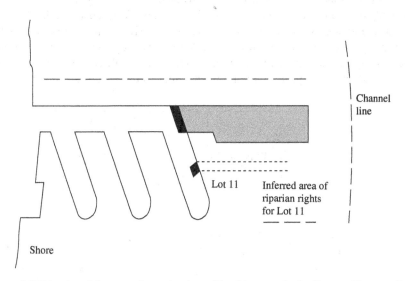

Figure 9.2 Riparian rights area determination with wide water body. Source: Hayes v. Bowman, Fla., 91 So.2d 795 (1957) "/U.s. Supreme court.

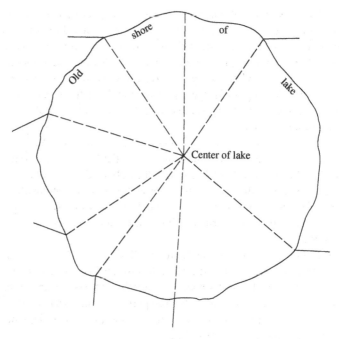

Figure 9.3 Round lake method for proportioning riparian rights.

measurement of the shoreline in question as well as measurement of some outer line such as the thread of the stream or the closing line of the cove.[49]

Where riparian rights on lakes are involved, different approaches often are required, since a closed figure is involved. In round lakes, most case law has suggested side lines drawn to a center point, creating pie slice–shaped areas (Figure 9.3) such as suggested in *Hansen* v. *Rice*.[50] For long lakes, the judicial direction has been to use an approach similar to that used for rivers with a thread or centerline being constructed and lines projected out perpendicular to the centerline (Figure 9.4), as

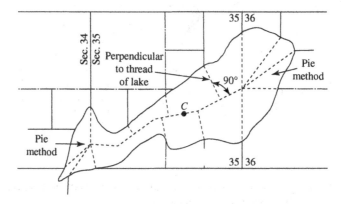

Figure 9.4 Long lake method for proportioning riparian rights.

suggested in *Rooney* v. *County of Stearns*.[51] Another approach for long lakes with straight shorelines is to use division lines run perpendicular to the shoreline. For large, irregularly shaped lakes, proportionate division of the line of navigable water, similar to that suggested for irregular shorelines in rivers, has been suggested.

A proportionality test may be used to evaluate the fairness of a proposed apportionment. This technique compares the ratio of the areas allotted to two adjacent riparian lots to the ratios of the lengths of the shorelines of the two lots. Such an evaluation is based on methods used to evaluate proposed division of continental shelf areas for international boundaries.[52] If the ratios are not similar, the proposed division may be considered inequitable.

Reviewing all of this, it may be seen that there are at least four basic cases—narrow streams, wide water bodies, irregular shorelines, and lakes—with each case having its own rules. With narrow rivers, the general rule is to use a division line perpendicular to the thread of the stream. With wider water bodies, the general rule is to use a line run perpendicular to the shoreline, offset line, or to some other baseline such as a bulkhead or channel line. For most lakes, the general rule is to use lines drawn to a common center point or centerline. For curving or irregular shorelines, proportional division of the centerline based on length of the shoreline is the general rule.

Because of variations in case law, there are other approaches for certain applications such as the Colonial Method used for division of tidal flats in some of the New England states (Figure 9.5). That method uses proportionate measurement between the low-water and high-water marks. Instructions for the use of that method as prescribed in the case of *Emerson* v. *Taylor*[53] are as follows: "Draw a base line from the two corners of each lot, where they strike the shore, and from those two corners, extend parallel lines to low-water mark, at right angles with the base line. If the line of the shore be straight there will be no interference in running the parallel lines. If the flats lie in a cove, of a regular or irregular curvature, there will be an interference

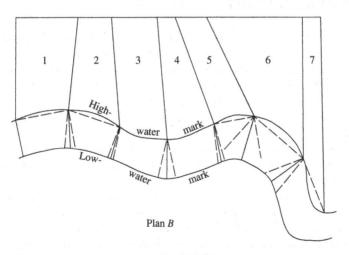

Plan *B*

Figure 9.5 Colonial Method for proportioning tidal flats.

in running such lines, and the loss occasioned by it must be equally borne or gain enjoyed equally by the contiguous owners . . . ."

9.15 EMERGENT OR OMITTED ISLANDS

Principle 15. *Emergent islands are generally considered to belong to the owner of the bed of the water body. In public land states, omitted lands (such as islands existing at the time of, but not surveyed in, the public land survey) remain in the ownership of the federal government until specifically conveyed. Islands emerging after the conveyance of the water body to the states with statehood, belong to the owner of the bed of the water body.*

The general rule regarding ownership of accreted lands also applies to emergent islands, where it is generally held that islands formed in a water body belong to the owner of the bed of water body. If an island forms on one side of a stream and the riparian proprietors own the bed of the stream, the island belongs to the proprietor on whose side of the stream it formed. If an island is formed in the middle of a stream and the riparian proprietors own the bed of the stream, the island is divided by the line through the thread of the stream. When the public of the state owns the bed of the stream or lake, the adjoining riparian owners are not entitled to an island that emerges from the bed of the stream. The right to new islands follows the right of the soil on which they were formed. In public land states, islands that existed at the time of the original public land survey but were overlooked in the survey are considered to be *omitted lands* that remain in the ownership of the federal government until validly conveyed by patent.

9.16 WATER BOUNDARIES OTHER THAN SEA

The International Boundary Research Unit at Durham University, England, has estimated that today there are in excess of 2000 boundary disputes dealing with water boundaries between countries. Some countries have resorted to shooting while others have resorted to law. Many of these boundary issues are centuries old, and the documentary evidence as well as physical evidence is scarce or not available. The United Nations and the World Court have attorneys and surveyors working worldwide trying to keep peace in the world.

At one or more times in their careers, there are several classes of water boundaries that attorneys and surveyors should be acquainted with: inland water boundaries of streams and lakes.

In dealing with these types of boundaries they are significant in that many are privately owned and landowners become very possessive of use by strangers.

Within this area of inland boundaries one can find meandered lakes, non-meandered lakes, navigable streams and rivers, and non-navigable streams and rivers.

Streams and lakes became an important issue because many of the early surveyors used these natural features as boundary-dividing lines between nations, states, counties, and individuals. One can see that the law has placed a high degree of legality of the accurate location of the features.

9.17 MAJOR RECOGNIZED AREAS

Some states are blessed with an overabundance of natural lakes and rivers, and some states have a scarcity. Each state has its own peculiarities. The natural divisions on inland water boundaries are meandered lakes of the GLO, non-meandered GLO lakes, private lakes, rivers as state boundaries, and rivers as private boundaries. Private lakes all in one ownership; private lakes in multiple ownerships. Each and every one of these has different rules for surveying and for applying the laws.

9.18 CONCLUSIONS AND RECOMMENDATIONS

In this final section, we wish to bring to the student and the professional who will deal with water boundary issues that this area of boundaries is steeped with ambiguity and uncertainty. This area of boundary creation, retracement, and litigation is still trying to find its way out of the "dark ages." One can find a proliferation of recommendations on how water boundaries should be divided. None are absolute. After decades, the courts are struggling with water issues. Our recommendation is to approach a water boundary problem with caution, but do an adequate job of research and do your professional best.

NOTES

1. Thomas C. Sandars, *The Institutes of Justinian* (London: Longmans, Green & Co., 1874).
2. George M. Cole, *Water Boundaries* (New York: John Wiley & Sons, 1997).
3. H. S. Truman, The Continental Shelf, Proclamation No. 2667, 59 Stat. 884, 1945.
4. Ronald Reagan, Territorial Sea of the United States of America, Proclamation 5928, Federal Register, Vol. 54, W05, 1989.
5. Nathaniel Bowditch, *American Practical Navigator* (Washington, D.C.: U.S. Government Printing Office, 1962).
6. *U.S.* v. *Louisiana, Texas, Mississippi, Alabama* and *Florida*, 3 63 U.S.I. 212 (1960).
7. Cole, supra note 2.
8. *Cinque Bambini Partnership et al.* v. *State of Mississippi et al., Miss.*, 491 So.2d 508 (1986).
9. *Phillips Petroleum* v. *Mississippi*, 484 U.S. 469 (1988).
10. *Martin* v. *Waddell*, 41 U.S. (16 Pet.)367 (1852).
11. M. Hale, *De Jure Maris* (1666). Reprinted in S. Moore, *A History of the Foreshore and the Law Relating Thereto*, 3rd ed. (1888).

12. *Attorney General* v. *Chambers*, 43 Eng. Rep. 486 (1854).

13. *Borax Consolidated Ltd.* v. *City of Los Angeles*, 296 U.S. 10 (1935).

14. 12 Am.Jur.2d, Boundaries, Sec. 36 Mass. Col. Ord. (1997).

15. Sandars, supra note 1.

16. Frank E. Maloney and Richard C. Ausness, "The Use and Legal Significance of the Mean High Water Line in Coastal Boundary Mapping," *The North Carolina Law Review*, December 1974.

17. Samuel B. Scott, *Las Siete Partidas* (New York: Commerce Clearing House, 1931).

18. *Luttes* v. *State*, 324 S.W.2d 167 (1958).

19. George M. Cole, "Sea Level Measurements and Their Applications," *Proceedings, SaGES Conference*, Mayaguez, Puerto Rico, 2011.

20. Hale, supra note 11.

21. *Barney* v. *Keokuk*, 94 U.S. 324 (1876).

22. *The Daniel Ball*, 77 U.S. (10 Wall.) 557, 563, 19 L.Ed 999 (1871).

23. *Howard* v. *Ingersoll*, 54 U.S. 381, 427 (1851).

24. Bureau of Land Management, *Manual of Instruction for the Survey of the Public Lands of the United States* (Washington, DC: U.S. Department of the Interior, 1973).

25. Cole, supra note 2.

26. George M. Cole, "Use of Hydrology for Determining Ordinary High Water in Non-Tidal Waters," *Technical Papers*, American Congress on Surveying and Mapping, Spring 1988.

27. George M. Cole, "The Significance of the Meandering of Water Bodies in Public Land Surveys," *Surveying and Land Information Systems*, American Congress on Surveying and Mapping 50, no. 3 (September 1990).

28. Lane J. Bouman, "The Meandering Process in the Survey of the Public Lands of the United States," in *Proceedings of the Water Boundary Workshop*, California Land Surveyors Association, 1977.

29. *Connerly* v. *Perdido Key Inc.*, Fla., 270 So.2d 390, 397. (Fla.App. 1 Dist. 1972).

30. *Trustees* v. *Wetstone*, Fla., 222 So.2d 10 (1969).

31. *Lopez* v. *Smith*, 109 So. 2d 176, Fla.2d DCA (1959).

32. Edward S. Bade, "Titles, Points, and Lines in Lakes and Streams," *Minnesota Law Review* 24:305, 1940.

33. *Stubblefleld* v. *Osborn, Neb.*, 31 N.W.2d 547., (1948).

34. Bade, supra note 32.

35. Bureau of Land Management, supra note 24.

36. George M. Cole, *Hydrology-Based Wetland Delineation*, doctoral dissertation, Department of Geography, Florida State University, 2007.

37. *San Francisco Savings Union* v. *Irwin*, 28 F 708 (1886).

38. Bureau of Land Management, supra note 24.

39. Sandars, supra note 1.

40. Cole, supra note 36.

41. *McDowell* v. *Trustees*, Fla., 90 S.2d 715 (1956).

42. *Board of Trustees* v. *Sand Key Associates Ltd.*, 512 So.2d 209, Fla.2d. DCA (1973).

43. *Yearsley* v. *Gipple*, 175 N.W. 641 (1919).

44. *Allen* v. *Curran*, 168 N.W. 761 (1918).
45. Cole, supra note 2.
46. *Campau Realty Co.* v. *Detroit, Mich.*, 127 NW 365 (1910).
47. *Hayes* v. *Bowman, Fla.*, 91 So.2d 795 (1957).
48. George M. Cole, *Study of Apportionment of Littoral Rights*, prepared for Mississippi Department of Marine Resources and Mississippi Secretary of State, 2003.
49. *Deerfield* v. *Arms*, 28 Am. Dec. 27, Mass. (1835)
50. *Hanson* v. *Rice*, 92 N.W. 982, Minn. (1903).
51. *Rooney* v. *County of Stearns*, 153 N.W. 858, Minn. (1915).
52. Jonathan I. Charney, "The Delineation of Ocean Boundaries," in *Rights to Oceanic Resources* (Boston: Martinus Nijhoff Publishers, 1989).
53. *Emerson* v. *Taylor*, 9 Me. 42 (1832).

CHAPTER 10

RETRACING AND "RESURVEYING" SECTIONALIZED LANDS

10.1 INTRODUCTION

As described in earlier chapters, the creation of boundaries is unique and demands people who are technically qualified. Yet the retracing of these once-created boundaries also requires a second unique type of person, who probably should be more attuned to the law of evidence than to being a "technology freak." From the beginning, it should be understood in the practice of land surveying that a resurvey and a retracement are not synonymous. The surveyor who elects to conduct an original survey should have different qualifications from those of a retracing surveyor. Retracements require people of different capabilities and possibly specialized knowledge and training.

Chapter 6—Creation and Retracement of General Land Office Boundaries—discussed the creation of the General Land Office (GLO) or Public Land Survey System (PLSS) boundaries. This chapter discusses the suggested rules for retracing those boundaries. It will discuss retracing metes and bounds boundaries as well. The majority of the modern surveyor's work, except for certain local subdivisions, is to "find the land to be surveyed" by finding the evidence of the original footsteps.

When the words "sectionalized lands" or "public lands" are mentioned, the public lands immediately should come to mind. But the surveyor may find that there are other lands that use that description.

Brown's Boundary Control and Legal Principles, Eighth Edition.
Donald A. Wilson, C.A. "Tony" Nettleman III, and Walter G. Robillard.
© 2024 John Wiley & Sons, Inc. Published 2024 by John Wiley & Sons, Inc.

We will address how to find the boundaries to be surveyed and what should be the proper rules and procedures for their retracement. We will concentrate on the rules and methods that should be used to find those original boundaries, with the emphasis on the GLO boundaries, but referencing metes and bounds boundaries as well.

The following principles are discussed in this chapter:

PRINCIPLE 1. An original survey creates original boundaries as well as the evidence of these boundaries. This created evidence and the subsequent written record become the "footsteps" to be followed in a retracement.

PRINCIPLE 2. In any retracement, whether by a federal agency or by a private surveyor, the description contained in the patent of the original boundaries is the controlling description the retracing surveyors should use to place the original boundaries on the ground.

PRINCIPLE 3. When retracing or dividing sections of land created under federal laws, federal rules of survey must be followed (1) when the final adjudication of problems is tried in the federal courts or (2) when a state has accepted or adopted the federal rules and when the original patents were federal in nature.

PRINCIPLE 4. Provided that a superior right is not interfered with or a fraud is committed, the boundaries of public lands, when approved, accepted, and patented, are unchangeable and unassailable by individuals or by the courts.

PRINCIPLE 5. Original township, section, and quarter-section corners stand as the true corners that they were intended to represent, regardless of where they are located. This is not true of closing corners.

PRINCIPLE 6. Plats and all original field notes become part of the grant or patent. When conducting a retracement, the surveyor should consult available and known plats and notes.

PRINCIPLE 7. Closing corners that are not actually located on the line that was closed upon will determine the direction of the closing line but not its legal terminus; the true position is the point of intersection of the two lines.

PRINCIPLE 8. After making allowances for natural changes, a monument, to be identifiable as an original monument, should not differ greatly from the field notes.

PRINCIPLE 9. When there is acceptable evidence indicating the location of an original corner, the corner should be considered as obliterated and should be relocated from that evidence.

PRINCIPLE 10. The original location of a corner may be identified by the evidence as well as by testimony of a witness who saw that corner. It is an exception to the hearsay rule, but the evidence as well as the testimony must be evaluated as having to meet the "preponderance of evidence" criterion, not the "beyond a reasonable doubt" criterion or standard that has been established for criminal matters.

PRINCIPLE 11. In a very limited number of situations, a corner's position may be located by common use of the monument and point.

PRINCIPLE 12. A corner position may be proven from evidence of found original line trees or from natural features found in close proximity to the corner and identified in the notes.

PRINCIPLE 13. Lost corners may be located by proportioning or by using a combination of evidence and measurements.

PRINCIPLE 14. The surveyor should strive to utilize all forms of evidence to prove a corner as an obliterated corner, rather than resorting to proportional measurements to set a lost corner.

PRINCIPLE 15. When the surveyor has exhausted all possibilities of locating corners of public land surveys from available evidence, only then, as a last resort, should the surveyor rely on measurements alone.

Guiding principles for proportionate measurements are as follows:

A complete discussion of the following principles are listed and discussed in the latest edition of *Manual of Surveying Instructions For the Survey of the Public Lands of the United States.*[1]

PRINCIPLE 15A. When proportionate measurements are used, the new values given to the several parts, as determined by the measurement, should bear the same relation to the record lengths as the new measurement of the whole bears to that record.

PRINCIPLE 15B. A single proportionate measurement should be made between two proven corners, and may locate one or more original points on a line.

PRINCIPLE 15C. Double proportionate measurement usually requires four original corners, and the location of the proportioned corners is positioned at the proportionate distances of lines run from the proportioned points. Primary weight is given to distance, not to angle or bearing.

PRINCIPLE 15D. Lost standard corners will be restored to their original positions on a baseline, standard parallel, or correction line by single proportionate measurement on the true line connecting the two nearest identified original standard corners. Proper adjustment should be made to secure the correct latitudinal curve. Closing corners are not to be used for either direction or measurement.

PRINCIPLE 15E. When the principal meridian or guide meridian was established by alignment in one direction only, lost township corners on such lines should be restored by single proportionate measurement. Where guide meridians were established as part of the scheme of township boundaries, all surveyed in the same system under a single contract, the township corners located thereon should be relocated by double proportionate measurements.

PRINCIPLE 15F. When lost regular township corners common to four sections and lost regular section corners between township lines were originally established with ties from four directions, the lost corners will be reestablished by double proportionate measurements.

PRINCIPLE 15G. All lost section and quarter-section corners on township boundary lines will be restored by single proportionate measurement between the nearest identified corners on opposite sides of the missing corner, north and south on a meridional line, or east and west on a latitudinal line, after the township corners have been identified or relocated. An exception to this principle will be noted in the case of any exterior whose record shows deflections in alignment between the township corners.

PRINCIPLE 15H. When a line has not been established in one direction from the missing township corner, the record distance will be used to the nearest corner identified in the opposite direction.

PRINCIPLE 15I. Where the intersecting lines have been established in only two directions, the record distances to the nearest identified corners on those two lines will control the position of the temporary points; then, from the latter, cardinal offsets will be made to fix the desired point of intersection.

PRINCIPLE 15J. All lost quarter-section corners on section boundaries within the township will be restored by first establishing or proving the section corners on each side; then the quarter corners will be relocated by single proportionate measurement on line. An exception occurs where the original lines had angular deflection.

PRINCIPLE 15K. Where a line has been terminated with measurement in one direction only, a lost corner will be restored by record bearing and distance, counting from the nearest regular corner, the latter having been duly identified or restored.

PRINCIPLE 15L. A lost closing corner on a standard parallel will be reestablished on the true line that was closed upon and at the proper proportional interval between the nearest regular corners to the right and left. The only corners that will control the direction of the line being closed upon are (1) standard township, standard section, and standard quarter corners; (2) meander corners terminating the survey of the standard parallel; and (3) closing corners in those cases where they were established originally by measurement along the standard line as points from which to start a survey.

PRINCIPLE 15M. When the north quarter corner on a closing section was originally set and lost, the lost corner will be reestablished on the closing line at a point at the proper proportionate interval between the nearest found or relocated corners to the right and left.

PRINCIPLE 15N. The method to be followed in the subdivision of a section into quarter sections is to run straight lines from the established quarter-section corners to the opposite quarter-section corners; the point of intersection of the lines thus run will be the corner common to the several quarter sections, or the legal center of the section.

PRINCIPLE 15O. Preliminary to the subdivision of quarter sections, the quarter–quarter or sixteenth-section corners will be established at points midway between the section and quarter-section corners, and between the quarter-section corners and the center of the section.

PRINCIPLE 15P. The law provides that where opposite corresponding quarter-section corners have not been or cannot be fixed, the subdivision of section lines shall be ascertained by running lines from the established corners north, south, east, or west, as the case may be, to the watercourse, reservation line, or other boundary of such fractional section, as represented on the official plat.

PRINCIPLE 16. Lost meander corners, originally established on a line projected across the meanderable body of water and marked on both sides, will be relocated by single proportionate measurement after the section or quarter-section corners on the opposite sides of the missing meander corner have been duly identified or relocated. Where a line was terminated with measurement in one direction only, a lost corner will be restored by record bearing and distance, counting from the nearest regular corner, the latter having been duly identified or restored.

PRINCIPLE 17. A meander line platted as the boundary of a riparian owner is seldom the true boundary; the body of water represents the true boundary. Lands acquired by the Swamplands Act and fraudulent surveys go to the meander line.

PRINCIPLE 18. After locating the positions of the meander corners on section lines, the record meander courses and distances are run, setting temporary angle points; the closing error from the last meander course to the section line meander corner is noted. The error of closure is balanced out by the compass rule.

PRINCIPLE 19. Where boundaries that would normally be straight lines were originally established in an irregular manner, the resurvey (retracement) will follow the irregular procedure.

PRINCIPLE 20. Where an original quarter-corner was not originally set, as in the case of sections closing on a correction line, sections closing on a range line with double or triple corners, or sections closing on a township line with double or triple corners, place the missing corner on the correction line (or range or township line) at a point between the found or relocated closing section corners a distance that is proportional to the measurements used for the acreage calculations on the original plat. The missing corner is usually set midway between closing section corners except in Section 6, where it is usually 40 chains proportional measure from the northeast or southwest closing section corner.

PRINCIPLE 21. The method to be followed in the subdivision of a section into quarter sections is to run straight lines from the established quarter-section corners to the opposite quarter-section corners; the point of intersection of the lines thus run will be the corner common to the several quarter sections, or the legal center of the section.

PRINCIPLE 22. Preliminary to the subdivision of quarter sections, the quarter–quarter or sixteenth-section corners will be established at points midway between the section and quarter-section corners, and between the quarter-section corners and the center of the section, except on the last half-mile of the lines closing on township boundaries, where they should be placed at 20 chains, proportionate measurement, counting from the regular quarter-section corner. The quarter–quarter or sixteenth-section corners having been established as directed in these principles, the centerlines of the quarter sections will be run straight between opposite corresponding quarter–quarter or sixteenth-section corners on the quarter-section boundaries. The intersection of the lines thus run will determine the legal center of a quarter section.

PRINCIPLE 23. The law provides that where opposite corresponding quarter-section corners have not been or cannot be fixed, the subdivision-of-section lines should be ascertained by running lines from the established corners north, south, east, or west, as the case may be, to the watercourse, reservation line, or other boundary of such fractional section, as represented on the official plat. In this, the law presumes that the section lines are due north and south, or east and west lines, but this is not usually the case. Hence, to carry out the spirit of the law, in running the centerlines through fractional sections, it will be necessary to adopt mean courses, where the section lines are not due cardinals, or to run parallel to the east, south, west, or north boundary of the section, as conditions may require, where there is no opposite section line.

PRINCIPLE 24. Other than creating new subdivisions for private landowners, today the majority of the modern surveyor's work is retracing the lines of the original surveys, some of which are ancient in nature.

10.2 AREAS OF AUTHORITY

Principle 1.   *An original survey creates original boundaries as well as the evidence of these boundaries. This created evidence and the subsequent written record become the "footsteps" to be followed in a retracement.*

The term *survey* may have different meanings to different people. A surveyor is usually asked, "Did you conduct a survey?" The response, usually is yes or no. The surveyor could reply, "Do you mean the noun, the verb, or the adjective ?" Surveying as a generic term may take on multiple meanings, depending on the circumstances and whomever you ask.

In conducting retracements and resurveys, two areas of authority may be found. The ultimate authority in a resurvey of public lands is the Bureau of Land Management (BLM), the administrative successor to the GLO, but through the authority of the US Secretary of the Interior. The ultimate authority in a retracement by a private surveyor lies in the authority granted to the registered surveyor by the registration agency of the respective state. Technically and legally, no private registered surveyor may conduct a *resurvey* of a parcel that is described in reference to a federal survey by an aliquot parcel description. Surveyors have authority only to conduct retracements, which in and of themselves are binding on no landowner or other surveyor, whereas a retracement should be considered nothing more than an opinion of the corners and lines based on the evidence of the original survey that is recovered. This can include tangible evidence as well as measurements, both original and subsequent.

The BLM is the only agency that Congress has authorized to conduct original surveys of the public domain land and is also authorized to promulgate rules and regulations to put into effect the land laws of the United States. These rules and guidelines apply only to the agency that created the rules. This agency issued various manuals and a pamphlet entitled *Restoration of Lost and Obliterated Corners.* The pamphlet summarizes the findings of federal courts on how to make resurveys and how to divide protracted portions of sections. The information contained in this pamphlet is also included in one of the *Manuals.* Although the instructions given on how to make original surveys are entirely within the authority of the BLM, the rules for resurvey are acceptable only when they agree with court findings. As noted, the pamphlet summarizes federal court findings, but not those of all state courts. Also, because it does not contain instructions on how to resurvey in special situations, other sources of advisory information must be sought.

In this context, no state legislature can enact survey laws for the retracing or reestablishment of public land survey corners or lines that are in conflict with the federal laws under which the original lines were surveyed or created. This also applies to the retracing of federally created corners and resurveying and replacing lost corners.

Principle 2.   *In any retracement, whether by a federal agency or by a private surveyor, the description contained in the patent of the original boundaries is the controlling description the retracing surveyors should use to place the original boundaries on the ground.*

10.3 RESURVEY OR RETRACEMENT

A few states have adopted the *BLM Manual* and the *Manual*'s definition of resurvey; thus, in those states it is binding. The *Manual* reads: "A resurvey is a reconstruction of land boundaries and subdivisions accomplished by rerunning and re-marking the lines represented in the field-note record or on the plat of a previous official survey. The field-note record of the resurvey includes a description of the technical manner in which the resurvey was made, full reference to recovered evidence of the previous survey or surveys, and a complete description of the work performed and monuments established. The resurvey, like an original survey, is subject to approval of the directing agency."

As discussed in an earlier chapter, an *original survey* creates boundaries, and without an original survey, there would be no boundaries. But once the boundaries are created, no lesser survey agency has authority to modify or alter the original boundaries, other than the agency that created them; and that agency has no authority to change these boundaries if *bona fide rights* granted previously are in any way impinged upon.

Since an original survey can never be re-created, these agencies and registered surveyors are authorized to conduct other classes of surveys. For other states, the definition of *resurvey* is only suggestive, in that resurveys are classified into two major categories. First, we should consider a *retracement survey* or *retracement*.

Once an original survey is conducted and accepted, that parcel can never be resurveyed. It can only be retraced and the original measurements defined to more modern or precise references.

According to the 1973 *Manual,* a retracement merely remeasures lines and identifies monuments or other marks of the prior established survey that is being retraced, as well as recovering any uncalled-for monuments, without the restoration of lost corners or re-blazing of the original lines. Perhaps the best definition would be to consider a *retracement* as merely an *investigation survey* to determine which original lines and corners can be found and identified. In public land states, the authority of the originally creating authority was first the US Treasury Department, then the GLO, and now the BLM. In metes and bounds states, case law decisions, as well as the various definitions of land surveying, become the critical criteria. In a *retracement,* the surveyor attempts to recover as much as possible of the original survey evidence of the original lines. Then, if the surveyor has been given the authority by the original creating agency or its successor, the surveyor surveys and locates the lost corners based on the accepted principles outlined and described in the various manuals. But in metes and bounds states, no such manuals exist, and the numerous *minimum technical standards* that surveyors should understand and adhere to mention only the technical aspects of the works, not the legal aspects. In other words, the minimum technical standards state that the surveyor should close surveys to a specific standard, but *absolutely no* technical standard states that the retracing surveyor has to be on the correct corner.

10.4 TYPES OF SURVEYS AND RESURVEYS

Although a private surveyor does not conduct resurveys of lands that were once public lands, it is important that all surveyors understand the distinctions among the various terms that concern surveys and resurveys.

These may be listed in descending order of control. The recognized classes are as follows:

I. Original surveys

II. Retracements

III. Resurveys

  A. Dependent resurveys

  B. Independent resurveys

First is the original survey. The law holds that the original survey does not describe boundaries, but it creates the boundaries. That *original* survey creates evidence and leaves this evidence behind for the subsequent surveyor to find when a *retracement* is conducted. Probably the best definition of a retracement can be found in the Florida decision *Rivers* v. *Lozeau*, 539 So. 2d 1147 (Fla. App. 1989), when the court wrote:

> [A] surveyor can be retained to locate on the ground a boundary line which has theretofore been established. When he does this, he *"traces the footsteps"* of the "original" surveyor in locating evidence of the boundaries. Correctly stated, this is a *"retracement"* survey, not a resurvey, and in performing this function, the second and each succeeding surveyor is a "following" or "tracing" surveyor and his sole duty, function and power is to locate on the ground the boundaries, corners and boundary line or lines established by the original survey. He cannot establish a new corner or new terminal point, nor may he correct errors of the original surveyor. He must only track the footsteps of the original surveyor. The following surveyor, rather than being the creator of the boundary line, is only its discoverer and is only that when he correctly locates it. (Emphasis added.)

A retracement is the intervening step between the original survey and the resurvey, because this retracement is conducted to recover the evidence created and left by the original surveyor. Then, predicated on the recovery of the evidence in the retracement, a decision is made about conducting either a *dependent* resurvey or an *independent* resurvey.

A *dependent resurvey* is first a retracement of all recoverable evidence of the original corners and lines, and then a reestablishment of lost or obliterated corners and lines in accordance with the proper rules, and then in accordance with the best available evidence and applicable rules of survey. The framework of the new survey is dependent on the recovery of the original corners and evidence of any lines. An *independent resurvey* casts aside the original survey and all of its boundary lines and creates all new monuments and corners and may include the establishment of new township lines without reference to the original survey.

During an independent resurvey of a township, parties with bona fide rights usually have their land delineated by a metes and bounds survey and designated by a tract number. In southern California, near the Mexican border, several townships have numerous tracts platted and surveyed without showing a relationship to the original survey. Also, as these tracts are shown with bearings of north, south, east, or west, and distances are in even multiples of 20 chains, the tracts could not possibly be based on the original survey. Without a corrective deed (patentee deeding the patent back to the government and the government deeding the tract to the patentee), title companies refuse to insure title by the tract number. In effect, this stops the development of new subdivisions because title insurance is not available.

In theory, once an independent resurvey is approved, the original survey no longer exists even though monumentation of the original survey may still exist. This is true for land always owned by the federal government, but not for patented parcels existing prior to the independent resurvey. In one township in California (Blaney Meadows), the government decided that the original survey was fraudulent and ordered an independent resurvey. During the independent resurvey, the old original corners were destroyed (although tied out), and the bona fide patent owner was given a new location. As the owner gained several hot springs, he did not complain.

The landmark retracement case, *Cragin* v. *Powell*,[2] contains a reference to a letter from the commissioner of the GLO, which reads:

> The making of resurveys or corrective surveys of townships once proclaimed for sale is always at the hazard of interfering with private rights, and thereby introducing new complications. A resurvey, properly considered, is but a retracing, with a view to determine and establish lines and boundaries of the original survey … but this principle of retracing has been frequently departed from, where a resurvey (so called) has been made and new lines and boundaries have often been introduced, mischievously conflicting with the old, and thereby affecting the areas of tracts which the United States had previously sold.

10.5 COURT OF PROPER JURISDICTION

Principle 3.   When retracing or dividing sections of land created under federal laws, federal rules of survey must be followed (1) when the final adjudication of problems is tried in the federal courts or (2) when a state has accepted or adopted the federal rules and when the original patents were federal in nature.

A surveyor retracing a federally created township is bound to follow the federal rules of survey that created that township. This may be a problem if the surveyor determines that the respective state has enacted legislation or follows rules that are different. Pragmatically, if property rights are created under one set of surveying rules, it hardly seems reasonable that a second set of surveying rules should be followed to relocate the lines. A young surveyor may encounter "old-time surveyors" who either refuse to accept this principle or do not understand it.

Once land passes from the federal government to private parties within a state, the jurisdiction over court trials passes from the federal courts to the state courts. Although most state courts recognize federal court rules as applicable in resurveys, this is not universally true. For example, the Missouri state court accepts that state's statute regarding restoration of a lost section corner, which is quite different from the requirements of federal courts, and in California courts, topography calls have much greater force than that given in federal courts.

10.6 FEDERAL PATENTS

A federal patent is the legal means by which the federal government grants lands from federal ownership into private ownership. A patent carries no warranty of title or any other warranties, so for all practical and legal purposes it may be equated to a quitclaim deed. For the most part, a patent acts as a quitclaim deed granted from the government to the individual. A patent from the federal government is senior to all other rights except prior-granted rights by some previously authorized entity. It is unusual, but not impossible, to determine that several patents were granted to the same parcel. In these instances, the senior patent must be determined, and it will control. When conducting research for a retracement, a surveyor will find it advisable and probably necessary to extract the title to the original patent to be certain to determine what is to be surveyed.

When conducting a retracement, it may be advisable that the surveyor "take" the description back to the original patent. Because of the numerous and varied ways that patents were granted to individuals, states, and corporations, it should be realized by the retracing surveyor that a patent may be granted to a state by one description, yet the state may convey the same land using a different description, thus having different boundaries to be retraced.

10.7 INTENT OF THE GOVERNMENT

In law, the actual proven intentions of the parties to a land transaction are paramount to other considerations, senior rights excepted. Because official government subdivisions were conducted and made under the statutes in effect at the time of surveys and because it is presumed that the instructions, rules, and regulations of the GLO were obeyed, any intent on the part of the government must be interpreted from the statutes, rules, and regulations in effect as of the date of the patent. By federal statute, "the boundary lines actually run and marked in the surveys returned by the surveyor general shall be established as the proper boundary lines of the sections or subdivisions for which they were intended." Thus, the original lines as run by the original surveyors are the best indicators of intent. It becomes axiomatic that because the retracement depends on the original surveys described by the plats and field notes, the surveyor who undertakes any retracement must base the retracement survey on the original notes and plats by which the descriptions were created.

10.8 SENIOR RIGHTS

From the inception of the land distribution program and its surveys, the federal laws never made provisions for the federal government to convey prior land rights that were not obtained. The federal land system was one of the few systems providing for recognition of valid land rights that were granted in areas of the public domain prior to the American Revolution. French, Spanish, some Dutch, and even English land grants were evaluated, and decisions were made as to their validity. There were several instances in which the validity of the grants was determined after the public land survey network was surveyed on the ground. In some instances, this placed a senior survey (GLO) over a senior grant (e.g., a Spanish grant). There were many instances in which the senior grants existed only on paper and not on the ground.

When a surveyor is asked to retrace either a township, with its sections, or a prior senior grant, the surveyor must examine the date of the respective surveys to determine which is senior or junior as well as the date of the grants.

> It is presumed that when a land grant is superimposed within a township, it is senior in title, but it may be junior in the survey of its defining lines.

Provisions were never made for the federal government to sell or convey valid land rights acquired by private citizens prior to establishment of the public domain. A government survey encroaching upon a private grant, such as a rancho land grant, cannot control for conveyance purposes, but the set monuments can be used as reference monuments. In encroached areas, all rights originating from the public domain cease at the boundary of the grant closed upon.

10.9 FOLLOWING THE FOOTSTEPS

Any surveyor who conducts an original survey or an independent resurvey (GLO) is the person who creates the footsteps to be followed in a subsequent retracement. The retracing surveyor (private) follows the footsteps of the creating surveyor. However, this term has presented difficulty in both application and interpretation. If one were to follow the dictum to the "letter of the law," it would mean that the retracing surveyor should first run the section lines with a compass, or solar compass or solar transit, and then use a 2-pole chain, or a 5-chain ribbon tape, to conduct the initial retracement.

Following the footsteps simply means following the *evidence* of the footsteps, which includes the physical evidence as well as methodologies of how the work was accomplished.

Courts and surveyors sometimes rely on the advice to "follow the footsteps of the original surveyor" or, as several courts have written, "follow the footprints," to justify their findings and, in many instances, attempt to discredit the work of other surveyors. No one can walk in the exact steps of the original surveyor; rather, it becomes a matter of "tracking the footsteps" by the evidence that is recovered, and then being able to say with a great degree of certainty, "This is where the surveyor walked."

A surveyor in private practice who is testifying in court will probably be asked, "Did you follow the footsteps?" This question has been asked by attorneys in both federal and state courts. The answer should be "No." No two surveyors can create the same footsteps because the "footsteps" are in reality the evidence of the original surveys. The retracing surveyor should attempt to recover the *totality* of the evidence of the original surveyor and then determine where the original surveyor stood. A reply should be: "I found the evidence of the original footsteps." The evidence should include the instructions, the field notes, the plats, and any other evidence, both documentary and physical, that can be found at the time of the retracement. To "follow the footsteps" implicitly would require using the same equipment, using the same methods for computations, and using the same methods of calculating the areas of the respective sections.

10.10   LINES MARKED AND SURVEYED

Principle 4.   *Provided that a superior right is not interfered with or a fraud is committed, the boundaries of public lands, when approved, accepted, and patented, are unchangeable and unassailable by individuals or by the courts.*

Congress intended that subsequent surveyors and all courts be restricted or limited in their treatment of public lands relative to inputting their own opinions as to whether they should be retraced. The foundation of the public land system and resurvey procedures depends on this principle, first approved by the courts and later enacted by statute:

> The boundary lines, actually run and marked in the surveys returned by the surveyor general, shall be established as the proper boundary lines of the sections and subdivisions for which they were intended.[3]

The burden of proof is now placed on the retracement surveyor to locate the lines as they were actually placed on the ground, not where they should have been. The resurvey becomes a question of evidence and the ability to convince the judge and/or the jury that the evidence presented is superior to all other evidence collected and presented.

10.11   ORIGINAL CORNERS

Principle 4.   *Provided that a superior right is not interfered with or a fraud is committed, the boundaries of public lands, when approved, accepted, and patented, are unchangeable and unassailable by individuals or by the courts.*

In many instances, one may find multiple plats for the same township, with each plat depicting different and conflicting information. A federal township plat does not become effective until it is signed by the surveyor general. Along with this principle, the footsteps of the township may not be readily identified in that the retracing surveyor must determine which one of the plats was used to identify the parcels that were patented.

The preceding principle is repeated because of its importance. When doing retracements in public lands the federal survey laws apply. There have been instances when states have attempted to enact guidelines that circumvent the federal laws, but these have proven to be inapplicable.

Principle 5. *Original township, section, and quarter-section corners stand as the true corners that they were intended to represent, regardless of where they are located. This is not true of closing corners.*

As discussed earlier, in the PLSS, the original corners control. Yet in almost all instances, those corners were identified by set monuments and their record evidence. This basic principle was identified by Congress on February 11, 1805, in the land act named for that date. This is a logical sequence of the following principle:

> The boundaries of the public lands, when approved and accepted, are unchangeable and unassailable through the courts.
> The monuments set at the time of the original survey are the best evidence as to where the original boundaries were established; as such, the monuments must remain unchangeable. Closing corners are an exception to previous principles.

10.12 ORIGINAL FIELD NOTES AND PLATS

Principle 6. *Plats and all original field notes become part of the grant or patent. When conducting a retracement, the surveyor should consult available and known plats and notes.*

In a now famous US Supreme Court decision, this principle was stated as follows:

> It is a well settled principle that when lands are granted according to an official plat of the survey of such lands, the plat, itself, with all its notes, lines, descriptions and landmarks, becomes as much a part of the grant or deed by which they were conveyed and controls so far as limits are concerned, as if such descriptive features were written out upon the face of the deed or the grant itself.[4]

Government field notes are the major source from which the descriptive information concerning corners, monuments, accessories, and line information can be found. Rarely will information concerning witness trees and natural objects appear on the face of the plat. The best evidence of how and where the lines were run appears in the notes. If there is a discrepancy in measurements between the plat and field notes, the plat usually must give way to the field notes, and the land department may properly correct the plat so as to conform to the field notes.[5] However, if a discrepancy affects the description, the plat controls.

For example, during a dependent survey, the original quarter corner, though out of position, was identified and called the quarter corner in the field notes. However, the drafter decided when platting the notes that the section would be skewed and, thus, renamed the original quarter corner the "witness corner," identifying the quarter corner at midpoint. Because the plat is what was patented, the plat controls.

Field notes should be used in conjunction with the plats, because they may explain why certain anomalies are on the plat. See Figure 10.1.

The problem: The surveyor did not understand how the variance in bearing occurred. The surveyor took the mean of 2 and 4 and applied the mean to 3, not realizing that the field notes showed it to be east and west. He indicated a mean bearing of N 89 27 W.

The solution: Had the surveyor consulted the field notes, he would have seen in both situations that when the closing lines were run in from the west the field notes indicated a closure or "falling" of "35 links North of the originally set corner. Now multiply 35 by 3 and divide by 7 and the result is the number of minutes of correction = 15 minutes correction, this 89 45 to the North line of Sec. 12.

10.13 CLOSING CORNERS

Principle 7. *Closing corners that are not actually located on the line that was closed upon will determine the direction of the closing line but not its legal terminus; the true position is the point of intersection of the two lines.*

A surveyor who is asked to retrace the lines of those sections that border the range lines or the township lines will encounter closing corners. It should be understood that a closing corner, even though it is a corner, should not be considered as having the same dignity as that of a standard section corner. Because these corners were set after the original township and range lines were created, the closing corner is important but not absolutely controlling. In the running of the closing corner, the surveyor intended that these corners be set on line between two original corners, when in many instances they were simply "estimated" by the original surveyor when he or she ran the original section lines.

Once a line is run and established, it cannot be altered at a later date. If the surveyor failed to place the closing corner on the true line closed upon, the original line cannot be changed to fit the new corner; the closing corner must be moved to the line closed upon. This rule applies to closing corners on standard parallels, correction lines, land grant lines, and the last-set double corner. When a new township is being surveyed or an old one is being resurveyed and double corners are set along the township lines, the second corner set must be considered a closing corner.

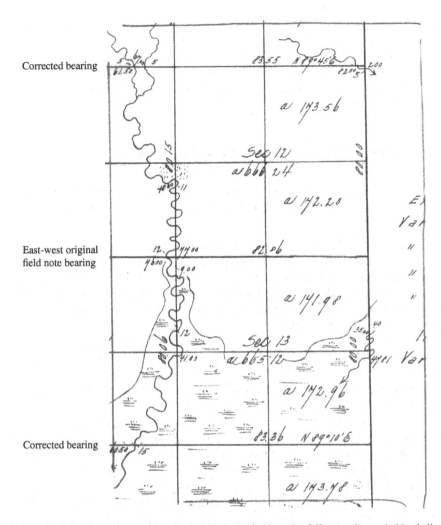

Figure 10.1 Portion of township plat in Mississippi. *Note the following lines:* 1. North line of Section 1 east and west although not indicated. Field 2. North line of Section 12 N 89 45 E 3. North line Section 13, no reference. What is it. Field notes not consulted 4. South line Section 13, N 89 10 E.

10.14 IDENTIFICATION OF CORNERS AND LINES

Any surveyor who undertakes to conduct a retracement or what some may call a resurvey must be trained in the area of evidence recovery and evaluation. A retracement or resurvey does not require the services or capabilities of a top technical expert, but, rather, a retracement requires the services of a person who understands evidence, including its significance and its legal implications. Evidence of the original surveys and their significance, and how this evidence relates to the original measurements as well as to modern measurements and not to technical aspects of modern surveying, is the most important element of a resurvey or a retracement.

A surveyor will appreciate that evidence can consist of verbal and real evidence, as well as hearsay evidence. All are equally important in determining where the original footsteps were left.

10.15 MONUMENTS AND THEIR IDENTIFICATION

Principle 8. *After making allowances for natural changes, a monument, to be identifiable as an original monument, should not differ greatly from the field notes.*

All surveyors should realize that monuments set at the corner points are like people. Time affects each differently and in its own way. (1) The character and dimensions of the monument in evidence should not be widely different from the record; (2) the markings in evidence should not be inconsistent with the record; and (3) the nature of the accessories in evidence, including size, position, and markings, should not be greatly at variance with the record.[6]

Before any attempts are made to conduct a retracement, the original field notes of surveys must be obtained from the proper authorities. If more than one "original survey" is available, it should be determined which is applicable by examining court records or dates of survey. Next, a complete chronology of monuments, descriptions, accessories, line objects, and calculations is made.

A variety of monuments have been used throughout the years; posts, stones, stone mounds, and dirt mounds with pits were most frequently used. When undisturbed, stones or stone mounds last for decades, and, on occasion, portions of wooden posts have been uncovered after 100 years. In one area, wooden posts, established in 1799, were recovered in the 1950s.

The type of monument found must correspond in size, species, and general character to that called for in the field notes. Deviations from the called-for information make the acceptance of location difficult, and a completely different call makes the corner suspicious. In one area, the original survey called for "wooden posts"; however, today each corner recovered is monumented by a marble post of unknown origin. Here, reputation and common usage come into play—is there better evidence?

Some of the problems that a retracing surveyor may encounter in evaluating evidence recited in field notes as to what is found on the ground may be a variation of materials or variations of distances, bearings, or other recorded evidence. For example, the field notes call for a sandstone monument, 6 by 12 inches, but the surveyor recovers a basaltic monument the same size and marked "as called for." Are these the same? Or a white oak post, 4 inches square marked with two notches is called for, but what is recovered is a red oak post, marked in the same manner. Is it the same post? Or the notes may refer to a bearing tree as being "12 Cyp., N 23 E, .33 ch.," but the surveyor finds a 23-inch cypress, with an old face blaze facing the corner point, whose bearing is S 22–½ W, 34 links. Is this the same tree? Is this the corner point?

By reading the signs of natural wearing and aging, such as the buildup of moss on stone or wooden corners, the approximate time of placement can be determined. Each stone in a mound is removed and examined to determine the nature of the ground under it and the presence of undecayed leaves. A change in coloration at ground level indicates aging.

The frailties of the creating surveyor become important. Did the creating surveyor estimate the size of the stone, or did he measure it? Did the creating surveyor read the wrong end of the compass? Did the creating surveyor measure to the bearing tree or "guesstimate" it? These are questions that will go unanswered forever. Yet the retracing surveyor must address these questions before an opinion can be formulated.

Few original posts or dirt mounds are recovered because of rotting, fire, or weathering. Redwood and red cedar are naturally long lasting, and white oak is more durable than red oak. "Lighter" pine has a large concentration of resin, which resists rotting but allows the wood to burn readily. In prairie areas, the pits and mounds have long since disappeared, yet the telltale signs of changes in vegetation, soil color, or texture help the surveyor to determine where the pits and mounds were once located.

Accessories such as bearing trees called for in the writings have equal dignity with the corner itself. Blazed bearing trees can be traced back to the time of the original survey by counting tree ring overgrowth. Because trees always grow larger, the reported size of the tree at the time of the original survey can be compared with its present size. In a few locations of very poor soil, however, growth may be slight, and some palm trees remain the same size. Identification of stumps to determine the species is also sometimes important. Bearing trees were usually scribed, and the scribing should be consistent with the field notes or the standard instruction of the time. If the notes for Section 26 call for the scribing to be in Arabic and in the resurvey Roman numerals XXVI are found, a problem arises. On the other hand, if the notes call for a "BT" on each bearing tree and if during a retracement of the entire township no "BT" is found on any tree, is this an error or an omission? Here, the surveyor must rely on his or her judgment to make a decision.

10.16 EVIDENCE OF CORNERS

Principle 9. *When there is acceptable evidence indicating the location of an original corner, the corner should be considered as obliterated and should be relocated from that evidence.*

A surveyor should never hesitate to consider every form or type of evidence to locate a corner monument. Because a corner is the controlling point for a line, courts and surveyors should understand that no form of evidence should be cast aside that could help to locate the original position. Courts consider a lost corner so serious that any shred of acceptable evidence may be utilized to help raise the dignity of a lost corner to an obliterated corner. Before a corner can be considered as lost, the surveyor must research all available records from such varied sources as utility companies, title companies, highway departments, private surveyors, and public agencies for any supporting data. In many western states, the railroads were built very soon after or at the same time as the original surveys were conducted. The right-of-way drawings depict many tie-outs to original section and quarter corners as related to mile posts.

Courts look at the standard of evidence needed to accept corners quite differently than *does* the surveyor and particularly the BLM. Whereas the BLM once required that the degree of proof be *beyond a reasonable doubt*, which is the most stringent, the courts as well as the new edition of the *BLM Manual* now require only the *preponderance of evidence* or creditable evidence.

10.17 USE OF TESTIMONY IN BOUNDARIES

Principle 10.   *The original location of a corner may be identified by the evidence as well as by testimony of a witness who saw that corner. It is an exception to the hearsay rule, but the evidence as well as the testimony must be evaluated as having to meet the "preponderance of evidence" criterion, not the "beyond a reasonable doubt" criterion or standard that has been established for criminal matters.*

Courts and the *Manual* accept that testimony of eyewitnesses to the original corners can be critical in determining the location of original corners. Yet they accept the fact that some people have *selective memory* or *selective recall* of facts. They also accept that some people may be prone to be less than truthful, even under oath. Generally, there is a modern trend, when accepting testimony, to identify the location of a corner position to use the "preponderance of the evidence" standard or requirement and not the "beyond a reasonable doubt" standard. Several states have addressed this, and the courts have universally held that *no* agency can set or establish a legal standard for the acceptance of evidence. Testimony is a form of evidence that is important in determining facts. When testimony is used, it must meet all requirements relative to the individual who is testifying and the presentation of the evidence.

The Federal Rules of Evidence [Rule 803(20). Hearsay Exceptions] and many state rules of evidence or statutes give special consideration to ancient testimony. In Hawaii, ancient testimony has been given the name *kamaaina*. The Hawaiian court in Pulehunui used *kamaaina* testimony to determine the boundaries of Pulehunui, a central Maui grant of 16,000 acres that was granted by name only.[7] *Kamaaina* was described as "a person familiar from childhood with any locality." Ancient testimony has been applied to land, evidence, and events.

When a surveyor receives testimony while in the field, he or she should consider the applicability of the field-related testimony to future use. Several states permit surveyors to take oaths for the establishment of old corners and recognized lines. The California law provides:

Every licensed land surveyor or registered civil engineer may administer and certify oaths: *(a)* When it becomes necessary to take testimony for the identification or establishment of old, lost or obliterated corners. *(b)* When a corner or monument is found in a perishable condition and it appears desirable that evidence concerning it be perpetuated. *(c)* When the importance of the survey makes it desirable to administer an oath to his assistants for the faithful performance of their duty. A record of oaths shall be preserved as part of the field notes of the survey and a memorandum of them shall be made on the record of survey filed under this article.

To be usable in the future, statements made under oath must contain facts that are the personal knowledge of the person giving the testimony. Statements of facts about what a person was told by third parties are considered hearsay evidence or double hearsay, or, as some courts have called it, hearsay on hearsay. A witness can testify that he or she personally observed a corner being destroyed when the county road was widened, and he or she can point out to the surveyor the location of the monument as he or she remembers it, but he or she cannot testify where a neighbor said it was located.

A surveyor should place little reliance on the testimony of people who are involved in the land dispute or on statements about boundaries or corners that originated after the dispute began. The *BLM 1973 Manual* (p. 130) describes reliable testimony as follows:

A corner will not be considered as lost if its position can be recovered satisfactorily by means of the testimony and acts of witnesses having positive knowledge of the precise location of the original monument. The expert testimony of surveyors who may have identified the original monument prior to its destruction and thereupon recorded new accessories or connections, etc., is by far the most reliable, though land owners are often able to furnish valuable testimony.

The testimony of individuals may relate to the original monument or the accessories, prior to their destruction, or to any other marks fixing the locus of the original survey. Weight will be given such testimony according to its completeness, its agreement with the original field notes, and the steps taken to preserve the original marks. Such evidence must be tested by relating it to known original corners and other calls of the original field notes, particularly to line trees, blazed lines, and items of topography.

10.18 COMMON USAGE

Principle 11. *In a very limited number of situations, a corner's position may be located by common use of the monument and point.*

The Federal Rules of Evidence, as well as most state rules of evidence, accept the fact that boundaries can be identified by long common usage. Several of the courts have extended this exception to include the termini of these lines, the corners, as being determinable by common usage, or as some courts call *repose*. In many areas, state and local highways were initially located along section lines with full knowledge of the individual landowners and local residents. Where, by local acceptance, such roads have been superimposed upon the section lines by common usage, local surveyors accept these points as being the best remaining evidence. In the absence of better or more conclusive evidence, a section corner that falls within road intersections may be considered as obliterated. Until recently, this privilege has not been extended to permitting people to testify as to ownership. But unfortunately, several state courts have extended this exception to the hearsay rule to permit a person to testify as to who they thought owned the land.

This certainly flies in the face of historic English law that requires a written document to prove one's claim to title.

When courts rely on the acceptance of a period of "long" prior usage to justify the acceptance of a corner as being the "true" corner without the supporting physical evidence, the courts usually justify their action by calling it a *corner of repose.*

10.19 USING RECORDED INFORMATION TO LOCATE ORIGINAL LINES

Principle 12.   *A corner position may be proven from evidence of found original line trees or from natural features found in close proximity to the corner and identified in the notes.*

The original surveyors, who, for the most part, were meticulous and professional, kept excellent field notes. Many of the original notes call for marked-line trees, natural features, and numerous other identifiable features. If the retracing surveyor is able to ascertain these original trees and other evidence, their positions may be used in numerous situations. Some of these would be as an aid to proportion lost corners or to use as possible positioning of the corners "intersecting" these original lines to locate the "most probable" corner location, short of proportioning. Many of the original notes contain references to natural features or line trees, and after positive identification, these may be utilized for the alignment of the section lines. In many older surveys, little evidence of these calls remains, and thus only by a careful and diligent search can it be said with certainty that the particular feature is the one called for.

10.20 PROPORTIONING: THE LAST RESORT

In football, "on fourth down and 50 yards to go," the quarterback asks the coach, "What do I do?" The answer probably will be, "When all else fails, you *punt.*" This same question will be answered by many surveyors, attorneys, and courts, who have failed to apply a primary requirement in considering corners. The failure to ask the initial question "Is the corner lost or obliterated?" can pose serious problems. Before any discussion is conducted about any corner, the surveyor must address this important question.

> Only when all of the evidence fails to prove a corner do you then proportion.

Any proportioning of lost corners should be considered as the last resort. It should be relied on only after an extensive search of the records, after conducting adequate measurements, after talking with people in local areas, and after much deliberation. In the case *U.S.* v. *John Citko*,[8] the federal judge in Wisconsin isolated the question to: Was the corner lost or obliterated? The United States held that the corner was lost, and thus it was proper to relocate it by single proportion. The defendants presented evidence based on witnesses' statements and the location of a rock mound, on the ground. Only when all of the evidence fails to prove a corner do you proportion. It has been found that many younger surveyors are quite innovative in locating or resurveying lost corners. Although in metes and bounds states there are no absolute guidelines, the GLO states have criteria that are strict yet tempered with flexibility. It is suggested that metes and bounds surveyors should understand the principles of locating lost GLO corners and apply the pertinent ones to their practices. This statement is particularly true now that many of the younger surveyors have ready access to electronic stations that permit easy measurements of angles and distances and make computations with little or no effort as well as global positioning system (GPS) units. Relying on measurements as the primary methodology for ascertaining corners will soon dull the surveyor's ability to recover the evidence and evaluate this evidence that proves the existence or nonexistence of original corners.

Once again, it should be stressed that no proportioning should be applied until it is absolutely determined that a corner is absolutely lost. It must also be stressed that no proportioning between found corners can be predicated on corners located outside the boundaries of the township in a GLO state or the parent parcel of a metes and bounds state. All proportioning must be tied to two or more proven corners found within the original unit.

How lost is lost? The case of *U.S.* v. *Doyle* states that "for corners to be lost, they must be so completely lost that they cannot be replaced by reference to any existing data or other sources of information and before courses and distances can determine boundary, all means for ascertaining location of the lost monuments must first be exhausted."[9] Too many times, surveyors, and others, give up too quickly and set a new point by proportioning. This, at the least, may put a cloud on the resulting titles or, worse, cause resulting titles to be void, since the determination does not fit the basic requirement of the law.

10.21 RELOCATING LOST CORNERS

Principle 13. *Lost corners may be located by proportioning or by using a combination of evidence and measurements.*

As pointed out, lost corners are serious in that a proportioned point is presumed to be placed in its original position. One must apply a large degree of experience and common sense when using proportioning. The rules or guidelines do not permit proportioning if superior evidence is available. Assuming that positive evidence of an original section line location is superior to a proportioned point, there are no limitations on what can be used. Also, assuming that a surveyor is able to positively identify an east–west fence line as being a section line by long-accepted reputation and testimony, that should be used. However, if a proportioned point would place the latitudinal point off the fence line, the surveyor would be well within acceptable survey practices to use the fence line as a latitudinal position and then rely on a proportioned longitudinal point.

> Courts prefer an obliterated corner based on "weak" evidence to a proportioned corner based on measurements alone.

Once it is determined that a corner is lost, the surveyor must decide whether it will be restored by federal or state rules. The first rules presented are an application of federal law. State laws that deviate from federal law are then identified. When a corner is positioned in accordance with the lost corner theory or proportioning, the surveyor must be convinced that no future surveyor or court will be able to locate the original corner in a different place. Proportionate measurement is always a rule of last resort and is always subject to question.

Principle 14.   *The surveyor should strive to utilize all forms of evidence to prove a corner as an obliterated corner, rather than resorting to proportional measurements to set a lost corner.*

With the availability of surveying equipment and computer programs that make the measurements of angles and distances relatively easy, many surveyors are lulled into relying on the subsequent measurements and calculations to set new corners that may conflict with possession lines and may cause boundary disputes. The key between a lost corner and an obliterated corner may just be an interpretation of the recovered evidence and the reliance and weight that the retracing surveyor places on it, as well as the tenacity of the retracing surveyor.

It is difficult to argue with a measured line or a computed distance, yet it is very easy to disagree with the interpretation of the evidence relied on to position a corner. For this reason, the surveyor should be experienced in the interpretation of evidence. The surveyor must accept the realization that he or she can be firmly convinced about the authenticity of a corner, based on the evidence, but may find himself or herself having to take a defensive position with the "modern surveyor" who is armed with the latest in surveying measuring equipment and computer technology. Few young surveyors realize that it does not take US$10,000 worth of surveying equipment to prove an original corner—the corner may be proved with a US$10.95 shovel, used at the correct location.

> Only when the recovered evidence fails to prove a corner position can a surveyor resort to proportioning.

10.22 PROPORTIONATE MEASURE OR PRORATION

Principle 15. *When the surveyor has exhausted all possibilities of locating corners of public land surveys from available evidence, only then, as a last resort, should the surveyor rely on measurements alone.*

The retracing surveyor must strive to recover sufficient evidence on which to predicate a decision about the authenticity of the corner. If, after serious consideration and contemplation, the surveyor is still convinced that the evidence is insufficient to determine the validity of the corner, the surveyor must resort to measurements alone.

Principle 15A. *When proportionate measurements are used, the new values given to the several parts, as determined by the measurement, should bear the same relation to the record lengths as the new measurement of the whole bears to that record.*

Existing original corners (except closing corners) cannot be disturbed; consequently, discrepancies between the new measurements and those of the record will not in any manner affect the measurement beyond the corners identified, but the difference will be distributed proportionately within the several intervals along the lines between the corners.[10] In a subdivision where there is a difference in measurement between the original surveyor and the recent surveyor, it must be concluded that the difference occurred in all parts of the line unless the contrary can be proved beyond any reasonable doubt. In government surveys, many corners were set on a single line, whereas others were set in checkerboard fashion with cross-ties. Single proportionate measure is applied to measurements along a single line, whereas double proportionate measure is applied to checkerboard surveys.

10.23 SINGLE PROPORTIONATE MEASUREMENT

Principle 15B. *A single proportionate measurement should be made between two proven corners, and may locate one or more original points on a line.*

The classic example of single proportionate measure is the restoration of a lost quarter-section corner, as shown in Figure 10.2. Section corners A and B are known; the record distance from A to the lost quarter corner Q was 41 chains; the record distance from Q to B was 40 chains; the present measure from A to B is 81.50 chains.

Quarter corner Q is restored in the exact proportion of the original record; that is, (41 ÷ 81) 3 81.50, or 41.26, chains is the distance from A to Q and (40 ÷ 81) 3 81.50, or 40.24, chains is the distance from Q to B.

United States Code, Title 43, Section 752, states in part: "and the length of such lines (section lines) as returned shall be held and considered as the true length thereof." The original reported length is presumed to be the correct length, and any discrepancy in measurement that exists is presumed to be due to the recent surveyor. In a resurvey, it is rarely found that the new measurement agrees with the old; hence,

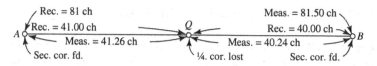

Figure 10.2 Single proportionate measure.

the new survey must be adjusted so that it will be in proportionate agreement with the old survey. In other words, the new chain must be adjusted in length so that it agrees with the original measurement. Actually, this is done by mathematics, as illustrated in the preceding example.

By law and by court ruling, original monuments, except closing corners, are fixed in position and cannot be moved. Proportionate measure cannot extend beyond any fixed point or fixed monument and is applicable only to lost points between original monuments.

Not all lines surveyed were straight lines. Corners on a standard parallel or east–west township lines were originally set on a curved line, and any restored corner that is located in the footsteps of the original surveyor must be on the same curved line.

10.24 DOUBLE PROPORTIONATE MEASUREMENT

Principle 15C. *Double proportionate measurement usually requires four original corners, and the location of the proportioned corners is positioned at the proportionate distances of lines run from the proportioned points. Primary weight is given to distance, not to angle or bearing.*

To restore a lost corner of four townships by double proportionate measurement, a retracement will first be made between the nearest known corners on the double proportionate measure meridional line, north and south of the missing corner, and upon that line a temporary stake will be placed at the proper proportionate distance; this will determine the latitude of the lost corner. Next, the nearest corners on the latitudinal line will be connected, and a second point will be marked for the proportionate measurement east and west; this point will determine the position of the lost corner in departure (or longitude). Then, through the first temporary stake run a line east or west, and through the second temporary stake a line north or south, as relative situations may determine; the intersection of these two lines will fix the position for the corner restored.[11]

During restoration of a corner by either single or double proportionate measurements, bearings or angles are of minor importance. Greater weight is placed on distance measurements, the reverse of what is usually true for federal metes and bounds surveys. In some states, the concept of distance controlling over direction may be in direct conflict with either state court holding or statute. On occasions, the instructions

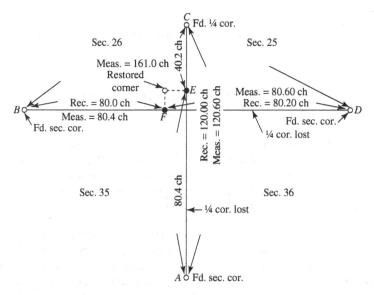

Figure 10.3 Double proportionate measure.

for double proportionate measurements have been misinterpreted. To obtain the inter-section point, the instructions direct that the lines must be in cardinal directions, not at right angles to the lines. In most instances, the difference is slight, but occasionally it becomes significant.

Double proportionate measurement is illustrated by Figure 10.3, which shows the conditions in a hypothetical problem. To restore the lost section corner common to Sections 25, 26, 35, and 36, as shown in Figure 10.3, the surveyor located the south-east (A) and northwest corner (B) of Section 35, and the west quarter corner (C) and the southeast corner (D) of Section 25. A straight line from A to C is run and found to measure 120.60 chains, whereas the original recorded government measure was 40 chains for each half-mile, or a total of 120.00 chains. The surplus of 0.60 chain is divided into three equal parts of 0.20 chain each. A temporary point is established at E, exactly one-third of the way between C and A, or 40.20 chains south of C on the line AC. On the east and west straight line between B and D is established a temporary point at F, the position of which is computed as follows:

$$\text{distance } BE = \frac{161 \times 80}{80 + 80.2} = 80.40 \text{ chains}$$

$$\text{distance } FD = \frac{161 \times 80.2}{80 + 80.2} = 80.60 \text{ chains}$$

From the proportionate point E a line is run due north (astronomical), and from the point F a line is run due west; their point of intersection determines the restored position for the lost section corner.

This is probably the main reason that surveyors who are trained in GLO principles will usually give primary weight to distances and secondary weight to angles or
bearings.

10.25 RESTORATION OF LOST STANDARD CORNERS ON STANDARD PARALLELS, CORRECTION LINES, AND BASELINES

Principle 15D.   *Lost standard corners will be restored to their original positions
on a baseline, standard parallel, or correction line by single proportionate measurement on the true line connecting the two nearest identified original standard corners.
Proper adjustment should be made to secure the correct latitudinal curve. Closing
corners are not to be used for either direction or measurement.*

The term *standard corners*, as used here, will be understood to mean all corners
that were established on the standard parallel during the original survey of that line,
including but not limited to standard township, section, quarter-section, and meander
corners. Closing corners or other corners purported to be established on a standard
parallel after the original survey of that line will not control the initial restoration of
lost standard corners.

In cases in which a closing corner was set on the standard parallel at the calculated
position of the closing corner and at the time of the survey of the standard parallel,
such a corner controls the direction of the standard parallel. For standard parallels,
the standard corners set varied from time to time; that is, sometimes they applied to
sections north of the line, at other times to sections south of the line.

Since corners originally set on standard parallels were set on a curved line, the
lost corners must be restored on the original curved line connecting the nearest identified standard corner on each side. Lost corners on curved lines are restored by
single proportionate measure on the curved line by first setting a temporary proportional corner on a straight line, then measuring over the proper correction for Earth's
curvature.

10.26 RESTORATION OF LOST TOWNSHIP CORNERS ON PRINCIPAL MERIDIANS AND GUIDE MERIDIANS

Principle 15E.   *When the principal meridian or guide meridian was established by
alignment in one direction only, lost township corners on such lines should be restored
by single proportionate measurement. Where guide meridians were established as
part of the scheme of township boundaries, all surveyed in the same system under a
single contract, the township corners located thereon should be relocated by double
proportionate measurements.*

When guide meridians were run as independent lines before the subdivision of
township lines, cross-ties did not exist and restoration work should proceed on the
basis of single proportionate means. However, under most contracts, guide meridians

were established as part of the general scheme for the township boundaries with cross-ties to other township lines. When this occurred, double proportion should be used to restore lost township corners.

10.27 RESTORATION OF LOST TOWNSHIP AND SECTION CORNERS ORIGINALLY ESTABLISHED WITH CROSS-TIES IN FOUR DIRECTIONS

Principle 15F.   When lost regular township corners common to four sections and lost regular section corners between township lines were originally established with ties from four directions, the lost corners will be reestablished by double proportionate measurements.

This rule is applicable to township corners not set on a baseline, principal meridian, and, occasionally, guide meridians. Most guide meridians were originally established as an integral part of the establishment of the township lines, with cross-ties in four directions; hence, lost township corners on guide meridians are usually restored by double proportionate means. The rule is also applicable to most lost section corners common to four sections and located between township lines. Where lost section corners between township lines were established by alignment in one direction only, such as corners along a sectional correction line or sectional guide meridian, single proportionate measure will be used.

Although the original east–west township lines were run on a curve, it is not necessary to rerun the lines on a curve where double proportion is used, as the final results will be identical whether curved or straight lines are employed. That is because the differences between the two are insignificant. Proportionate methods cannot extend beyond an identified original corner regardless of whether it is a quarter corner, section corner, meander corner, or a rarely set sixteenth corner. Once a corner is found undisturbed, its position is fixed and cannot be altered by proportionate means.

10.28 RESTORATION OF LOST CORNERS ALONG TOWNSHIP LINES

Principle 15G.   All lost section and quarter-section corners on township boundary lines will be restored by single proportionate measurement between the nearest identified corners on opposite sides of the missing corner, north and south on a meridional line, or east and west on a latitudinal line, after the township corners have been identified or relocated. An exception to this principle will be noted in the case of any exterior whose record shows deflections in alignment between the township corners.

The strength of this principle lies in the fact that township lines were established before the subdivision of sections within a township. The corners that were originally set on a line without cross-ties should be reestablished on the same line without cross-ties. The east–west township lines were curved lines, and the relocated lost corners must also be placed on a curved parallel of latitude. Single proportion cannot extend beyond a township corner that must be reestablished by double proportionate means.

10.29 RESTORATION OF LOST TOWNSHIP AND SECTION CORNERS WHERE THE LINE WAS NOT ESTABLISHED IN ONE DIRECTION

Principle 15H. *When a line has not been established in one direction from the missing township corner, the record distance will be used to the nearest corner identified in the opposite direction.*

This principle is applicable where double proportion normally would be used but cannot be applied because a line was not established in one direction. In Figure 10.4, the northwest section corner, the north quarter corner, the southeast section corner of Section 1, and the east quarter corner of Section 36 are all known. The townships to the east were not surveyed. Corner *X*, the lost township corner, is to be relocated. Run a line exactly 40 chains the record distance, due east from point *E*, and set a temporary point at *A*. Because due east means due east in the direction called due east by the original surveyor, the prolongation of line *F–E* would be considered due east. Next,

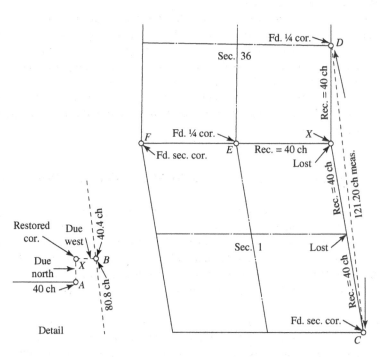

Figure 10.4 Restoration of a lost township corner originally established in three directions.

set a point proportionate measure at *B* on a straight line from *D* to *C*, or 80.80 chains from *C* proportionate measure. Run a line from *B* due west and run a line from *A* due north. Their point of intersection at *X* is the lost corner.

10.30 RESTORATION OF LOST CORNERS WHERE THE INTERSECTING LINES HAVE BEEN ESTABLISHED IN ONLY TWO DIRECTIONS

Principle 15I.   *Where the intersecting lines have been established in only two directions, the record distances to the nearest identified corners on those two lines will control the position of the temporary points; then, from the latter, cardinal offsets will be made to fix the desired point of intersection.*

This principle is applicable in Figure 10.5, where the north and east quarter corners are known and the northeast section corner (also township corner) is to be set. The other townships were never surveyed. Run a line 40 chains due east from *C*, and set a temporary point *A*. Run a line 40.0 chains (record distance) due north from *D*, and set a temporary point at *B*. Run a line due west from *B* and a line due south from *A*. The point of intersection *X* is the lost corner.

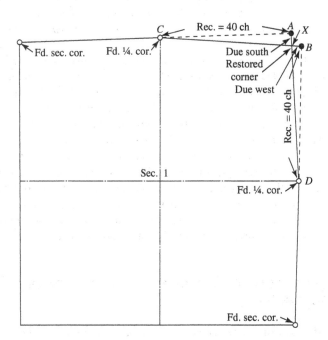

Figure 10.5 Restoration of a township corner originally established in two directions.

10.31   RESTORATION OF QUARTER-SECTION CORNERS IN REGULAR SECTIONS

Principle 15J.   All lost quarter-section corners on section boundaries within the township will be restored by first establishing or proving the section corners on each side; then the quarter corners will be relocated by single proportionate measurement on line. An exception occurs where the original lines had angular deflection.

The importance of one line over another is recognized by this rule, in that section corners must be relocated before the restoration of a lost quarter corner. Lost regular quarter corners, whether located on standard parallels, sectional correction lines, or any other line, are restored by single proportionate measure between the section corners located on each side of the missing quarter corner. Where the original line run was a curved parallel of latitude, the relocated quarter corner must also be placed on the original curved line. If the quarter corner was originally placed on a line that showed angular deflections, the restored corner must be relocated with the angular deflection.

10.32   RESTORATION OF QUARTER-SECTION CORNERS WHERE ONLY PART OF A SECTION WAS SURVEYED ORIGINALLY

Principle 15K.   Where a line has been terminated with measurement in one direction only, a lost corner will be restored by record bearing and distance, counting from the nearest regular corner, the latter having been duly identified or restored.

In applying this principle, some surveyors may tend to give it more credence than it deserves. This should be considered as the last resort, yet it places a technical responsibility on the retracing surveyor. Because the lost corner is dependent on the bearing and the distance, there is no leeway in applying the distance. Under this rule, the distance controls, but there is latitude in the bearing on which the line is to be measured.

There are other questions that should be addressed: Is the bearing of the line supposed to be the cardinal or true bearing that is indicated in the notes or the plat? Or can it be the mean of the two lines that lie on either side of the line? If a chaining index is determined, can the surveyor apply this chaining index to the record bearing that is to be placed on the ground? Unfortunately, these questions cannot be answered, because there are no definite answers to them. The ultimate decision will be made by the retracing surveyor, who must ultimately defend any methods used.

10.33   RESTORATION OF A CLOSING SECTION CORNER ON A STANDARD PARALLEL

Principle 15L.   A lost closing corner on a standard parallel will be reestablished on the true line that was closed upon and at the proper proportional interval between the nearest regular corners to the right and left. The only corners that will control the direction of the line being closed upon are (1) standard township, standard section, and standard quarter corners; (2) meander corners terminating the survey of the standard parallel; and (3) closing corners in those cases where they were established originally by measurement along the standard line as points from which to start a survey.

Figure 10.6 shows a standard parallel and Sections 2 and 3, whose lines closed upon the parallel when surveyed from the south toward the north. The corner common to Sections 2 and 3 is to be restored by proportionate means. Because the lost corner was originally set on the standard parallel, it must be reset upon the line at a proportionate distance from the next regular corner to the right and left. Closing corners were not set at the time at which the standard parallels were run and cannot be considered as determining the direction of the parallel; only those set as part of the standard parallel when it was run originally may be used. In the preceding example, the record distance from the closing corner to the nearest regular corner to the east was 30 and 10 chains to the quarter corner to the west. Recent measurements revealed a 1-chain surplus between the found regular corners, which is divided into 0.75 and

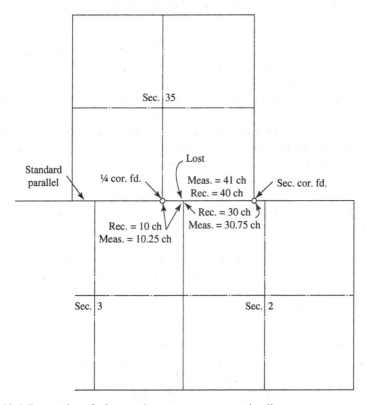

Figure 10.6 Restoration of a lost section corner on a correction line.

0.25 chains by proportion so that the new distance to the regular section corner to the east becomes 30.75 chains.

10.34 RESTORATION OF A LOST NORTH QUARTER CORNER IN A CLOSING SECTION

Principle 15M.   *When the north quarter corner on a closing section was originally set and lost, the lost corner will be reestablished on the closing line at a point at the proper proportionate interval between the nearest found or relocated corners to the right and left.*

The north quarter corner of a closing section on a standard parallel was not normally set.

Principle 15N.   *The method to be followed in the subdivision of a section into quarter sections is to run straight lines from the established quarter-section corners to the opposite quarter-section corners; the point of intersection of the lines thus run will be the corner common to the several quarter sections, or the legal center of the section.*

This principle is probably the most discussed. Although this principle is established in law, the Land Act of 1805, practicing surveyors deviate from it considerably. Over the years, numerous short cuts were taken to locate the *legal center of the section*, yet none were sanctioned by the law. In retracing past surveys that were conducted improperly, if the modern surveyor conducts his or her work properly, deviations would be found in possession lines and surveyed lines. However, the modern surveyor is obligated to conduct the modern work according to the law and principles. There can be no deviation. If, in conducting a modern survey in accordance with proper rules and principles, the surveyor determines possession lines, and surveyed lines disagree on the center of the section, the only obligation the retracing surveyor has is to report these variances to the client, not to ascertain property rights to the mislocated fences.

Principle 15O.   *Preliminary to the subdivision of quarter sections, the quarter–quarter or sixteenth-section corners will be established at points midway between the section and quarter-section corners, and between the quarter-section corners and the center of the section.*

Principle 15P.   *The law provides that where opposite corresponding quarter-section corners have not been or cannot be fixed, the subdivision of section lines shall be ascertained by running lines from the established corners north, south, east, or west, as the case may be, to the watercourse, reservation line, or other boundary of such fractional section, as represented on the official plat.*

10.35   RESTORATION OF LOST NONRIPARIAN MEANDER CORNERS

The original surveyors also set a special type of corner that does not identify specific sections. In later surveys, these *meander corners* were set at points where meander lines were run to identify the exterior of topographic features, usually water bodies, cross-section lines, or township or range lines. In most instances, no bearing tees were set, but the point was monumented. Meander corners can be used for line direction as well as proportioning.

The wording of the original surveyors in the field notes becomes critical when it comes to whether a line is a meander line or not. The application of survey principles to meander lines and non-meander lines is totally different. Read the field notes, patents, and all related documents is the best advice that can be given.

The following principle applies to meander corners.

Principle 16.   Lost meander corners, originally established on a line projected across the meanderable body of water and marked on both sides, will be relocated by single proportionate measurement after the section or quarter-section corners on the opposite sides of the missing meander corner have been duly identified or relocated. Where a line was terminated with measurement in one direction only, a lost corner will be restored by record bearing and distance, counting from the nearest regular corner, the latter having been duly identified or restored.

The actual shoreline of a body of water is considered the correct terminus of a line; the meander corner controls the direction of the line. The restoration of a lost meander corner would be required infrequently.

10.36   RESTORATION OF RIPARIAN MEANDER LINES

Principle 17.   A meander line platted as the boundary of a riparian owner is seldom the true boundary; the body of water represents the true boundary. Lands acquired by the Swamplands Act and fraudulent surveys go to the meander line.

The purpose of meandering any body of water was to obtain information that would aid plotting the body of water on maps. The shoreline itself is the best evidence of the true location and will govern.

When a survey was fraudulent or grossly inaccurate in that it purported to bound tracts of public lands on a body of water when in fact no such body of water existed at or near the meander line, the false meander line, not an imaginary line to fill out the fraction of the normal subdivision, marks the limits of the grant of a lot abutting thereon, and, upon discovery of the mistake, the government may survey and dispose of the omitted area as a part of the public domain.[12] Swamplands conveyed to the state under the Swamplands Act are generally limited by the meander line, as no definite riparian waterline exists.

10.37   RESTORATION OF NONRIPARIAN MEANDER LINES

Principle 18.   *After locating the positions of the meander corners on section lines, the record meander courses and distances are run, setting temporary angle points; the closing error from the last meander course to the section line meander corner is noted. The error of closure is balanced out by the compass rule.*

This principle is attained in the field by moving each temporary angle point or location on a bearing of the closing error a distance in proportion to the length of the closing error as the sum of the courses to the angle point is to the sum of all the courses. Although this method of locating meander points is valid in GLO states, the principle can be applied to a metes and bounds description when a surveyor determines several corners in a description as lost. Where the angle points in the meander lines are considered corner points, either approach applies.

The method used is illustrated in Figure 10.7. The resurvey, beginning on the west side of the section at the found or reestablished meander corner, is run eastward on the bearing and distances of the original notes. After setting temporary stakes at each angle point, the closing bearing and distance from the last temporary stake to the true meander corner is noted, or, as illustrated, N 20° E, a distance of 30 feet. From each temporary stake, the true corner is set on the bearing of the closing line a proportionate distance. The first temporary point would be moved 30 times 36/103, or 10.48 feet, as illustrated. Exactly the same results are obtained by using the compass rule of adjustment.

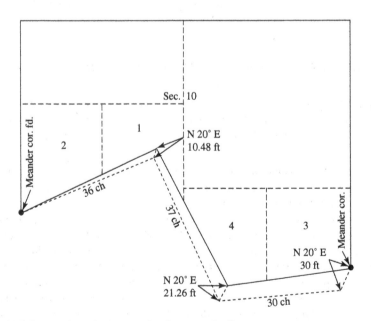

Figure 10.7 Restoration of a lost nonriparian meander line.

10.38 RESTORATION OF IRREGULAR EXTERIORS

Principle 19.   Where boundaries that would normally be straight lines were originally established in an irregular manner, the resurvey (retracement) will follow the irregular procedure.

Where a township line was partially surveyed and later completed by surveying from another direction, an angle point at the junction of the two lines would be perpetuated in resurvey methods. Recent retracements of township lines for the purpose of resurveying a township frequently uncover deflections in alignment along the township lines. New corners set upon such lines have an original-record angular alignment and must be resurveyed, taking the angular deflections into account.

10.39 LOST CORNER RESTORATION METHODS

A summary of the usual methods by which surveyors should restore lost corners within a standard township surveyed in accordance with *Manuals* after 1855 is outlined in Figure 10.8. This diagram may not be applicable where irregular original procedures were used or where early instructions deviated from those approved in published *Manuals*. For example, corners on a sectional correction line acquire the status of a township line; lost corners are restored by single proportionate measurements. The letter D indicates that double proportionate measurement is used. The letter S indicates that single proportionate measurement is used. The letter $S9$ indicates that single proportionate measurement on a curved parallel is used. The letter C indicates a closing corner that is to be adjusted to the line closed on. The most important rule to remember is: proportionate measurement is a rule of last resort; don't use it merely because it requires less work.

10.40 RESURVEY INSTRUCTIONS ISSUED IN 1879 AND 1883

On November 1, 1879, the GLO of the Department of the Interior issued a circular signed by J. M. Armstrong, Acting Commissioner, which stated: "There being no special law in regard to the reestablishment of lost corners, the rule (as given) is to be considered merely as an expression of the opinion of this office." With regard to lost section corners, the circular said: "Missing section corners in the interior of townships should be reestablished at proportionate distances between the nearest existing original corners NORTH and SOUTH of the missing corner." Double proportioning was not used!

In the pamphlet entitled *The Restoration of Lost and Obliterated Corners*, published in 1883, the instructions were to use double proportioning to relocate lost interior section corners. We can only wonder how many lost section corners were set by single proportionate measurements during the period between 1879 and 1883.

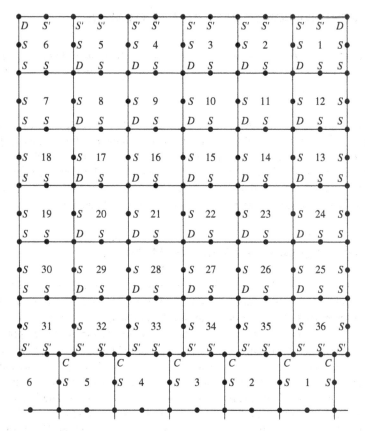

Figure 10.8 Chart showing the normal methods of restoring lost corners. *S*, single proportion; *S′* single proportion on a curved parallel; *D*, double proportion; *C*, closing corner.

10.41   HALF-MILE POSTS IN FLORIDA AND ALABAMA

Two states, Florida and Alabama, were treated to the unique situation of half-mile posts set or established at 40.00 chains along east and west or north and south lines, as described in Section 6.14. Part of the time, the half-mile posts happen to fall where the quarter corner should be. Sometimes the quarter corner was set; at other times it was not. In all instances, the half-mile post was placed on the random line, and if it happened to fall where the quarter corner should be, it became the quarter corner.

In a retracement, the half-mile post, when found, is always used as a control point but not always as the quarter corner.

Some notes may indicate that the original surveyor returned and corrected the half-mile post to the true quarter corner position without new ties to bearing trees. Thus, if the quarter corner post is lost, the bearing trees relocate the half-mile post, not the quarter corner. The half-mile post, being a fixed point on the original survey, can be used as a proportioning point to restore adjoining lost section corners.

Proportionate measurements can never extend beyond original points identified. If an adjoining lost section corner is relocated by double proportionate measurement, the quarter corner is located by single proportionate measurements utilizing the half-mile post.

The evidence of the half-mile post should never be destroyed by the retracement surveyor, nor should the surveyor assume, without checking the field notes, that the half-mile post is the quarter corner—often it is not.

SUBDIVISION OF SECTIONS

10.42 GENERAL COMMENTS

The present federal rules for subdividing sections are clear; the date of their first application is questionable. Certainly, these rules could not have been formed prior to the time the Land Office was empowered to sell land in portions of a section. In 1805 the law provided for land to be sold in quarter sections, in 1820 in half-quarter sections (80 acres), and in 1832 in quarter–quarter sections (40 acres). Prior to 1820, the practice of dividing sections by protraction on township plats did not exist. In 1879, J. M. Armstrong of the Land Office issued a pamphlet on how to restore lost corners and subdivide sections. His method of dividing a section into quarter–quarter sections agrees with what is accepted today. In older areas, where the government did not protract portions of sections and where land was later sold by aliquot parts, the surveyor should consult state court findings to obtain subdivision guidance.

Prior to the issuance of the pamphlet *Restoration of Lost and Obliterated Corners and Subdivision of Sections*, and even afterward, many surveys conducted by private parties or self-proclaimed surveyors were accomplished by measuring lines north, south, east, or west in increments of 1320 feet (20 chains). Such procedure is, of course, entirely wrong; no portion of a section can be laid out correctly without first resurveying the entire section.

10.43 SUBDIVISION BY PROTRACTION

Sections of land, being the minimum area normally surveyed, are too big to completely serve the purpose of disposal of public lands. Where an entryman may patent a quarter of a quarter of a section, the section must be subdivided further to allow proper delineation of the land boundaries; the process is known as *subdivision by protraction*. Upon the original township plats are shown dashed lines that indicate the intent of the government in disposing of parcels less than a section in size. Together with the federal land statutes or regulations of the BLM and plats subdivided by protraction, the surveyor can determine the proper procedure for the survey of portions of sections.

Regular sections of land, as shown on the official plats, have straight dashed lines connecting opposite quarter corners to indicate that quarter sections are

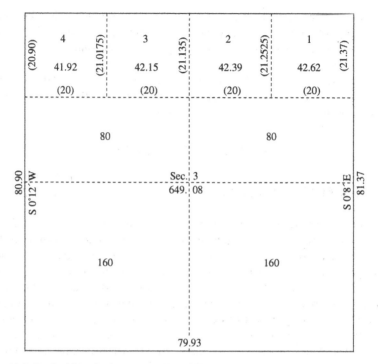

Figure 10.9 Dashed lines indicating smaller divisions are used only when irregular conditions exist.

to be subdivided by running lines connecting opposite quarter corners. Further dashed lines to indicate smaller divisions of land are omitted except where irregular conditions exist, as shown in Figure 10.9. All quarter–quarter section lines not shown on the plat are assumed to be of regular size, as required by statutes. Irregular-sized parcels and quarter–quarter parcels containing irregular measurements are given lot numbers on the plat to distinguish them from regular parcels.

10.44 ESTABLISHING THE NORTH QUARTER CORNER OF CLOSING SECTIONS ON A STANDARD PARALLEL AND OTHER QUARTER CORNERS NOT ORIGINALLY SET

Principle 20. *Where an original quarter-corner was not originally set, as in the case of sections closing on a correction line, sections closing on a range line with double or triple corners, or sections closing on a township line with double or triple corners, place the missing corner on the correction line (or range or township line) at a point between the found or relocated closing section corners a distance that is proportional to the measurements used for the acreage calculations on the original plat. The missing corner is usually set midway between closing section corners except* *in Section 6, where it is usually 40 chains proportional measure from the northeast or southwest closing section corner.*

By an older principle, when a township was being subdivided, the last section lines, run west to intersect the west line of the township, did not have to meet the township line at the same point as the formerly set section corner. Double corners existed: the original section corner for sections west of the township line and the newly set closing section corner for sections east of the township line. By such a rule, the quarter corners for the sections east of the township line—6, 7, 18, 19, 30, and 31—were never set. The quarter-section corner found is applicable only to the sections west of the township line. In the event of triple corners, the quarter corner on the township line was not applicable to either section other than as a line point.

The north quarter corner of Sections 1 to 6, inclusive, was seldom set where the north line of the township was a correction line. Prior to establishment of the missing quarter corner, the closing section corners found must be tested for alignment. If found off the line closed upon, they must be moved up to the line closed upon. From the acreage figures on the original plat, the distances used for computing the original acreage can be reconstructed. For the purpose of illustration, Section 3 shown in Figure 10.9 gives the original measurements of a section whose north quarter corner was never set. Figures in parentheses are those computed by the rules of protraction. Since 1 acre is 10 square chains, the acreage of Lot 1 is $20 \times (21.37 + 21.2525) \div 2 = 426.2 \div 10 = 42.62$ acres. Where the original width of Lot 1 was 20 chains, the acreage was determined by the sum of the lengths of the east and west lines. Conversely, where the sum of the east and west sides equals the acreage, 20 chains was the assumed original width.

Because each of the lots in Figure 10.9 can be proved by computation to be 20 chains wide, the missing quarter corner would be set midway between the found or relocated closing section corners to the east and west.

10.45 ESTABLISHMENT OF CENTERLINES AND CENTER QUARTER CORNERS

Principle 21.   The method to be followed in the subdivision of a section into quarter sections is to run straight lines from the established quarter-section corners to the opposite quarter-section corners; the point of intersection of the lines thus run will be the corner common to the several quarter sections, or the legal center of the section.

Boundary lines that have not actually been run and marked shall be ascertained by running straight lines from the established corners to the opposite corresponding corners.[13] Federal laws are very specific and unambiguous; the legal center of a regularly created section is at the exact point of intersection of the two lines running from opposite quarter-section corners. This rule does not require a section to be divided into two halves of equal area. Although a monument found near the legal center may be controlling for property rights because of occupancy rights, it is never the legal center of the section.

Of course, if the *original surveyor* used an irregular procedure and set a monument at the center of a section, the monument, wherever set, will be the center of the section. Also, if half of a section was created by one original survey and the remaining half by a dependent survey, the rule may not apply because of the irregular original procedure. From 1849 to 1851, Justin Butterfield, Commissioner of the GLO without authorization, issued special instructions that directed that the center-quarter corner would be located at the midpoint of the line connecting the east quarter-section corner and the west quarter-section corner. If it can be shown that during those years these instructions were followed, the corner so set may be the legal center of the section.

When in a retracement the surveyor finds a use and occupation monument inconsistent with the legal center of the section, a dilemma arises: Should it be cast aside? Certainly, the monument can never be the legal center of the section, but property rights may have been lost or gained by unwritten means, or conveyances may have been completed to the monument. When an improperly placed monument is found, the surveyor must investigate the various possible legal ramifications that present themselves.

For some unexplainable reason, the center of the section seems to enter into the discussion in conversations with surveyors at annual meetings, parties, and other occasions. At times, the discussions become quite opinionated. The young surveyor must make a distinction between "Where is the center of the section?" and "What is the center of the section?" In nearly all sections that have been surveyed, a retracing surveyor, after locating the intersection of the lines as dictated by the law, will find a fence corner or monuments at positions that disagree with the intersecting lines or at a location other than the intersection. The distance may range from a few inches to chains. There are surveyors who are willing to accept these monuments as the center of the section. This problem has plagued surveyors and the courts for over 200 years. The problem still has not been solved.

In a recent Oregon decision, *Dykes* v. *Arnold*, 129 P3d 257 (2006), the Supreme Court totally disregarded the law and created a "new center of the section some 75 feet away from the true intersection of the two lines by declaring that the "new center of the *section*" had been created by a doctrine of repose. The question that persists is: Why did they exceed their authority to take this route when they could have declared the fence corner a "property corner"? Are future surveyors who will subdivide the section into smaller parcels required to use this new center section to further subdivide?

These monuments may ultimately be considered as property corners, but there is no manner in which they may be considered as the center of the section. The law is specific and unyielding. On several occasions, some surveyors have attempted to justify the acceptance of these monuments by calling them by such names as "as-built corners" or "accepted center section corners." Their acceptance or rejection are not technical questions but legal questions.

10.46 ESTABLISHMENT OF QUARTER–QUARTER SECTION LINES AND CORNERS

Principle 22. *Preliminary to the subdivision of quarter sections, the quarter–quarter or sixteenth-section corners will be established at points midway between the section and quarter-section corners, and between the quarter-section corners and the center of the section, except on the last half-mile of the lines closing on township boundaries, where they should be placed at 20 chains, proportionate measurement, counting from the regular quarter-section corner. The quarter–quarter or sixteenth-section corners having been established as directed in these principles, the centerlines of the quarter sections will be run straight between opposite corresponding quarter–quarter or sixteenth-section corners on the quarter-section boundaries. The intersection of the lines thus run will determine the legal center of a quarter section.*

"In every case of the division of a quarter section the line for the division thereof shall run north and south . . . on the principles directed and prescribed by the section preceding."[14] The preceding section states: "And the boundary lines which have not been actually run and marked shall be ascertained by running straight lines from the established corners to the opposite corresponding corners."

This rule is applicable in most states, but there are several exceptions. In Louisiana and Mississippi, the original surveyors laid out some of the townships with the error of closure being divided equally between the last two closing half-miles (halfway between the two section corners). In several other states, particularly in Florida and Alabama, the error of closure was sometimes *placed* in the south half mile of the southern tier of sections. Of course, in situations like this, the usual rules of retracement will not apply.

10.47 FRACTIONAL SECTIONS CENTERLINE

Principle 23. *The law provides that where opposite corresponding quarter-section corners have not been or cannot be fixed, the subdivision-of-section lines should be ascertained by running lines from the established corners north, south, east, or west, as the case may be, to the watercourse, reservation line, or other boundary of such fractional section, as represented on the official plat. In this, the law presumes that the section lines are due north and south, or east and west lines, but this is not usually the case. Hence, to carry out the spirit of the law, in running the centerlines through fractional sections, it will be necessary to adopt mean courses, where the section lines are not due cardinals, or to run parallel to the east, south, west, or north boundary of the section, as conditions may require, where there is no opposite section line.*

In those portions of the fractional townships where no such opposite corresponding corners have been or can be fixed, the boundary lines should be ascertained by running from the established corners due north and south or east and west lines, as the case may be, to the watercourse, Indian boundary line, or other external boundary of such fractional township.[15]

This statute is interpreted by the federal courts, as noted. In a few states, the courts say that the east–west centerline of a section should be parallel with the south boundary of the section. The federal rule is preferred.

10.48 SENIOR RIGHT OF LINES

The principle of the precedence of one line over another of less original importance has long been recognized. Like property rights, surveys also have senior and junior rights or dignity. For example, if Township 1 is surveyed and accepted prior to the later survey and acceptance of Township 2 and it is found that Township 2 encroaches on Township 1, then as the senior survey, Township 1 will receive full measure. All closing corners, representing a junior survey, are adjusted to the line closed on.

This has led to the legal question, "May a surveyor cross a township line to proportion lost corners?" The answer is a resounding *no*. (See Section 10.50.)

10.49 GROSS ERRORS AND ERRONEOUSLY OMITTED AREAS

When originally establishing sections of land by survey, gross errors did occasionally occur. This was particularly true of subdivisions adjoining rancho land grants, Indian treaty lines, and the meander of lakes or waterfronts. Where the federal government sold land not its own (overlaps or rancho land grants in particular), the error can be resolved by senior rights or by prolonged occupancy with color of title.

Areas omitted present a complex problem, as occupancy rights against the federal government cannot ripen into a fee title. Omissions of small areas (failure of a closing corner to close on the line called for) cause no problem, as it is a well-established rule that erroneously located closing corners may be adjusted to the original line closed on.

In the event of omission of islands that existed as of the date of the original survey, the US government can claim the islands as unsubdivided public domain land. Many present-day surveys by the BLM are for the purpose of mapping omitted islands in rivers and lakes.

In several instances, the original surveyors included meanders of nonexistent lakes or exaggerated meanders of existing lakes (presumably, to get extra pay for mileage of lines run). Figure 10.10 shows the meanders of Moon Lake (T.12 N., R. 9 E., Fifth Principal Meridian, Arkansas) as originally reported. Investigation on the ground disclosed that the greater part of the tract was covered with oak, maple, cottonwood, hickory, sycamore, hackberry, cypress, and willow, and that no lake could have existed as of the date of Arkansas's admission to the Union. Arkansas had acquired the surrounding fractional sections by the Swamplands Act and was claiming the Moon Lake area as based on riparian rights. In rejecting the riparian claim of the adjoiners, the court commented as follows:

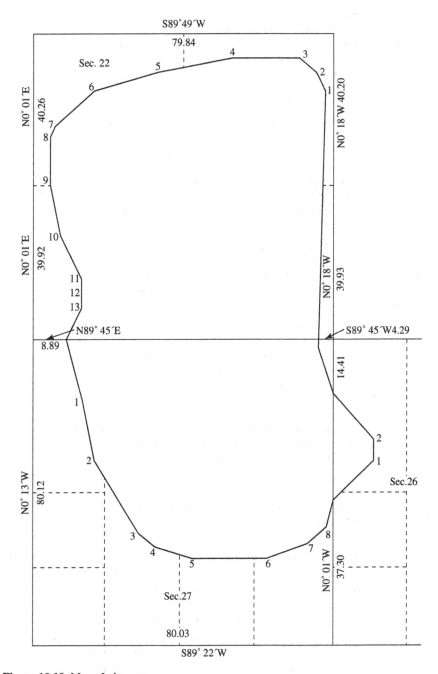

Figure 10.10 Moon Lake case.

First. Where, in a survey of the public contain a body of water or lake is found to exist and is meandered, the result of such meander is to exclude the area from the survey and to cause it as thus separated to become subject to the riparian rights of the respective owners abutting on the meander line in accordance with the laws of the several States. *Hardin* v. *Jordan*, 140 U.S. 371; *Kean* v. *Calumet Canal Co.*, 190 U.S. 452, 459; *Hardin* v. *Shedd*, 190 U.S. 508, 519.

Second. But where upon the assumption of the existence of a body of water or lake a meander line is through fraud or error mistakenly run because there is no such body of water, riparian rights do not attach because in the nature of things the condition upon which they depend does not exist and upon the discovery of the mistake it is within the power of the Land Department of the United States to deal with the area which was excluded from the survey, to cause it to be surveyed and to lawfully dispose of it. *Niles* v. *Cedar Point Club*, 175 U.S. 300; *French-Glenn Live Stock Co.* v. *Springer*, 185 U.S. 47; *Security Land & Exploration Co.* v. *Burns*, 193 U.S. 167; *Chapman & Dewey Limited Co.* v. *St. Francis Levee District*, 232 U.S. 186.

If in the making of a survey of public lands, an area is through fraud or mistake meandered as a body of water or lake where no such body of water exists, riparian rights do not accrue to the surrounding lands, and the Land Department, upon discovering the error, has power to deal with the meandered area, to cause it to be surveyed, and lawfully to dispose of it.

The fact that its administrative officers, before discovery of the error, have treated such a meandered tract as subjected to the riparian rights of abutting owners, under the State laws, and consequently as not subject to disposal under the laws of the United States, cannot stop the United States from asserting its title in a controversy with an abutting owner; and even as against such an owner, who acquired his property before the mistake was discovered and in reliance upon action and representations of Federal officers carrying assurance that such riparian rights existed, the United States may equitably correct the mistake and protect its title to the meandered land. The equities of the abutting owners, if any, in such circumstances are not recognizable judicially, but should be addressed to the legislative department of the Government.[16]

10.50 RELOCATING CORNERS FROM OTHER TOWNSHIPS OR FROM INTERIOR CORNERS

As has been pointed out, surveyors are innovators when it comes to using exotic, and possibly unauthorized or unaccepted, methods of locating or proving corners. In the basic scheme of authority in the subdivision of a section, sixteenth corners are located and surveyed from section and quarter-section corners. Thus, they are superior to the sixteenth corners, and the sixteenth corners should not be used to locate section corners or quarter-section corners. If, in the course of a survey, a question is presented as to whether an original corner is lost or obliterated, the measurement from a sixteenth corner may be used, along with all of the other evidence, as a means of attempting to prove the corner in question, but not to set it.

Applying the principle that each township is a separate and independent entity, one cannot cross that invisible township or range line to resurvey a corner in an adjacent township. The surveyor should rely on original corners to set or position lost corners; that is, the surveyor should not use a sixteenth corner to set a quarter corner. Except for those rare instances when an original surveyor set an interior corner within a section, the retracement surveyor should not go from within a section to set a corner along a section line. If, in the course of a survey, an original corner position is thought to be on the verge of a lost corner and additional evidence is needed to prove it to be obliterated, a measurement from a found monument within the section may be taken into consideration.

However, a found interior subdivision corner should not be accepted without being referenced to a found or proven original corner or without evaluating the field notes of the surveyor who subdivided the section.

10.51 PROCEDURES FOR CONDUCTING RETRACEMENTS

A surveyor should not undertake a retracement of a section, a township, or a metes and bounds survey without first having an understanding of what is required. A retracement survey should be planned in advance to ensure that nothing is overlooked. The steps suggested for this procedure are as follows:

1. If possible, obtain necessary original field notes and township plats pertaining to the area being surveyed. If the state has more than one source, determine which source has the "official" field notes. For example, original California field notes were burned in the San Francisco fire and copies were made from the government copies. Thus, California now has copies of copies and the government copies are "official" documents. Included should be original surveys and any subsequent dependent or independent resurveys as well as official retracements. In several metes and bounds states, all official records were burned when Sherman made his infamous "march to the sea" during the Civil War in 1865. In many metes and bounds states, the original surveys were private in nature and the original notes were never turned over to the respective states.

2. Search known records for subsequent surveys conducted by private or public parties, such as road, railroad, and court-ordered surveys.

3. Contact old residents concerning ancient land boundaries. This may include taking affidavits to preserve testimony from elderly witnesses.

4. Examine recorded and known pertinent deeds of landowners along the lines to be surveyed and any recorded documents that may show easements or encumbrances.

5. Make a diligent search for all necessary corners and apply the rules of evidence to determine whether a corner is original, obliterated, or lost. If it is an obliterated corner, it should be refurbished with a durable monument with the surveyor's identification numbers on it. If the corner is lost, reposition it according to the proper rules of survey.

6. Set new monuments for new positions and replace any monuments that have deteriorated.

7. Subdivide the section and set any required subdivision corners according to applicable rules of survey, whether federal or state.

8. Prepare and file a record of survey indicating the dignity of all points recovered or set in the retracement, and identify all new monuments set. Furnish the client with a written report of the methods used, monuments set, and basis of decisions made.

The responsibility of the retracement surveyor is to follow the evidence of the footsteps of the original surveyor as nearly as possible. However, these instructions should equate to "following the evidence of the footsteps." This will require that the retracing surveyor go into the field armed with as much of the documentary evidence as possible and then try to determine or ascertain the location of the original footsteps through the recovered evidence. In light of modern technological advances, it is virtually impossible, surveywise, to follow the footsteps of the original surveyor. Using calculation to determine positions of the original lines, these original lines can be retraced. The retracing surveyor must always keep in mind the precision of the original surveys. The degree of positional tolerance from lines that are all measured to the full chain and the angles measured to the full degree are not absolute in position but they have "areas of certainty."

After recovering and evaluating the evidence available, the surveyor determines that there is no exact certainty as to the precise location of the exact corner point, but its corner can be located only within an area of certainty; for example, "The location falls within a circle of 9 feet in radius, based on the evidence." What some people would consider as a weakness is not, in that we can say: "The corner can be anyplace within this 6-foot circle but certainly is not some 300 feet over there."

10.52 INTERPRETATION OF ALIQUOT DESCRIPTIONS

Surveyors seldom survey or retrace more than a fractional portion of a section or sections. For the surveyor to understand what is required, the client usually furnishes the surveyor with a description, usually referring to an aliquot portion. Yet the description furnished is often far from a true aliquot description. Some of these descriptions may be as follows: "the east 80 acres of the NE ¼," "the west 1320 feet of the NW ¼," "the north 10 chains of the NE of the NW," "the NE ¼ lying north of Highway 6." In fact, one of the worst descriptions, which has no legal meaning, is "the NE ½ of the section"; there is no basis of validity for this description. These descriptions should not be considered as an aliquot description but are in all probability a modified (quasi) metes and bounds description. One could also encounter a description that calls for "lot 2, containing 80 acres." There could possibly be a conflict between "lot 2" and "80 acres."

In examining the original plat, the closing Section 3 would depict two lots, with Lot 2 depicted as consisting of four parcels containing a total of 153.98 acres and Lot 1

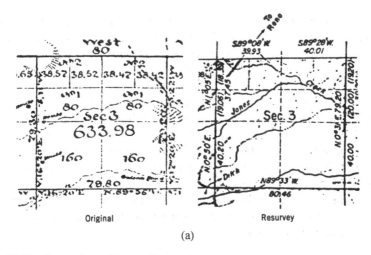

Original   Resurvey

(a)

Figure 10.11a Comparison of survey lines and areas.

depicted as two parcels containing 160 acres (see Figure 10.11a). However, after a dependent resurvey and a subsequent retracement, it was determined that the closing lines were short, and the prorating of the distance of the two lots indicated that Lot 1 actually contained 156.24 acres and Lot 2 contained 150.6 acres. The question that the retracing surveyor must address is: What do I survey? Lot 1 or 2 or the area? It depends. Even though the original government surveyor did not physically run the boundary line between Lots 1 and 2, they were fixed legally, and the bona fide rights were established in accordance with the original survey. Accepting the fact that, although there are fewer acres present within the actual boundaries of Lots 1 and 2 than were indicated in the original survey, the answer would be: "First come gets the total area and the second gets the remainder." Thus, the bona fide rights of one would be jeopardized. The correct answer in locating the original boundary between Lots 1 and 2 would be to apply the principle that each lot takes its proportionate share of the deficiency, and the dividing line between the two retraced lots would be proportionate to the original. In this situation, the retracing surveyor must examine prior deeds, the chain of title, and possible unwritten rights.

There has been considerable manipulation with the federal method of describing land parcels. Federal laws permit legal subdivisions to a ¼ ¼ or 40 nominal acres. One can find many descriptions that will describe a parcel to 1 ¼ acres by subdivision. The law does not permit it. Also those who write descriptions play "loose" describing less than 40 acres. In some states descriptions such as NE ½ of the NE 1/4. This description, when challenged in the courts has been held valid in some and void in others.

Over the years, confusion and ambiguity have resulted from improper identification of portions of sections. A description that conveys the east 80 acres of the northeast quarter of a section does not convey the eastern half of the northeast quarter. The official plat of the section usually indicates a return of 640 acres, with all the lines 80 or 40 chains in the quarter section. Theoretically, this would place 160 acres in the northeast quarter and 80 acres in the eastern half of the quarter section. Since in a resurvey a section is rarely found to be the size reported, a description conveying the east 80 acres may or may not convey 80 acres for the following reasons.

If in the retracement of a section the surveyor finds that the section measures exactly the same as the record—an unlikely event—the east 80 acres and the eastern half of the northeast quarter would be identical. If in the retracement the section was found to be long in measurements and the northeast quarter contained more than 160 acres, only 80 acres would be conveyed. If the original patentee patented the eastern half, the seller of the east 80 acres would retain a strip of land. On the other hand, if the section measured short in size and the patentee had obtained only the eastern half, sale of the east 80 acres would not be a sale of 80 acres, since the seller did not own 80 acres. Persons describing land should refrain from conveying an area if it is intended that an aliquot portion be conveyed.

10.53 ACCORDING TO THE GOVERNMENT MEASURE

The words "according to the government measure" or "according to the government survey" become controlling elements in a description. When these words are discovered, the retracing surveyor must accept the realization that all the elements and the principles of retracing and locating a US government parcel must be applied. That also requires that the retracing surveyor relate back to the original measurements that are indicated on the original plat and in the original notes. If the phrase "according to the government measure" is included as a portion of a description, the interpretation resorts to the original measure as indicated on the official plat, and the conveyance is intended to be a proportionate part of the original measure. With the use of these terms, the retracement takes on an entirely new approach, in that the retracing surveyor should refer to as many of the original documents as possible. The insertion of these words gives any description an entirely new meaning. They may become the controlling factor if questioned.

DIFFERENCES BETWEEN STATE AND FEDERAL INTERPRETATIONS

10.54 APPLYING STATE LAWS

It is an accepted principle that, once lands were divested from the United States into private ownership, the responsibility of retracement became that of the states. As long as the federal government retains any rights in any lands, the federal rules of survey must apply. Since the public surveys were completed, many states have enacted statutes that conflict with federal laws. If the state has final authority in a jurisdiction, the state law would be followed so far as property rights are concerned. Some of the deviations from federal laws are discussed in the following sections.

Missouri Statute Law

Missouri statute law (Section 60-290) provides that a lost section corner should be reestablished at the point of intersection of straight lines connecting the nearest corner found on each side of the lost corner; double proportionate measurement is not used. In the case of *Simpson* v. *Steward*, the court observed that double proportioning was one method of restoring a lost corner and the state rule was another.[17] As neither rule would restore a corner exactly where the original was, either rule would be equally valid, and in Missouri the state statute is enforced. Recently, this statute was repealed. Court opinion will be needed to clarify the status of the old and new situations.

Wisconsin Law

In 1862, a law was passed by the state of Wisconsin which provided that whenever a surveyor is required to make a subdivision of a section, as determined by the US Survey, except where the section is fractional, the surveyor should establish the interior quarter-section corner at a point that is the same distance from the east quarter-section corner as it is from the west quarter-section corner, and the same distance from the north quarter-section corner as it is from the south quarter-section corner.[18] In 1867, the law was repealed, and new legislation provided that whenever a surveyor is required to subdivide a section or smaller subdivision of land established by the US Survey, the surveyor should proceed according to the statutes of the United States and the rules and regulations made by the secretary of the interior in conformity thereto. In Wisconsin, the problem of the center of the section is now resolved by statute law.

10.55 TOPOGRAPHY

For the most part, the identification of topographic calls was one of the responsibilities of the original surveyors. There are times when the topographic calls harmonize well with actual conditions, and there are times when the retracing surveyor questions whether the original surveyor was even in the state, let alone the township, in that no harmonious relationship can be made of the results. The retracing surveyor will find mountains where a stream is indicated on the original map, or the retracing surveyor is positive that the original surveyor was there, in that all of the features are checked but are out of position by 20 chains. In such a situation, the retracing surveyor is presented with a major problem to ascertain just what happened.

The presumption is that the survey was conducted as indicated in the field notes and depicted on the plat. Since no court has the authority to determine that the survey was erroneous, the retracing surveyor can only report the facts. In California, if topographic calls can be identified, the courts generally accept them, even though they may be substantially off distance calls. Reestablishing a corner by proportionate methods is truly a rule of last resort.

According to *Hanes* v. *Hollow Tree Lumber Co.*, "The government survey cannot be impeached by proof of fraud, negligence, or mistake on the part of the Government surveyor."[19] In a California court, as in some other states, it is never wise to say that the original survey was fraudulent; the surveyor should merely point out that topography calls and other items are not in agreement with the field notes.

10.56 BOUNDARIES BY AREA

Surveyors retracing boundaries from descriptions that refer to area may be confused as to the proper methods that were used to create the original boundaries. The surveyor will find that two basic approaches are accepted; which one should be used cannot be stated with certainty in a book such as this. A description of the "north ½ of the SE ¼" may be approached in several ways. In applying the federal approach, the north half is determined by taking one-half of the distance of all lines involved. This division may or may not give each divided portion an equal area. Under the state approach, the dividing lines are determined using equal area as the factor for locating the boundary lines. Not all states use this method; many of the GLO states use the federal method. It is important that a surveyor understand what method is used. In a California case, a description was written for Section 30. The original plat indicated the east half as containing 80 acres and the west half as containing 98.98 acres. A single patent was issued by the United States for the entire southwest quarter. Subsequently, the owner sold the east half. The buyer took possession of approximately 90 acres, or half of the total area that was based on a survey, not on the 80 acres depicted on the plat. In the ensuing litigation the court awarded half of the area, noting that the owner could sell the parcel by any description that he or she wanted to use. This decision probably would have been different had the deed made reference to "according to the government survey."

This should be a warning to surveyors who are asked to identify or survey boundaries of such parcels. Aliquot portions should be surveyed according to the federal rules of survey, but after lands have been patented into private ownership, the state rules may apply and equal area will be considered, unless the magic words "according to the government survey" are recited. In the opinion cited here, if the United States had patented the two parcels to two separate persons, the decision would have been different. In that specific case, the federal rules would have applied.

By the federal aliquot part rule, one-half is seldom one-half of the area of a whole; by state common law, one-half is one-half the area of a whole. In *Wood* v. *Mandrilla*,[20] in Section 30, T 20 S., R 24 E. Mount Diablo Meridian, the original township plat pictured the east half as 80 acres and the west half as 98.98 acres. One patent was issued for the entire southwest quarter, and the owner sold the east half. The buyer took possession of half the area, about 90 acres, instead of the 80 acres shown on the government plat.

According to the court, the seller has the right to sell by either the federal method or by proportion of the area; the court awarded half the area. Aliquot parcels obtained by patent from the federal government should be surveyed in accordance with federal rules, but after a patent has been issued and the land is divided by fractional parts, the state rule of equal area may apply. In the *Wood* v. *Mandrilla* case cited here, if the east half had been patented from the federal government separately from the west half, federal rules would apply.

10.57 ESTABLISHING CORNERS

A few states have passed laws whereby the county surveyor could "establish" a corner or boundary line by following the procedure required by law. If the county surveyor's findings were not appealed within a definite period of time, the findings became final. Very often, the county surveyor located centers of sections or sixteenth corners in violation of the law. When it could be shown that the center of the section was not on the intersection of lines connecting opposite quarter corners, the question arose: Is the corner located by the county surveyor the legal center of the section, or should the true center be located? In general, where the county surveyor sets the corner in accordance with statutory requirements and the period of time stated in the statute has passed, the county surveyor's corner becomes the legal center of the section for parties properly notified. Such proceedings are much like a court case. Unless it can be proven that legal process was not followed, no future survey can supersede the one established by law.

10.58 SECTIONS CREATED UNDER STATE JURISDICTION

Not all sectionalized lands were surveyed and created under federal laws (see Figure 10.11b).

In West Texas, many sections were originally surveyed under the authority of Texas; these lands were never owned by the federal government. This also applies to a major portion of Northern Maine and throughout Northeastern United States. In California, much confusion was created when a law was passed making it necessary for the assessor to assess parcels not greater than 1 square mile. Most of the owners of large ranchos filed maps projecting section lines through the rancho as they would have existed if sections had been surveyed. Later, lands were sold by the theoretical sections (never monumented). In a few instances, the sections were monumented by variable methods, depending on the surveyor.

If the sections of land were not created by federal rules of the BLM, or its predecessors, and were not under federal jurisdiction, the foregoing principles are not necessarily applicable. State laws must be examined to find proper procedures. In keeping with the majority opinions and rules now accepted by many states, state courts, and knowledgeable surveyors, the federal rules for subdividing sections of land that were originally created under the federal laws apply when retracing those lands:

The description of the original parcel may determine which law(s) to apply in a retracement.[21]

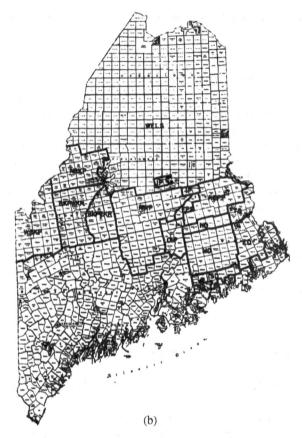

(b)

Figure 10.11b Maine was divided into townships of irregular and regular shape. Later surveys were rectangular, with 6 miles on a side. Since these townships were created under state law, federal rules, laws, or customs are not controlling. For resurvey procedure, state court interpretation is controlling.

10.59 PRESUMPTIONS AND REALITIES FOR GLO SURVEYS

A surveyor retracing the lines of aliquot portions of GLO surveys not only should have an understanding of the theoretical township or sections but should also understand the actual or realistic township and sections. The following facts may apply both to the theoretical sections and to those as actually surveyed (Figure 10.12). In an actual presumed, theoretical, legal, statutory section that was anticipated by the original surveys, we should anticipate the following:

1. All bearings are recited in the notes and depicted on the plats as north, south, east, and west, or otherwise as indicated.
2. All sides are 80 chains in length and not 5280 feet or as indicated in the notes and depicted on the plats.

3. All other distances indicated are the true surveyed field distances.

4. All angles at the corners are right angles.

5. Each section contains the exact number of acres indicated.

6. Quarter corners are equidistant between section corners.

7. Quarter corners are on a straight line between section corners.

8. The legal center of the section is the geometric center.

9. Opposite sides of the section are parallel.

10. The section contains exactly 640 acres, and the aliquot parts contain the exact aliquot share of the entire section.

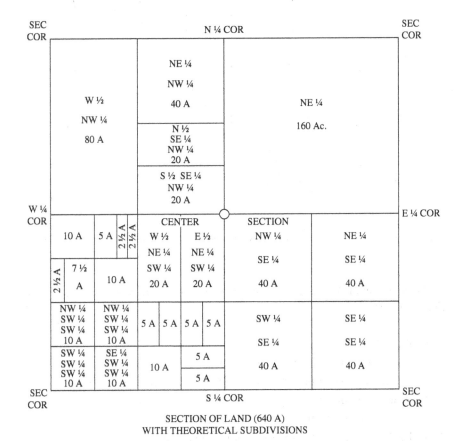

SECTION OF LAND (640 A)
WITH THEORETICAL SUBDIVISIONS

Figure 10.12 Idealized interior section. Shown are standard aliquot part designations (with nominal areas) within an interior section, according to the US Public Lands Survey System. Those sections against the north and west boundaries of a township, those quarter–quarters (or "forties") that occur against those north and west boundaries are designated as lots and not aliquot pans of sections. In addition, other specialized situations may also require that certain parcels of land have lot or tract designations. For sections that more nearly resemble real-world conditions, see Figure 10.13.

TYPICAL SECTION AS LAID OUT

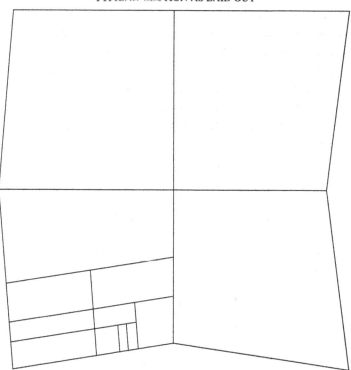

Figure 10.13 Typical "regular" section. *Each* section must be considered a *special* case.

The true section found as a result of a resurvey or retracement is as follows (Figure 10.12):

1. The sides of the section are not 80 chains long.
2. Quarter-section sides are not 40 chains long.
3. Quarter-section corners are not equidistant between section corners.
4. Quarter-section corners are not on a straight line between section corners.
5. No one line is on a true cardinal bearing.
6. The legal center of the section is not the geometric center.
7. No angle between lines is 90°.
8. Opposite sides of the section are not parallel.
9. The section does not contain exactly 640 acres, nor do any of the subdivision parts contain their exact theoretical acreage.
10. The actual field measured distances and bearings do not agree with those reported originally.

10.60 CONCLUSIONS

Principle 24.   *Other than creating new subdivisions for private landowners, today the majority of the modern surveyor's work is retracing the lines of the original surveys, some of which are ancient in nature.*

Today, the majority of the services performed by land surveyors is finding previously run parcels of land. Granted, some original surveys are conducted for the creation of new subdivisions, but this does not constitute the major portion of a surveyor's work. Most surveyors conduct retracements of previously created parcels of land, both federally created parcels and private parcels. Few surveyors concentrate on creating new land parcels or townships. The federal government conducts few original surveys in the GLO states—Alaska being the exception. Some surveyors' entire work is running previously created lines and then redescribing them for today's terms for today's landowners. A few firms may still prepare new subdivisions for the sale of the newly created lots or parcels, but in many areas this is the exception. Since the future of the land surveyor may be in the area of retracements, it is imperative that the modern surveyor have a complete understanding of the differences between the methods of retracement and an original survey.

NOTES

1. *Manual for the Survey of the Public Lands of the United States, Bureau of Land Management, U.S. D,I.* printed by the Government Printing Office, 2009.
2. *Cragin* v. *Powell*, 128 U.S. 691, 2 S.Ct. 203 (La. 1888).
3. U.S.C.A., Title 43, Sec. 472.
4. Supra note 1.
5. *Harrington* v. *Boehamer*, 134 Cal. 196 (1901).
6. Bureau of Land Management, *Manual of Instructions for the Survey of Public Lands* (Washington, DC: U.S. Department of the Interior, 1973), 310.
7. Pulehunui, 4 H 239 (1879).
8. *U.S.* v. *John Citko et ux*, 517 F.Supp. 233 (1981).
9. 468 F.2d 633 (1972).
10. *Restoration of Lost or Obliterated Corners* (Washington, DC: U.S. Government Printing Office, 1976), 6.
11. *Restoration of Lost or Obliterated Corners* (Washington, DC: U.S. Government Printing Office, 1976), 13.
12. *Manual*, 366.
13. U.S.C.A., Title 43, Sec. 752.
14. U.S.C.A., Title 43, Sec. 752.
15. U.S.C.A., Title 43, Sec. 752.
16. *Lee Wilson & Co.* v. *U.S.*, 245 U.S. 24 (1917).

17. *Simpson* v. *Steward*, 281 Mo. 228 (1919).

18. Wisc. Stat Ch. 120, Sec. 4.

19. *Hanes* v. *Hollow Tree Lumber Co.*, 191 C.A.2d 658 (1961).

20. *Wood* v. *Mandrilla*, 167 C.A. 607 (1914).

21. *Bryant* v. *Blevins*, 884 P2d 1034 (Cal. 1994).

CHAPTER 11

LOCATING SEQUENTIAL CONVEYANCES

11.1 INTRODUCTION

In Chapters 11 and 12—Locating Sequential Conveyances and Locating Simultaneously Created Boundaries, two contemporary conveyance systems are discussed
separately. In theory, they could be discussed in a single chapter. But we chose to
discuss the two systems, sequential and simultaneous, in separate chapters to encour-
age students to recognize their importance. Retracing surveyors need to know the
distinctions between the two.

In a retracement, the surveyor is frequently confronted with problems of resolving
conflicts of evidence discovered when required to place a deed description on the
ground. Relying on previous decisions, courts have given surveyors guidance on the
resolution of such problems. Three basic elements that have been identified are as
follows:

1. An occupancy right that has ripened into a legal right extinguishes or becomes
   superior to all written title to occupied land.
2. As between private parties in a land dispute, a senior right is superior to a junior
   right.
3. Written intentions of the parties are paramount.

Because they seldom determine the validity of occupancy or property rights, surveyors are concerned primarily with elements 2 and 3. The intent of the parties in a

Brown's Boundary Control and Legal Principles, Eighth Edition.
Donald A. Wilson, C.A. "Tony" Nettleman III, and Walter G. Robillard.
© 2024 John Wiley & Sons, Inc. Published 2024 by John Wiley & Sons, Inc.

conveyance is first determined from the writings and then by the order of the importance of the deed elements. Usually, a call for a prior survey controls; monuments control over direction, distance, or area; under some circumstances, distance controls over direction; and area is usually, though not always, the least important element. Although these are common-law rules, specific state statutes may give a variant order of importance.

As can be determined from reading various court decisions, the applicable rules for survey, resurvey, or retracement of conveyances are not simple, nor will they be found to be uniform from state to state.

Surveyors are often asked to determine the intent of the various parties. There is no correct answer. Any list may be altered to meet the requirements or the facts presented.

The following is a collection of decisions reflecting how various courts have determined and resolved the intent of parties. When locating a parcel from a written description, the surveyor will see that there are three general classes of surveys or resurveys or, more particularly, retracements: (1) locating metes and bounds descriptions where the parcels have been located in sequential order; (2) locating lots within subdivisions where the lots were created simultaneously; and (3) locating parcels that are predicated on sectionalized federal government lands. Each of these will be treated separately.

A surveyor will find metes and bounds surveys in all the states as well as the various territories. As will be seen with sequential surveys or conveyances, one will seldom apply proportionate measurements, but in a retracement of parcels created simultaneously, proportionate measurement may be the rule unless the contrary can be shown.

The following principles are discussed in this chapter:

PRINCIPLE 1. Ambiguities in patents issued by a state are usually resolved in favor of the state.

PRINCIPLE 2. Having knowledge of sequential principles is important to surveyors and attorneys in understanding the basics of title and boundaries. Applying these principles in a faulty manner may lead to confusion and unnecessary litigation if they are not readily understood.

PRINCIPLE 3. As between private parties, a junior grant in conflict with a senior grant yields to the senior grant.

PRINCIPLE 4. A grantor that sells a portion of his property automatically becomes junior to a grantee purchasing such property.

PRINCIPLE 5. When establishing the boundaries of a property in accordance with a written deed, the boundaries must be established in accordance with the written terms of the deed. Parol evidence may not be taken to determine the terms of a deed but may be used only to explain ambiguous terms of a deed.

PRINCIPLE 6. When a conveyance is reduced to writing, it is assumed to contain all the terms intended and is construed, if possible, to give effect to all the terms mentioned in the conveyance.

PRINCIPLE 7. Where a property description calls for a plat or map and the parties acted with reference to the map, the plat or map becomes a part of the description as much as if it were expressly recited in the deed itself.

PRINCIPLE 8. A call for an adjoiner is not always a correct criterion for determining senior rights. A title search back to the original formation of the conveyance is necessary for a correct solution.

PRINCIPLE 9. Excepting senior rights of others and a valid unwritten right of possession, the intentions of the parties to a deed, as expressed by the writings, are the paramount considerations in determining the order of importance of conflicting title elements.

PRINCIPLE 10. Title lines established by estoppel, agreement, prescription, or other unwritten means are local in character and cannot be used to establish lines of the written deed.

PRINCIPLE 11. Where lines are actually located and marked on the ground as a consideration of the transaction and called for by the deed, the lines so marked show most clearly the intentions of the parties and are presumed paramount to other written considerations, senior rights, and clearly expressed contrary intentions being excepted.

PRINCIPLE 12. Monuments called for in a deed, either directly or by a survey, or by reference to a plat that the parties relied on, are subordinate to senior rights, clearly stated contrary intentions, and original lines actually marked and surveyed, but are presumed superior to direction, distance, or area.

PRINCIPLE 13. When there is a conflict between sequential title and sequential surveys, the title controls the surveys.

PRINCIPLE 14. If a description of land contains an error or mistake, and if the error or mistake can be isolated, the error or mistake is placed where it occurs.

PRINCIPLE 15. Bearing and distance are presumed superior to surface, and only where bearing and distance more clearly show the intent do they control other elements.

PRINCIPLE 16. Except where area expressly states the intentions of the parties to a deed, area is presumed as subordinate to other considerations.

PRINCIPLE 17. A grant or conveyance is to be interpreted in favor of the grantee except that a reservation in any grant, and every grant by a public officer or body, as such, to a private party, is to be interpreted in favor of the grantor.

PRINCIPLE 18. Certain and definite statements will prevail, and erroneous or ambiguous terms may be rejected, but such rejections should be as few as possible.

PRINCIPLE 19. Unless stated otherwise, a distance is measured in a straight line along the shortest measurable distance, usually horizontal.

PRINCIPLE 20. When a curve factor is given as a whole number, the curve factor with the whole number was probably the controlling assumed figure originally and should remain as such. When radius, tangent, and degree of curve are all odd, a prorated adjustment between fixed monuments is indicated. Government (city, county, and state)-relocated curved street lines are not acceptable for lot location unless based on original curve stakes.

PRINCIPLE 21. When two factors are in conflict and nothing else indicates which of the two is correct, the first stated is preferred.

PRINCIPLE 22. When numbers are shown both as figures and spelled out as words, the words will control unless the contrary can be proved.

PRINCIPLE 23. The scale unit of the map is implied to refer to all distances without a character mark on the map.

PRINCIPLE 24. A particular intent will by presumption control a general one that is inconsistent with it.

PRINCIPLE 25. Unless the contrary is stated, it is presumed that every bearing given in a present-day metes and bounds description refers to the same basis at the same point, and that the bearing of every line is constant throughout its length.

PRINCIPLE 26. (a) Where the bearing of a known line is given in a metes and bounds description, the bearing as given is assumed to be correct; successive courses are surveyed relative to the given bearing, whether or not the given bearing is astronomically correct. (b) Where a land description refers to a map and no basis of bearing is stated, it is implied that the map bearings are to be used, and all bearings in the description are referred to the same basis. (c) Where no basis of bearing is given or implied by a call for a map, true or magnetic bearings are to be used, depending on the presumption in the particular state.

PRINCIPLE 27. If an original line of a deed can be identified, and if it was described originally by a measured magnetic bearing, the difference between the original record bearing and the present measured magnetic bearing is the correction to apply to other record bearings of the same description.

PRINCIPLE 28. If a line can be found in the vicinity whose magnetic bearing was determined reliably at the time of the deed, determine on that line the change in magnetic bearing from the date of the deed to the present time. Next, correct the bearings of the lines being surveyed by the amount of correction noted.

PRINCIPLE 29. In the absence of a direct method of determining declination changes between given dates and reliable local data relative to declination changes from a given date, apply the magnetic declination corrections as given in National Geodetic Survey tables or website.

11.2 DEFINITION OF SEQUENTIAL CONVEYANCES

When a portion of a tract of land is sold, two or more parcels are created, a new parcel and the remainder of the parent parcel. Because the new parcel must receive all the land described, it is called the *senior deed*, and the remainder, at the time of conveyance, becomes the *junior deed*. *Sequential conveyances* are those written deeds in which junior and senior rights exist between adjoining parcels. In general, sequential conveyances came into being because of a lapse of time between successive conveyance instruments. For example, the state of Virginia granted a patent for a parcel of land to Jones, and at a later date granted an adjoining patent to Smith. If there happens to be an overlap of Smith's parcel on Jones's parcel, the party first in time (Jones, the senior deed) receives all that is described in the patent, and the adjoiner or second conveyance in time (Smith, the junior deed) receives the remainder. If a gap exists, that portion, in theory, belongs to the state. In Texas, Virginia (including the Virginia Reserve in Ohio), Kentucky, and several other southern states, warrants were issued for an area of land. The person with a warrant went upon the land, selected the area that he or she desired, and had it surveyed, described, and platted, as required by law. The description usually included calls for monuments, trees, topographic features, measurements, and area. The person then applied for a patent, and if all the legal requirements were met, the patent could be granted. Since each parcel was created in sequence, usually as indicated by the date of the patent, these were sequential conveyances.

As expected, in states issuing indiscriminately located sequential patents, gaps, vacancies, and overlaps sometimes existed between parcels. For a period of time in Texas, such vacancies belonged to the first person applying for them, and in oil fields, vacancy hunters often became wealthy. Later, this law was changed, and the schools became the beneficiaries of vacancies. Each state now determines how or whether vacancies may be conveyed, and each surveyor must be acquainted with the laws of his or her state. Because of the numerous conflicts in deed rights, many states with sequential patents shortened the time necessary for occupancy rights to ripen into a fee right from the 20-year common law.

11.3 SIMULTANEOUS CONVEYANCES

When several parcels of land are created at the same moment in time, from the same source tract, such as lots in a subdivision, several parcels in a will, or sections in a township, all parcels have equal legal standing; they were all created at the moment of filing the subdivision map, at the moment of death of the testator, or, in the case of a US township, at the time the plat was approved. Sequential rights (senior rights) to lots in a filed subdivision rarely exist. Usually, each party is entitled to his or her proportion of any excess or deficiency discovered. However, sequential rights may exist between separate subdivisions.

In a resurvey of a deed, if no significant error of closure exists and all monuments fall in their measured positions, the resurvey merely amounts to replacing old monuments or setting new monuments at corners where none were set originally. However, when measurements between original monument positions do not agree with those called for in the conveyance or when there is a significant error in closure, correcting these discrepancies becomes a problem. The treatment of the discrepancy varies depending on whether the conveyance is sequential or simultaneous. In this chapter, errors in measurements found in sequential conveyances are discussed.

11.4 POSSESSION

Any unwritten right or possession that ripens into a fee right extinguishes prior title rights. In such an event, all the rules for construing the order of importance of conflicting deed elements are meaningless; hence, an unwritten right ranks first in the order of importance of elements making up the "right of title and possession" or, in the language of the layperson, "ownership." Normally, a surveyor does not decide who has unwritten possession rights; he or she merely notes the evidence of encroachments or adverse uses and evidence of other surveys. The final decision is a legal one to be made by the courts.

11.5 SEQUENTIAL PATENTS

Principle 1. *Ambiguities in patents issued by a state are usually resolved in favor of the state.*

In several states, some of the original land titles were derived from the state, and many of these states issued patents in a sequential manner. As a part of the patent procedure, a survey of the land parcel selected, a written description, and a plat are usually required. The question now arises: What are the junior and senior rights between the state and the person acquiring the patent?

Patents are quitclaim deeds. A state or governmental body gives no warranty. In the event of an ambiguity, the patent is to be construed in favor of the government, not in favor of the person who holds the patent or his or her successors. This is just the reverse of what is accepted as proper during a land sale between private parties. In *Creech* v. *Johnson*, the court ruled that "where the terms of a patent from the State are uncertain or doubtful, it is to be strictly construed against the grantee."[1] In *Fordson Coal Co.* v. *Napier*, the court held that "construction most against party claiming under uncertain (patent) survey should be adopted in case of doubt in establishing boundaries."[2]

11.6 IMPORTANCE OF KNOWLEDGE

Principle 2. *Having knowledge of sequential principles is important to surveyors and attorneys in understanding the basics of title and boundaries. Applying these*

*principles in a faulty manner may lead to confusion and unnecessary litigation if
they are not readily understood.*

Like any other profession in retracing completed surveys, there are certain basics
that do not have to be identified. These principles are a result of standards that seem
so basic that they are imposed by unwritten dictates. An example would be "A doctor
must wash his or her hands before treating a patient "or" Surgical instruments must
be sterilized."

This axiom also applies to surveyors. No surveyors should be reminded that they
should calibrate their equipment. That should be a given.

In a 1987 Virginia decision, a dissatisfied landowner, through his attorney, filed
a negligence, per se action against his retracing surveyor.[3] The defendant landowner
challenged the decision from the Circuit Court of Albemarle County (Virginia),
which, in a jury trial, entered judgment for the plaintiff surveyor in the surveyor's
suit to recover for the unpaid balance for a boundary survey. The landowner had
the action removed from the district court, denied liability, and counterclaimed for
damages arising from the surveyor's alleged negligence.

The court reversed the decision after concluding the surveyor was negligent as
a matter of law, and the circuit court erred in denying the landowner's summary
judgment motion on the issue of liability. The court entered final judgment for the
landowner on the surveyor's claim, entered judgment on the issue of liability in the
landowner's favor as to the landowner's counterclaim, and remanded for trial on
the issue of damages on the counterclaim.

Land surveyors, like other professionals, are governed by certain standards, which
have ripened into rules of law. In the absence of evidence of contrary intent, a distinct
order of preference governs inconsistencies in the description of land: (1) natural
monuments or landmarks; (2) artificial monuments and established lines, marked or
surveyed; (3) adjacent boundaries or lines of adjoining tracts; (4) calls for courses
and distances; and (5) designation of quantity.

11.7 JUNIOR AND SENIOR RIGHTS BETWEEN PRIVATE PARTIES

Principle 3.   *As between private parties, a junior grant in conflict with a senior grant
yields to the senior grant.*

"First in deed and last in will" is the often-quoted principle. One who conveys
an interest in land to one person cannot at a later date sell the same intentionally to
another. For every title policy, abstract, or boundary location of a junior deed, the title
examiner or surveyor should research and investigate the rights of the senior deed.
The availability and quality of records to prove seniority of deeds vary considerably
throughout the United States. In states in which title insurance policies are issued on
land, the problem is simplified because the facts of seniority are usually disclosed by
the title policy. The surveyor who is locating land from a title policy description need
not devote research time to title matters other than those described or called for by

the policy. The title company assumes responsibility for the correctness of the title as written. The surveyor's only responsibility is the interpretation of the description, not the title.

In the absence of a reliable abstract company, the facts of seniority are sometimes difficult to determine, and in some localities deed records are meager. In some southern counties, in Massachusetts, in Chicago, and in many rural areas, fires destroyed most of the records. As early as 1872 in Vermont, the court held: "The line in dispute has two descriptions in the deed; one, its course North 34 degrees West, and the other, on said Morgan's line and Simon Colgon's north line. As between courses and distances on the one side, and abuttals on the other, in case of disagreement, abuttals, when identified, must, as a general rule, control."[4]

11.8 JUNIOR AND SENIOR RIGHTS BETWEEN PRIVATE PARTIES; EXCEPTION

Principle 4.   *A grantor that sells a portion of his property automatically becomes junior to a grantee purchasing such property.*

The previous principle stated that the earlier in time that the grantee took title to his property, the more senior he will be in comparison with other grantees that took title to their properties later in time. There is one major exception to this rule, known as the Grantor-Subdivider exception.

Imagine a grantor, Tom, owns a tract of land that measures 1000 feet by 1000 feet. Tom subdivides his property is half, retaining the east half for himself and selling the west half to the grantee, Jerry. The legal description from Tom to Jerry reads "the east 500 of Blackacre." Years later, the property is surveyed at the width of the property is actually only 900 feet. Would it be fair for Tom to sell "the east 500 feet" to Jerry, then come back years later and only leave Jerry with 400 feet, instead of the 500 feet stated in his legal description? No. Therefore, if a grantor sells a portion of his property and retains the remainder, then the grantor (owner of the parent parcel) becomes junior to any grantees to whom the grantor conveyed land.

11.9 DEEDS MUST BE IN WRITING AND DEEMED TO BE WHOLE

Principle 5.   *When establishing the boundaries of a property in accordance with a written deed, the boundaries must be established in accordance with the written terms of the deed. Parol evidence may not be taken to determine the terms of a deed but may be used only to explain ambiguous terms of a deed.*

Parol testimony is oral testimony. To allow an owner to express an opinion as to what the terms of a deed were is equivalent to permitting land to be transferred by parol means. Explanations of deed terms are sometimes necessary and proper, as in the following deed: "Beginning at the southwest corner of Love's intersection; thence, etc." An oral explanation that a Mr. Love formerly lived at the intersection of the present Tavern and South Grade Roads and that the intersection was then known as Love's Intersection is proper, but to dispute the fact that the point of beginning was at Love's intersection is improper. Only the physical location of Love's intersection can be explained.

Words may not be added to or subtracted from a deed, nor may they be varied. The person who drew up the deed would not be permitted to claim that additional land was to be included in the deed. A deed that is applicable to either of these properties is void and may not be identified by verbal means. An existing word may be explained. "My house and lot" is a valid deed if it can be explained that "my house" can apply to only one house, but such a deed is void if verbal explanation is necessary to distinguish between two houses that might be "my house." In other words, the addition of a term to differentiate between the two houses is not permitted. Any deed may be void or voidable if the description does not describe a property.

Parol evidence may be resorted to for the purpose of applying the description contained in writing to a definite piece of property and to ascertain its location on the ground, but never for the purpose of supplying deficiencies in a description otherwise so incomplete as not to definitely describe any land. The description itself must be capable of application to something definite before parol testimony can be admitted to identify any property as the thing described.[5] When the parties to an agreement reduce the terms to writing, the agreement is to be considered as containing all those terms, and therefore there can be, between the parties and their representatives, no evidence of the terms of the agreement other than the contents of the writing. This does not exclude other evidence of the circumstances under which the agreement was made to explain an extrinsic ambiguity or to establish illegality or fraud. For proper construction of an instrument, the circumstances under which it was made, including the situation of the subject of the instrument and of the parties to it, may also be shown, so that the judge may be placed in the position of those whose language he or she is to interpret.

Limitations on the Principle

Although in all states deeds must be in writing to be valid or enforceable at law, there are exceptions in a few states as to the necessity that all the terms of the deed have gone upon the land and made a physical survey of the same, giving it a boundary that is actually run and marked, and the deed is thereupon made, intending to convey the land that they have surveyed. Such land will pass, although a different and erroneous description may appear on the face of the deed.[6] It was further observed: "This is regarded as an exception to the rule, otherwise universally prevailing, that in the case of written deeds the land must pass according to the written description as it appears in the instrument."

11.10 DIRECTION OF THE SURVEY

In surveying a lot within a subdivision, any corner identified and found to be original has equal standing as a starting point. If four corners of a block are found, each corner has an influence on the location of a lot within the block, and no definite direction of survey is called for. Most metes and bounds surveys are written so that the deed proceeds in a definite direction from one point of beginning and returns to the same point of beginning.

11.11   TERMS OF THE DEED

Principle 6.   When a conveyance is reduced to writing, it is assumed to contain all the terms intended and is construed, if possible, to give effect to all the terms mentioned in the conveyance.

To be valid, deeds must be in writing; to permit the addition of terms to a deed or the omission of terms from a deed would defeat the purpose of requiring a deed to be in writing. California, one of the later states admitted to the Union, enacted many of the common laws of older states into statute laws, as in the following example:

> In the construction of an instrument, the office of the judge is simply to ascertain and declare what is in terms or in substance contained therein, not to insert what has been omitted, or to omit what has been inserted; and where there are several provisions or particulars, such a construction is, if possible, to be adopted as will give effect to all.[7]

When working with any interest in real property, the law requires that in order to be able to prove or have a cause of action, the complainants must have a document proving their action. As an example, one must have a deed to prove title or any "arrow" within the bundle of rights. The extent of land ownership one is able to claim is that defined in their title deed.

11.12   CALL FOR A PLAT

Principle 7.   Where a property description calls for a plat or map and the parties acted with reference to the map, the plat or map becomes a part of the description as much as if it were expressly recited in the deed itself.

Whenever a map or plat is called for, the monuments, distances, bearing, and data of the plat become part of the description: "According to Map 1205," "as shown on Record of Survey No. 1272," "as shown on the plan of lot 6 in range the 3rd in Boston Narragansett Town No. 5, being second division," "as shown on Map of survey No. 12566, Virginia Military District, Ohio," or "according to Road Survey No. 1020" are phrases that include a map or plat as part of a deed. It makes no difference whether the map was legally recorded or not, provided that the map or plat was referred to and is identifiable as the exact one referred to. A call for an adjoiner may include any call for a map or plat not mentioned in the deed; thus, "to Smith's west line" may include any plat called for in Smith's deed, which determines Smith's west line. In *Ferris* v. *Coover*,[8] the court stated that a map, referred to in a grant for the purpose of identifying the land, is to be regarded as a part of the grant itself, as much as if incorporated into it.

Limitations on the Principle

The importance of a plat is dependent on whether the parties acted with reference to the map and also on the data given on the plat. If there is a clear error in the call for a map or it is apparent that the parties did not act with reference to the map cited, the map will be treated as surplusage. When there is a call for a plat, it is presumed that the parties did act with reference to the plat; the contrary must be proved, not surmised. The data on the map determine its importance. If the plat shows an accurate survey with calls for monuments, much greater control is afforded to the map than would be given to a picture drawing with little or no information on it.

11.13 INFORMATIVE AND CONTROLLING TERMS

An *informative term* in a deed is considered a "modifier" and is one that adds information about the terms of the deed, but if the informative term is in conflict with the term that it modifies, this term yields. Informative terms act as aids to distinguish the *controlling term* in a manner similar to adjectives that assist in distinguishing like objects or nouns from one another.

"A house," "a red house," and "a red house with a green roof" may all refer to the same house, but in the last-described house there is more certainty of its being located correctly because of the additional descriptive terms. "Beginning at the southeast corner of lot 2; thence N 12° 10′ E, 200.00 feet to a concrete monument" contains two informative terms, "N 12° 10′ E" and "200.00 feet," and a controlling term, "a concrete monument." The "N 12° 10′ E, 200.00 feet" as given here is an aid to distinguish the concrete monument from similar monuments and even conflicting monuments, but the added terms should act as a modifier of the term "a concrete monument." In a deed in which the controlling term cannot be distinguished or located, the informative term may then become the controlling term, as in the foregoing, "N 12° 10′ E, 200.000 feet" could become exact if the monument were lost.

In determining the title interest of a client, conflict may occur between the written instrument, the subdivision map, and the measurements on the ground, or within the written instrument itself. When a conflict occurs, the surveyor or attorney should be able to recognize the deed term that is controlling and the one that is informational but not controlling. Which conflicting term should be followed is discussed in the normal order of importance.

ORDER OF IMPORTANCE OF CONFLICTING TITLE ELEMENTS

11.14 GENERAL COMMENTS

Surveyors prefer to work with and are comfortable with mathematical equations because equations give a positive and presumably correct and comfortable answer for a set of facts. In listing the order of importance of title elements, the list cannot be compared with a mathematical formula; the order of importance must have flexibility to fit solutions most nearly deemed to be correct by the court. The following outline of the order of importance of conflicting deed elements, which is often quoted, is seldom understood and has many exceptions.

The retracing surveyor should understand and become comfortable with relying less on measurements and more on evidence when it comes to a question of boundary location. In recent years, the various government agencies as well as the courts have made possible a greater latitude for accepting "minimum" corner evidence. From the 1930s until 2009, the distinction of a lost corner and an obliterated corner, so far as degree of proof was concerned, was the difference between "beyond a reasonable doubt" and creditable. The US *BLM Manual*, required the higher degree of proof "beyond a reasonable doubt," the degree of proof usually reserved for criminal convictions. With the most recent edition of 2009, the Bureau of Land Management (BLM) changed the degree of proof to "creditable" equal to or probably lower than the usual "preponderance of evidence." Several courts have written that they would prefer a corner based on lesser degree of evidence than on measurements.

Order of Importance of Conflicting Elements That Determine Land and Boundary Location

A. Right of possession (unwritten conveyance)
B. Senior right (in the event of an overlap)
C. Written intentions of parties
    1. Call for a survey or an actual survey on which the conveyance is based
    2. Call for monuments
       a. Natural
       b. Artificial
    3. Call for adjoiners
    4. Call for direction and distance
    5. Call for direction or distance
    6. Call for area (quantity)

Right of possession should never be mentioned in a conveyance. Senior rights may or may not be apparent from the wording of a conveyance. The importance of all other items within a conveyance must be interpreted in light of the intentions of the parties as of the time of the conveyance. The following discussion is mainly an explanation of what the courts, in the event of conflicting elements, have declared expressed the intent of the parties.

Although surveyors should be knowledgeable about category A, they should only report the facts and not advise clients on legal conclusions.

The order of importance of conflicting deed elements listed here, while generally true, can vary from state to state, and within the same jurisdiction it can vary under different circumstances. In most states, both distance and direction are subordinate to monuments or adjoiners, but when it becomes necessary to choose whether direction or distance is the controlling consideration, variable and conflicting court opinions exist. For example, in the sectionalized land system of the United States, the BLM was given the authority to establish rules and regulations to put the system into operation. In the BLM methods used to relocate lost corners, distance during the application of proportionate measurement becomes controlling over direction, and this has been accepted by most courts. However, in the interpretation of metes and bounds descriptions, federal courts and many state courts have declared that direction is superior when the two are in conflict. In California and Colorado, a statute lists distance as preferred to direction. Thus, to conduct a proper retracement, surveyors must know the state statutes and when, how, and why the courts have reacted to different situations within their state or states of practice.

It should always be kept in mind that while the order of conflicting elements may serve to resolve differences between calls, for a variety of reasons, strictly speaking, the order of conflicting elements applies to the resolution of ambiguities within a written description. This set of rules is not intended to resolve all conflicts or conflicts between written evidence and physical evidence. For instance, called for, found, and proven monuments may control a course or a distance for a variety of reasons, but this rule applies only when both are recited in the same description and are in conflict with each other.

11.15 SENIOR RIGHTS

Where two parties are given title to the same parcel of land, and where possession is not a consideration, the party with senior rights has the right of possession according to common law. Patrick L. Brown purchased land from Ashley Bishop on March 16, 1950, and his title reads: "The westerly 50 feet of lot A." Willard Woods purchased land from Ashley Bishop on March 17, 1950, and his deed reads: "The easterly 50 feet of lot A." Because of the earlier time, Patrick Brown is said to be senior, and he receives all of the land coming to him, as shown in Figure 11.1. Willard Woods could not buy more than Ashley Bishop's remainder and is junior in character. The overlap in Figure 11.1 in title belongs to Brown, and if there is no possession, Brown has the right of possession. If, however, Woods had possession up to his title line for a prolonged period of time as required by law, his possession might give him the right to go to court and obtain title to the overlap. Whether senior rights are investigated by the surveyor or not depends on the custom within the state and the terms of the contract under which the surveyor is working. In many states, especially in the West,

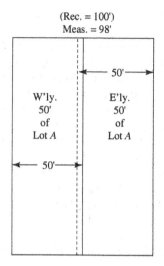

Figure 11.1 Senior rights and overlaps.

title companies issue title policies with senior considerations stated in the description furnished. In such states, it is advisable for surveyors to work with title policies and deeds. In some states, little or no title information is available, and it is extremely difficult to trace senior rights. An abstract and attorney's title opinion are common in the Midwest, and in these areas, opinions and abstracts are used by surveyors. In many of the East Coast states, where grants were made in sequence by the state, it may become necessary for the surveyor to determine the dates of entry and/or survey to understand junior and senior rights.

11.16 CALL FOR AN ADJOINER

Principle 8. A call for an adjoiner is not always a correct criterion for determining senior rights. A title search back to the original formation of the conveyance is necessary for a correct solution.

A call for an adjoiner (bounded by Jones on the east: to the line of Smith as described in Book 1021, page, etc.) may be a call for a senior right, but not always. For example, Richard Johnston's deed reads in part: "Bounded on the south by the land of Stanley Burne as described in volume 7, book of deeds, page 42, etc." Stanley Burne's deed reads in part: "thence N. 0°27'00" E., 301.27 feet to the southerly line of Richard Johnston's land as described in … etc." Each calls for the other. In such an event, an extensive title search is necessary to determine who has prior rights. They do exist! The cause of such ambiguity is usually a subsequent change in the words or the form of a description after the first conveyance. Later owners cannot legally obtain more than was described in the original document, except by the process of lawful prolonged possession, which ripens into a title right.

11.17 WRITTEN INTENTIONS OF THE PARTIES TO THE DEED

Principle 9. *Excepting senior rights of others and a valid unwritten right of possession, the intentions of the parties to a deed, as expressed by the writings, are the paramount considerations in determining the order of importance of conflicting title elements.*

The primary and fundamental principle to which all others relate and must yield is that the intentions of the parties gathered from the whole instrument, taken in connection with the surrounding circumstances, must control.[9] A deed should be construed according to the intentions of the parties, as manifested by the whole instrument.[10] Principles given to determine the order of importance of conflicting elements are not conclusive but are principles of evidence or principles of construction adaptable to surrounding circumstances. A call that would defeat the parties' intentions is rejected regardless of its comparative dignity.[11]

The following principles of construction, given to determine the control between conflicting elements within a deed, are rebuttable presumptions subordinate to the preceding principle. Like all principles based on rebuttable presumptions, when the contrary is shown, the presumption is overcome and the principle does not apply. In court decisions involving land boundary disputes, the written intentions of the parties who were a part of the original transaction are the paramount considerations; the written intentions control all other points. When land is first conveyed, the transfer cannot be by parol means; only properly written and signed documents in accordance with the laws of the state can be used. To determine the intent of an instrument from the parties of the transaction by oral statements is tantamount to permitting transfer of titles by parol means. This is a violation of the statute of frauds.

The intent must be determined from the written instrument itself, not from the mistaken ideas of one of the parties. Where a party believes that he or she has a right to a disputed parcel of land and that right is not based on a written deed, only a court or a true title owner can transfer paper title. In such cases, the surveyor should advise the client to seek legal advice.

If two elements in a deed are in conflict, before a proper location can be made, it becomes necessary to decide which one was intended and which one was informational. A deed written "N 20°E a distance of 310 feet to Boulder Creek" presents a conflict because Boulder Creek is 410 feet away. What was intended? Here, the court rule is that the natural monument, the creek, more clearly shows the intent than does the informative term "310 feet." An additional problem is whether the line goes to the side line of the creek or to the thread or center of the creek. In this case, it would be improper to ask the buyer or seller what the intentions were; the document as signed is the best evidence of intent. In most states, if the stream is navigable, the line terminates at the side of the creek; if it is not navigable, the line ends at the thread or center of the creek. This principle is applicable in every state in one form or another.

To determine the intent of a deed, sometimes an explanation of the terms of a deed or an understanding of the surrounding circumstances existing when the deed was written is necessary. Thus, a deed beginning "at a well-known sycamore tree in Alpine" must be investigated and parol evidence taken to explain where the tree is. A deed stating "starting at a blazed pine tree; thence running a line through a second blazed pine tree to the Cuyamaca Park line" is indefinite without extrinsic evidence. Here, terms are not being added to or subtracted from a deed; an explanation of the terms existing in the deed is being sought for clarification or meaning. Nowhere is a statement made that any particular element written in a deed is controlling; the element most effectually expressing the intent of the parties is to be adopted. As early as 1858, it was stated:

> The rules adopted in the construction of boundaries are those which will best enable the courts to ascertain the intentions of the parties. Preference is given to monuments, because they are least liable to mistake; and the degree of importance given to natural or artificial monuments, course and distances, is just in proportion to the liability of the parties to err in reference to them. But they do not occupy an inflexible position in regards to each other. It may sometimes happen, in case of a clear mistake, that an inferior means of location will control a higher.[12]

The intentions of the parties to the deed must be gathered from all the terms of the deed, each term taken in the light of all other terms. A call for a monument, although normally controlling, may be rejected where all the other terms in the deed indicate that the call for the monument was inserted in error.

Exception to the Principles of Intent

The intentions of parties to a conveyance can never overcome the rights of a senior claimant who is not a part of the conveyance proceedings or litigation. Thus, a call in a deed for "2172 feet to Hyde Road" cannot go to the road if the seller does not in fact own the road. Considerations in the senior deed (adjoiner) may limit the call for Hyde Road (a record monument) to a line determined by items of lesser standing, such as measurements or area. The principle of intent is applicable only to the parties of the conveyance.

11.18 AIDS TO INTERPRET THE INTENT OF A DEED

The following maxims of jurisprudence are frequently quoted in cases involving land disputes:

1. When the reason for the principle ceases, so should the principle itself.
2. Where the reason is the same, the principle should be the same.

3. One who grants a thing is presumed to grant also whatever is essential to its use and enjoyment.
4. Between rights otherwise equal, the earliest is preferred.
5. Particular expressions qualify those that are general.
6. An interpretation that gives effect is preferred to one that is made void.

General Acceptation of Terms

The terms of a writing are presumed to have been used in their primary and general acceptance, but a local, technical, or otherwise peculiar signification may be shown to be the intent, in which case the agreement must be construed accordingly. The terms represent their meanings at the time of writing.

Least Likely Mistake

If a deed contains conflicting description elements, that description element is to be adopted which is the least likely to be affected by mistakes.[13] If the conflict is significant, the deed may be either totally void or voidable.

Date of Execution

Unless there is something in the instrument to the contrary, a conveyance referring to natural objects and boundaries speaks as of the date of its execution or as the date of a survey, if a survey is called for.

11.19 CONTROL OF UNWRITTEN TITLE LINES

Principle 10.   Title lines established by estoppel, agreement, prescription, or other unwritten means are local in character and cannot be used to establish lines of the written deed.

Any boundary line determined by unwritten means can only be considered local in time and character. An agreement with a neighbor to fix a disputed unknown line cannot be used as a basis to establish lines of a previously written deed. In other words, unwritten title lines stand alone, since they are established on an individual basis. They cannot be used to locate or control other lines of title or boundaries.

11.20 LINES MARKED AND SURVEYED

Principle 11.   Where lines are actually located and marked on the ground as a consideration of the transaction and called for by the deed, the lines so marked show most clearly the intentions of the parties and are presumed paramount to other written considerations, senior rights, and clearly expressed contrary intentions being excepted.

Limitations on the Principle

The lines marked and surveyed have force only where (1) the lines run were considered as the lines of the transaction, (2) the lines can be identified, (3) the lines run do not encroach on a senior right, (4) the lines run are not for the purpose of meandering a body of water, and (5) in most states the lines run are called for in the deed.

Proof that a survey was intended as a part of the transaction may be determined from a statement in the deed itself or from a law. In some states, such as Kentucky, Virginia, and Texas, and for federal lands, a general law required a survey prior to issuance of a patent. As a matter of law, it would be presumed that a survey was made for such patents unless the contrary could be shown.

If an owner incurs the expense of a survey and then describes lands in accordance with the lines laid out, it can only be presumed that he or she intended to convey to the lines delineated upon the ground and not to erroneous informative calls for bearings and distances. Measurements taken upon the ground are aids to locate where the lines are if lost, but the lines themselves, when identified upon the ground, represent the original lines intended for the conveyance and are controlling for that conveyance survey.

All the rules of law that have been adopted for guidance in locating disputed boundary lines have been to the end that, in so doing, the steps of the surveyor who originally created the lines on the ground may be retraced as nearly as possible.[14] If a deed describes the land conveyed by courses and monuments and boundary lines of other tracts, and then declares that the description given is to be according to a survey theretofore made by a person named, such survey is incorporated into the deed and becomes a part of it, and the grantee acquires title only to the land contained within the exterior boundaries of the survey.[15]

Numerous other cases can be cited in which the principle of acceptance of the original survey is paramount to boundaries, distances, angle, or area. A call for monuments in a metes and bounds description does not necessarily prove that a survey has been made. Metes and bounds descriptions that contain calls for monuments may have been written without a survey.

Proof of the original survey is usually embodied in the location of the original monuments set to mark the original survey. If, in identifying the lines as run by the original surveyor, there is a discrepancy in course and distance over monuments, the evidence of the actual original location must be by a preponderance of the evidence. Preliminary or conditional lines are not binding.

Where a surveyor runs lines to meander a body of water or stream, the lines run are not considered lines of the conveyance; they are considered lines to aid in the plotting of the feature itself. The boundary of the meandered monument itself is the true line.

Of course, if a survey is called for in a conveyance and it is shown that the survey was not made or was incomplete, the rule for the control of an original survey does not apply. In *Swift Coal and Timber Co.* v. *Sturgill*,[16] it was proved that the surveyor ran the first three courses only and protracted the remainder. Although the first three courses called for the top of a ridge, the monuments set were on the side of a ridge.

The next courses called for a second ridge and gave bearings and distances. If the top of the second ridge had been adopted, the area would be 17,000 acres instead of 12,000 acres, as called for. Because the top of the ridge could not be followed in the three courses surveyed, the court reasoned that the call for the second ridge did not mean the top of the ridge, and the bearings and distances were adopted. Although the court did not say so, it should be remembered that the original patent came from the state of Kentucky, and, in case of doubt, patents issued by a state are construed against the grantee.

In Kentucky, where an original survey and plat were required for a patent, the question of the performance of the original survey was questioned. The court observed:

> It is true the rule is that the calls of a patent for course and distance must give way to known or established objects found on the ground. But after all, the rules that have been laid down on this subject are for the purpose of establishing the actual location of the lines and corners of the original survey, and they have little application where the lines were not run out in the original survey, but were simply laid down by the surveyor by protraction as was evidently the case in the patent before us. When the lines were not in fact run, we have little to guide us except the calls of the patent and the plot of the land accompanying the original survey. The plot accompanying the original survey is potent evidence in the determination of the general shape of the tract of land intended to be patented. To follow the lines of the other surveys in the case before us on the east to John Murphy's survey on the west to King's survey, and then on around with still other surveys would be to make this tract include five times as much land as the grantee paid for and give the tract an entirely different shape from that which was evidently contemplated in the grant.[17]

Note again that in a Kentucky patent an uncertainty is to be construed against the patentee, not the state. In *Givens* v. *United States Trust Co.*,[18] course and distance were given control over calls for adjoiners, wherein, if the adjoiner lines were followed, it would have increased the area 10 times over that stated in the patent. These decisions are understandable in that the courts did not rely upon a single element but considered all elements collectively or in their totality.

Where there is a conveyance of land from one private person to another, the call for adjoiners usually controls course and distance, even though a large excess of area is conveyed over that stated in the deed. An ambiguity in a patent from the state is resolved in favor of the grantor (the state), whereas an ambiguity in a conveyance between private parties is construed against the grantor (seller).

11.21 CORNER DEFINITIONS

The following terms are frequently used to identify the status of a corner:

A *corner* is a point of change of direction of the boundary of real property. It may be marked by a monument, fence, or other physical object, or it may not be marked at all. A call for a point is assumed to be a call for a corner, but the call for a course without the call for a point is not necessarily assumed to be a call for a corner.

An *existent corner* is one whose position can be identified (1) by an existing original monument or (2) by accessories acceptable to the original monument position.

An *obliterated corner* is one at whose point there are no remaining traces of the original monument, replacement monument, or its accessories, but whose location may be recovered (1) by competent testimony, (2) by some acceptable record evidence, (3) by improvements built at the time the original position was known, or (4) by a monument proven to be a replacement of an original monument.

A *lost corner* is a point of a survey whose position cannot be determined by the original monument or by acceptable evidence as to where the original monument was, even though a monument may be present.

An *accessory to a corner monument* is any bearing object, bearing tree, pit, or object recited in the original field notes as having a definite relationship to the corner. An accessory assumes the same dignity as the original monument if called for in the writings. A proven original corner and an accepted obliterated corner enjoy the same legal standing.

A *memorial* is any durable article deposited alongside, near or beneath the monument that will serve to identify the corner location in case the monument is destroyed. Metal, charcoal, glass, stoneware, and marked stones were common memorials in the sectionalized land system.

11.22 CONTROL OF MONUMENTS

Principle 12. *Monuments called for in a deed, either directly or by a survey, or by reference to a plat that the parties relied on, are subordinate to senior rights, clearly stated contrary intentions, and original lines actually marked and surveyed, but are presumed superior to direction, distance, or area.*

The surveyor must determine whether the monuments called for in the deed were in place at the time the deed was conveyed. The biblical importance of monuments was recognized in Deuteronomy: "Thou shalt not remove thy neighbor's landmark, which they in old times have set" (Deut. 19:14) and "Cursed be he that removeth his neighbor's landmark" (Deut. 27:17). "Where parties agree upon definite monuments that fix the boundaries of a parcel of land conveyed by one to the other, such monument should unquestionably control."[19]

Whenever an original survey is made, the surveyor either finds a monument in place or sets a monument. The distance and direction measured between the monuments are dependent on the skill and accuracy of, which is dependent on the equipment used by, the surveyors doing the measuring; if an error occurs, the error is due to the inability of humans to measure properly or to copying errors. The monuments are fixed in position and, if found undisturbed, are not in error. In deeds written without benefit of a survey and including calls for monuments, the presumption is that the parties intended to go to the monuments; otherwise, the calls would not have been inserted. Because an original monument is considered more certain in fixing the location of a line or corner, it is given preference over distance, direction, or area.

An uncalled-for monument cannot be considered controlling when in conflict with superior elements: In describing the boundaries of the land granted, "Monument[s] mentioned in a deed control both the courses and distances given in the deed, if there is a conflict, without regard to whether, in fact, the monuments were seen by the parties to the deed or not."[20]

Because odd or confusing language is sometimes used to express a finding, the entire report of a court case should be read or a wrong impression can be formed. For example, one court's opinion was: "It is a long since settled principle of construction in respect to location, that metes and bounds in the description of the premises granted control courses, distances, and quantities, when there is any inconsistency or conflict between them."[21] Apparently, the court erroneously used "metes and bounds" to mean monuments, as further reading of the case indicates that the dispute was about the location of a hickory tree. A surveyor should never quote findings in any case without reading the entire case first. When quoting a case, it is imperative that the surveyor compare the facts in each case with those in hand.

In deciding whether an iron pipe represented the "stake" referred to in a deed, the lower court found that the burden of proof was not met and rejected the iron pipe. On appeal, the upper court cited the general reputation of the pipe and prior acceptance by prior grantors, and concluded that the burden of proof had been met. In commenting on monuments, the judge said:

> A monument, when used in describing land, can be defined as any physical object on the ground which helps to establish the location of the line called for. It may be either natural or artificial and may be a tree, stone, stake, pipe or the like. Just as in contracts and wills the intentions of the parties governs the interpretation of deeds. It is for that reason that monuments named in deeds are given precedence over courses and distances, because the parties can see the tree, stone, stake, pipe or whatever it may be, which is referred to in the deed, but would require equipment and expert assistance to find a course and distance.[22]

Many surveyors are guilty of using nonspecific terms such as stakes, posts, staffs, or IPs (probably meaning "iron pipe" or "iron pin"). To eliminate future conflict, it is imperative that each monument be identified as to its exact nature, including size, material, and identifying marks.

In writing descriptions, the use of abbreviations should be avoided. All words should be fully written out to avoid future interpretation problems and possibly costly litigation.

Limitations on the Principle

For a monument to be controlling, it must be (1) called for in a document, (2) identifiable, and (3) undisturbed. If the monument is obliterated, it is controlling if its former position can be identified (a) by reliable witness evidence, (b) by surveyor's notes, (c) by improvements, and (d) sometimes by hearsay and reputation.

In written conveyances or documents, uncalled-for monuments cannot be considered as controlling. If it is the intent of the parties to have a monument controlling, it should be so stated in the deed. If a monument has deteriorated beyond recognition, either visual or by witnesses' evidence, the monument itself is no longer controlling. Once a monument is disturbed, its value as a control point ceases, but if a monument is merely obliterated and its former position can be identified, the former position will control.

Superiority of monuments over distance, angle, and area is so frequently accepted by the courts in all states that the danger of applying the principle comes from a contrary intent. If numerous other inferior terms in a deed refute the call for a monument, and if the other terms taken together indicate a contrary intent, the reason for the principle is nullified.

Control of Monuments Shown on a Reference Plat

Where a deed refers to a plat and the parties acted with reference to the plat, all the monuments shown on the plat have equal dignity with those referred to in the writings. If the call for the plat is inconsistent with portions of the deed, the plat is controlling only where the parties acted with specific reference to the plat itself.

Obliterated Monuments

If a monument is merely obliterated, but its former position can be identified by a preponderance of the evidence by the testimony of landowners, competent surveyors, or other qualified local authorities or witnesses or by acceptable evidence, the position so identified is controlling. Proof sufficient to convince a judge and jury that an obliterated monument has been restored to its original position rests with the surveyor.

11.23 CONTROL BETWEEN CONFLICTING MONUMENTS

Principle 13.   When there is a conflict between sequential title and sequential surveys, the title controls the surveys.

Monuments in the form of fences or boundary improvements built soon after the deed was written and in accordance with the original survey may become controlling, especially where several surveyors would locate the boundary lines in different places or where the true survey lines are uncertain. This possibly could include monuments set after the description was written. It is suggested that the surveyor not yield to requests to do such an act. Usually this practice is requested by individuals who wish to save money by placing monuments after lots are sold.

Where there are conflicts between monuments called for and no senior right is interfered with, the monument most clearly showing the written intentions of the parties is controlling. Unless a contrary intent is indicated by the deed wording, the following order of importance is presumed:

a. *Identified lines or monuments of a survey called for in the chain of title.*

b. *Natural monuments (boundaries) called for in title documents and still in their original locations.*

c. *Artificial monuments as described in title documents.*

d. *Monuments set after the deed was written, and not occupying the spot of an original monument, are not controlling except where the deed calls for a survey to be made.*

e. *Where two monuments, otherwise equal, are in conflict, the one in harmony with distance, angle, or area becomes controlling.*

f. *Monuments in the form of fences or boundary improvements built soon after the deed was written and in accordance with the original survey may become controlling, especially where several surveyors would locate the boundary lines in different places or where the true survey lines are uncertain.*

11.24 EXPLANATION OF THE PRINCIPLES

Any monument that marks a senior right is superior to any other monument listed here. For example, George Miller's deed calls for artificial monuments (surveyor's stakes) located so as to include all of Green River and several feet on each side. Ted Harper's deed, which is junior to George Miller's deed, reads to "Green River" and overlaps Miller's deed. The area of interference (the overlap of Harper's deed on Miller's deed) belongs to Miller. Normally, the artificial monuments would yield to the natural monument (Green River); it does not do so in this case because of a senior right. The first part of the principle, "and no senior right is interfered with," should not be overlooked, and, if applicable, should control or at least be examined.

The preceding principles have force in the event that monuments are called for. A specified artificial monument in conflict with an uncalled-for natural monument would be controlled by the artificial monument unless the intent is clearly indicated by other written words. A portion of a deed reading "west 200 feet to an iron pin; thence S 1°02′ W, 200 feet to an iron pin; thence east 700 feet; thence to the point of beginning" would be interpreted by the found pins even though the course indicates something different. S 10°02′ W, 200 feet were found to be adjoining and parallel with an uncalled-for river. If the deed had read "west 200 feet to an iron pin marking the bank of Red River, etc.," the deed would be interpreted by the natural monument (river).

Normally, parcels described in deeds are controlled by the monuments called for, but if other factors show that the call for the monument was inserted in error, and the other factors more clearly show the intent of the parties, the reason for the control of the monument ceases. Only when called-for monuments most clearly show their intentions—and they usually do—are they controlling.

The results of the authorities may be stated thus: the lines run and marked on the ground are the true survey and, when they can be found, will control the call for a natural or other fixed boundary and conclusively establish the survey; however, when a younger survey calls for an older as an adjoiner, and no lines are found to have been marked for the younger on the side on which the older is called for, the line of the older becomes the division line between the two tracts; in other words, the younger is to be laid so as to adjoin the older. If a line is inconsistent with the older survey, the actually run and marked line on the ground for the younger survey may prevail.[23]

The general rule is that where the lines of senior and junior surveys conflict, the lines of the senior survey control, particularly where the junior is bounded with express reference to the senior. A junior survey *cannot* be used to control a senior survey.

Natural Monuments

A stake placed on the shore of a lake or on the bank of a stream and called for is to be used for line (direction) purposes and in some instances for proportioning, whereas the more certain monument, the water, is the determining natural monument that establishes the termination of the line. Thus, in Figure 11.2, the original surveyor set a stake on a bank near the ocean to designate A's and B's parcels. Because the waterline represents the limit of ownership of the subdivider, and because the surveyor could not conveniently set a stake at a submerged location, it is assumed that the stake set on the bank was intended only for the line and that the water as called for by the plat or in the description is intended to be the true termination of the line.

In a relevant court case, it was noted:

> Generally, in interpreting boundaries of land, resort must be had first to natural objects or landmarks, next to artificial monuments, then to adjacent boundaries, and thereafter to course and distance, and, whenever a natural object is distinctly called for and satisfactorily proved, it becomes a landmark to which preference must be given, because the certainty which it affords excludes the possibility of mistakes.[24]

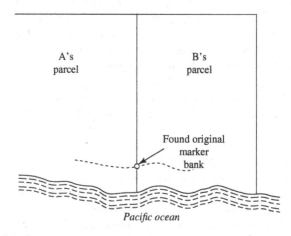

Figure 11.2 Original monuments set on the shore control the direction of a line but not its terminus.

In this court report, there is an obvious error. If the call for an adjacent boundary is a call for a senior right, resort is first to the senior right.

Artificial Monuments

Monuments set prior to a deed and referred to can be considered in interpreting the description for retracement purposes in that any referenced found monuments are presumed superior to other monuments, provided that no senior right is interfered with. But artificial monuments set after the deed is written are presumed subordinate to other monuments. Only those monuments called for or considered a part of the deed are presumed controlling. All others are supplemental or extrinsic evidence, and their acceptance or rejection is a matter for the trial judge. But the surveyor and attorney should fully understand the local *Rules of Evidence*.

Record and Artificial Monuments

In the event of a gap between a call for an adjoiner (record monument) and a call for artificial monuments, court decisions as to which controls have varied. In Figure 11.3, Brown's parcel is senior and Jones's parcel is described as " ... thence N 89° E a distance of 200.00 feet to an iron pin located in Brown's property line: thence S 1° E 200 feet along Brown's property line to an iron pin located in the northerly side of sixth street; thence ... " If there is an overlap as shown on the right in Figure 11.3, the area of interference goes to the senior deed (Brown). However, when a gap exists between parcels, the courts have varied in their decision.

If the gap is small, the courts generally give the adjoiner line control; however, if the gap is large, the iron pins generally control on the theory that the gap belongs to the owner of the original parcel from which the two were carved. To avoid liability, whenever a surveyor finds a gap between artificial monuments and an adjoiner, as illustrated, he or she should disclose the facts on a plat presented to the client or the

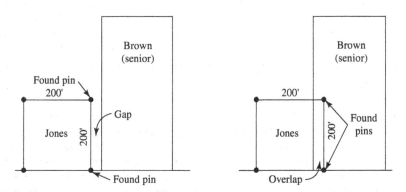

Figure 11.3 Overlaps and gaps.

client's attorney. If the state issued two patents and there was a gap between them similar to that shown in Figure 11.3, the gap is a vacancy and belongs to the state (ambiguities in patents are construed in favor of the state) or as provided by state law. Several other situations may be found in which the "found acreage" belongs to one of the adjoining owners under a legal doctrine, one of which has become known as the *Strip and Gore Doctrine*.

Uncalled-for Monuments and Boundary Improvements

One of the more serious problems retracing surveyors encounter is what weight to give to "Topsy" corners. This term was coined to represent the character "Little Topsy" in *Uncle Tom's Cabin*. When asked where she came from, Topsy's response was "I don't know. I guess I just growed." Uncalled-for monuments that are supposed to identify corners just seem to appear. Yet many people want to give them the same authority and dignity as if they resulted from a survey and a description. Usually, no one knows who set them or where they came from, nor when they appeared. These found uncalled-for monuments should be approached with suspicion and control nothing.

> A found monument that cannot be related to a title document should be given little or no credence.

Monuments set after a deed was written do not control a boundary, although they may be used as evidence for possible prescriptive points. Monuments in the form of fences or improvements built soon after the deed was written and in accordance with the original survey may become controlling where several surveyors would locate the property lines in different places or where the true survey lines are uncertain. When there is certainty in the location of the boundaries of a parcel of land and when several surveyors would all locate the property in precisely the same place, improvements such as buildings and fences are usually treated as encroachments, but if the survey lines are uncertain from lack of control of known fixed monuments, and several surveyors might place the lines in different places, the fences and improvements are probably better evidence of the original lines of the original parties. The courts accept the most certain evidence to fix the limits of a property. Where the survey lines are uncertain, the courts are loathe to change existing fences without just cause, as in the following case:

> Where the call of both lots in distance from the line of a street is rendered uncertain by difficulty in ascertaining the true line of the street, some surveyors sustaining plaintiff's line, and others the boundary claimed by the defendant, but there is not uncertainty as to the possession given to plaintiff's grantor by the original owner of both lots, and as to the fence which such grantor was then permitted to erect and maintain, and the subsequent grant by the original owner of the remainder of the land to defendant's grant was bounded by plaintiff's lot, which was then fenced, such boundary must be

deemed a monument which will control the call for distance in the description of the defendant's lot.[25]

In time, most monuments become lost or obliterated. Wood rots, iron rusts, trees die, and land erodes. With the destruction of monuments, land boundaries may not be lost. Monuments and boundaries must be located from the best available evidence. Often, possession is all that remains. The most difficult task of the surveyor is in the evaluation of fences. In the western states, this task may be easier, for the surveys are more recent and many witnesses may be available to testify. In the eastern states, where possession extends back hundreds of years, only reputation and hearsay may remain.

11.25 IMPORTANCE OF THE WORD "TO"

To is a word of exclusion. In a land description, "to" flashes a warning to the surveyor, attorney, or title person—look for a call that excludes informative calls of distance, angle, or area. "To a stone," "to a stake," "to the corner of lot 16," and "to the point of beginning" are all examples of the use of the word *to* where the distance, area, or course given yields by presumption to the object or point called for. Surveyors often fail to locate points properly where there is a call for an adjoiner such as "S 76°21' E, a distance of 327.21 feet to the southeast corner of that land conveyed to Jones in Book 1276, page 16 of Official Records." This means in effect that Jones's boundaries must be located before the survey can be completed. The cost of the survey increases because boundaries of more than one parcel of land must be determined.

11.26 DIGNITY OF RECORD MONUMENTS

Because a record monument is a monument called for in a land conveyance, its dignity depends on whether it is a senior right (a call for a senior adjoiner), a natural monument, or an artificial monument. In court reports, especially in the syllabus, certain statements may fail to take into consideration the true nature of a record monument. For example, a statement that artificial monuments control a call for an adjoiner (a record monument) can be quite true, provided that the call for the adjoiner is not a senior right. Referring to the discussion accompanying Figure 11.3, courts are not always in agreement as to the best solution when this occurs. The proper procedure may depend upon the circumstances of each case and the law in the particular jurisdiction. This last sentence conflicts with the statements accompanying Figure 11.3.

11.27 CONTROL POINT OF A MONUMENT

The exact location of the point on a monument that one measures from or to has various meanings under different circumstances. Where the deed term is "to a stake," the measurement is to the center of a stake. A call for a ditch, wall, tree, fence, road, or nonnavigable stream, unless the contrary is stated, usually means the intent is to the center of the object called for. Where a call is to go to a structure such as a building, house, mill, or wharf, the call usually terminates at the side of the structure.[26] When a line terminates at a water boundary, the point of termination is usually defined by state law and varies from state to state.

A problem may be encountered if the corner monument cannot be occupied, such as a tree, or when it falls at a point that cannot be occupied, or in water. Then nothing remains except to take "offset points" and calculate the line length and angle. If required to do this, the surveyor should show his or her calculations in the field book for reference.

11.28 UNCALLED-FOR MONUMENTS

As noted previously, an unmentioned monument, set after a parcel of land has been conveyed, has no value whatsoever for determining where the written deed should be located. However, such a monument may have value to determine unwritten rights. Cited in court cases, usually involving reformation of a deed, are situations in which uncalled-for monuments, set after a conveyance is made, determine a boundary location. In New Hampshire, the court said: "Where a monument does not exist at the time a deed is made, and the parties afterwards fairly erect such a monument, with intent to conform to the deed, such monument will control."[27] A similar finding has been approved in other states, including Maine.[28]

Cases of this type are based on what is called an *unwritten agreement*, where (1) the adjoiners do not know where their true boundary line is located, (2) the adjoiners agree to a location and set a monument to identify it, and (3) the adjoiners intend the line so marked to be the true line. Later, one of the parties discovers that the written deed line does not agree with the unwritten agreement line and then wants the written deed line to control.

Surveyors have no authority to reform a deed without approval of the adjoiner; only a court can do this. If a surveyor does uncover a situation in which all the elements of the unwritten agreement can be proved, all the surveyor can do is call attention to the situation and advise the client to seek legal advice or attempt to suggest to the adjoiners that they consider executing an agreement deed. At all times, the obligation of the surveyor is to call attention to differences between the conveyance writings and occupancy. Reforming a deed is a legal matter that should be referred to the legal profession.

If the evidence shows the monuments were placed in the ground soon after the original survey, by the same individual who conducted the original survey, infrequently the courts have been known to accept these as original monuments.

This problem of uncalled-for monuments probably has caused extensive litigation from inexperienced surveyors. When *any* monument is found, whether called for or not, inexperienced surveyors feel obligated to accept any monument "close by" when the finding of these uncalled-for monuments indicate the possible need for additional research and field investigations.

11.29 ERROR OR MISTAKE IN A DESCRIPTION

Principle 14. *If a description of land contains an error or mistake, and if the error or mistake can be isolated, the error or mistake can be placed where it occurs.*

In Maryland, the usual rule in the event of conflict is that monuments control over course and distance, and course (direction) controls over distance.[29] In the case of *Wood* v. *Hildebrand* involving monuments, course, and distance, the court observed:

> When the course and distance conflicts, the whole description is to be considered to determine which conforms to the intentions of the parties, and there is ordinarily no rule by which preference is to be given to one element as against the other.

> We take the view that the distance of 225 feet was in error. The scale of the original plat shows this on its face, and the course of the closing line N 15 & ¼ degrees W confirms it. The correctness of this is likewise confirmed by the call to the shore of St. Nicholas Creek.... This solution is confirmed by still other facts; it accepts the correctness of the back line of the entire tract, as shown upon the plat, both as to scale, figures and actual measurement, and it gives effect to the call of the Western boundary. This solution requires a closing line of 228 feet instead of 338 feet.[30]

In this case, distance was corrected because all other factors indicated that it was the only item in error. The error was placed where it occurred.

In Kentucky, a deed from the state did not close, and in accordance with the rule that course controls distance, the last two courses had to be extended 38 poles and 62.7 poles to reach the point of beginning.[31] This theory was rejected by the court, and the court noted that if the seventh course were changed from south 77° west 80 poles to north 77° west 80 poles, the deed would close within 40 feet. Furthermore, the original plat indicated that this was true and was the proper method for resolving the conflict. The error was placed where it occurred, or at least where the court determined it occurred.

11.30 CONTROL OF BEARING AND DISTANCE

Principle 15. *Bearing and distance are presumed superior to surface, and only where bearing and distance more clearly show the intent do they control other elements.*

In the *priority of calls* these two elements rank very low on the scale of acceptance. A course can be helpful in looking for evidence, in that the end of a course would indicate a search area to look for corners, etc.

Bearing and distance quoted are more often informative rather than controlling terms. Dimensions give way to the objects called for; hence, they are frequently more or less in meaning. Where there is a call for a monument, bearing and distance are elements to be used to pinpoint the area of investigation to recover monuments called for. Only when the called for monument is missing can the elements of bearing and distance control.

When either bearing or distance must yield, the courts have had diverse opinions as to which should yield. In metes and bounds descriptions, where monuments are not called for, both bearing and distance are essential for the determination of a line, and neither need yield to the other. In the case of a call for the adjoiner (record monument), bearing may be held and distance may yield. Thus, "N 10° W a distance of 25 feet to the line of Jones's property" means that you go "N 10° W" to the line of Jones's property, and if the distance is more or less than 25 feet, the distance must yield. A Texas court held:

> Though courts have agreed that amongst a diversity of calls, preference, everything else being equal, is to be given first to the natural objects called for, next to the artificial objects, and lastly to course and distance indicated, yet, when a discrepancy among the calls is established, and the circumstances in proof show that course and distance are the more certain and reliable evidence of the true locality, then course and distance will prevail.[32]

11.31   CONTROL OF EITHER BEARING OR DISTANCE

Where there are several provisions—that is, bearing, distance, and a call for a monument—such a construction is to be adopted, if possible, as will give effect to all. Frequently, bearing, distance, and a call for an adjoiner are in conflict, and where in conflict, neither bearing nor distance is considered superior to the other. The construction is adopted that will give control to the largest number of terms and still recognize the presumed paramount control of a call for an adjoiner. There are three cases: (1) distance yields, (2) bearing yields, and (3) both distance and bearing yields. A description including all three cases is "thence N 10° E a distance of 50 feet to the southerly line of Maple Street; thence south 89°51′ E a distance of 50 feet along the southerly line of Maple Street; thence S 00°09′ W a distance of 50 feet to a 1-inch iron pipe; etc." (see Figure 11.4). In the first call given, the line extends N 10° E to the southerly line of Maple Street whether the distance is 50 feet or not. In the second call, the distance is 50 feet along Maple Street whether the bearing is 89° 51′ E or not. In the third call, you extend to the 1-inch iron pipe whether it is 50 feet or S 00°09′ W or not. In each case bearing, distance, or both yielded to the call for a monument, but in each case only that element(s) yielded that was in conflict with the monument.

This analysis seems logical and convincing, but let the evidentiary facts change. Suppose that a 1-inch iron pipe is found at the end of the third course, and the courses are run in reverse. The N 0°09′ E 50.00 feet is run and it intersects the street line exactly. Next, 90° is turned and the line falls exactly on Maple Street. After the distance of 50 feet is measured along Maple Street, an angle is turned to the point of beginning of the deed and it is found that the distance to the point of beginning is exactly 50 feet but the bearing is S 11° W instead of S 10° was called for. Now what

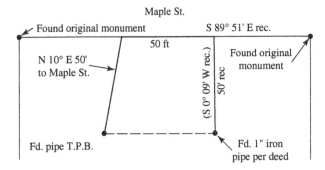

Figure 11.4 Description with both distance and bearing yields.

controls: the bearing of S 10° W or the distance of 50 feet? In this situation, the only inconsistent element is the bearing of 10°, and therein lies the error. Hence, all conditions must be analyzed before a distance or bearing is considered as superior or is rejected. The first analysis was correct in theory, but it had to be modified when the cause of the error was discovered and isolated. Before giving bearing or distance control, the possibility of a mistake or error must be investigated and eliminated. The surveyor has an obligation to examine all reasonable avenues to resolve a discrepancy.

Situations have occurred in which direction was declared controlling over distance, and others in which distance was declared controlling over direction. Where an error must be placed in either of two adjacent lines, the courts have varied in opinion. To apply a pro rata application of bearing or distance defeats the responsibility of the surveyor or the courts to isolate the error and apply it where it is located.

In Figure 11.5, two monuments, one at point C and the other at the beginning of the land description, are fixed in position and the last two courses read "thence West 300.00 feet; thence South 300.00 feet to the Point of Beginning." When running these last two courses, the end of the last course fails to reach the beginning point by an error of closure of S 45° E, 50 feet. Where is the error placed? Three possible solutions exist: (1) Let direction control, (2) let distance control, or (3) place the error in the last course.

There is no hard-and-fast rule about what will control. One surveyor will have the distance control, while a second surveyor will have the bearing control. The retracing surveyor should give serious thought as to which he or she will accept and not just make the assumption one will control over the other.

Control by Direction

In a New Jersey case, the court held: "Course governs distances in a survey; and it is error to instruct the jury that a certain line must stop at the distance given in the description."[33] According to the court's ruling, if this situation occurred in New Jersey, the problem would be solved by running west from point C to a point located north of the beginning point and then running south to the beginning point. This is

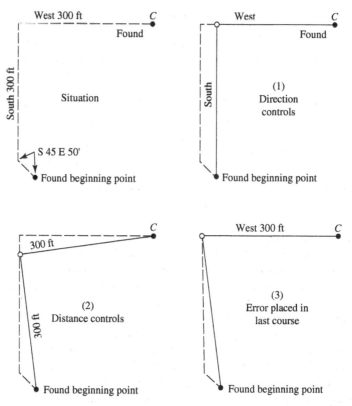

Figure 11.5 Possible solutions to placing an error.

in general agreement with Texas, Georgia, Illinois, Maryland, Kentucky, Ohio, and federal common law as well as some eastern states.

In Kentucky in 1957, one corner was found, the next was lost, and the next was found.[34] The two courses in between the found points were run on deed bearings even though one line was shortened from 100 to 50 poles and course controlled distance. In 1915, wherein distance was given preference to course (meaning direction), the court stated:

"It may be laid down as a universal rule, that course and distance yield to natural and ascertained objects. But where these are wanting and the course and distance cannot be reconciled, there is no universal rule that obliges us to prefer one or the control of distance over angle." The retracing surveyor will find court decisions all over the spectrum and should use his or her own determination.

Control of Distance Other

"Cases may exist in which the one or the other may be preferred upon a minute examination of all the circumstances."[35]

Research has revealed that the second solution in Figure 11.5, that of keeping the two distances 300 feet each and allowing the directions to fall where they may, has been the minority finding in states other than those containing sectionalized General Land Office (GLO) lands.

Error Placed in the Last Course

The third solution, that of placing the error in the last course that recites to the point of beginning, has also been used, especially when the error of closure is minor. In litigation involving metes and bounds situations, the majority of state courts have given bearing control over distance. Although this may be so, the true rule is "to say that distance shall yield to course, or vice versa, where there is a conflict would seem to be entirely arbitrary; and the true rule seems to be that one or the other shall be preferred according to the manifest intent of the parties."[36] As previously illustrated, all factors, such as possessory acts, evidence, and prior decisions of specific jurisdiction, should be analyzed before a final decision of the proper solution is made. The surveyor should always be prepared to testify in court as to alternative solutions, but if asked, he or she should be able to give an opinion as to the correct solution for his or her area of practice. This is a modification of the rule that when balancing a survey, when an error is determined, the surveyor may place the correcting error in the line or angle at which the error occurred, if it can be positively ascertained.

11.32 DISTRIBUTION OF ERRORS IN SEVERAL BOUNDARY LINES

In Kentucky, where course is preferred to distance, a situation occurred in which the point of beginning was identified as a tree and the remaining points called for were stakes, none of which could be found or identified. Neither the plat nor the patent mathematically closed, and the question of where to place the error of closure arose. The court found that where a "call patent" (meaning wording by bearing and distance without set monuments) exists, the error of closure was placed in the last course as follows:

> There are instances where the plat may be resorted to correct an error in the patent calls, but, before this can be done, it must first appear that the mistake is in the patent (calls) as issued. [The error] is just as apt to be a mistake in the surveyor's plat as in the patent calls. There being no difference between the calls in the patent and the calls in the surveyor's certificate, the surveyor's plat is of equal dignity with his other certified work but not superior. In such cases, the correct method of locating the patent is to commence at the beginning corner, where, as here, it is clearly established, follow the calls, course and distance of the patent, and close the last line so as to make a complete boundary.[37]

Here the error was placed in the last course.

In a previous case in the same state in 1907, the syllabus states: "Where corners marking three successive lines of a survey are lost, the distances of the three lines will

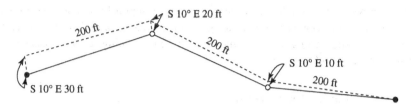

Figure 11.6 Compass adjustment rule.

be altered proportionately to make them in conformance with the known parts of the boundary rather than to radically alter the course of one of the lines."[38]

It is rare that altering only distances proportionately causes a traverse to close mathematically, as in this case. In the event that two court cases within a state are in disagreement or are inconsistent, the most recent decision controls or is binding.

Where there are a number of angle points in a boundary and an error of closure that cannot be isolated, as shown in Figure 11.6, the error of closure, according to the *Manual of Instructions for the Survey of Public Lands*, is disposed of by the compass rule adjustment. The lines are run and temporary stakes are set. The bearing and distance of the closing error are determined in the field by measurement.

At each angle point, the temporary corner is moved in the direction of the closing bearing a distance proportional to the angle point's distance from the starting point.

11.33 CARDINAL DIRECTIONS

A client's deed reads: "Beginning at a point (as defined below); thence west 20 rods; thence south 20 rods; thence east 20 rods; thence north 20 rods to the point of beginning." The question in each of the following circumstances is: What determines the direction of west, south, east, or north? Are astronomic bearings used, or are there modifiers that predicate the use of some other basis of bearing?

Situation 1. The point of beginning is a rock mound, which is found. Adjoiner deeds use the same type of calls and there are no senior rights. Astronomic bearings are indicated, except in those states in which old deeds are presumed to have used magnetic bearings.

Situation 2. The point of beginning is described as the southeast corner of Section 10 in a given township and range of sectionalized land. When the original patent is researched, it is found that the original patentee had the southeast quarter of the southeast quarter.

In this situation or a similar one, the courts have held that words are presumed to be used with their ordinary meanings and that west, south, east, and north, when unmodified, are relative to true north. However, extrinsic evidence may be taken to explain a local or an unusual meaning or the custom in the area.[39] In a number of cases, the decision has been to interpret "west" as parallel with other lines.[40]

In California, when the word *due* was placed in front of the word *north*, the basis of bearings was changed to astronomic north, as indicated in the case of Richfield Oil Corp.[41] Prior decisions did not differentiate between geodetic and astronomic bearings.

11.34 UNRESTRICTED GENERAL TERMS

In deeds, general words and phrases such as "northerly" or "about 20 chains" are sometimes used, and if the terms are not restricted by other words, they may become exact. In *Bosorth* v. *Danzien* unrestricted "northerly" became due north,[42] and in *Wise* v. *Burton* unrestricted "about 20 chains" became exactly 20 chains.[43] If the deed containing "northerly" had been written "northerly to a 2-inch iron pipe," then northerly could mean anything from 45° east to 45° west of north, to the found 2-inch iron pipe. The problem is created when the pipe is missing and cannot be found.

It is probably safe to assume that deeds containing unmodified ambiguous terms such as "about" or "northerly" were prepared by the unqualified. The use of these terms simply meant that the scrivener did not know the true distance or bearing. Such deeds keep courts and surveyors busy. In descriptions, a scrivener will use a general term such as *near* with the assumption that it will provide certainty to the location of another point. If a question arises as to the legal interpretation of a word, the courts usually rely on *Black's Law Dictionary*. Black describes *near* as follows:

> NEAR. The word as applied to space is a relative term without positive or precise meaning, depending for its signification on the subject-matter in relation to which it is used and the circumstance under which it becomes necessary to apply it to surrounding objects. *Case-Fowler Lumber Co.* v. *Winslett*, 168 Ga. 808, 149 S.E. 211, 213.

In all probability, if the term *near* is used in a description, it is ambiguous and probably should not be used unless accompanied by specifics of distance and other specific references.

11.35 DIRECTION OF SURVEY

"In attempting to trace the description on the ground, the court should follow the footsteps of the surveyor rather than the reverse course."[44] In the absence of certainty of the point of beginning and in the presence of certain monuments farther along, the surveys are sometimes backed in, but this procedure is to be avoided except where the intent is clarified. The presumption is that the survey is to be made in the direction of the deed; the contrary must be justified.

> If an insurmountable difficulty is met with in running the lines in one direction, and is entirely obviated by running them in the reverse direction, and all the known calls of

the survey are harmonized by the latter course, it is only the dictate of common sense to follow it.[45]

There is no hard and fast rule for closing a survey, but such rules as are employed are but rules of construction in aid of an effort to relocate lost lines as they were located in the original survey, which is always the problem for solution, and rules of construction must give way to competent evidence disproving their applicability to a given case. One rule of construction often recognized and applicable here is that reversing calls is as lawful and persuasive as following their order.[46]

There may be times when a conveyance has no direct call for monuments, and its point of beginning is **an** adjoiner's deed. In New York in 1913, the issue was whether the starting point was intended to be at the line of occupancy or at the premises conveyed. The point determined by the adjoiner's deed was correct, and the occupancy point was rejected. The court said: "The beginning corner of a survey is not usually of any more importance in determining the location of the survey than any other corner therein."[47]

11.36 AREA OR SURFACE

Principle 16. *Except where area expressly states the intentions of the parties to a deed, area is presumed as subordinate to other considerations.*

"South 5 acres of lot 13" is a description in which area is the sole controlling factor that establishes the north line of the survey, and in the absence of other calls, area is the controlling element. In a conveyance reading "5 acres, no more, nor no less, described as follows," or "exactly five acres," area probably will prevail if an ambiguous description follows. But if the perimeter description is without error, the area becomes more or less. Occasionally, area is the deciding factor where alternate lines can be drawn from the written instructions, and area computations fit one of the alternate lines, as in the case in which the judge observed:

> Quantity ... of itself is no description. It does not give boundaries of course (bearing and distance), therefore, metes and bounds will prevail where there is conflict. But when boundaries are doubtful in themselves, quantity often becomes the controlling fact. It often makes the metes and bounds certain. The quantity, taking the eastern edge as the boundary, approximates the quantity mentioned in the deed, being eleven acres more than the deed calls for. Taking the western ridge for the boundary is added 680 acres, nearly 2/3 as much more, to the deed. Such a discrepancy should not be disregarded in arriving at the proper location of the disputed boundary under the circumstances disclosed in the case.[48]

In the case in which the original surveyor established the first 13 calls in a patent and then, at the instructions of the seller, inserted adjoiner calls for the remainder of the land, the land conveyed follows the lines of the original 13 calls and the adjoiners described, even though 600 acres were conveyed instead of 150 as supposed (construed in favor of grantee).[49] In New York in 1838, when the land was described by monuments, course, and distance, and stated 200 acres, strict measure, and no more, the monuments, course, and distance controlled.[50]

11.37 POINT OF BEGINNING

Although every deed must have a point of beginning, it is not to be assumed that the beginning corner is of any more importance than any other corner recited in the description. A monument called for and found at the second corner often fixes the position of the second corner independently of the first corner. Both corners have equal standing, and in all conveyances all recited corners have equal dignity.

11.38 CONSTRUED MOST STRONGLY AGAINST GRANTOR

Principle 17.   A grant or conveyance is to be interpreted in favor of the grantee except that a reservation in any grant, and every grant by a public officer or body, as such, to a private party, is to be interpreted in favor of the grantor.

Because language used in the instrument was selected by the grantor, who should have been more familiar with the property, the deed should be construed most strongly against the grantor. This rule applies only where two or more meanings are possible, but not where one of the parties misunderstands a clearly expressed written meaning.

The northerly 50 feet of lot 10, as shown in Figure 11.7, implies that the 50 feet is measured at right angles to the northerly line, thus giving the buyer the maximum possible area. In Figure 11.8, the westerly 50 feet shown actually includes more than 50 feet at the angle point, as parallelism is implied by the "westerly 50 feet."

In New York, a similar situation occurred in a deed reading "two miles on each side of a certain creek." The court found that the line should be run so that at every point the line will be 2 miles from the stream in some direction, and not by parallel lines running from points 2 miles distant from the stream on a line traversing it.[51]

In Vermont, the court observed that "if the intentions of the parties upon the face of the deed be ambiguous, the construction is to be most strongly against the grantor."[52] The reverse was held where the grant was a patent from the state. In that the words "near Cumberland Gap" could not be construed to be "in Cumberland Gap," it was construed against the grantee.[53]

11.39 ERRORS AND AMBIGUOUS TERMS

Principle 18.   Certain and definite statements will prevail, and erroneous or ambiguous terms may be rejected, but such rejections should be as few as possible.

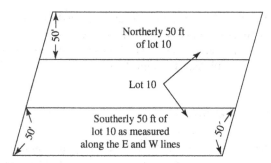

Figure 11.7 The northerly 50 feet of a lot are measured at right angles to the northerly line.

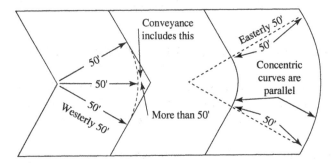

Figure 11.8 At an angle point, the westerly 50 feet conveys more than 50 feet because of implied parallelism.

Errors in any description may be corrected or rejected if there is sufficient information in the deed to indicate where the errors occur. A change of bearings from N 10° E to N 10° W, as proved by a mathematical closure, may be corrected. A general statement refuted by a particular call may be rejected, but only that part refuted may be rejected. A call in a survey or deed may be rejected for inconsistency when, in the description, enough still remains to ascertain the land with certainty.[54]

11.40 COORDINATES

Coordinates are computed from measurements of distance and angles by various formulas. The coordinate values are a product of calculations, which themselves have inherent errors. The quality of the coordinates is only as good as the least significant number of the measurements from which they were calculated. In the order of importance of conflicting deed elements, coordinates cannot be presumed to rank higher than the method used to determine them; they are presumed subordinate to monuments. Whether coordinates will outrank other measured distances or angles may depend on the accuracy of the method used to determine the coordinates and the proximity of the control points from which the coordinates were calculated. Although coordinates can be established with precision, not all can. If a monument is found and the coordinates of the monument are determined precisely by an acceptable method, and the monument is later lost, the coordinates so established will probably form the best available means of reestablishing the former position or providing an area of search to look for monuments. Similarly, if an original monument is set and the surveyor carefully determines the exact coordinates of the monument, the coordinates, an informational call, will probably be the best means of restoring the corner, if lost; but if the monument is not lost or disturbed, the monument itself is presumed correct irrespective of whether the coordinates were determined correctly. Coordinates should be considered an informational aid to assist in replacing a lost monument, not a means to determine where a found undisturbed monument should have been.

11.41 DIRECT LINE MEASUREMENT

Principle 19. *Unless stated otherwise, a distance is measured in a straight line along the shortest measurable distance, usually horizontal.*

Distances recited in descriptions are presumed to be straight-line point-to-point (corner-to-corner) distances. In all probability, this was not the method used in creating the original description. The modern methodology is to run random lines using traversing methodology and then compute the resultant straight line, which is then placed in the description, without any explanation.

Ambiguous conditions arising from this rule are shown in Figure 11.9. Where a distance is intended to be along a line or curve, the fact should be so stated. "Beginning at a point in Orange Avenue 200 feet easterly from the NW corner of lot 2"

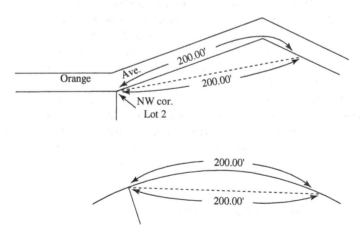

Figure 11.9 Deeds reading "200.00 feet from the northwest corner of lot 2" may be ambiguous.

is indefinite, but "beginning at a point in Orange Avenue, 200 feet easterly from the NW corner of lot 2 as measured along the road" is certain, as is "along and with the right-of-way line."

> The law does not declare in favor of a straight line between monuments, where the language employed in the deed shows that a different line was intended. If, in a deed, the boundary line on one side of the land conveyed is described as running from a given monument easterly to a creek parallel with the southerly line of another tract of land, and such line of the other tract of land is not a straight line, but meanders, then the boundary line described in the deed will run parallel with the other line in its meandering, and not straight, and parallel with its general course.[55]

11.42 TREATMENT OF CURVES

Principle 20.   *When a curve factor is given as a whole number, the curve factor with the whole number was probably the controlling assumed figure originally and should remain as such. When radius, tangent, and degree of curve are all odd, a prorated adjustment between fixed monuments is indicated. Government (city, county, and state)-relocated curved street lines are not acceptable for lot location unless based on original curve stakes.*

As curve data are fixed by computation from two known factors, the problem is to fix the most likely control. The usual situation is when the given delta differs from the delta as measured by the resurvey. Almost always, the original surveyor assumes one of three factors after the delta has been fixed: (1) the radius is assigned a whole number, (2) the curve is assigned an even degree of curve, or (3) the tangent is assigned a definite amount. Where the radius is an odd number, the chances are that the radius was computed from other data. All even degrees of curvature have odd radii; a check must be made to determine whether the radius was derived from an even degree of curvature (always suspect this where a curve adjoins a former or existing railroad right-of-way). Where the tangent is an even whole number (centerline or side line), the chances are that this was the control intended.

Frequently, in rerunning street lines, the highway surveyors locate curves with changed radii, delta, or both. Such relocation can be used for defining the recognized street boundaries, which may have arisen from prescriptive methods but cannot be used to locate original lines. It is far safer to relocate lots from known lot corners or to block corners and work toward the curve in question.

11.43 FIRST STATED CONDITIONS

Principle 21.   *When two factors are in conflict and nothing else indicates which of the two is correct, the first stated is preferred.*

Although only two elements of a curve are necessary to determine a curve mathematically, three are usually given. If the three—the radius, delta, and tangent—are in conflict and all are odd in measurement, the first two stated are preferred. If, however, by reasonable analysis one of the three is found to be in error, the one in error should be disregarded. If a correct closure is obtained by using the delta and tangent, the radius given is in error.

11.44 WRITTEN AND CHARACTER NUMBERS

Principle 22. When numbers are shown both as figures and spelled out as words, the words will control unless the contrary can be proved.

The probability of spelling out a word in error is much less than the chance of writing a number erroneously, especially where a decimal point is misplaced.

11.45 UNIT IMPLIED

Principle 23. The scale unit of the map is implied to refer to all distances without a character mark on the map.

The mapping custom of surveyors has been to omit the sign indicating which unit has been employed. Where the scale of the map is given, a comparison between the enumerated distances and the distance scaled will reveal the unit. On a map with a scale of 1 inch equals 50 feet, unless the contrary can be shown, it is implied that all numbers unmarked by a unit are in feet.

11.46 FEET AND INCHES

Modern surveying practice calls for measurements to be made in feet and decimal parts of feet. Some of the older survey plats did use inches without stating such facts on the map or placing the customary inch symbol on the inches intended. Only by an examination of the map as a whole can the intent be determined. The use of 11", meaning 11 inches, is a clue for proving the intent to be inches, whereas the use of numbers above 12, indicating decimal parts of a foot, is a contrary proof.

11.47 GENERAL AND PARTICULAR PROVISIONS

Principle 24. A particular intent will, by presumption, control a general one that is inconsistent with it.

The validity of a deed and or description is a question for the court to determine.

Most deed descriptions have a general statement to identify the locale or vicinity, and a particular statement to identify the land from all other parcels in the vicinity. "All that portion of Section 10, Township 15 South, Range 3 East, SBM, more particularly described as follows" is a general description that will be followed by a particular description. If in the particular description the land was clearly described as lying within both Sections 10 and 11, the particular facts are controlling and that portion in Section 11, if owned by the grantor, would be conveyed even though not mentioned in the general description. This principle is based on the fact that a particular thing described in detail is much less apt to be in error than a general statement written without detailed thought.

A deed reading "Thirty-one acres in the east side of lot seven in Section fourteen, township seven south, range nine west, together with all improvements thereon" and followed by a particular description, "the land hereby conveyed being bounded on the south by the Bay of Biloxi; on the east by the Scale property; on the west by the lands of Martin; and on the north by the north line of lot number seven" is to be construed by the particular description even though only 17.3 acres were conveyed.[56]

But if a contrary intent is shown by the wording, the principle may be overcome. In a will, the term "the house and lot known at No. 114 Tenth Street" controlled a particular description that did not include all the land that the house occupied.[57] The general description showed the intent of the testator, that of willing a house and including all things necessary to enjoy the use of the house. The principle that a general description is not controlling over a specific description applies to cases where the specific description is not ambiguous.[58]

Directory calls are those that merely indicate the neighborhood wherein the different calls may be found, whereas *locative calls* are those that serve to fix the boundaries.[59] *Particular calls* are special locative calls; *general calls* are descriptive or directory. General calls are merely to direct a person's attention to the vicinity or neighborhood, whereas locative calls are made with care and exactness. General calls cannot be given much credit when in conflict with a particular locative call.[60]

In some states, a deed that recites a conveyance by street number is either void or voidable.

BASIS OF BEARINGS

11.48   DEFLECTION METHOD VERSUS COMPASS BEARINGS

Direction determinations in early surveys were made employing the magnetic compass. The results were either magnetic bearings or the declination was "set off" as calculated to give a true bearing. Starting around 1840, the solar compass was the accepted instrument for federal land surveys. Later surveys, beginning about 1870 and mostly since 1900, were made using an instrument to measure the angle between lines. From the angles, bearings were computed based on a determination of direction on one or more of the lines. The angular method of determining bearings (commonly

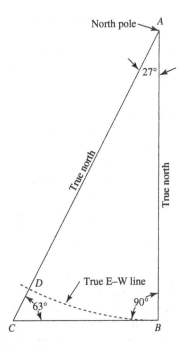

Figure 11.10 Convergence of meridians.

called the *deflection method*) differs fundamentally from the compass method. To illustrate this difference, an extreme situation is shown in Figure 11.10, where a parcel of land (ABC) apexing at the North Pole is being surveyed. Assuming that points A, B, and C are fixed in position by monuments and that the deflection method is being used, the surveyor measures the angles at A, B, and C. At point B, a star observation is made, and it is determined that the direction of line BC is due west. The resulting description is "Beginning at the north pole; thence due south 981.30 feet; thence west 500.00 feet; thence N 27° E a distance of 1101.33 feet to the point of beginning." Obviously, the N 27° E is true north, and the convergence of meridians is not taken into account. However, the figure will close mathematically. If we were to survey the same parcel by the compass method, the resulting description would be "Beginning at the north pole; thence due south 981.30 feet; thence west 500 feet; thence north 1101.33 feet to the point of beginning." This description will not, of course, close mathematically. Furthermore, the straight line west (CB) does not have a constant bearing; at point C, looking back at point B, the bearing is N 63° E, not east. Technically speaking, all lines, other than true north lines or lines along the equator, change bearing as a person travels along the line.

The differences between the compass and the deflection angle method of describing land boundaries diminish as the land is located farther from the poles and as parcels diminish in size. Within a small area of the continental United States, the differences are usually negligible.

In modern descriptions, the following presumption exists:

Principle 25.   *Unless the contrary is stated, it is presumed that every bearing given in a present-day metes and bounds description refers to the same basis at the same point, and that the bearing of every line is constant throughout its length.*

This presumption is exactly right if the original surveyor or deed author used the deflection method to compose his description. The presumption is equally valid for descriptions written from data obtained from a compass survey, provided that the area of the survey is small and not near the North or South Pole.

Seldom in modern surveys does a surveyor find it necessary to make astronomical observations. Where a new parcel is carved from an older subdivision, the bearings as given on the older map are assumed to be correct; new lines are deflected from the older lines, and bearings are computed. If a deed calls for a line and defines that line as having a definite bearing, succeeding lines are surveyed relative to the line called for.

In the eastern states, especially in the original colonies, use of the compass was prevalent before the development of the deflection method. Many early deeds failed to state whether their basis of bearing was magnetic or true. In some states, the presumption is that magnetic was used in older deeds; in others, true north is the presumption.

Often, where there is a marked difference between magnetic north and true north, the facts as observed on the ground will disclose which is correct. If the original surveyor used the deflection method, succeeding surveyors, in order to follow the footsteps of the original surveyor, must use the deflection method to reestablish deed lines. If a question exists between surveyors, the question may be answered only by application of rules of evidence.

Three conditions arising in descriptions lead to the following principle for determining the basis of deflected bearings:

Principle 26.   *(a) Where the bearing of a known line is given in a metes and bounds description, the bearing as given is assumed to be correct; successive courses are surveyed relative to the given bearing, whether or not the given bearing is astronomically correct. (b) Where a land description refers to a map and no basis of bearing is stated, it is implied that the map bearings are to be used, and all bearings in the description are referred to the same basis. (c) Where no basis of bearing is given or implied by a call for a map, true or magnetic bearings are to be used, depending on the presumption in the particular state.*

Referring to Figure 11.11, a description reading "Beginning at the NW corner of Pueblo Lot 1204, from which the northeast corner bears N 89°50′ E; thence S 40° E, 850 feet to the true point of beginning; thence, etc.," implies that the angle to be turned from the north line of Pueblo Lot 1204 is 50° 109, not 50°409 as figured from the true astronomical bearings. Should the deed read "beginning at the northwest corner of Pueblo Lot 1204 as shown on the map of View Crest; thence S 40°E, 850 feet to the true point of beginning; thence, etc.," Principle 2 applies. Where the View Crest map

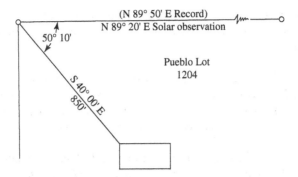

Figure 11.11 Pueblo lot 120.

shows the bearing of the north line of Pueblo Lot 1204 to be N 89°50′ F, an angle of 50°10′ would be turned from said line. "Beginning at the northwest corner of Pueblo Lot 1204; thence S 40° E, 850 feet to the true point of beginning; thence, etc." makes it necessary to use astronomical or magnetic bearings (depending on the state) where no bearings are given on the map of the Pueblo lands.

Compass Bearings

Early original surveys were conducted using staff compasses. This presumption is given credence by the written descriptions that usually refer to bearings to fractions of a degree; that is, 43 ¼ or 62 or 21 ½. To retrace these early bearing references, the retracing surveyor must take into account the changes in magnetic declination that have occurred throughout the years.

The magnetic bearing of a line is subject to daily variations, annual variations, magnetic storm variations, and local attractions. Where an old description has magnetic bearings based on a compass survey, the problem today is to determine the difference between magnetic north as it existed at the date of the original deed and as it exists today. It is well known that two compasses may give substantially different bearings for the same line at the same time. Considering this, the most positive way to determine the difference between magnetic north as it existed at the time of the survey and as it exists today is according to the following principle.[61]

A closed traverse consisting of compass bearings does not have an error of closure.

Principle 27. *If an original line of a deed can be identified, and if it were described originally by a measured magnetic bearing, the difference between the original record bearing and the present measured magnetic bearing is the correction to apply to other record bearings of the same description.*

Original lines are not always identifiable. Lacking an original line to determine the difference between the present and the former declination, the following principle should be used.

Principle 28. *If a line can be found in the vicinity whose magnetic bearing was determined reliably at the time of the deed, determine on that line the change in magnetic bearing from the date of the deed to the present time. Next, correct the bearings of the lines being surveyed by the amount of correction noted.*

In the absence of direct methods of determining declination variations with time, tables published by the government (*Magnetic Declinations in the United States in 1945*, Samuel A. Deel, U.S. Government Publication 664, Coast and Geodetic Survey) should be consulted. As an alternative, the website may be consulted for computations back to 1900. Prior to 1900, tables may be produced, but at this time the site is not able to automatically make the correction for values prior to 1900. The website is www.ngdc.noaa.gov. The annual change of declination may be small, but over a period of time it can become significant. The change may accumulate for many years in one direction and then change to the opposite direction. Because of the many uncertainties in applying theoretical corrections, the following should be considered as a principle of last resort.

Principle 29. *In the absence of a direct method of determining declination changes between given dates and reliable local data relative to declination changes from a given date, apply the magnetic declination corrections as given in National Geodetic Survey tables or website.*

11.49   SEQUENTIAL CONVEYANCES IN TEXAS

The land tenure systems employed within Texas can only be described as "unique." These land tenure systems have developed from a sort of "mixing pot," combining elements from all six flags of Texas: the Kingdom of Spain, the Kingdom of France, the Republic of Mexico, the Republic of Texas, the Confederate States of America, and the United States of America.

In all other states, the date of sequential conveyance used is when the land was subdivided from its parent parcel. This is known as the "date of survey." The same system is used in Texas, but the Texas-specific process used to award land to settlers often confuses out-of-state professional surveyors.

In Texas, the awarding of state lands to settlers worked as follows: first, the settler received a "certificate" that entitled him to a certain acreage, sometimes within a certain county. Next, the settler would go find a "vacant" parcel of land and claim it as his own. Third, the settler would hire a Registered Professional Land Surveyor (RPLS) to carve out that specific acreage in the area the settler desired. Finally, the settler would have to meet whatever requirements were set forth by the law under which the settler was claiming title (i.e., cultivation, fencing, settlement for a period of time). Only years later, when all of these elements had been met, could the settler gain clear title to the parcel of land. No matter, the date surveyors use to determine junior/senior rights is the date the parcel was surveyed.

11.50 SUMMARY, INTERPRETATION OF THE PRINCIPLES, AND CONCLUSION

In court decisions, certain elements are presumed to be true until the contrary is proven; thus, if a letter is duly written, sealed, addressed, stamped, and placed in a mailbox, it is presumed to be delivered unless the contrary can be proven. The principles stated in this chapter can be considered in the same light; they are presumed to be correct until a contrary intent can be proven.

When interpreting conflicting terms within a deed, courts apply the rule or rules of construction that show most clearly the intent of the original parties to the deed as well as attempting to make the description valid. To state definitely that one construction always controls another is to err. The foregoing principles aid in interpreting what the courts declared to be the normal manner in which intent is expressed. To determine the intent of each term of a deed without considering each term in the light of all other terms is to err. The intent is to be gathered from all the terms of the deed, the circumstances under which the deed was written, and the facts on the ground. Many of the principles given here have force when only one other term is in conflict. Thus, a call for a monument is normally given precedence over calls for distance, angle, and area; however, where several other terms in the deed contradict the call for the monument, and where the acceptance of the monument voids all other considerations in the deed, the principle that monuments control is overcome by a contrary intent.

Figure 11.12 shows the facts found on the ground for a deed reading "... to a 1-inch iron pipe being the true point of beginning: thence, N 1°10′ E, 200 feet to a stone mound; thence, N 89°50′ E, 200 feet to a 1-inch iron pipe; thence, S 10°10′ W, 200 feet to a 1-inch iron pipe; thence, S 89°50′ W, 200 feet to the point of beginning." Considering the first call alone, "N 1°10′ E, 200 feet to a stone mound," the stone mound found at that point would be considered as correct, whereas the stone mound found at N 10°10′ E would be rejected in accordance with the rule stated in Section 5.21. However, the remainder of the deed shows a contrary intent. A mathematical closure and the finding of three other monuments in agreement with one another all indicate that a typographical error caused the N 10°10′ E to become N 1°10′ E. Only by considering all the terms of the deed can a proper intent be arrived at.

The theory of majority probability (advanced by William C. Wattles in his book *Land Survey Descriptions*), whereby all factors are balanced, giving each its proper weighted value, forms a logical approach to a complex deed problem. Mathematical correctness, location of monuments, location and age of lines of possession, superiority of one call over another, previous survey records, the seniority of adjoiner deeds, common customs of other surveyors, and all other factors must be examined, weighted, and balanced to arrive at a proper location.

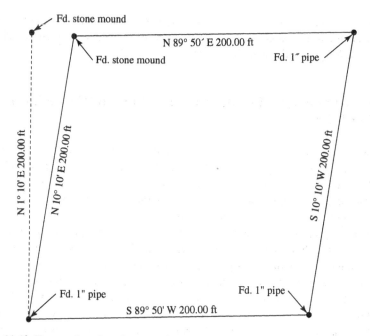

Figure 11.12 Facts as found on the ground.

BIBLIOGRAPHY

Black's Law Dictionary, 5th ed. St. Paul, MN: West Publishing Co., 1982.

Love, John. *Geodaesia* (1687). Reprint by Walter G. Robillard, Atlanta, GA, 2000.

Bureau of Land Management. *Manual of Instructions for the Survey of the Public Lands*. Washington, DC: U.S. Department of the Interior, 1973.

Leybourn, William. *The Compleat Surveyor* (1659). Reprint by Walter G. Robillard, Atlanta, GA, 2002.

Wattles, William C. *Land Survey Descriptions*. Los Angeles: Title Insurance and Trust Company, 1956.

NOTES

1. *Creech* v. *Johnson*, 116 Ky. 441 (1903).
2. *Fordson Coal Co.* v. *Napier*, 261 Ky. 776 (1935).
3. *Spainhour* v. *B. Aubrey Huffman & Associates, Ltd.*, 237 Va. 340, 377 S.E.2d 615 (Va. 1989)
4. *Bundy* v. *Morgan*, 45 Vt. 46 (1872).
5. *Cushing* v. *Monarch Timber Co.*, 75 Wash. 678 (1913).
6. *Clark* v. *Aldridge et al.*, 162 N.C. 326 (1913).
7. C.C.P. Sec. 1858 (1983).

8. *Ferris* v. *Coover*, 10 C 589 (1858).

9. *Cates* v. *Reynolds*, 228 S.W. 695, 143 Tenn. 667 (1920).

10. *Mills* v. *Catlin*, 22 Vt. 98 (1849).

11. *Miller* v. *Southland Life Insurance Co.*, 68 S.W.2d 558 (1934).

12. Ferris supra note 8.

13. *Vance* v. *Fore*, 24 Cal. 436 (1864).

14. *Morris* v. *Jody*, 216 Ky. 593 (1926).

15. *Hudson* v. *Irwin*, 50 Cal. 450 (1875).

16. *Swift Coal and Timber Co.* v. *Sturgill*, 188 Ky. 694 (1920).

17. *Bryant* v. *Struck*, 151 Ky. 97 (1912).

18. *Givens* v. *U.S. Trust Co.*, 260 Ky. 762 (1935*); Albertson* v. *Chicago Veneer Co.*, 177 Ky. 285 (1917).

19. *Norbery* v. *Todd*, 236 N.W. (Mich.) 826 (1931).

20. *Anderson* v. *Richardson*, 92 Cal. 623 (1892).

21. *Friend* v. *Friend*, 64 Md. 321 (1885).

22. *Delphey* v. *Savage*, 227 Md. 373 (1961).

23. *Quin* v. *Heart*, 43 Pa. 337 (1862).

24. *Earhart* v. *Rosenwinkel*, 108 Ind. App. 281; 25 N.E.2d 268 (1940).

25. *Powers* v. *Bank of Oroville*, 136 Cal. 486 (1902).

26. *Hoff* v. *Tobey*, 66 Barb. 347 (N.Y. 1866).

27. *Lerned* v. *Morrell*, 2 N.H. 197 (1816).

28. 38 Me. 99 (1854).

29. *Wilson* v. *Inloes*, 6 Gill 121 (1847).

30. *Wood* v. *Hildebrand*, 185 Md. 56 (1945).

31. *Speed* v. *Cheech*, 307 Ky. 765 (1948).

32. *Booth* v. *Upshur*, 26 Tx. 64 (1861).

33. *Curtis* v. *Aaronson*, 49 N.J.L. 68 (1958).

34. *Hoskins' Adm'x* v. *Louisville Cooperage Co.*, 297 S.W.2d 846 (1957).

35. Ibid.

36. *Green* v. *Pennington*, 105 Va. 8011 (1906).

37. *Combs* v. *Combs*, 238 Ky. 362 (1931).

38. *Morgan* v. *Renfro*, 124 Ky. 314 (1907).

39. *Reed et ux* v. *Tacoma Building and Savings Ass'n.*, 2 Wash. 198 (1889).

40. *Burgess* v. *Healey*, 73 Ut. 316 (1901).

41. *Richfield Oil Corp. et al.* v. *Crawford et al.*, 39 Cal. 2d 729 (1952).

42. *Bosworth* v. *Danzien*, 25 Cal. 296 (1864).

43. *Wise* v. *Burton*, 73 Cal. 166 (1887).

44. *Birk* v. *Hodgkins*, 159 Cal. 576 (1911).

45. *Ayers* v. *Watson*, 137 U.S. 584 (1890).

46. *Cornett* v. *Kentucky River Coal Co.*, 175 Ky. 718 (1917).

47. *Green* v. *Horn*, 207 N.Y. 489 (1913).

48. *Winans* v. *Cheney*, 55 Cal. 567 (1880).

49. *Rock Creek Property Co.* v. *Hill*, 162 Ky. 321 (1915).

50. *Suffern* v. *McConnell*, 19 Wend. 175 (N.Y. 1838).

51. *Jackson* v. *Dennis*, 2 Caines 177 (N.Y. 1804).

52. *Mills* v. *Catlin*, 92 Vt. 98 (1917).

53. *Creech* v. *Johnson*, 116 Ky. 441 (1903).

54. *Vose* v. *Handy*, 2 Me. 322 (1823).

55. *Woodard* v. *Pratt*, 32 Cal. 219 (1867).

56. *Carrere* v. *Johnson*, 1439 Miss. 105 (1929).

57. *Gilbert* v. *McCreary*, 87 W.Va. 56 (1920).

58. *Haskell* v. *Friend*, 196 Mass. 198 (1907).

59. *Cates* v. *Reynolds*, 228 S.W. 695, 143 Tn. 667 (1920).

60. *Stafford* v. *King*, 30 Tex. 257 (1867).

61. Authors' Note. The authors, having worked many decades with compasses, have found that early bearings were less trustworthy than were the recited distances; the courts take the opposite view.

CHAPTER 12

LOCATING SIMULTANEOUSLY CREATED BOUNDARIES

12.1 INTRODUCTION

The area of simultaneously created boundaries has been, and still is, a fruitful area for professional surveyor examinations and litigation. It has been found that both state registration questions and court decisions have inadvertently misapplied basic principles that surveyors must understand in order to practice in the daily world of boundary retracement.

Prior Chapter 11 have covered one type of parcel creation—sequential. This is one of junior/senior creation and rights. The second form of real property interest creation is a simultaneous creation. This could be likened to the *big bang* theory whereby there was nothing originally and then, upon the occurrence of certain required events, *it becomes a fact*.

Definition. A simultaneously created boundary results when several smaller parcels of land are created from one large parcel or from several combined large parcels at the same legal instant by the same person, persons, or agency and by the same instrument. All parcels have equal standing, and no such portion can be said to have prior rights or seniority over any other portion.

This concept of simultaneously created boundaries falls into three separate and distinct areas: those created under federal laws (the General Land Office [GLO] system),

Brown's Boundary Control and Legal Principles, Eighth Edition.
Donald A. Wilson, C.A. "Tony" Nettleman III, and Walter G. Robillard.
© 2024 John Wiley & Sons, Inc. Published 2024 by John Wiley & Sons, Inc.

those created under state laws, and those created under combinations of metes and bounds descriptions, wills, and partition proceedings.

The following relevant principles are discussed in this chapter:

PRINCIPLE 1. The boundaries and corners of lot lines within a subdivision are determined by the exterior lot lines of an older subdivision or by the boundary lines of a metes and bounds description.

PRINCIPLE 2. The micro boundaries of the interior lot lines located within the senior macro lines should not extend beyond the macro boundaries of the parent parcel.

PRINCIPLE 3. In a subdivision that is a product of a replat, the retracing surveyor should consult both plats and then realize different rules may apply under certain conditions.

PRINCIPLE 4. Lost corners on exterior (macro) boundaries cannot be retraced from micro corners found on inside lots. These micro corners can be used as collaborative evidence with other evidence to prove obliterated corners but not to proportion from for lost corners.

PRINCIPLE 5. A subdivider who subdivides another's land cannot convey good title to land improperly monumented. After a period of time, however, a person may acquire title to an improperly monumented strip of land through unwritten methods of title transfer.

PRINCIPLE 6. A subdivider who describes the boundaries of a subdivision incorrectly but owns all the lands monumented conveys title to the land described improperly.

PRINCIPLE 7. If interior lots of a subdivision were originally created by protraction, the lots are located by measurements from the subdivision boundaries; if interior lots of a subdivision were created originally by survey with monumentation, the original survey and monuments control lot locations.

PRINCIPLE 8. The intentions of the parties to a subdivision are paramount to all other considerations.

PRINCIPLE 9. Once a lot, street, or block line within a subdivision is established by the original surveyor and the land is sold in accordance with the original plat, the lines originally marked and surveyed are unalterable except by re-subdivision. Within the subdivision, boundary lines may be an exception.

PRINCIPLE 10. In a lot and block description, subdivision monuments called for on the plat, or monuments set by others and known to perpetuate the position of the original monuments called for, if properly identified and undisturbed, control the position of the original lot lines.

PRINCIPLE 11. Monuments other than the original monuments or replacements of the original monuments may become title monuments by prescription, agreement, estoppel, or other means; however, these monuments cannot be used to determine original lines of the subdivision or to control such lines.

PRINCIPLE 12. Original monuments set on the ground, except where the intent is clearly otherwise, control facts given on a plat.

PRINCIPLE 13. After due allowance has been made for weathering, deterioration, and other disruptive forces, the monument and its markings should be substantially the same as the record describing the monument. In the absence of visual evidence of a monument, location of a former monument can be determined by competent witnesses who saw the monument and remembered its location.

PRINCIPLE 14. Within a subdivision, distance, direction, and area are presumed subordinate to the intent of the subdivider; the lines are as marked and surveyed, and original monuments control. Where excess or deficiency is found, it is distributed in proportion to linear measurements between original found monuments.

PRINCIPLE 15. Natural monuments, where called for, are presumed to control street lines.

PRINCIPLE 16. After natural monuments, artificial monuments that represent the actual lines run by the original surveyor at the time of making the plat are presumed to control street lines irrespective of whether the courses, distances, and street improvement agree with the plat.

PRINCIPLE 17. In the absence of natural monuments or evidence of lines actually run by the original surveyor, improvements, such as curbs and paving, which were installed simultaneous to the original platting in accordance with the original survey monuments, may be presumed controlling.

PRINCIPLE 18. In the absence of artificial monuments and evidence of the lines as marked and surveyed, where a street is plotted as being the continuation of a nearby street or beginning at a nearby street, the line of the nearby street is presumed to control.

PRINCIPLE 19. In the absence of evidence covered by the foregoing principles, the exact width of the street as given on the plat and the distances and angles are presumed to govern street location. Occasionally, a measurement index is applied.

PRINCIPLE 20. If the street width is not given on the plat, the width as scaled on the map will govern as a last resort.

PRINCIPLE 21. Offset monuments set by the city engineer or other officials to perpetuate the position of the original monuments of the original surveyor control street lines. Offset monuments not based on adjacent original monuments are afforded control only in proportion to the accuracy with which they were set in accordance with the foregoing rules. City engineers' monuments long acquiesced to are presumed to be correct; the contrary must be proved.

PRINCIPLE 22. Excess or deficiency existing in a straight line between fixed monuments within a subdivision is distributed among all the lots along the line in proportion to their record measurements.

PRINCIPLE 23. Excess or deficiency occurring within a block should not be prorated among other blocks.

PRINCIPLE 24. Excess or deficiency existing in a straight line between fixed monuments within a subdivision is distributed among all the lots along the line in proportion to their record measurements.

PRINCIPLE 25. Excess or deficiency existing between fixed monuments on a curve is distributed among the lots along the curve in proportion to their record linear measure.

PRINCIPLE 26. Excess or deficiency cannot be distributed beyond any undisturbed original monument.

PRINCIPLE 27. When the end lot measurement is not given, all the excess or deficiency is presumed to be given to the end lot.

PRINCIPLE 28. Where excess or deficiency is given to an irregularly shaped lot at the end of a block, the method is called the remnant rule. Few jurisdictions accept it.

PRINCIPLE 29. In the absence of physical evidence on the ground and in the absence of measurements given on the plat, as a last resort scaling of the plat may be used.

PRINCIPLE 30. Where the boundaries of a subdivision were staked correctly originally, the foregoing principles for the establishment of lots are applied.

PRINCIPLE 31. Lots abutting on a subdivision boundary line cannot extend beyond the title interest of the original subdivider except for a right of title resulting from lawful possession.

PRINCIPLE 32. Where the original subdivider failed to subdivide all the land as shown by found original monuments, the surveyors should not extend the lot lines beyond the limits of the original monuments. Title to the unsubdivided land probably remains in the original subdivider.

PRINCIPLE 33. A considerable excess or deficiency existing in blocks adjoining a subdivision boundary is prorated only as a last resort.

PRINCIPLE 34. It is safer to locate lots from the interior of a subdivision than from the boundary line of a subdivision.

12.2 DEFINING SUBDIVISIONS

To have a need for determining a simultaneous boundary, the primary requirement is that there first must be a subdivision being divided into two or more parcels created from a parent parcel, with the following conditions being fulfilled:

1. All the divisions of land parcels are indicated and shown on a map or plat.
2. The resulting map or plat is approved and filed with or by the agency that either created the parcels or by whomever is responsible for compliance with the respective subdivision laws. This will not apply if there is no requirement for recording.
3. The parcels of land are sold by reference to the map or plat, and no parcel is sold or conveyed until the map is either approved or filed.
4. If the subdivision map is not being recorded, then it must be specifically mentioned in the conveyance documents for the information to be controlling,

   This order may also apply if the jurisdiction does not require recording of the map or plat. If lots are sold in accordance with a plan or map, and no reference to the lot exists other than the map, no one lot can be said to be senior to any other lot. All lots being created at the same moment of time—that is, at the moment of filing the map or at the moment of the first transfer of title—are of equal standing. A deficiency in a block found to exist within a platted subdivision is divided among the several lots in proportion to frontages as indicated on the plat, without regard to the sequence of their sale by the proprietor.[1] But if lots are sold in sequence and the lots are added to the map after they are sold, senior rights may exist. Locating subdivision lots on the ground presents essentially the same legal elements as those presented for metes and bounds descriptions, except for certain modifications caused by an absence of senior rights. In a metes and bounds description, the owner of the senior title receives all that is coming to him or her, and the junior title holder has the remainder. Within a subdivision, any excess or deficiency is divided among several lots in accordance with principles that have evolved from case law. Excepting federal subdivisions, lands subdivided within a state are governed by the state laws, and relocation of boundaries must be done in accordance with the rules or laws of the state. Lot and block conveyances calling for a plat offer a minimum of written language on the face of the deed; however, this does not necessarily mean that a lot and block description is the simplest conveyance to establish on the ground. Many older subdivisions and some poorly surveyed modern subdivisions present difficult situations to the person attempting to reestablish the true deed lines. Litigation arising from ambiguous lines and figures on maps has produced many principles of common law. The exterior boundaries of a subdivision may or may not have senior conveyance considerations, depending on the original deed of the subdivider. If the deed defining the boundary of the subdivision is junior to the adjoiner, all lots adjoining the senior deed are junior in character to the adjoiner. Lots within a subdivision may be junior to an adjoiner of the subdivision but not to another lot within the same subdivision.

SUBDIVISION BOUNDARIES AND CORNERS

12.3 ALIQUOT PART SUBDIVISION

Principle 1.   *The boundaries and corners of lot lines within a subdivision are determined by the exterior lot lines of an older subdivision or by the boundary lines of a metes and bounds description.*

In attempting to retrace subdivision boundaries without relying on stakes that were set for the original subdivision, the surveyor must examine the original description of the subdivision and the lots within the subdivision proper. This is actually a two-step process. There have been times when the creating surveyor set monuments, intending them to identify the corners of the subdivision and the lots. An example of this is in a subdivision where boundaries and/or corners represent an aliquot part of a section. Figure 12.1 shows that the original surveyor who created the exterior boundaries of the subdivision was in error when he set the exterior boundaries, apparently because the surveyor did not properly conduct a true and correct aliquot part survey.

Monuments were set, and the respective lots were located within the erroneously located subdivision. Figure 12.1 shows that the creating surveyor failed to correctly locate the east–west centerline. Apparently, the original subdivision corner was a "stubbed corner" at 1320 feet (not 20 chains GLO). This scenario presents two valid possibilities or solutions to this problem. The ultimate answer depends on how the descriptions of the subdivision and the lots were written in the subsequent deeds.[2]

The surveyor must understand the relationships in order to determine the correct solution. The two solutions are:

1. The subdivision and respective lots refer to aliquot portions only and reference to lots, without reference to the monuments set.
2. The subdivision and respective lots refer to the monuments set in the survey of the subdivision.

12.4 CONTROLLING BOUNDARIES

Principle 2.   *The micro boundaries of the interior lot lines located within the senior macro lines should not extend beyond the macro boundaries of the parent parcel.*

When the exterior macro boundaries of a parcel are created, they become the limiting perimeters of the individual interior lots.

This principle should have great significance for the retracing surveyor in the retracement of boundaries. It requires a basic understanding of the sequential

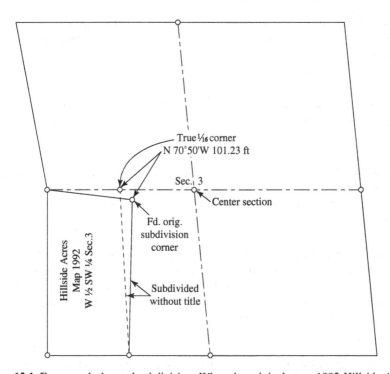

True ¹⁄₁₆ corner
N 70°50'W 101.23 ft

Sec. 3

Center section

Fd. orig.
subdivision
corner

Hillside Acres
Map 1992
W ½ SW ¼ Sec. 3

Subdivided
without title

Figure 12.1 Erroneously located subdivision. When the original map, 1992 Hillside Acres, was recorded, the northwest corner indicated a "2-inch iron pipe set at a fence corner in a boundary of long standing." Upon retracing the original subdivision boundaries, the situation depicted in the figure resulted. In locating the west half of the southwest one-quarter of Section 3, the following was found: an area of land beyond the boundaries of the west half was included in the description of the monuments. Now the surveyor must conduct additional research to determine if the original subdivider owned the east half; that is, was there common ownership of the entire parcel? If this was the case, one avenue for solving the problem could be followed; if not, a second solution would be followed. Without reference to the monuments and with only a reference to an aliquot subdivision, the description controlled. The lot owners who fell outside the east boundary of the west one-half had title problems and possibly could be evicted if the original subdivider did not own the east half. Actions for breach of warranty would be proper. However, if the original descriptions called for set monuments and the original landowner did not own the east half, affected landowners could consider a quiet title action under color of title. If the original subdivision owner owned the east half, there could possibly be a title problem involving junior–senior interest. It is suggested that any surveyor who finds such a situation as this should instantly consult legal counsel and prepare a detailed report and analysis for the client before attempting to undertake any solution.

application as well as the simultaneous creation when applying the reasoning. First, the exterior boundaries of the parcel control all lots within the boundaries. With this in mind, if one of the sequential interior corners of a boundary is dependent on an exterior boundary, the exterior boundary should be located first; then the interior corner (boundary) is identified.

Principle 3.   *In a subdivision that is a product of a replat, the retracing surveyor should consult both plats and then realize different rules may apply under certain conditions.*

The law permits a landowner to prepare new plats of former platted subdivisions as long as former vested rights are not violated. The replat cannot change any of the former lots that had already been conveyed. If a surveyor is asked to undertake such a project, he or she should approach it with caution.

Principle 4.   *Lost corners on exterior (macro) boundaries cannot be retraced from micro corners found on inside lots. These micro corners can be used as collaborative evidence with other evidence to prove obliterated corners but not to proportion from for lost corners.*

Corners placed along exterior boundaries control those boundaries, and the corners of the individual subdivision lots are separate. They should not be the controlling evidence from which to locate lost corners; they may be used as corroborative and supporting evidence to give added weight to these lost corners after they have been located by other methods or means.

12.5 SUBDIVISION MACRO BOUNDARY WRONGLY MONUMENTED

Principle 5.   *A subdivider who subdivides another's land cannot convey good title to land improperly monumented. After a period of time, however, a person may acquire title to an improperly monumented strip of land through unwritten methods of title transfer.*

Referring to Figure 12.1, a subdivider who inadvertently subdivides and sells another person's land may be able to solve the problem and clear title to the mislocated lots by purchasing or exchanging the land in question and making the title granted by applying the doctrine of after-acquired title, which is recognized in most states. It should be noted that in Figure 12.1, if the monumented lines are accepted, a triangular portion of land will result at the midsection line that still is in the ownership of the subdivider.

Limitations on the Principle

A subdivider who subdivides another's land may later purchase the adjoining land and, by after-acquired rights, automatically clear the title to the land subdivided. In the case of Hillside Acres (see Figure 12.1), title to the triangular strip of land on the east side could not pass because the subdivider did not have title to the land. Under certain circumstances, where a person occupies land with color of title, pays taxes on the land, and so on, he or she may acquire a fee title by adverse rights.

If you, as a surveyor, are hired to survey the east half of the southwest quarter of Section 3, and you find the encroachment of Hillside Acres on the east half as here, you should note the encroachment on your map whether or not you considered the land lost by adverse rights. The surveyor should point out to the client what rights the adjoiner has to encroachments of long standing and also advise the client to consult an attorney. Along land grant lines, especially where the grants are several miles long, subdivisions are frequently found to overlap or not touch the true grant lines. When the cost of running out the true line is excessive or difficult, surveyors often fail to establish the line properly.

12.6 SUBDIVISION BOUNDARIES INCORRECTLY DESCRIBED

Principle 6. *A subdivider who describes the boundaries of a subdivision incorrectly but owns all the lands monumented conveys title to the land described improperly.*

A landowner petitioned the county court to lay off a town site completely within the north 40 acres of his property. The town was laid off in error in that the south boundary extended into the south 40 acres in his direction. Although a plat was filed depicting the town to be located properly, the court held that all sales of lots created and their conveyance was to the lots and land as monumented.[3]

CONFLICTING ELEMENTS IN DESCRIPTIONS

12.7 GENERAL COMMENTS

The control-afforded conflicting elements *within* a subdivision, but not necessarily the boundaries of a subdivision, are expressed in the following rules. The boundaries of a subdivision, described in the map title or subdivider's title, are controlled either by the rules for metes and bounds surveys or by the rules of construction for a larger subdivision from which the new one was carved. The rules presented in the following pages are to be interpreted as applying to lots within a subdivision or to a portion of the lots within a subdivision, but not to the boundary lines of a lot abutting on a subdivision, unless specifically included.

12.8 ORIGINAL METHOD OF CREATING LOTS

Lots placed within a subdivision may be created in several ways. They may be created in an original survey with the corners monumented or they may be created by protraction, on paper, without benefit of actual survey. The retracing surveyor must know which method was used, because the procedures for retracement of the lots are different.

Principle 7. *If interior lots of a subdivision were originally created by protraction, the lots are located by measurements from the subdivision boundaries; if interior lots of a subdivision were created originally by survey with monumentation, the original survey and monuments control lot locations.*

To determine *what* must be done may require extensive field and office search—a requirement, regardless of costs.

12.9 INTENTION OF THE PARTIES

Principle 8. *The intentions of the parties to a subdivision are paramount to all other considerations.*

As one practices and testifies concerning boundary and title problems, attorneys can be expected to discuss and argue this principle in the belief that they understand what they are doing. Surveyors may be asked to testify as to what the intent of the parties was when a description was written or a plat was prepared. That question is impossible to answer, in that only the original parties to the original deed know, with certainty, what they intended. This can be solved when writing the deed in the supplemental clauses. If the parties state what the intention is, there will be no question at a later date. Such statements as "It is the intent of this deed to convey the land only to the south right-of-way line and that line will be the true and correct boundary" or "It is the intent that no riparian rights be conveyed," which detail the intent at the time of the conveyance, will solve future problems.

By most state laws, the subdivider must subdivide and survey the land before offering any one lot for sale. The intent of size and shape of all lots is thus platted and caused to be surveyed by the subdivider in accordance with current subdivision laws. The monuments set by the original surveyor to show the lines as marked and surveyed express the intent of the subdivider and become the paramount control for resurvey within a recorded subdivision. This is not true for the exterior boundary lines of a subdivision, where prior conditions, rights, surveys, or deeds may dictate otherwise.

One must be extremely cautious not to confuse intent of the parties with what the parties intended. The intent of the document is what the parties stated in the document, not what they meant to say or some unexpressed intention. In addition, the meaning and intending clause is not the parties' *intent*. Most courts are in agreement that the meaning and intending clause is an aid to trace the title, or to supplement the parties' previous language, not to express their intention.

12.10 FINALITY OF ORIGINAL LINES

Principle 9. *Once a lot, street, or block line within a subdivision is established by the original surveyor and the land is sold in accordance with the original plat, the lines originally marked and surveyed are unalterable except by re-subdivision. Within the subdivision, boundary lines may be an exception.*

In the GLO states, this principle is fixed by federal law. The Land Act of February 11, 1805, states emphatically that the original lines and corners are unalterable by retracing surveyors. In fact, not even courts can alter or change a section line or a section corner.[4] If land boundaries are to remain unalterable and fixed in position, no subsequent retracing surveyor should be permitted to relocate them based on more precise methods of surveying. This also applies to private subdivisions. The lines, corners, and monuments set by the creating surveyor, regardless of how imprecisely they were located, must remain fixed in place. No subsequent surveyor has the authority to "correct" any errors that are found. To do so would wreak havoc on possessions, structures, and other improvements within the subdivisions. Neighborhoods that have enjoyed a long history of peace will be thrown into total disorder.

If land is to remain fixed in position and not altered by every resurvey, the principle must stand. It would indeed be folly to alter the lines of a survey and the location of the improvements thereon just because the original surveyor failed to set his or her monuments in the measured position called for. The entire foundation for the stability of the land depends on this rule.

All rules of law that have been adopted for guidance in locating disputed boundary lines have been directed to the end that in so doing the steps of the surveyor who originally projected the lines on the ground may be retraced as nearly as possible. No rule that has been adopted to accomplish that end is more firmly established than that courses and distances are controlled by marked and fixed monuments.[5]

Limitations on the Principle

This principle does not apply to the lines of later surveyors or to subdivision boundary lines other than free lines. If a later surveyor establishes lines and purports them to be the original lines of the original surveyor, they can be accepted only if they are in fact the original lines. Later surveyors are not empowered to place the lines in any position other than those established by the original surveyor, and if the later surveyors establish the lines erroneously, the lines must be reestablished in their original position. Where the original surveyor of a new subdivision establishes the boundary lines of the new subdivision erroneously, and the subdivider did not own the adjoining land nor has he or she acquired title to it since the new subdivision was filed, the boundary lines erroneously established cannot be considered unalterable unless title by after-rights, prescription, agreement, estoppel, or other unwritten means has set in. When not in agreement with the original subdivision determining the boundaries, the boundaries of the newer subdivision may be altered because of the permanence of the lines of the older subdivision.

The boundary lines of older subdivisions established in error often become the true boundary lines by prescriptive title. If the boundaries of a subdivision have been in error and improvements have been maintained along the erroneous lines for the statutory limit, it is probably safe to assume that the adjoiner would be barred from asserting rights to the true line. This, of course, will depend on the specific facts of each situation.

12.11 CONTROL OF ORIGINAL MONUMENTS WITHIN SUBDIVISION BOUNDARIES

Principle 10.   *In a lot and block description, subdivision monuments called for on the plat, or monuments set by others and known to perpetuate the position of the original monuments called for, if properly identified and undisturbed, control the position of the original lot lines.*

The retracing surveyor must thoroughly understand the difference between existing (original) and obliterated corners. The principal difference is the evidence accepted to prove them. There is no difference in their legality. In descriptions of the type "lot 2, block 3, map 2701," nothing other than data given on the map is called for by the writings. When the map is the sole written consideration, original monuments are paramount. Monuments, as placed by the original surveyor of a subdivision, represent the true location of the lines as run by the original surveyor, and as such have prior rights over any informative call for angles, distance, or surface. Errors in the measurements of distance, angles, or surface occur, but the location of original monuments, if undisturbed, is certain and conclusive as to the original location of the lines run by the original surveyor. Calls on plats for distance, angles, or surface are informative terms to aid in the location of monuments; the monuments mark the lines as run.

A plat is a subdivision of land into lots, streets, and alleys, marked on the earth and represented on paper. The monuments or marks placed on the ground by the surveyor in making a survey constitute the survey, and the courses and distances are only evidence of the survey. Although evidence based on courses and distances from other known points is admissible as evidence to fix a corner, where no corner is found, it is not admissible to change the location of an original corner of the survey when found.

The survey is composed of the monuments placed on the ground by the creating surveyor and the maps and plats prepared to describe that survey. Courses (bearings) and distances must be considered as evidence as to where the corners, with their monuments, are. They are "finger pointers."[6] These courses and distances can be used as evidence from other known points to fix or to help find a corner; however, when found and identified, they cannot control a found original corner and change its position.[7] When it comes to a difference between the found corners on the ground and the course and lines depicted on a plat, the plat is nothing more than a subdivision of land, marked on the earth and represented on paper.[8] Numerous other court cases can be cited that recognize the control of original monuments. In Iowa, the surveyor set original monuments and a few years later set other monuments in the correct position as indicated by his field notes.[9] An original monumenting error caused a loss of about 4 feet in Cottage Grove Avenue, a gain of 4 feet in University Avenue, and a shifting of all lots by about 4 feet. As the earliest purchasers had found the original stakes and relied on them and constructed improvements in accordance with them, the court ruled that the original stakes must control. The original surveyor had no right to alter the position of the lots after the new owners had acted with respect to the stakes set even though they were not in accordance with his plat and field notes.

When a monument is found but not identified as an original monument set by the original surveyor or as a replacement of an original monument, the control given to the monument is often subordinate to distance and angles given on the original map. Having no locative value, an uncalled-for monument cannot be considered an original monument and cannot be given priority over distance or angle as given on the original map. In older subdivisions in which all the original stakes are gone as well as the records of any replacement of original stakes, the burden of proving that existing city engineers' monuments or tie points are replacements of original stakes is impossible. If the tie points are in agreement with street and building improvements, and the improvements represent monuments built on the original lines, the tie points can be accepted.

12.12 TITLE MONUMENTS

Principle 11.   Monuments other than the original monuments or replacements of the original monuments may become title monuments by prescription, agreement, estoppel, or other means; however, these monuments cannot be used to determine original lines of the subdivision or to control such lines.

A retracing surveyor may find several classes of title monuments. The original title may be modified or altered by unwritten legal principles, including agreement, estoppel, and others. A monument found but not located in accordance with the written deed or map may be a correct title monument, yet it may not be an indicator of the original lines of a subdivision. When a monument within a subdivision becomes a title monument by agreement, estoppel, adverse rights, judicial proceedings, or any other means, and where the monument is not an original, the monument, although binding on the adjoiners of that particular line, has no effect on the location of the original lot lines of the original surveyor; that is, it is local in character. It also cannot be used to control the replacement of other original monuments.

12.13 CONTROL OF MONUMENTS OVER PLATS

Principle 12.   Original monuments set on the ground, except where the intent is clearly otherwise, control facts given on a plat.

The lines marked on the Earth represent the true full-scale map of the subdivision; the lines as marked on paper are a shorthand representation of what the surveyor purported to do. When there is an inconsistency between the map and the facts on the ground, the map must yield to the facts on the ground.[10] When facts cannot be established on the ground—that is, the lines were never run on the ground or are lost completely—the data on the map are the best available evidence.[11]

According to most state laws, where land is described by lot and block numbers and several parcels are created simultaneously by the filing of a map, the land must be surveyed before the sale of any one lot. In such cases, the lines as marked by the original surveyor represent the intentions of the subdivider and control resurveys.

12.14 CERTAINTY OF MONUMENT IDENTIFICATION

Principle 13. *After due allowance has been made for weathering, deterioration, and other disruptive forces, the monument and its markings should be substantially the same as the record describing the monument. In the absence of visual evidence of a monument, location of a former monument can be determined by competent witnesses who saw the monument and remembered its location.*

Time takes its toll on monuments; some alter more quickly and differently than others. From the standpoint of permanence, stability, visibility, and certainty of identification, concrete monuments rank high, iron stakes intermediate, and wood stakes low. Pine stakes rot and disappear after a short life span; in moist locations, five years is the most that can be expected for positive identification. Redwood is well known for its resistance to decay and termites. In the Eureka Lemon Tract, many of the original 2-inch by 2-inch redwood lot corner stakes were found after a 60-year existence, and, from the excellent state of preservation, another 60 years could be expected. In the National Ranch, 3-inch by 3-inch by 4-foot scribed redwood posts, set for corner locations, were found well preserved after 82 years. In the southern United States, "lighter wood" posts were identifiable 150 years after placement by GLO surveyors.

Iron stakes in some locations are not nearly as permanent as is generally supposed. At one location, a 3/4-inch ungalvanized iron pipe was rusted away almost completely at the end of four years, whereas at another location little rusting took place in the same length of time. At the end of 25 years, 2-inch ungalvanized pipes set at Sunset Cliffs were completely rusted, whereas the redwood cores, driven in the pipes, were in good condition.

Of all devices invented by human beings, the bulldozer disrupts more stakes than any other. Monuments yield to the power of the dirt-happy dozer operator, pushing the earth around at will. In one subdivision, after all improvements were in, not a single lot stake or boundary stake was left—construction work destroyed them all. It is imperative that work in a subdivision be conducted so as to eliminate this problem.

12.15 RECORD DESCRIPTION OF MONUMENTS

Retracing surveyors have found much to complain about when attempting to identify original monuments set by the creating surveyors. The liberal use of abbreviations such as IP, IPS, and WS has made the job of retracement most difficult. With the advent of electronic data collectors, we may find once again that descriptions of monuments set at corners are less than acceptable. Surveyors are often careless in describing monuments. The description of pipe sizes is an example. Common sizes of pipes do not measure the same as the stock designation; for example, a 1 ½-inch pipe measures 2 inches in outside diameter (O.D.). In reporting the size of a set pipe, many take the actual outside measurement, but others give the stock designation. If the O.D. of a 1½-inch pipe is measured, it should be noted as 2 inches O.D., not as a 2-inch pipe. Early survey records were notoriously poor in monument descriptions, which today makes the job of retracement less certain and more difficult. Most modern laws have rectified this by requiring descriptions of all monuments set.

12.16 PRINCIPLES FOR PRESUMED CONTROL BETWEEN CONFLICTING MONUMENTS WITHIN SUBDIVISIONS

Early case law seldom referred to the term *natural monument.* Case law did refer to a *natural boundary*, but somehow, through years of legal decisions, courts have coined the term *natural monument.* Excluding possession considerations, two or more original monuments called for in a description that are in conflict with one another are given control in the following order of presumed importance:

1. Original natural monuments (boundaries).
2. Original artificial monuments set within a subdivision.
3. Original monuments set to mark the boundary lines of a subdivision will yield to senior rights in the event of an overlap. In the event of a gap between the subdivision boundary and an adjoiner, the lines marked and surveyed will control over the call for an adjoiner.
4. Uncalled-for monuments may become controlling by common report and/or acceptance.
5. A series of boundary improvements built soon after the original stakes were set, in agreement with one another and long acquiesced to by adjoining owners, are sometimes better evidence of original survey lines than are measurements of angles and distances from other points.
6. Where two monuments, otherwise equal, are in conflict, the one in harmony with distance, angle, or area becomes controlling.
7. Judicial decisions affecting that particular subdivision. A major portion of the surveyor's work when surveying a lot within a subdivision is the relocation of monuments set by the original surveyor or the relocation of points set to perpetuate original monument locations. Land surveyors can be considered detectives who have specialized in the art of finding and relocating monuments and lines run by the original surveyors. In the absence of found monuments, other factors, such as distances given on a plat, become controlling and will be discussed under the establishment of streets and lots.

12.17 EXPLAINING PRINCIPLES

Except along the boundary lines of subdivisions, senior rights do not exist, thus accounting for the greater importance of monuments within subdivisions.

Any principle given to determine the control between different types of monuments is merely a rule of construction to put in force the intentions of the parties, and any rule of construction is rebuttable and may be overcome by contrary evidence. In conveyances, natural monuments in the form of lakes, rivers, and the like are given preference because of the certainty of their identification. Artificial monuments representing the lines originally marked and surveyed are given preference over record monuments, such as the line of the adjoiner, because of the greater certainty of location. Along subdivision boundaries, artificial monuments cannot overcome senior rights unless prescriptive rights have ripened into a fee.

Control of Artificial Monuments

If artificial monuments set or used by the original surveyor represent the lines as actually run by the original surveyor, the artificial monuments are presumed superior to record monuments. Thus, in Figure 12.2, the original surveyor assumed in error that the iron pin found at the northwest corner of Olive and A streets was the true corner and started a subdivision from the erroneous point. To relocate a lot within this subdivision, the surveyor should use the pin as his or her starting point and overlook the call for the record monuments (street line). The street as originally staked is 76 feet wide and is unalterable except by conveyancing. In a New Jersey case, very similar in nature, the opposite was held. The court reasoned that the unmarked street was a natural monument '(*sic*) that should control the artificial monument.

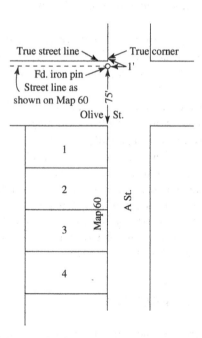

Figure 12.2 Case in which an artificial monument is superior to a record monument.

Meander Lines

Meander lines are quite unusual, in that they were never intended to be considered as boundary lines. Initially, they were to represent the natural curvatures of the water body or other feature that was being surveyed. The primary purpose of a meander line was to permit the calculation of area to the proximity of the water body or other boundary. Artificial monuments set by a surveyor to meander a natural monument, such as a lake, river, or ocean, must yield to the more certain monument—that is, the waterline. Meander lines are run for the purpose of plotting the natural monument being meandered and to permit the determination of area. Meander lines should not be considered as marking the boundaries or limits of the natural monument or of lots adjacent to it. "Where a meander line of land included in a patent is run to locate a shoreline approximately and to afford the means for calculating the area of the land granted, the shoreline constitutes the real boundary."[12]

This presumption may not be applicable in true metes and bounds descriptions.

Uncalled-for Monuments

Monuments that are not called for are uncertain; to be controlling they must be proved. If they are found in the proper position for angle and distance, they may automatically become the correct corner markers. If they are not in the measured position, they cannot be accepted without substantial evidence showing that they are in the position of the original stakes or that they have been accepted by common report.

Unfortunately, in many older subdivision surveys a statement explaining what monuments were set was not required by law; yet we know that the lands were surveyed and points were set. A statement on a map "Surveyed by Wheeler in 1880" is conclusive proof that the land was surveyed, and it must be presumed that monuments were set.[13] Notes taken at the time of early surveys amounted to private property, and as such have been lost as the years advanced. In some instances, a thorough resurvey reveals consistent types of corner material set by the original surveyor. In Rancho Mission, 4-inch by 4-inch redwood posts about 3 ½ feet long, set 1 foot in the Earth and stenciled with black paint indicating the corner numbers shown on the original map, were found as corner monuments. In a superior court case involving lot line location in Wadsworth's Olive Grove, 2 ½-inch by 2 ½-inch by 18-inch redwood stakes were accepted as the original stakes even though they were not mentioned on the original map. In both of these examples, the principle that a monument should be called for before it is controlling is defeated because of the preponderance of evidence proving what was actually set, even though it was not described on the maps.

This principle is sustained by the court in its findings:

In an action to quiet title involving a disputed boundary between lots, referring to a map uncertainly locating the line, and no field-notes of the survey for the map are indicated, it is proper for the court to receive the best evidence obtainable to determine the location as a matter of fact, and in so doing, was important to inquire as to the location of such stakes and monuments as were commonly recognized, accepted and used in lieu of lost or destroyed original monuments, and in the absence of more certain evidence, these stakes and monuments would be sufficient to support a finding as to the location of the boundary.[14]

In this case, the court accepted parol descriptions of the old monument found but not described on the original map. One surveyor testified: "In making surveys here in the city [Los Angeles] the last 13 years I have had occasion to retrace a great many of those Hancock corners. There were a few of them left and that was one of the last to go." Although this is considered to be hearsay, it is *considered* an exception to the hearsay rule.

Common Report

By reputation, certain markers are commonly accepted by surveyors and others as being correct, even though they cannot be positively proved to be correct. As time progresses and the records of replacement of original property monuments are lost, surveyors become more and more dependent on the acceptance of monuments whose history is lost in antiquity. If people accept a monument, and especially if numerous surveyors accept a monument, the monument is said to be the true marker by *common report*, or what contemporary courts call "repose." Monuments that can be proved to be neither correct nor incorrect and are commonly accepted as being correct may be accepted by common report, but a monument that can be positively proved to be incorrect, although accepted by many as being correct, cannot be accepted by common report. Mere measurements from distant objects do not disprove a monument. But finding an original monument not in agreement with a ½-inch pin accepted by common report certainly refutes the pin. For a monument to be accepted by common report, the following factors should be present: (1) the monument is commonly accepted and has the reputation of being correct, (2) the monument is in a position that represents a spot or place marked by an original surveyor or a spot that could have been marked by an original surveyor, and (3) the monument cannot be disproved. Several courts have also referred to this doctrine as the *doctrine of repose*.

Improvements as Monuments

A resurvey made after the monuments of the original survey have disappeared is for the purpose of determining where the monuments were, not where they ought to have been.[15] If the original monuments of an original survey are lost and a series of old fences, old buildings, or other ancient boundary indicators are in agreement with one another, it is usually presumed that the improvements were built on the original lines of the original surveyor and stand as monuments representing the original lines. Such improvements are considered as being the best remaining evidence of the original survey. This rule has more force when it can be shown that the improvements were made soon after the original stakes were set. Many tie points are set in accordance with this principle. When there is uncertainty about city tie points that represent the position of the original monuments and several surveyors are uncertain about the true location of the original monuments, the courts will frequently settle on a fence of long standing as being the better evidence of a boundary line. Acceptance of improvements is a principle that should not be applied if there is better evidence of the original positions of the monuments, and such evidence as fences should have no greater weight than any other element of evidence. Associate Justice Cooley, in 1878, stated in his opinion:

> This litigation grows out of a new survey recently made by the City Engineer. According to this survey the practical location of the whole plat is wrong, and all the lines should be moved four or five feet east.... When an officer proposes thus dogmatically to unsettle the landmarks of the whole community, it becomes of highest importance to know what was the basis of his opinion. The records in this case fail to give an explanation. Nothing is better understood than that few of our early plats will stand the test of a careful and accurate survey without disclosing errors. This is true of the government surveys as of any others, and if all the lines were now subject to correction on new surveys, the confusion of lines and titles that would follow would cause consternation in many communities. Indeed the mischiefs that must follow would be incalculable, and the visitation of the surveyor might well be set down as a great public calamity.

> But no law can sanction this course. The surveyor has mistaken entirely the point to which his attention should have been directed. The question is not how an entirely accurate survey would locate these lots, but how the original stakes located them. No rule in real estate law is more inflexible than that monuments control course and distance ... the city surveyor should therefore, have directed his attention to the ascertainment of the actual location of the original landmarks, and if those were discovered, they must govern. If they are no longer discoverable, the question is where they were located; and upon that question the best possible evidence is usually to be found in the practical location of the lines, made at the time when the original monuments were presumably in existence and probably well known. As between old boundary fences, and any survey made after the monuments have disappeared, the fences are by far the better evidence of what the lines of a lot actually are.[16]

Also, Justice Hand noted: "In case of a disputed boundary line in a town, city or village, where the monuments from which the town, city or village was platted are lost or destroyed, the courts ought not to disturb boundary lines between lot owners which have been acquiesced in for years and upon which the lot owners have erected improvements."[17]

12.18 INTRODUCTION TO PROPORTIONING

Principle 14. *Within a subdivision, distance, direction, and area are presumed subordinate to the intent of the subdivider; the lines are as marked and surveyed, and original monuments control. Where excess or deficiency is found, it is distributed in proportion to linear measurements between original found monuments.*

Few landowners or courts fully understand proportioning of excess or deficiency. The one basic principle underlying all proportioning is that proportioning can occur only between found original corners of lots. Although in many states direction is considered more certain than distance, excess or deficiency discovered in subdivisions is distributed in proportion to linear measurements. This is probably because frontage on a street is considered as paramount. Even though most surveyors know that distance is low in priority and may be a mere estimate, it is often the most important item in a layperson's mind. Usually, explanations must be given to clients.

In summary, the primary objective of retracing a boundary is to locate and prove, by the totality of best evidence available, where the original corners were placed and the remains of the monuments set to identify these corners. It does not matter whether the original corner monument was destroyed or whether the corner itself is difficult to position. The surveyor is permitted to use originality, and the courts will permit flexibility in recovering and interpreting evidence to prove the original positions.

ESTABLISHMENT OF STREETS

12.19 GENERAL COMMENTS

Boundaries of streets are established by the following methods, listed in their usual order of importance: (1) by natural monuments, (2) by artificial monuments and lines actually run at the time of making the plat, (3) by improvements, (4) by the line of nearby streets where called for, (5) by the data given on the plat, and (6) as a last resort, by proportional measure. As the lines of streets and blocks are usually one and the same thing, the establishment of street lines normally establishes block corners.

12.20 ESTABLISHMENT OF STREETS BY NATURAL MONUMENTS

Principle 15. *Natural monuments, where called for, are presumed to control street lines.*

Figure 12.3 shows a portion of Bird Rock subdivision with Ocean Boulevard existing between Blocks 20 and 21 and the ocean. This defines the boundary of Ocean Boulevard as being riparian and extending to the ocean, regardless of the distance given or scaled on the original map. To define it otherwise in this case would defeat the intentions of the subdivider, who has shown clearly by the plat that the intention was for the street to extend all the way to the ocean.

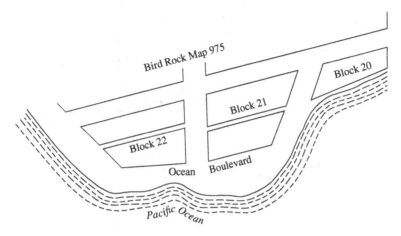

Figure 12.3 The boundary of Ocean Boulevard is riparian and extends to the ocean.

12.21 ESTABLISHMENT OF STREETS AND ALLEYS BY ARTIFICIAL MONUMENTS AND LINES ACTUALLY RUN AT THE TIME OF MAKING THE PLAT

Principle 16. *After natural monuments, artificial monuments that represent the actual lines run by the original surveyor at the time of making the plat are presumed to control street lines irrespective of whether the courses, distances, and street improvement agree with the plat.*

Identified original monuments set by the original surveyor and found undisturbed will control the street line as shown in Figure 12.4. Street A was found to measure 50 feet between monuments instead of 40 feet as indicated on the original map. The adjoining lots cannot each receive 5 feet extra by narrowing the street; the artificial monuments found definitely establish the street line. Similarly, if the record measurement is greater than the actual width indicated by monuments, the streets cannot be widened at the expense of the adjoining lots. In this case, the subdivider's intentions are clearly shown by the markers set at the time of the subdivision. The fact that the mapmaker erroneously noted 40 feet instead of 50 feet does not alter the facts on the ground.

In the city of Decatur, Illinois, an alley was located by the city crew and again by the county surveyor; neither found an original stake, but both agreed that by measurement the alley should be farther north and that this location would include the

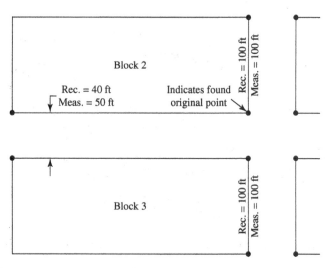

Figure 12.4 Streets are not of record width when original monuments indicate otherwise.

enclosure and parts of the buildings of the defendant. The opinion of the court was as follows:

> A. T. Risley, a former county surveyor, testified that when surveying he found a number of stakes on the line of the alley as now located, and recognized some of the original corners in Block 2; that the alley in block 2 was on a line with the alley in block 1, and that the line is substantially the same now that it was then.... That the last time he [Niedermeyer] re-built the fence, nine or ten years ago, he found one of the original pegs in the ground. All the witnesses testified that the fence had been on the same line ever since it was first put up; that no change had been made in the location of the alley since it was first fenced on both sides.... This evidence is uncontradicted.... As has been repeatedly held by this court, the true boundary lines of a city lot are where they are actually marked by the monuments placed by the surveyor to indicate where they are to be found, and the most satisfactory evidence of the place where the lines were located is afforded by the original stakes.[18]

Owners of the land south of Cheltenham Avenue in St. Louis could not claim 20 feet of the surplus in the subdivision merely because they were entitled to a prorated share of the entire subdivision.[19] Stones placed in Cheltenham Avenue by the original surveyor fixed the position of the street; all surplus belongs to the land north of the street.

Street improvements not built in accordance with the original stakes set by the original surveyor cannot change the street alignment; the original control is conclusive and cannot be altered. This does not preclude a city or other governing agency from claiming the streets by prescription; it merely means that a street claimed by prescription, if properly described by written title, must be described in accordance with the original lines as marked and surveyed by the original surveyor if such lines can be proved. The city must meet the same burden as that met by a private citizen when proving title by prescription.

12.22 ESTABLISHMENT OF STREETS BY IMPROVEMENTS

Principle 17. *In the absence of natural monuments or evidence of lines actually run by the original surveyor, improvements, such as curbs and paving, which were installed simultaneous to the original platting in accordance with the original survey monuments, are presumed controlling.*

The duty of the surveyor is to relocate lines as they were originally run by the first surveyor; curbs located properly when the original stakes were available may be the best evidence as to where the original lines were run and, as such, are controlling. The court observed:

> In determining the line of a street, measurements upon such street are of more value than measurements taken elsewhere; and if they or the places where they were cannot be located, the boundaries of the street as actually opened and used should be ascertained; and if such location has been generally acquiesced in by the public, by lot owners and by the municipality, in the *absence of more certain evidence*, it will beconclusive.[20]

When the reason for the principle ceases—that is, the improvements were not built in accordance with the original stakes—so does the principle. An engineer's failure to place the construction stakes in accordance with the original stakes does not allow the street to be moved in position to cover a construction blunder. When a street is long used by the public and the location of the improvements is not in accordance with the original monuments, the public may acquire a right, by prescription, to that portion improved. This *does not* alter the original street lines; they are unchangeable. Where by reasonable analysis it can be shown that the original street lines and the present improvements are not in agreement, the original block corners may not be in the side lines of the streets as obtained by prescription.

Home Avenue was constructed and staked by locating the street from Swan's Addition, which was a re-subdivision of a portion of Wadsworth's Olive Grove. When attempting to locate Home Avenue as shown on Wadsworth's Olive Grove, the surveyor measured the record distance as shown on said map and set the side line markers for Swan's Addition. He failed to note that the measurements extended to the centerline of the street and thus located the street incorrectly by one-half its width. The Superior Court ruled that the starting of a survey from the side line of Home Avenue as established by Swan's map was improper for the establishment of a lot in Wadsworth's Olive Grove. Home Avenue had been built and improved for more than 20 years in its incorrect position; yet the improvements alone, unless built in accordance with the original stakes, are insufficient to control original lines. If there had been a lack of evidence to show where Home Avenue should have been with respect to Wadsworth's Olive Grove, the court might have ruled differently.

12.23 ESTABLISHMENT OF STREETS BY THE LINE OF A NEARBY STREET

Principle 18.   *In the absence of artificial monuments and evidence of the lines as marked and surveyed, where a street is plotted as being the continuation of a nearby street or beginning at a nearby street, the line of the nearby street is presumed to control.*

The call for the line of a street is a call for a record monument and, as such, is presumed to control over distance and angle. Figure 12.5 shows part of the Frary Heights subdivision, where Palm Street is shown as a continuation of the street of the adjoining subdivision. In the absence of finding original monuments within the subdivision on Frary Heights, the continuation of the line of Palm Street as found in the adjoining subdivision is proper. This principle is defeated where found original monuments indicate that the street line is not a continuation of the adjoining street.

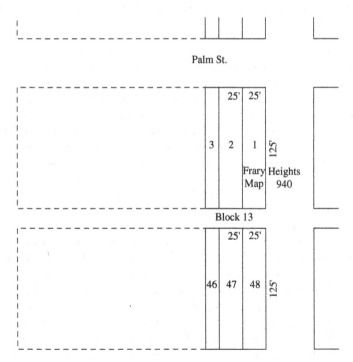

Figure 12.5 Where a street is plotted as the continuation of a nearby street, the line of the nearby street controls.

12.24 ESTABLISHMENT OF STREETS BY PLAT

Principle 19.   *In the absence of evidence covered by the foregoing Principles, the exact width of the street as given on the plat and the distances and angles are presumed to govern street location. Occasionally, a measurement index is applied.*

In the absence of monuments, streets are given the width called for on the plat, regardless of any excess or deficiency that may exist within a subdivision. In Figure 12.6, two original 2-inch by 2-inch hubs found two blocks apart measured 2 feet in excess of the record distance. The proper procedure is to give Blocks 1 and 2 exactly half of the surplus (1 foot each) and give the street exactly 50 feet, as called for by the map. The rights of the public to a street are thus protected by the courts so that deficiency or excess cannot exist within a street except where the original monuments set by the original surveyor indicate otherwise. In a few jurisdictions, a measurement index is applied.

Measurement Index

Not all surveyors and courts are in agreement with the theory that excess or deficiency is not to be prorated in the width of the street. If the original surveyor, establishing monuments within a subdivision, used a long or short tape, the streets as well as the blocks would be equally long or short. The uniform excess or deficiency is referred to as the *measurement index*. In Long Island City, New York, the chain used to tie a point was 100.08 feet instead of 100.00; the custom is to prorate 0.04 foot in a 50-foot street. In San Diego, California, Horton's Subdivision has a fairly uniform surplus of ½ foot to a 300-foot block, yet streets are always staked at exactly 80.00-foot widths, and all surplus is given to the blocks.

The measurement index may also be considered in conducting a GLO retracement. If a surveyor is not able to recover two original or obliterated corners, if in remeasuring between those two corners the surveyor determines a difference between the original measurement he or she obtained and the original measurement, this factor may be considered in looking for additional evidence not possible in surveying lost corners.

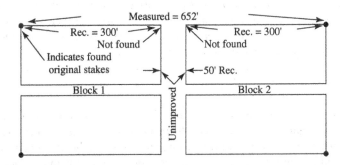

Figure 12.6 In the absence of monuments, streets are given the width called for in the plat.

A measurement index can be applied where there is a uniform overage or shortage in many blocks within the same subdivision, but it cannot be applied where there are erratic measurements that do not indicate a uniform error. In a subdivision where the blocks are supposed to be 300 feet each and the actual measurements indicate 300.25, 300.35, and 300.30 feet, it might be assumed that the original tape was in error. If the measurements were 299.75, 300.25, and 300.00 feet, no measurement index exists.

Only one court case was found in which excess and deficiency were distributed within a public way, wherein a 43.8-foot shortage was divided between two lots and an alley.[21] The weight of reason is against such a procedure. Before the establishment of any subdivision, street widths that are acceptable to the public are determined by the governing agency, and the subdivision is accepted by the public agency on the condition that the streets are of a certain definite width. The size of lots is determined by the whims of the subdivider, the desires of the purchasers, and the minimum area requirements of the planning agency. The tendency of all subdividers is to make the streets of minimum width so as to have a maximum amount of land for lots. The streets are definite and fixed, whereas the lots are variable. This area is particularly difficult, in that streets may be platted but not located on the ground. In doing their research on a retracement, surveyors will often overlook the streets that were not built.

12.25 ESTABLISHMENT OF STREETS WHERE WIDTH IS NOT GIVEN

Principle 20.   If the street width is not given on the plat, the width as scaled on the map will govern as a last resort.

The complete absence of any other acceptable evidence on the ground makes it necessary to use scaling as the only means available to determine street width. Scaling should be considered as the very last resort.

12.26 ESTABLISHMENT OF STREETS BY CITY ENGINEERS' MONUMENTS

Principle 21.   Offset monuments set by the city engineer or other officials to per-petuate the position of the original monuments of the original surveyor control street lines. Offset monuments not based on adjacent original monuments are afforded con-trol only in proportion to the accuracy with which they were set in accordance with the foregoing rules. City engineers' monuments long acquiesced to are presumed to be correct; the contrary must be proved.

Simply because a city engineer or the engineer's assistants set a monument does not prove the monument to be correct. Where a monument is set by measurements of angle and distance from other known original monuments, the monument set cannot be considered as controlling except where it is in agreement with the data given on the map. This is well illustrated in the case in which the original town of Sacramento was surveyed by Sutter in 1848 or 1849. In 1878, the city engineer established the street lines by starting at points many blocks apart and prorating in the intervening streets. Evidence showed that the Sutter survey and the stones set in the 1878 city survey were not in agreement. A surveyor in 1911 established a lot with respect to the 1878 stones and overlapped a fence. The court observed:

> In this action in ejectment and for damages for the unlawful detention of land, which involved the location of the boundary line between two lots in a city of Sacramento, upon the dividing line of which a fence has existed for probably 40 years, or more, it is held that in view of the meager character of the evidence of the real boundary line as fixed by the original survey of the city [Sutter survey] and in accordance with which the deeds of the parties were made, and of the disputed strip in the enclosure of the defendant, the court was justified in finding that the plaintiffs had failed to establish any title to the property in controversy.[22]

Here, the court accepted a fence as better evidence than uncertain monuments. Note that, although the stones set in the later survey were not acceptable for the location of lot lines within a block of the subdivision, they might be acceptable for determination of the street lines. Streets can be acquired by usage of the public even though not based on an original survey.

In a case in which the city had passed an ordinance making all surveys not conforming to the new points established by the city engineer null and void, the court observed:

> An ordinance of a city providing that a certain monument shall from that date of the ordinance be the initial point of the town survey and of all locations of lots and streets, and that all surveys made thereafter, that shall deviate from such initial point shall be null and void, is void, and not admissible in evidence to show enlarged rights against the city. The city council cannot change the location of streets by such a resolution, or affect the rights of land owners under grants previously made, nor can it lay down rules of evidence by which the courts are to determine the location of the points or lines of the survey.[23]

In this case, a landowner was attempting to move his fence some 16 feet to include a portion of the street that was long recognized by improvements. The shift in the lot position was brought about by using the point of beginning, as mentioned previously.

> Where a deed from a city bounded the land granted by a street which had been previously located and surveyed by a city surveyor, and the grantee had for many years held possession of and fenced the lot nearly according to such location and survey, and the street was in use by the public accordingly, the fact that a new survey of the street is afterwards made, changing its line so as to exclude therefrom a strip of land adjoining the lot granted, cannot entitle the grantee to remove his fence to the line of the street as fixed by the new survey, so as to include such strip in his lot, but he is restricted to the lot as bounded by the line of the street as surveyed *originally*.

In determining the boundaries of city lots the line as originally located on the surface must govern; and a line shown by monuments as platted by the city authorities and acquiesced in for many years must control course and distance, and cannot be overturned by measurements alone.[24]

In summarizing the three court cases cited here, it can be said that city engineers' monuments are not controlling where they were not set correctly with reference to the original stakes. When, however, city engineers' monuments have long been acquiesced in and used by surveyors and the public, they will be *presumed to be correct* and may be prima facie evidence except when the contrary can be shown.

ESTABLISHMENT OF LOTS WITHIN SUBDIVISIONS

12.27 EFFECT OF MATHEMATICAL ERROR

Principle 22. *Excess or deficiency existing in a straight line between fixed monuments within a subdivision is distributed among all the lots along the line in proportion to their record measurements.*

In Block 9 of Mission Beach (Figure 12.7), the length of the north line of lot D is indicated as being 66.58 feet, but the scaled distance is 96 feet, more or less. Mathematically, the lot will not close by 30 feet, more or less. If each lot were to receive a portion of the surplus, the lot corner common to D and C would shift some 24 feet west of that pictured, and the intent that lot lines are about 90° to the north line and parallel with each other is defeated. The proper method is to give lots *A*, *B*, and

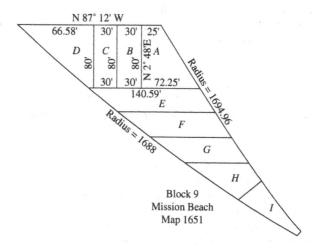

Figure 12.7 Excess or deficiency within a block caused by mathematical error is given to the lot in which the error occurs.

C their record distance, which will automatically correct the mathematical error of lot *D* by establishing the length of the north line of lot *D* at about 96 feet. This refers to the principle that "an error should be placed where the error occurs, if it can be so ascertained."

12.28 EXCESS OR DEFICIENCY

Excess or deficiency occurs whenever a recent precise measurement between original monument positions differs from the accurate measurement originally reported between the same monument positions. Excess or deficiency often implies that the original surveyor or scrivener of the conveyance made an error or blunder, or relied on erroneous measurements. Whatever the cause, the excess or deficiency belongs to someone or several people; the problem is, who is entitled to it? As an error may have been made, the courts are reluctant to accept the principle that an error should be divided among several people. As a last resort, excess or deficiency is prorated. For example, Mr. Stringfellow dies and leaves all his land to his two sons—the west 1320 feet to the older and the east 1320 feet to the younger. After the land is measured, it is discovered that there are 2710 feet instead of 2640 feet as assumed. A whole was inherited by the two; each received equal quantities of land; each is entitled to half of the excess.

Mr. Burne dies and leaves the west 1320 feet to his older son and the remainder to his younger son. In this case, according to the written record, no matter how much excess or deficiency exists compared to the measurements reported originally, the younger receives a remainder and the older receives 1320 feet.

Mr. Brown sells the west half of his lot and retains the east half. Two equal rights are created simultaneously and each receives half regardless of whether the area or measurements differ from the record. Each party has proportional rights.

Within a block, 10 equally sized lots are protracted (not monumented) and the surveyor indicates that he set only the block corners. Each lot is entitled to a proportionate share of the entire block and excess or deficiency is divided among the lots.

Accretion forms in front of several properties and belongs to all the properties. The excess is distributed among the properties by some proportionate means. There is no hard-and-fast rule as to what is correct.

12.29 PRORATION: A RULE OF LAST RESORT

Before prorating an excess or a deficiency, an investigation of the possible causes of the excess or deficiency should always be made. In a block shown to be 240 feet wide, a deficiency of 15 feet was found.[25] The lower court found that the eastern lots should get full measure because of acquiescence, but this was rejected by the upper court. In ordering a new trial, the Supreme Court observed that the shortage probably did belong to the western lots (for a different reason) and suggested that (1) the eastern lots in question, although included in the subdivision, were sold by metes and bounds prior to the filing of the subdivision and thus had senior rights, and (2) the street west of the western lots was on the centerline of a section, and, at the time the city improved the street, it probably included part of the western lots. In other words, the upper court ordered a new trial for the purpose of discovering where the shortage occurred. Proration was not to be applied if the cause of the shortage could be discovered.

12.30 EXCESS OR DEFICIENCY CONFINED TO A BLOCK

Principle 23.   *Excess or deficiency occurring within a block should not be prorated among other blocks.*

Excess or deficiency in the land platted into lots, blocks, and streets along with an absence of original markers does not always indicate that intervening streets should be located by proration. If possible, each block should be treated as distinct and the shortage or surplusage therein apportioned among the lot owners.[26] Proration is a principle of last resort, and once a street is established by one of the means previously discussed, the street lines are unchangeable. Just because Block 1 in Figure 12.8 is 1 foot short and Block 2 is 1 foot long does not mean that the location of B Street can be moved 1 foot east in order to give Blocks 1 and 2 each exactly the record measure. All the shortages existing in Block 1 are prorated among the lots of Block 1, irrespective of the surplus in an adjoining block. Only if B Street were not established by improvements, possession, or found points could Blocks 1 and 2 each receive an equal frontage between streets. Under this principle, the surveyor is prohibited from going beyond the limits of the specific block. This is the basic principle of retracements.

12.31 EXCESS OR DEFICIENCY DISTRIBUTION WITHIN BLOCKS

When an excess or deficiency is found in measuring a line, it must be assumed that the error of measurement occurred not in any one part of the line, but in all of the

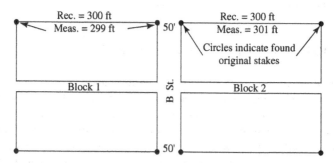

Figure 12.8 The shortage in Block 1 is prorated among the lots of Block 1.

line. The original surveyor probably had a chain that was either too long or too short, and the error would occur in all parts of the line. Where a subdivision shows 10 lots of equal frontage, it is logical to assume that any small excess or deficiency should be distributed among all lots equally and not given to the end lot.

Excess or deficiency distribution is applicable where several parcels were created simultaneously by proceedings in partition, by protraction, or by subdivision. Proration of surplusage and shortage cannot apply to metes and bounds descriptions where a senior grant exists, to blocks with undimensioned end lots, to errors the location of which can be established, to lines where monuments fix the location of the error, or to lines where points are fixed by improvements made in accordance with the original stakes.

> Where land is plotted into lots, blocks and streets, and surplus exists, the surplus is to be divided among the lots.[27]

> Where in platting a village it turns out that by mistake the blocks are not so long as the plat represents, the deficiency must be apportioned between the lots of the block according to their apparent size as shown by the map.[28]

There are many more examples in each state or jurisdiction. To fully understand each local application, the surveyor must research each specific area where retracements are to be made.

12.32 SINGLE PROPORTIONATE MEASURE

Principle 24. *Excess or deficiency existing in a straight line between fixed monuments within a subdivision is distributed among all the lots along the line in proportion to their record measurements.*

In block 1 of Hambleton, Weston and Davis's subdivision in Chicago, Illinois, a shortage of 2.68 feet occurred. By record measurement from one end of the block, a brick wall of a building was on its own lot (the north half of Lot 5 and the south half of Lot 6), but if the 2.68 feet was divided by proportion among the lots, the building encroached on the adjoiner. Assuming that proration applied, the adjoiner used the brick wall to support several girders of his new building. In approving the action, the court observed:

> In the absence of any agreement or question of title by adverse possession, where a block has been platted into lots and the lots sold, a shortage in the block will be prorated among the several lots. Complainant, therefore, when it received its deed, the property being described as the north half of lot 5 and the south half of lot 6, and not by metes and bounds, did not get paper title to 509 frontage, but only received paper title to the north half of lot 5 and the south half of lot 6 after the same had been prorated by reason of the deficiency.[29]

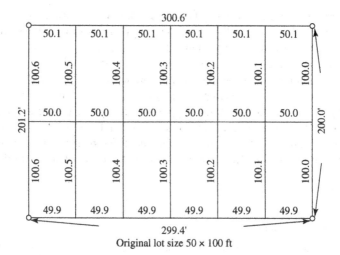

Figure 12.9 Errors of measurement are divided proportionately between lots.

This principle should apply to the distribution of small errors and small differences between present measurements and original measurements. When gross errors occur, a different rule often, but not always, applies. When the block is abutting on a subdivision line, extreme caution should be exercised in locating lots by proration since the principle frequently does not apply.

Figure 12.9 shows the application of proration to a block where the block corners are fixed by monuments. Angles at the block corners must be measured to show the angles for any one lot, but the angles have no influence on the distribution of excess or deficiency, for they are based entirely on linear measurements.

12.33 SINGLE PROPORTIONATE MEASURE ON CURVES

Principle 25. *Excess or deficiency existing between fixed monuments on a curve is distributed among the lots along the curve in proportion to their record linear measure.*

In Figure 12.10, the length of the curve along Le Roy Street was found by actual measurement to be 8.40 feet short of the original measurements. The beginning and end of the curve were fixed by street improvements and by monuments set by the city engineer; no trace of the original markers could be found. Because the beginning and end of the curve were fixed by presumption and could not be altered, Lots 4, 5, and 6 must take all the deficiency. The fixed monuments preclude Lot 7 from being short in measurement.

Proration along a curve is controlled by the same rule that governs single proportionate measure; that is, each segment of the curve has the same proportionate

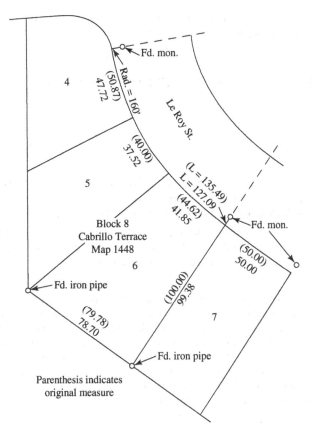

Figure 12.10 Proportionate distribution of errors along a curve.

length of the total curve as is shown on the original map. The new length of Lot 6 is calculated as follows:

$$127.09 \times 44.62 \div 135.49 \times = 41.85 \text{ feet}$$

The lot frontage then becomes 2.77 feet short of the record measurement. Frequently, the city engineer or surveyor establishes a curve that is not based on the original curve points because the curve data given on the map are out of agreement with the stakes set on the ground or because the stakes as set are not a true curve. Any survey locating lots from a curve alignment not in agreement with the original stakes has made use of a poor foundation for the survey. Surveyors should approach this area with caution.

Corners placed on curves are difficult, if not impossible, to measure their position if they become destroyed. The elements of curves can be described by numbers, but one cannot survey along curved surfaces. The elements of curves can be surveyed—the radius, central angle, chord lengths—but there is no known method of measuring a curved line that identifies the curve itself.

12.34 DISTRIBUTION OF EXCESS AND DEFICIENCY BEYOND A MONUMENT

Principle 26. *Excess or deficiency cannot be distributed beyond any undisturbed original monument.*

Undisturbed original monuments are fixed in position and cannot be overridden by proration even when the monuments are found in the middle of the block. Using the same block as that shown in Figure 12.9, the distribution of excess and deficiency is modified by the found monuments shown in Figure 12.11. Judge Marshall, in 1897, commenting on the establishment of lot corners from known corners found within a block as compared to remote block corners, observed:

> The *unvarying rule* to be followed in such case is to start at the nearest known point on one side of the lost corner, on the line on which it was originally established; to then measure to the nearest known corner on the other side of the same line; then if the length of the line is in excess of that called for by the original survey, to divide it between the tracts connecting such two known points, in proportion to the lengths of the boundaries of such tracts on such line, as given in the survey .... The method always followed in reestablishing corners is to measure the line connecting the nearest known corners, on the same line, on either side of the lost corner, and then divide the excess, if any be found, as before stated.[30] (Emphasis added.)

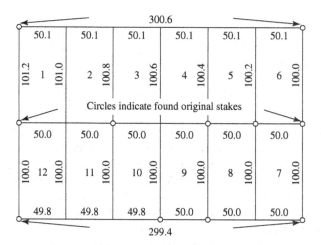

Figure 12.11 Errors of measurement are distributed proportionately between found original monuments.

Although proportioning cannot extend beyond an original monument, the original monument can be used to determine the direction of the line it controls. This will permit an alternate means of locating lost corners.

12.35 ESTABLISHMENT OF LOTS WHERE THE END LOT MEASUREMENT IS NOT GIVEN

Principle 27.   When the end lot measurement is not given, all the excess or deficiency is presumed to be given to the end lot.

When the original subdivider failed to give a dimension to the last lot in the block, it can be assumed that the subdivider intended to place all excess or deficiency in the last lot. The fact that the last lot scales differently from the amount of land on the ground does not alter the principle. This situation frequently occurs along the boundary of a subdivision where the subdivider was uncertain as to his true boundary line. Figure 12.12 shows the condition in Block 13 with the general note that all regular lots are 100 by 100 feet. Because Lots 3 and 4 are true remnants of lots, all excess or deficiency is given to Lots 3 and 4, regardless of the scaled distance. Lot 3, which scales 100 feet, will receive 150 feet. In New Jersey, a map showed 50 lots: 48 were regular with a width of 25 feet each, 2 were irregular and undimensioned; the court ruled that the regular lots would receive 25 feet each and the undimensioned lots would receive what was left over even though they did not get the amount of land that the map appeared to give them.

12.36 REMNANT PRINCIPLE

Principle 28.   Where excess or deficiency is given to an irregularly shaped lot at the end of a block, the method is called the remnant rule. Few jurisdictions accept it.

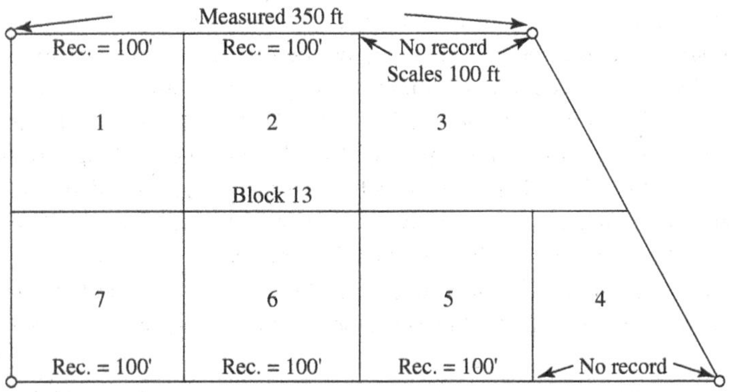

Figure 12.12 Remnant lots without dimensions receive the remainder.

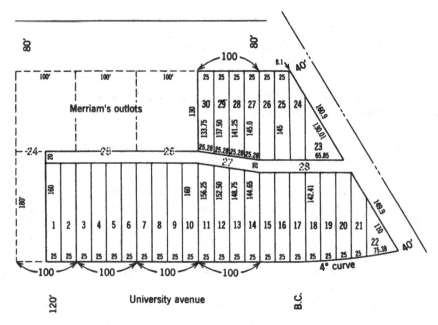

Figure 12.13 Hughes's addition to the City of St. Paul. Source: Barrett v. Perkins, 113 Minn. 480 (1911)/The President and Fellows of Harvard University/CC BY 4.0.

The remnant rule as a means of disposing of excess or deficiency has not been ignored in court discussions; the unsoundness of the principle has led to its rejection in most states.[31] The case most widely discussed is examined next in detail.

Figure 12.13 shows the essential parts of Hughes's addition to the city of St. Paul. An argument arose over the apportionment of approximately 25 feet of deficiency along University Avenue. In the lower court, the defendant alleged, among other matters, that at the time of the platting of the land, it was at first intended to embrace in one lot the land included in Lots 21 and 22 in the plat; that said plat was first made with that intention, and the distance along the southwesterly line of that portion of the plat was noted thereon as being 75.38 feet, but before the plat was completed, it was decided to divide that irregular portion into two lots, numbering the same 21 and 22, respectively, making the width of lot 21, 25 feet. That width of 25 feet was noted on the plat, but the parties making the plat neglected to change the length of the southwesterly line of Lot 22 by deduction therefrom of the 25 feet included in Lot 21, and allowed the length of the southwesterly line of Lot 22 to appear as 75.38 feet instead of 50.38 feet. This was not the finding of the lower court, and 75.38 feet was awarded to Lot 22 even though Lot 22 scales 50 feet and there is a 25-foot shortage in the block. In reversing the lower court, the state supreme court's lengthy opinion was as follows:

> The action was brought to determine the boundary between Lots 18 and 19, but in fact involves the boundary between all the lots in this part of the plat. All the lots, except 22,

have, according to the plat, a frontage of twenty-five feet on University Avenue, and Lot 22 a frontage of 75.38 feet. The distance between the outer boundary of Lot 1 and Fairview Avenue, the end of the plat, is insufficient to supply the number of lots given on the plat with the dimensions stated. So that beyond question there was a mistake in the preparation of the plat, or in the original survey, and the purpose of the action is to locate and correct it. The whole controversy in the court below narrowed down to the question whether Lot 22 was or was not erroneously designated on the plat as a seventy-five foot lot. The trial court in effect found that this lot was entitled to a frontage of 75.38 feet; and, if that conclusion be sustained by the evidence and the rules of law controlling the question, the case is at an end.

We are of the opinion that the learned trial court erred in the conclusion stated. The evidence is insufficient to sustain it. The court relied, in reaching its conclusion, chiefly upon the testimony of witness Armstrong, county surveyor of Ramsey County. This witness, though apparently fair and candid, seems to have fallen into the initial error of adopting as correct the notation on the plat that Lot 22 had a frontage of 75.38 feet on University Avenue.

But conceding for present purpose the correctness of the conclusion reached by the witness, his evidence served only to demonstrate the fact, concurred in by all the parties, that there was a mistake somewhere in the plat. There is not enough land within the platted tract to supply all the lots of the dimensions given on the plat, and this situation is not controverted. Giving to Lot 22 a frontage of 75.38 feet results in moving each of the other lots twenty-five feet West, and in the end to completely extinguish or eliminate Lot 1. In other words, there is a deficiency of land, and all the lots cannot be accounted for. In such a case the most the court is authorized to do, in the form of correcting the apparent mistake, is to apportion the deficiency among the several lots, and not eliminate one of them entirely, as the trial court in effect did in the case at bar.

The owner of Lot 1 has as much, and it would seem a greater, right to have his property remain a part of the plat, as the owner of Lot 22; the greater right, because Lot 1 was first laid out by the owner of the plat, and beyond controversy, with the intention that it should be and remain a lot of the subdivision of the dimension indicated. The rule requiring an apportionment of either an excess or deficiency of land in such cases is well settled . . . . The trial court was therefore in error, even from the standpoint of the testimony of the witness Armstrong, in reaching the conclusion, in substance and effect that Lot 22 was entitled to remain with a frontage of 75.38 feet.

However, we are not to be understood as holding that the rule stated should have been applied to the facts in the case at bar. The rule applies more particularly to tracts of land subdivided into smaller tracts of specified and uniform dimensions, and not be a situation like that here presented. Here the owner of a definite tract of land intended, and his intention is manifest, to lay out as many lots of the uniform width of twenty-five feet as the tract would contain. This he proceeded to do, laying off twenty-one twenty-five foot lots, after which there remained an irregular or triangular piece to be disposed of as a remnant. This he supposed was 75.38 feet at its base, and so noted it upon his plat. This, we are satisfied from the record was a clear mistake. But it is not necessary to extend the opinion by a discussion of the matter, it is not important.

There is a deficiency of land to make up the number of lots with indicated frontage, and the rule in such a case is that the deficiency must fall upon the last or irregular tract; the remnant of the whole after laying out the lots of a uniform size. Baldwin v. Shannon, 43 N.J.L. 596. In a situation like this the owner of the plat must be deemed to have intended to constitute the irregular remnant of a lot by itself, regardless of its dimensions, and a purchaser thereof takes the whole remnant, whether of greater or less area than that indicated by the plat. It cannot be enlarged at the expense of the owners of other lots, nor, if of greater area than shown by plat, diminished in their favor. Though the rule might in a given case work a hardship to the owner of the remnant lot, yet in the case at bar the combined injury to the owners of the other lots, arising, if defendant be sustained, from the displacement of their improvements and the total elimination of Lot 1, far outweighs any disclosed damage or hardship to defendant. The rule furnishes a definite and safe method and guide for the determination of mistakes of this nature, and we adopt and apply it to the fact in the case at bar; and if on a new trial the fact remains the same, the relief stated should be granted.

The decision of the state supreme court was correct, but the reason was wrong. In most states, this situation would have been judged by the rule of a blunder: "Where a mistake occurs, leave it in that place." In this case, the shortage could be explained by scaling on the map. The following situation illustrates how a surveyor may be trapped by blindly following the remnant rule.

Shown in Figure 12.14 are the conditions found to exist along the former city boundary of San Diego. Blocks east of the boundary are all 4.8-foot long, more or less; blocks west of the boundary are all 4.8-foot short, more or less. Evidence shows that at the time Boundary Street was improved, the street improvements were located 4.8 feet too far to the west. The shift in position seems to have come about because when the line was run between monuments several miles apart, it was found that the line was not straight. The 4.8-foot excess and the 4.8-foot deficiency belong to the remnant lots in Blocks 1, 2, 5, and 6, as this is where the error does in fact occur. In outward appearance, the remnant rule is correct. But in Block 12, where all lots were fully dimensioned, the error should be in Lots 10 and 11 (the city moved the line) and not the remnants.

In restoring lost corners in sectionalized lands, the Minnesota court rejected the remnant rule and adopted the apportionment principle. Figure 12.15 shows the conditions in *Goroski* v. *Tawner*.[32] In three adjoining lines, a deficiency of about 15 chains existed in the last half mile. These sections are on the northern tier of a township and according to original survey instructions they are remnants; that is, errors of measurement were placed in the last half-mile. Regular forties were created except for the top tier.

The east-quarter corner of Section 1 was lost. In restoring the lost corner, proportional measurement was used, and the concept of placing the deficiency in the irregular closing lot was rejected. Thus, within the same state inconsistencies in principles can be found. It can be argued that in this case, the state was following federal rules because it was sectionalized lands. A state does not have to accept federal apportionment rules for lands wholly under its jurisdiction.[33] It may choose to accept or reject any federal rule for restoring lost positions that does not involve federal ownership. In Michigan, the rule of apportionment of deficiency between lots of a block must stand irrespective of the end lot being fractional except where

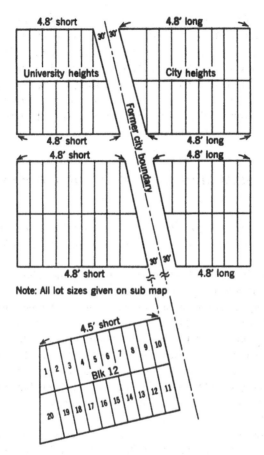

Figure 12.14 Conditions along the former San Diego city boundary.

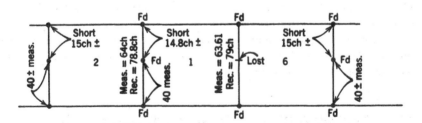

Figure 12.15 *Goroski* v. *Tawner*, 121 Minn. 189 (1913)/U.S.Supreme court.

the contrary is proved.[34] The majority opinion on the remnant rule is well expressed by Judge Cassoday.

Had the plat given the specific dimensions of each lot of the several lots fronting on Jefferson Street except Lot 1, and given no dimensions of that, then such absence of the dimensions of that lot would have evinced an intention that it should include whatever should be left after setting off the several lots of which the specific dimensions had thus been given, whether the same should be more or less, but where, as here, the specific dimensions of each and all of the several lots fronting on Jefferson Street are given upon the plat, and there is no lot in the block of which the specific dimensions are not thus given; there seems to be no substantial reason why such excess should be given wholly to one lot merely because its dimensions, as given upon the plat, differ from those of other lots.[35]

12.37 ESTABLISHMENT OF LOTS WHERE NO LOT MEASUREMENT IS GIVEN

Principle 29. *In the absence of physical evidence on the ground and in the absence of measurements given on the plat, as a last resort scaling of the plat may be used.*

Some early subdivisions failed to mention how big the subdivided lots were. On the map on which 10 lots, each scaling 50 feet, were shown, only 450 feet was found to exist on the ground. As several houses had been built by measuring 50 feet from each end of the block, a tangled legal snarl ensued. One of the unoccupied lots was finally purchased, and the block was re-subdivided to adjust the houses within the land occupied. Scaling in this case indicated that there were 10 lots of equal size, about 50 feet each. Facts on the ground showed the existence of only 450 feet, or 45 feet per lot. Scaling could not be relied on to establish any fact other than that the lots were of equal size; the physical evidence on the ground was better evidence of the intent of the lot sizes.

Paper shrinks or expands with changing moisture conditions. A map reproduced by the dry process or the wet process will give a print different in scale from the original. A map drawn on a scale of 1 inch equals 100 feet cannot be scaled sufficiently accurately to determine whether a distance intended to be 50 feet is either 48 or 52 feet. Because of these facts, scaling must be considered poor evidence and is to be relied on only as a last resort.[36]

The size of the drafting medium (paper) as well as its composition can also affect shrinkage and expansion, depending on conditions. Extreme changes in moisture and temperature may produce dramatic changes.

12.38 ESTABLISHMENT OF LOTS WITH AREA ONLY GIVEN

Where a subdivision shows area only, the proration of any excess or deficiency presents almost insurmountable difficulties. Area is a computation depending on two linear measurements and directions, any of which can be altered. Sometimes area errors can be localized by the existing streets or lines of possession. Location of property lines from area data is considered a last resort to be applied after all other means have failed. "Where in an instrument of conveyance of real property the quantity of land to be conveyed is stated, the statement is not to be viewed as a factor in establishing boundaries unless the more particular description given is indefinite and uncertain."[37]

12.39 NEW YORK RULE FOR ESTABLISHMENT OF LOTS

In New York and some other states, the custom has been to describe land by both lot and block in a subdivision and by a metes and bounds description. If excess or deficiency is found in the block, there is ambiguity, as two locations can exist. One is based on proportional measurements and the other on record measurements from one end of the block. By the New York rule, if both descriptions are used, the first buyer gets exactly the dimensions called for (by metes and bounds and by the map) and later buyers can receive only a remainder (see Figure 12.16).

The facts in this case are undisputed. Both plaintiffs and defendants derive their title from a common source. On June 15, 1886, Ann C. Morton was the owner of certain property shown on a map, entitled "Building Lots for Sale at Middle Village." This property was situated in block VI, as shown on the map, and included lots 1, 2, 3, 4,

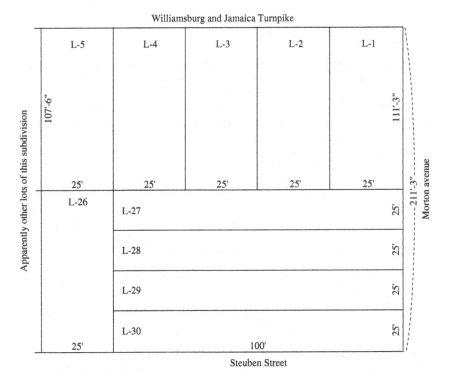

Figure 12.16 *Mechler* v. *Dehn*, 196 N.Y.S. 460 (1922)/U.S.Supreme court.

and 5, and also lots 26, 27, 28, 29, and 30 in that block. On that day she conveyed to one Jacob H. Gebhard said lots 1, 2, 3, 4, and 5, describing them by lot numbers according to the map, and also by metes and bounds, commencing "at a point on block VI formed by the westerly side of Morton avenue where it intersects with the southerly side of the Williamsburg and Jamaica turnpike road." The description then continues along the turnpike road 125 feet, then southerly 107 feet 6 inches, then easterly 125 feet, to Morton Avenue, and "thence northerly, along the westerly side of MortonAvenue, 111 feet, 3 inches to the point or place of beginning."

On February 14, 1898, Ann C. Morton conveyed to Joseph Burmel lots 26, 27, 28, 29, and 30, on the corner of Morton Avenue and Steuben Street, in block VI on the map in question, reciting that said lots contained 12,500 square feet, more or less, and that said lot is shown on a diagram contained in the deed, "and agreeable to said map," but no such diagram appears. This deed contains no description by metes and bounds.

It appears from actual survey, and is undisputed, that the distance between the Williamsburg and Jamaica turnpike (now called Metropolitan Avenue) and Steuben Street, shown on the map as 211 feet 3 inches, is in fact only 206 feet and about 3 inches; in other words, that the actual distance between the turnpike and Steuben Street is 5 feet and 7/8 of an inch short of the distance shown on the map. The defendants now occupy the strip in dispute.

Upon these undisputed facts the trial court found that the description of the plaintiff's boundary on the westerly line of Morton Avenue as 111 feet and 3 inches was erroneous, and was only 106 feet 2 inches; that defendants were the owners of the remaining 100 feet on the westerly side of Morton Avenue; and that they did not unlawfully enter into possession of the plaintiff's premises.

The learned trial court states the general rule to be that where land is conveyed by reference to a plan or map, and there is more or less in the entire tract which has been divided than the map shows, no grantee is entitled to any preference over the others, and the excess should be divided among, or the deficiency borne by, all of the lots in proportion to their area, citing numerous cases from other jurisdictions. But he says that there is an exception to this general rule: that where a map shows a plotting of a considerable tract, and the creation of lots of regular width and depth, if a few of them are of irregular dimensions, they are deemed to be the remnant of what remained of the entire tract after plotting the regular lots, and if there is a shortage in the entire frontage, the irregular sized lots must bear it: that this is held upon assumption that the owner intended to get as many regular sized lots as possible, and that whatever remained of his frontage was to go into the irregular-shaped plot, and hence, if he had more or less than the map showed, the difference in the frontage would affect only the irregular lot, citing *Baldwin* v. *Shannon*, 43, N.J. Law, 596, and *Barrett* v. *Perkins*, 113 Minn. 480, 485, 130 N.W. 67.

Following this exception to the general rule as stated in his opinion, the learned trial court found that the entire deficiency should be taken from the plaintiff's lots, because they constituted the irregular lots in Block VI. There appears to be no decision in this state which precisely covers the situation presented in this case at bar, nor do the ordinary rules which have been adopted in this state seem to have any direct application.

I am unable to agree with the learned trial court in his final conclusion, which adopts the rule of law as to irregular lots laid down by the courts in New Jersey and Minnesota. I do not believe that this rule is sound in principle. A party who maps a plot of land into lots, whether regular or irregular, in size, and gives them certain dimensions on the map, believed by him to be accurate, cannot, in my opinion, be held by any such strained construction to intend the irregular lots as a remnant, regardless of its dimensions. On the contrary, I think he believes and intends that each lot shown on the map, regular or irregular, has, or should have, the dimensions ascribed to it. Especially is this true, in my opinion, where the so-called irregular lots constitute, as in the case at bar, approximately half the block. A literal application of this doctrine might result, in case of a large deficiency, in the loss of all, or nearly all, of the irregular lots, although held under a conveyance prior to that of the regular lots.

In the case at bar the original grantor clearly intended to convey in the deeds to plaintiffs' and defendants' predecessors in title 211 feet 3 inches of land fronting on Morton Avenue, 111 feet 3 inches to plaintiffs' and 100 feet to defendants' grantor. Her grantees intended likewise, in accepting such conveyance, to secure that amount of land. Had there been no map, concededly, as stated in the opinion of the trial court, plaintiffs' predecessor in title, having the first conveyance, would obtain title of the entire 111 feet 3 inches. How, by any legal principle, can the mere existence of the map and reference thereto, in the descriptions of the deeds, affect plaintiffs legal right, except by arbitrarily correcting the mistake and apportioning the deficiency, without any sound legal reason for such apportionment.

I do not think we are at liberty to presume from the mere reference to the map in the descriptions that the grantor intended to convey those lots in accordance with the frontage shown on the map, if it was there, or, if not, then in ratable proportion. I am unwilling to carry legal presumptions to any such extent. In my opinion, the mere statement of such a proposition shows its unsoundness.

The use of the words "more or less" cannot operate to modify or alter the grantor's intention, shown by the map and the descriptions. It certainly could have no such effect upon the first conveyance to plaintiff's grantor, because, measuring from the point of beginning, the lots as laid out on the map contain the frontage called for by the deed, and that amount of frontage existed and had not then been conveyed. When the subsequent conveyance of the remaining frontage on Morton Avenue was made, the grantor intended, of course, to convey 100 feet, believing that that frontage remained. But it is equally certain that she could not, and did not then intend to, convey more than she had. Therefore she conveyed to the defendants' grantor only what was left, or about 95 feet. I do not see that the fact that the lots are shown on a map and their description in the deed made by reference to that map in any degree alters this result. To correct the original mistake by an apportionment of the deficiency would be a purely arbitrary determination by the court, which could in no sense be called judicial.

Plaintiffs' grantor had a right to rely on the description in his deed and on the map, and also on the fact that, as this was the first conveyance on that frontage, the entire 111 feet 3 inches called for by his deed was there on the ground. No survey was necessary to determine that fact, nor was he bound to know that there was a deficiency in the block. On the other hand, defendants' grantor knew when he received his conveyance that he could obtain only what was left. He could then do one of two things, either determine for himself the correctness of the map and description by a new survey, or accept the deed and rely on the convenants against his grantor for damages for any deficiency. He did the latter. Why, then, should plaintiffs now be called on to make up any portion of this deficiency? In my opinion, therefore, the judgment should be reversed, and judgment directed for the plaintiffs for the relief asked for in the complaint.[38]

In these results, the court recognized certain principles: (1) that proportionate measurement is a rule of last resort and (2) that the metes and bounds description and the senior deed indicated who should receive a *remainder*. In other jurisdictions, the courts reach as far as possible to discover who should receive the remainder and thus relegate proportionate measurements to a rule of last resort. Although the court could have invoked the remnant rule in this case, it did not (the results would have been reversed).

12.40 SUMMARY OF PRORATION RULES

From these situations and from others, the following general principles have evolved and can be relied on by the retracing surveyor:

Proportionate measurement is used to distribute discovered excess or deficiency as compared to the record where:

1. A whole is divided simultaneously into dimensioned parts by protraction; that is, the parts platted are not monumented on the ground, and no part is designated to receive a remainder. (See remnant rule.)
2. Equal parts are created by a will or partition; that is, each child receives a proportion of a whole.
3. Proportionate parts are created by the conveyance; that is, the west one-half and the east one-half of Lot 12, Map 1313.
4. To restore a lost monument position as nearly as may be to its original position where there are tie-outs from several directions, all of equal value (not applicable in sequence conveyances).

Proportionate measurement cannot be used:

1. To alter senior rights or in sectionalized lands to alter the superiority of one line over another.
2. To distribute a mistake that can be proved to exist in one place, that is, the rule of a blunder.
3. To alter legal rights obtained by possession.
4. To alter an acceptable original monument position.

Proportionate measurement:

1. Begins and ends at the nearest original monument position on each side of the area of excess or deficiency.
2. Is a rule of last resort.
3. Varies in interpretation in different jurisdictions.

12.41   ESTABLISHMENT OF LOTS ADJOINING SUBDIVISION BOUNDARIES

The foregoing discussion has been confined to lots within a subdivision but not adjoining the boundary of a subdivision. Lots abutting a subdivision boundary frequently require special attention where there was an error in the establishment of the original boundary line and at a later date the true boundary was recognized. The question of adverse rights often enters into these cases.

12.42   ESTABLISHMENT OF LOTS ADJOINING A SUBDIVISION CORRECTLY ESTABLISHED

Principle 30.   Where the boundaries of a subdivision were staked correctly originally, the foregoing Principles for the establishment of lots are applied

If there are no serious errors within the subdivision and corner monuments are found where they are indicated, then one can rely on retracing the various lots within the subdivision based on the conclusive evidence found, which, by law, *is without error.*

12.43   ESTABLISHMENT OF LOTS OVERLAPPING THE TRUE SUBDIVISION BOUNDARIES

Principle 31.   Lots abutting on a subdivision boundary line cannot extend beyond the title interest of the original subdivider except for a right of title resulting from lawful possession.

This principle is best explained by the following illustrations. The map of Middletown Addition filed in 1871 shows a picture plan and a general note giving the street widths, block sizes, and lot sizes. On the original map, the boundary line of the subdivisions was shown as in Figure 12.17. When the boundary was run out from known existing monuments on the ground, about 12 feet of the subdivision was not in existence. Lots 8, 9, 17, and 18 cannot extend beyond the true boundary because the subdivider did not own the land and could not extend beyond his ownership, nor can the deficiency be distributed among all the lots in the block, as the original lot lines are unchangeable. The proper procedure in this block is to give each lot in the block 25 feet and place all the excess or deficiency in the lots adjoining the subdivision boundary.

Limitations on the Principle

As noted previously, title to land cannot be transferred by estoppel, adverse rights, agreement, and other unwritten means; without a written instrument, only the lines are affected.

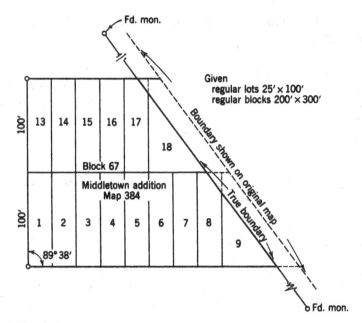

Figure 12.17 Subdividing another's land in error.

12.44   ESTABLISHMENT OF LOTS NOT TOUCHING THE TRUE BOUNDARY OF THE SUBDIVISION

Principle 32.   Where the original subdivider failed to subdivide all the land as shown by found original monuments, the surveyors should not extend the lot lines beyond the limits of the original monuments. Title to the unsubdivided land probably remains in the original subdivider.

When surveying College Park Unit 2, Map 2218, the surveyor failed to subdivide all the land owned by the subdivider by the amount shown in Figure 12.18. The original stakes for Lot 6 were found, and no doubt existed as to the original location of Lot 6. The fact that there was a strip of land varying in width from 10.6 to 12.94 feet, not subdivided by the subdivider, does not give Lot 6 a title interest to the strip (Figure 12.19).

Limitations on the Principle

This rule cannot apply to small, insignificant errors or cases where the original subdivider's monuments cannot be found. When a subdivider indicates by the title that all the land was subdivided and no monuments on the ground indicate otherwise, it must be presumed that all the land was intended to be subdivided.

In Washington, 4½ feet of surplus was found to exist between the street and the subdividers' true boundary line. The court ruled that when the lots, as marked on the plat, occupy the entire space between the north and south boundaries of the tract, the excess must be apportioned among the lots.[39]

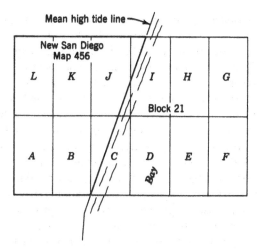

Figure 12.18 Gap between subdivisions.

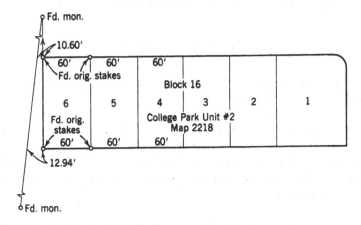

Figure 12.19 Land below the high-tide line cannot be privately owned.

12.45 PRORATION OF EXCESS AND DEFICIENCY IN BLOCKS CLOSING ON SUBDIVISION BOUNDARIES

Principle 33. *A considerable excess or deficiency existing in blocks adjoining a subdivision boundary is prorated only as a last resort.*

Considerable error in the closing block of a subdivision may indicate an error in the original location of the subdivision boundary. If an error existed in the original boundary location of the subdivision, and if at a later date the boundary was moved to fit the true deed location of the subdivider, it is not advisable to move the lots by proration to fit the new boundary. Lots once established are unalterable.

There must be reasonable proof showing that an error probably existed in the original subdivision boundary location. If there are 10 blocks in a subdivision, 9 of which measure very close to the original record and the tenth, existing next to the subdivision boundary, is found to be 30 feet short, the inference is that the original boundary line was established erroneously and later moved to the true location. However, if the subdivision consists of only one block and the block is found to be 30 feet in error, proof is lacking to show on which end of the block the error might have occurred, and in such a case the only equitable solution is to prorate the error. Blocks abutting on subdivision boundaries cause no end of grief to surveyors and landowners. Although a lot within a block that adjoins a subdivision boundary may be insured by a title company without fear of liability provided that it is described by lot number instead of size, the location and size of such a lot on the ground may be in serious doubt. Two possible solutions exist: (1) prorate the error or (2) give the error to the end lot. If the error was brought about by a relocation of an incorrect original boundary line, the error would be applied where it occurred, that is, next to the boundary line. But if the error was not due to the boundary line being moved from its original location and no evidence exists to localize the error in one lot, the error would be prorated. This is a question of proof, the burden of which is placed on the surveyor.

12.46 LOCATING LOTS FROM BOUNDARY LINES

Principle 34. *It is safer to locate lots from the interior of a subdivision than from the boundary line of a subdivision.*

Proven interior lots and street locations within a subdivision are safer starting lines for staking the lots within a subdivision. The frequency of errors in locating boundaries is far higher and is dangerous to use. Any lot or street within a subdivision should be located from the data on the map itself and not from a relocated boundary line of an adjoining subdivision.

Limitations on the Principle

In rare instances, the original lot lines of some of the early subdivisions were not staked on the ground, and in such cases the lots can be relocated only from the boundaries of the subdivision.

12.47 OBLITERATED AND LOST SUBDIVISIONS

Even after understanding all the preceding principles pertaining to boundary location procedures, and even after extensive practice in their application, the surveyor will find numerous perplexing situations created largely by conflicting evidence. Most surveyors can agree on measurements; they are either right or wrong. But when it comes to evaluating witness evidence, old documents, and physical evidence, many theories evolve. The mere fact that there have been thousands of court cases pertaining to the meaning of deed terms is sufficient proof that variations in interpretation occur.

In most cities, there are numerous subdivisions of lots and blocks. If the city is old, chances are that the original subdivision data are not available. No one can say with any degree of certainty what was originally set and usually no one can remember who the original surveyor was. The mere loss of original monument positions cannot and does not void a land title. Land is located based on the best evidence available. In older areas, custom, occupancy, tradition, and related maps are often the only methods available for locating lot lines. Surveyors with years of experience have background knowledge that tells them what is right. They are sought out by attorneys not because of their superior measurement methods but because of their firsthand knowledge of what has been. When retracing lots within ancient subdivisions, a new surveyor should do so with caution. Until the surveyor understands the unwritten history of similar subdivisions, he or she may not possess the qualifications to undertake such a task. The obligation of the surveyor is not to correct any errors made by the original surveyor; the obligation is to locate lot boundaries in conformity with what was actually done, and then to report the findings.

PROCEEDINGS IN PARTITION

12.48 GENERAL COMMENTS

A court decree resulting from proceedings in partition vests title in the various parties at the same time and by the same decree; none can be said to be senior to the other. Unless there is a definite statement limiting a party to a definite quantity of land, all are presumed to have a proportionate share of any surplusage or shortage.

12.49 ESTABLISHMENT OF LINES DETERMINED BY PROCEEDINGS IN PARTITION

In many proceedings in partition, a map or plan is made showing the division between the various allotments, and where the allotments are described by the map, the division lines between the various parcels are determined by the same principles as given previously for subdivisions. The map is treated like any other filed or recorded subdivision map. If the allotments are described in written language and there exists a gap or overlap between the allotments, the gap or overlap is prorated among the allotments.

The procedure for determining boundary lines resulting from proceedings in partition is clearly stated:

> *First:* If the monuments or marks on the bounds for the corners of the several allotments can be found, such marks or monuments must govern, and distance and bearing must be disregarded.

Second: If the monuments or marks on the ground are lost or obliterated, parol evidence may be introduced in connection with the record to show their location.

Third: If no monuments were set, except theoretically or on paper, the proper location of these monuments will be determined by prorating the distances as given in the records, according to the length of frontage of the several allotments.

Fourth: If the actual computed sum of the lengths of the several allotments given exceeds the length of the tract partitioned, it will be construed that the decree means that, upon the hypothesis that the entire length of the whole tract is as stated, then the length of each assignment shall be (proportionately larger than) as given; but if it be less, the assignments of allotments must lose in like proportion.[40] Sometimes there is no map; however, the rules still apply.

12.50 ESTABLISHMENT OF BOUNDARIES OF ALLOTTEES OF WILLS

All heirs receiving land under a will have equal standing, and unless there are contrary words, each is to share in any surplusage or shortage. Where by the terms of a will the testator intended to give all her land to her heirs, and the tract is found to contain excess as shown by the sum of the acres given to the different heirs, the surplus will be prorated among the heirs proportionate to the named acreage of each.[41] Again, usually there is no map, but the same rules still apply.

12.51 DEED DIVISIONS

Sometimes two or more people owning jointly would divide a tract of land by deed, describing their respective shares in a single deed, called a *division deed.* Unless there is a statement giving one party a remainder, the parcels so created are created simultaneously.

In many instances, when parties ask for an equitable division of land, they believe the court has the authority to subdivide a parcel by area. But in the strictest legal sense, a court of equity will ask that the land be sold and then the division of the prorated shares be distributed in the correct percentages.

12.52 COMMENTS

We *do not* consider the information contained in these chapters as legal advice for surveyors, but if placed in the hands of attorneys, this information may be helpful. Sandwiched between many complex survey concepts, the surveyor will find legal doctrines with which each surveyor should become familiar as being necessary to practice boundary creation and retracement.

This book encourages each and every surveyor to become acquainted with surveying principles that are necessary to perform his or her work and let his or her coworker, the attorney, make the legal conclusions.

NOTES

1. *O'Brien* v. *McGrane*, 27 Wis. 446 (1871).
2. *Rivers* v. *Lozeau*, 529 So. 2d 1147 (Fla. App. 1989).
3. *Pyburn* v. *Campbell*, 158 Ark. 321 (1923).
4. *Cragin* v. *Powell*, 128 U.S. 6791 (La. 1888).
5. *Morris* v. *Jody*, 216 Ky. 593 (1926).
6. *Riley* v. *Griffin*, 16 Ga. 151 (1854).
7. *Gordon* v. *Booker*, 97 Cal. 586 (1892).
8. *McDaniels* v. *Mance*, 47 Iowa 504 (1877).
9. *Tomlinson* v. *Golden*, 157 Iowa 237 (1912).
10. *O'Farrel* v. *Harney*, 51 C. 125 (1875).
11. *Heaton* v. *Hodges*, 14 Me. 66 (1836).
12. *Curtis* v. *Upton*, 175 Cal. 322 (1917).
13. *Burke* v. *McCowen*, 115 Cal. 481 (1896).
14. *Andrews* v. *Wheeler*, 10 C.A. 614 (1909).
15. *Diehl* v. *Zanger*, 39 Mich. 601 (1878).
16. Ibid.
17. *Westgate* v. *Ohlmaher*, 251 Ill. 538 (1911).
18. *City of Decatur* v. *Niedermeyer*, 168 Ill. 68 (1897).
19. *Williams* v. *City of St. Louis*, 120 Mo. 403 (1894).
20. *Orena* v. *Santa Barbara*, 91 Cal. 621 (1891).
21. *Coop* v. *Lowe Co.*, 263 P. 485 (Utah 1927).
22. *Perich* v. *Maurer*, 29 Cal. 293 (1865).
23. Orenasupra note 20.
24. *Daiser* v. *Dalto*, 140 Cal. 167 (1903).
25. *Waldorf* v. *Cole*, 61 Wash. 251 (1963).
26. *Anderson* v. *Worth*, 131 Mich. 183 (1902).
27. *Coppin* v. *Manson*, 144 Ky. 634 (1911).
28. *Quinnin* v. *Reimers*, 46 Mich. 695 (1881).
29. *Nilson* v. *Kahn*, 314 Ill. 275 (1924).
30. *Lewis* v. *Prien*, 98 Wis. 87 (1897).
31. *Barrett* v. *Perkins*, 113 Minn. 480 (1911).
32. *Goroski* v. *Tawner*, 121 Minn. 189 (1913).
33. *McKinzie* v. *Nichelini*, 43 Cal. App. 194 (1919).
34. *Quinnin* v. *Reimers*, 46 Mich. 605 (1991).
35. *Pereless* v. *Magoon*, 78 Wis. 27 (1890).
36. *Blaffer* v. *State*, 31 S.W.2d 172 (Tex. Civ. App. 1930).

37. *Wheatley* v. *San Pedro*, Los Angeles & S.L.R.R., 169 Cal. 505 (1915).

38. *Mechler* v. *Dehn*, 196 N.Y.S. 460 (1922).

39. *Booth* v. *Clark*, 59 Wash. 229n (1910).

40. *McAlpine* v. *Reicheneker*, 27 Kan. 257 (1882).

41. *Bennett* v. *Simon*, 152 Ind. 490 (1890).

CHAPTER 13

LOCATING COMBINATION DESCRIPTIONS AND CONVEYANCES

13.1 INTRODUCTION

The authors do not know of any state that identifies for attorneys or surveyors how land descriptions should be written. Many of the suggestions were created in private writings. Essentially, there are few restrictions on who can prepare land and/or boundary descriptions to the point that a landowner may prepare his or her own descriptions. This fact has led to unnecessary litigation over the validity of land descriptions.

> When creating boundaries, a scrivener should strive to use a consistent form of descriptions.

If there is any one profession that requires its practitioners to be well versed in many aspects and specifics or requirements, it is undoubtedly the surveying profession. Surveyors are not only asked to create boundaries and descriptions but are required then to survey and locate these boundaries on the ground at a later date and then found. This found evidence in all probability was created by some earlier persons.

In the everyday course of a retracement of boundaries of parcels of land, there is no one approach that can be explained. Because in most states it is legal for any person to prepare a deed, we find people of all social classes, educational backgrounds,

Brown's Boundary Control and Legal Principles, Eighth Edition.
Donald A. Wilson, C.A. "Tony" Nettleman III, and Walter G. Robillard.
© 2024 John Wiley & Sons, Inc. Published 2024 by John Wiley & Sons, Inc.

and capabilities writing descriptions that may defy future retracement. We often find descriptions that go clockwise for one portion and then counterclockwise for the remaining portion. Or we may discover that a parcel is partially described by a metes or bounds description or a metes *and* bounds description and also by reference to a plat, recorded or unrecorded. In a General Land Office (GLO) state, we may find an aliquot description for a parcel only to be "more particularly described" by courses and monuments. In this chapter, we consider some of the problems created by the scrivener of a description who uses various and possibly confusing combinations of sequential and simultaneous descriptions.

The following principles are discussed in this chapter:

PRINCIPLE 1. When an owner conveys a portion of a lot in a subdivision by an "of" description, the owner conveys that portion of the lot called for by measuring from the side line of the street. Vacations or openings of streets do not alter original lot lines.

PRINCIPLE 2. When an owner conveys a portion of a metes and bounds survey by an "of" description, the owner conveys from the boundary line or causes the measurements to be made from the boundary line, even though the boundary line is in a street.

PRINCIPLE 3. Words in a land boundary description should not be abbreviated. To do so may cause future problems for the retracing surveyor.

PRINCIPLE 4. A scrivener of a land/boundary description should not change from one style to another but should use the same method throughout the entire description.

PRINCIPLE 5. When an owner conveys an "of" description by lineal measurements, the presumption is that the measurement is at right angles to the boundary line from which the measurement is made, unless specified otherwise, as well as being a horizontal measurement.

PRINCIPLE 6. When a fraction of the whole is conveyed, such as the west half, it is presumed that the conveyance is based on area measurement unless stated otherwise.

PRINCIPLE 7. When the easterly and westerly lines of a lot are shown as parallel on the original map and in fact are nearly parallel, and the easterly half and westerly half are conveyed, the dividing line between the easterly and westerly halves is made on the mean bearing of the two lines.

PRINCIPLE 8. When the easterly and westerly lines of a lot are not parallel or north and the lot is divided into east and west halves, the dividing line should be made to run north and south.

PRINCIPLE 9. When a deed reads "east one-half of lot 1" and the second deed reads "lot 1, except the east one-half," it is commonly assumed that the west line of the east one-half is parallel with the east line of the lot, provided that the east line of the lot is nearly in a cardinal direction.

PRINCIPLE 10. In the absence of other qualifying terms, a given area of land on the side of a tract will include such quantity in the form of a parallelogram. If a given area of land is to be laid off on a given line, the shape is presumed to be a square. If a given area of land is to be laid off in the corner of a given parcel of land, the shape is presumed to be as nearly a parallelogram with equal sides as the circumstance permits.

PRINCIPLE 11. When writing a legal boundary description, avoid deed calls from two directions.

PRINCIPLE 12. The scrivener should write each description as an individual creation, and each description should assume a continuity of words and definitions.

PRINCIPLE 13. When a retracing boundary surveyor encounters a problem, he or she should try to give consideration to all elements of the description, but should be able to explain why specific elements were not adhered to.

"OF" DESCRIPTIONS

13.2 "OF," "IN," AND "AT" DESCRIPTIONS WITHIN SUBDIVISIONS AND ADJOINING STREETS

Land conveyed by an "of" description, such as the "easterly 50 feet of Lot 2," as shown in Figure 13.1, presents an ambiguity. Ownership of a lot within a subdivision usually extends to the centerline of the street; hence, the easterly 50 feet of Lot 2 could be measured from the centerline of the street. But in the minds of the public, the side

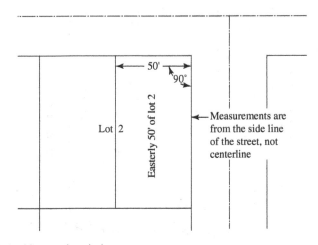

Figure 13.1 Ambiguous description.

line of the street, not the centerline, is the limit of private ownership, thus causing the 50 feet to be measured from the side line. In keeping with the principle "Where two meanings can be construed from a written instrument, the meaning giving the greater advantage to the grantee is usually used," and in keeping with the intent of the sale as understood by most people, the measurements are made from the side line of the lot in accordance with the following principle.

Principle 1. *When an owner conveys a portion of a lot in a subdivision by an "of" description, the owner conveys that portion of the lot called for by measuring from the side line of the street. Vacations or openings of streets do not alter original lot lines.*

This is explained in the following decision:

> However clear it may appear that the owner of a lot holds title to the center of the adjoining street, subject to the public easement, and that the boundary of the lot is technically, therefore, the center of the street, in view of the fact that the owner of such lot or land has no right to the possession or occupancy of any portion of such street, the word "lot" as generally and customarily used does not include that portion of the street. In the absence of any circumstances indicating that a more unusual and technical meaning of the word "lot" was contemplated and intended by the grantor, it will be presumed that the grant of a fractional part or of a given number of feet of a certain lot or parcel of land conveys the given fractional part or number of feet of that portion of the lot or parcel of land which is set apart for private use and occupancy.[1]

In Figure 13.2, a condition is shown where care must be used in conveying the east or west 25 feet of the ownership shown. Lot lines never change because of a

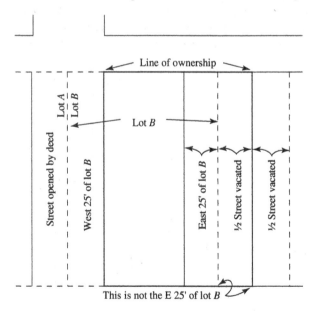

Figure 13.2 Lot lines never change because of a street opening or closing.

street opening or closing. Because of the vacation of the street, the east 25 feet of the ownership is properly described as the westerly one-half of the street now vacated and not the easterly 25 feet of Lot B. The west 25 feet of Lot B describes a portion of a street opened, whereas the east 25 feet of the west 50 feet properly describes the 25 feet immediately adjoining the west side of the ownership. The surveyor must be extremely careful if a subdivision has been repeated.

13.3 "OF" DESCRIPTIONS WITHIN METES AND BOUNDS DESCRIPTIONS AND ADJOINING STREETS

Principle 2. *When an owner conveys a portion of a metes and bounds survey by an "of" description, the owner conveys from the boundary line or causes the measurements to be made from the boundary line, even though the boundary line is in a street.*

In metes and bounds descriptions, the reverse principle to that given for subdivisions usually is implied. When the "easterly 50 feet of the following described property" is conveyed, as in Figure 13.3, the property conveyed is the easterly 50 feet of that part described, even though part of the land described may lie within a public street (Figure 13.4).

When writing "of," "in," and "at" descriptions, the original creator should extend the writings beyond the specific word. This can be included in the habendum clause of the description.

In Figure 13.3, the problem is that of the "in" description. A parcel of 1 acre is identified as "1 acre in the SE ¼." The courts have held that the boundaries are indefinite and ambiguous.

Principle 3. *Words in a land boundary description should not be abbreviated. To do so may cause future problems for the retracing surveyor.*

When writing boundary descriptions, some individuals resort to abbreviating some words, thinking this action will save paper and time. Any person who writes boundary descriptions should make it a practice to write all words fully without resorting to abbreviations. Scriveners believe that having a legend explaining what the abbreviations mean will suffice; this is not true.

Several years ago, a plaintiff brought an action in Memphis when his retracing surveyor encountered the abbreviation "CM" on a plat. In doing his fieldwork, he uncovered a concrete monument in the general locality and used that as a control point to which the landowner constructed a large movie theater complex. Several months later, a survey of the contiguous lot indicated a possible overlap of the adjacent lot of some 25 feet. When the surveyor who conducted the original survey was contacted, he informed the landowner that "CM" indicated a "chisel mark" and not a concrete

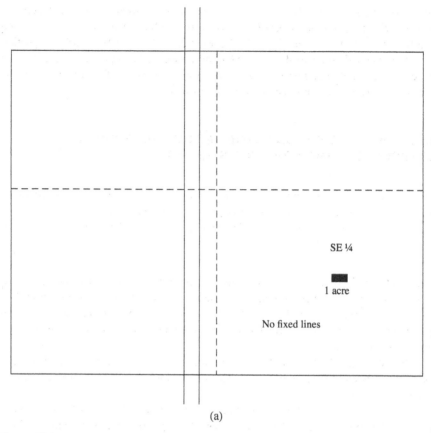

(a)

Figure 13.3a (a) "In" the SE ¼. (b) At the SE ¼.

monument on the curb line, which was found. Many surveyors use the abbreviation "IP" randomly to indicate iron pin or iron pipe without making the distinction. In conducting research, surveyors will find that title professionals are prone to take a well-written description and reduce it to abbreviations.

Principle 4. *A scrivener of a land/boundary description should not change from one style to another but should use the same method throughout the entire description.*

There are several accepted methods for describing land parcels and boundaries. Each is suitable with particular original evidence. One of the earliest forms is a description consisting of calls for adjoining parcels. This is probably one of the oldest forms in use today. This method of describing a parcel by referring to the contiguous parcels that concerned the parcel being described is probably a result of the early method of transferring land by *"livery of seisen"*: when the scrivener starts

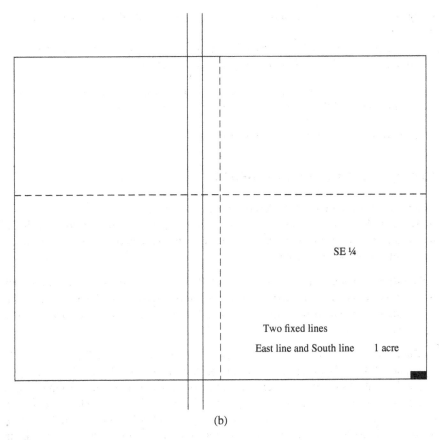

(b)

Figure 13.3b *(Continued)*

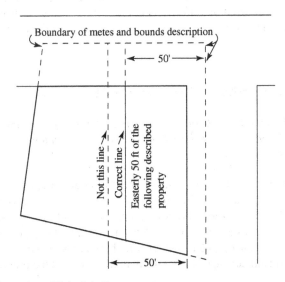

Figure 13.4 Illustration of Principle 2.

using one form, he or she should not adopt a second or third form in that same description.

The scrivener will find this form of description is one of the easiest to create. For example, a description may be partially written as follows:

> Bounded on the North by Issac Smith, and on the East by the road to Queensbury, then one line North 28° West, a distance of 124.23 feet to Corner 4, thence ... "

This form of description should be discouraged.

13.4   DIRECTION OF MEASUREMENT

When writing or retracing "of" descriptions, the retracing or creating surveyor should make a detailed research on the exact boundaries from which the description should be written and then ultimately retraced. Possession lines and boundary lines for measurements may not be the same.

Principle 5. *When an owner conveys an "of" description by lineal measurements, the presumption is that the measurement is at right angles to the boundary line from which the measurement is made, unless specified otherwise, as well as being a horizontal measurement.*

Figure 13.5a shows a typical case where the "northerly 50 feet" of a lot and the "southerly 50 feet as measured along the easterly and westerly lines" of a lot give different areas. As in all measurements of this type, where alternate meanings can be interpreted, the greatest advantage is given to the buyer. However, where the method of measurement is specified as in the "southerly 50 feet" of Figure 13.5a, the method as specified should be used since alternate meanings do not exist. In Figure 13.5b, the correct property lines for curves and angle points in lines where lineal measurement "of" descriptions are used are shown. It is interesting to note that where an angle point exists, as shown in Figure 13.5b, more is conveyed at the angle point than the distance called for.

13.5   PROPORTIONAL "OF" CONVEYANCE

Principle 6. *When a fraction of the whole is conveyed, such as the west half, it is presumed that the conveyance is based on area measurement unless stated otherwise. (See Figure 13.6.)*

This principle is not in harmony with the federal statutes that specify the method to be used for sectionalized land. Thus, the north one-half of the northwest one-quarter of Section 6 might be considerably less in area (especially in a closing section on a correction line) than the south one-half of the northwest one-quarter. In general, proportionate conveyances under state laws are based on acreage; under federal laws for sectionalized land, acreage is not considered; distance is the criterion.

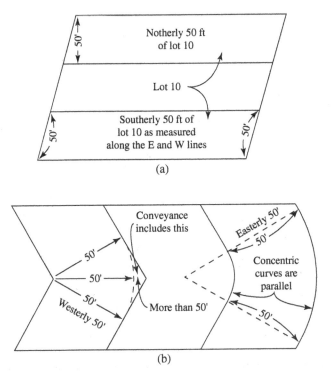

(a)

(b)

Figure 13.5 Illustration of Principle 5.

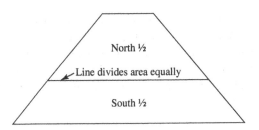

Figure 13.6 Presumption that conveyance is based on area measurement.

If the scrivener insists on using this form of description, his or her method of proportioning should be identified in the *habendum clause* of the description.

Occasionally, in sectionalized land where there is an odd-shaped lot bordering on a lake or land grant line and the land is divided *after it has passed under state jurisdiction*, a question arises as to the intent of the sale where half is sold. The south one-half of Lot 4, as shown in Figure 13.7, would be divided, according to state laws, into two parts of equal area. But by the federal principle, the south half would be determined by a line extending west from the midpoint of the easterly line of Lot 4. Any reference such as "according to federal government survey methods" implies that

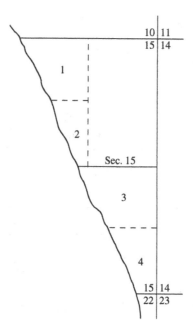

Figure 13.7 Odd-shaped lot.

the lot would be divided by linear measurements. A few states have adopted statute laws similar to federal laws, and thus it becomes mandatory that the federal rules be applied. Where division by acreage is intended, it is advisable to state "one-half the acreage of Lot 4."

"A deed calling for the north one half of a lot facing the Au Gress River and not indicating a division line, is to be divided so as to make the parcels equal in area; a division such as to give each one half of the river front, is erroneous."[2] Strictly speaking, any line can divide a lot into two halves, especially where the deeds state the northerly one-half and the southerly one-half. In a written description of a proportional or fractional conveyance, the direction of the dividing line should always be given.

Proportional conveyances, where the method of locating the dividing line is specified as "the south one-half as measured along the easterly line," are divided in accordance with the method specified. "Along a line" cannot be an area measurement nor does it imply equal areas; the line would be divided in half and the distance principle applied.

13.6 EXCEPTION BY ONE-HALF BY AREA

In New York, where the terms *east-half* and *west-half* imply equal area division, the rule may lose its effect when it appears that at the time the deed was formed some fixed boundary (such as a fence) divided the property somewhere near the center, so the words referred more properly to one of such parts than to a mathematical division, which had never been made.[3]

13.7 INDETERMINATE PROPORTIONAL CONVEYANCES

Indeterminate proportional conveyances are those in which the direction of the dividing line is not given or implied. Any area may be divided in half by a multitude of lines, and when the direction of the dividing line is not given, the conveyance may be indeterminate, as shown in Figure 13.8. A scrivener who fails to define the direction of the dividing line when writing a proportional conveyance is dedicating to the future a headache for the parties on either side of the dividing line and an unsolvable problem for surveyors. Such descriptions should and could be avoided by a few carefully inserted words in a document so that a clear and concise intent is conveyed rather than a dual meaning left to the fighting instincts of future owners. Under certain conditions, the direction of the dividing line is revealed by the geometric shape of the entire parcel or by the wording of the deed.

13.8 ANGULAR DIRECTION OF THE DIVIDING LINE IN "OF" DESCRIPTIONS

Principle 7. *When the easterly and westerly lines of a lot are shown as parallel on the original map and in fact are nearly parallel, and the easterly half and westerly half are conveyed, the dividing line between the easterly and westerly halves is made on the mean bearing of the two lines.*

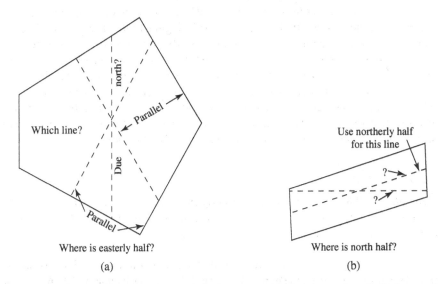

Where is easterly half?

(a)

Where is north half?

(b)

Figure 13.8 (a) Finding the eastern half of a parcel. (b) Finding the northern half of a parcel.

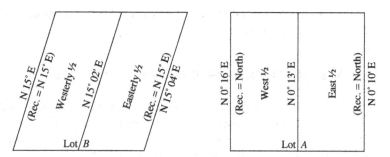

Figure 13.9 Illustration of Principle 7.

Figure 13.9 illustrates this principle as applied to two conditions. Lot A was originally shown as having due north lot lines that proved to be north 0°10′00″ east and north 0°16′00″ east when actually surveyed. Because two parcels were sold, the east one-half and the west one-half, the problem of the surveyor is to divide the area into two equal areas and use a line equitable to both parties. Because the original lot lines were shown as parallel, the dividing line should be as nearly parallel with the sides of the lot as possible or, in other words, on the average bearing of the two sides. This principle can be overcome by other factors more clearly showing the intent, such as stakes set at the time of the division and being parallel with either the west or the east line. This principle cannot be applied where the deeds read "the east one-half" and "all except the east one-half" since it is then inferred that the east one-half is intended to be a parallelogram and the west one-half is what is left. Use of the terms "east one-half," "west one-half," "north one-half," and "south one-half" should be avoided unless the original lot lines were due east, west, north, or south. Easterly, westerly, and so on, are the proper terms, which do not imply that the dividing line must be astronomic north or west. In the case of lot B in Figure 13.9, where the original lot lines were shown as parallel and where the easterly and westerly halves were conveyed, there is the implication that the dividing line should be as nearly parallel with the easterly and westerly lines as possible or on the average bearing of the lines.

This principle is of little value when the supposedly parallel lines are in fact considerably out of parallelism. In one such case, the court ruled that after the grantor had sold the easterly half, he could sell only his remainder. The easterly half was laid off by parallel lines, and the westerly half received the remainder.

Nonparallel Lines

Principle 8. *When the easterly and westerly lines of a lot are not parallel or north and the lot is divided into east and west halves, the dividing line should be made to run north and south.*

A deed reading "the east one-half" of the land shown in Figure 13.10 would be staked by turning 75° from the northerly property line at a point that will divide the lot into two equal areas. The record map bearing, S 75°00′00″ E, determines the basis of

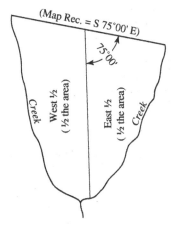

Figure 13.10 Dividing line based on bearings.

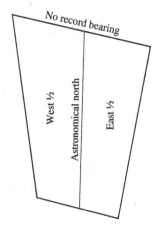

Figure 13.11 Dividing line based on astronomical observation.

bearings (where a deed calls for a map, the data shown on the map become a portion of the written description), and since the line is to be north, the angle of 75° properly defines north according to the map referred to. Where no record bearing exists on the map called for, as in Figure 13.11, only an astronomical observation can properly divide the lot into the east one-half and west one-half. In neither figure is there an implication that the dividing line is to be parallel to a particular line.

East Half of Lot and the Lot Except the East Half

Principle 9. *When a deed reads "east one-half of lot 1" and the second deed reads "lot 1, except the east one-half," it is commonly assumed that the west line of the east one-half is parallel with the east line of the lot, provided that the east line of the lot is nearly in a cardinal direction.*

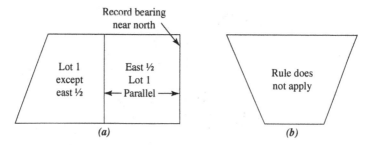

Figure 13.12 (a) Parallel rule applies. (b) Parallel rule does not apply.

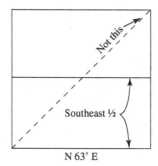

Figure 13.13 Application of parallel lines.

Figure 13.12a shows the application of this rule while Figure 13.12b shows a case where this principle does not apply. If the deeds were written "the easterly one-half and all except the easterly one-half," the common practice would be to make the dividing line parallel with the easterly line. As shown in Figure 13.13, the southeast one-half was ruled to be delineated by a line parallel to the N 63° E line, not by the diagonal line.[4]

13.9 ACREAGE "OF" DESCRIPTIONS

The same general rules of Section 13.5 that apply to proportional descriptions (the north one-half of lot 1) apply to acreage conveyances. The direction of the dividing line, as in proportional descriptions, should always be stated when a new deed is written.

A description reading "the north 10 acres" of the land shown in Figure 13.8b allows two interpretations, each equally valid. Similarly, the "easterly 10 acres" of the land shown in Figure 13.8a is an indeterminate area. But there are certain general principles for acreage descriptions that can be stated.

Principle 10. *In the absence of other qualifying terms, a given area of land on the side of a tract will include such quantity in the form of a parallelogram. If a given* *area of land is to be laid off on a given line, the shape is presumed to be a square. If a given area of land is to be laid off in the corner of a given parcel of land, the shape is presumed to be as nearly a parallelogram with equal sides as the circumstance permits.*

In a Texas case, Chief Justice McClendon stated:

> Where land is described generally by acreage out of the corner or off a side of a larger tract, the courts will construct a survey of the designated acreage, by lines drawn parallel with the designated line or lines of the larger tract; not, however, because the parties have so stated in their writings, but because the writing is silent on the subject, and the presumption that they so intended is deduced from what men ordinarily do under like circumstances.[5]

The conditions were shown in Figure 13.14. The south 75 acres and the north 80 acres of a 155-acre parcel were leased within two days of each other. Area A, containing an oil well, was claimed by the lessee of the 75-acre parcel on the grounds that the lessor stated that his intentions were to lease the north 80 acres and then lease the remainder. But the court ruled that the south 75 acres must be laid between parallel lines; hence, the lessee of the south 75 acres had no interest in Area A.

A description of "nine acres in the southwest corner of the SE ¼ of the SW ¼ of Section 1, T8N, R4W" was construed in accordance with the following: "A conveyance of a definite quantity of land in or off of a specified corner of a designated tract is, under a well-settled rule of construction, the grant of a corner quadrangle, of equal sides, extending to the corner."[6]

Under the rule that where a deed describes a tract of land as so many acres in a certain corner of a specified section, it will be taken to embrace the given number of acres in the form of a square in the corner of the section designated.[7]

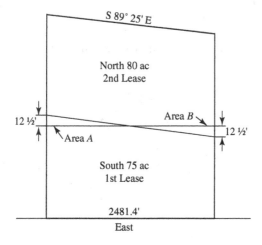

Figure 13.14 *Woods v. Selby Oil & Gas Co.*, 2 S.W.2d 89s, 12 S.W.2d 994 (1929).

Also, a description reading "15 acres more or less off the southwest corner of the NE ¼" was construed to mean "exactly 15 acres to be taken in a square body located in the southwest corner of the land."[8] The same principle was applied in a New York case.[9]

13.10 AMBIGUITY

All too often, ambiguous statements or phrases in deeds are difficult to interpret and require field information. In Figure 13.15, the "east 50 feet" is indeterminate where a curve return cuts off the corner. Better practice is to write the "east 50 feet as determined by a line parallel with the most easterly line of the lot and its northerly extension." "Beginning at a point 50 feet from the most northerly corner of Lot B," as shown in Figure 13.16, indicates two possible points. In Figure 13.17, the southwest one-quarter can be interpreted in two ways. The rear 15 feet and the west 50 feet in Figures 13.18 and 13.19 can be solved only by physical evidence on the ground or parol evidence of witnesses. Figure 13.20, the case of the double exception, can and should be avoided by the deed author, who should never use double exceptions without checking for dual meanings. "Lot B, except the north 50 feet, except the west

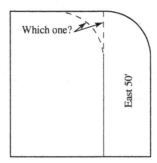

Figure 13.15 Avoid "east 50 feet" where there is a corner cut off by radius return.

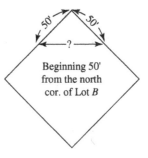

Figure 13.16 "50 feet from north corner" may mean two possible points.

Figure 13.17 Where is the SW one-quarter?

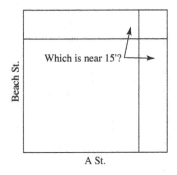

Figure 13.18 Ambiguity of "near."

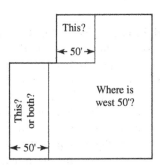

Figure 13.19 Ambiguity of "west 50 feet."

75 feet," is clear until the second exception is made. What does the "except the west 75 feet" refer to, the north 50 feet or lot B? Deeds being interpreted most strongly against the grantor set up a condition in which a double exception may cause the grantor to part with more land than was intended.

In Figure 13.21, ambiguity results where a deed reads the "east 20 feet of lots 9 and 10," owing to the dual meaning of (1) "lot 10 and the east 20 feet of lot 9," or (2)

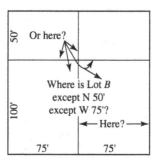

Figure 13.20 The double exception.

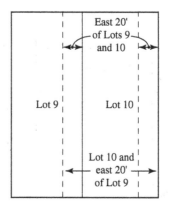

Figure 13.21 State the whole lot first.

the "east 20 feet of lot 9 and the east 20 feet of lot 10." It is better to state the whole lot or lots first, and then follow with the part lots. In Figure 13.20, the sale would be construed most strongly against the grantor unless the contrary could be proved.

The land conveyed to a person and the land described to a person can be entirely different points, as shown in Figure 13.22. Where land abuts a road in a subdivision, by presumption, title to the road vests in the owner of the abutting land. The land described in this instance would be Lot 27, excluding the road, whereas the land conveyed would include Lot 27 and the road up to the centerline.

"Thence along the west line of lot D to the mean high-tide line; thence 150 feet along the mean high tide line; thence … " is ambiguous, as shown in Figure 13.23, because the mean high-tide line changes with erosion of the shoreline. The point of beginning of the next course could change with every storm.

Interpretation of intent is difficult in deeds containing double calls of the form "thence N 10° E a distance of 200 feet to a point in the north line of lot 13, said point being 120 feet westerly from the northeasterly corner of said lot and also being 110 feet easterly from the northwesterly corner of said lot." If there is surplus or deficiency, both "being" clauses cannot be right. If the "N 10° E, 200 feet" is in agreement with one of the two distances, it would be controlling. But if the "N 10° E, 200 feet"

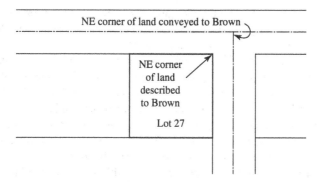

Figure 13.22 Land described versus land conveyed.

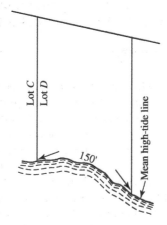

Figure 13.23 Distances along a body of water may change with erosion.

is not in agreement with either tie to the lot corners, a prorate is indicated, provided that no senior right is interfered with.

Deeds of the form "thence N 10° E a distance of 200 feet to a point in the northerly line of lot 13, said point being S 89° W, 120 feet from the northeasterly corner of said lot; thence N 89° E, 120 feet to the northeasterly corner of said lot; thence, etc.," would be construed by the tie distance. The N 10° E, 200 feet becomes more or less in character because of the extra emphasis placed on the N 89° E, 120 feet.

OVERLAPS AND GAPS

13.11 CALLS FROM TWO DIRECTIONS

Principle 11. *When writing a legal boundary description, avoid deed calls from two directions.*

This condition arises where portions of lots are sold from two directions, as the east 50 feet and the west 50 feet of a 100-foot lot, without calls for adjoiners. Often, a surplus or deficiency exists within a lot, and where the foregoing form is used, there is usually an overlap or gap. The proper way to describe the parcels is "the easterly 50 feet of lot 1" and "lot 1 except the easterly 50 feet."

When an overlap exists between two parcels, the senior or first owner receives what is coming to him or her, and the junior owner has the remainder. Where there is a gap and the two parcels do not meet, neither has title to the surplus, because the original grantor did not sell it. Thus, suppose that an owner sold the west 12 acres and the east 8 acres of a parcel reported to be 20 acres by original government measure. An accurate survey revealed that instead of there being 20 acres as supposed, the said 20 acres was found to be 22¼ acres, owing to surplus in the section. Because the original owner had not sold the surplus 2½ acres, title was vested in him.

ESTABLISHMENT OF PROPERTY DESCRIBED BY BOTH METES AND BOUNDS AND SUBDIVISION DESCRIPTIONS

13.12 DOUBLE DESCRIPTIONS

Ambiguity resulting from double descriptions of the same parcel of land can and should be avoided by deed authors. "Lot 4, block 2, according to map 1240 being also the following described land; beginning at the southeast corner of block 2; thence west 150 feet to the true point of beginning; thence, north 100 feet; thence west 150 feet to the true point of beginning; thence, north 100 feet; thence west 50 feet; thence, south 100 feet; thence, east 50 feet to the true point of beginning" is a description that invites trouble. The block shown in Figure 13.24, being long in all directions, causes the two parcels described to be in different locations. If the seller owned all of Block 2 at the time of the sale, certainty of location would be impossible (see Section 13.13).

In Louisiana, a square described as No. 2670 on a map and also described as bounded by Paris Avenue, Hamburg, Manuel, and Fowy Streets was controlled by

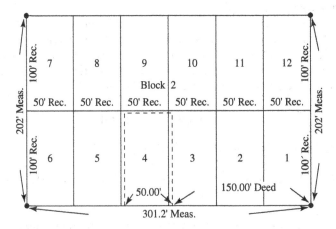

Figure 13.24 Ambiguity from double description.

the monuments called for and the possession of the owner rather than the erroneous square number.[10]

In some areas of the United States, particularly older regions, double descriptions are more prevalent. Occasionally, three or four rewrites of the same land create numerous ambiguities. If it can be shown that the new description was intended by the scrivener to be the same as the old description, the new description should be interpreted in that light. If this cannot be shown, the grantor then remains as the owner of any unconveyed land. But the grantor cannot convey that which he or she does not own; hence, any land included that did not belong to the grantor is not conveyed.

13.13 NEW YORK DOUBLE DESCRIPTIONS

In New York, where there is excess or deficiency in a subdivided block, and the land is described by *both* lot and block, and by a metes and bounds description using the same dimensions shown on the plat, the deed is interpreted by the metes and bounds description; that is, any surplus or deficiency is not prorated. The senior deed gets exactly what is coming to it by its metes and bounds description, and the junior deed is plus or minus in character.

13.14 NATURAL PHENOMENA AND BOUNDARIES

On more and more occasions, surveyors are being asked to solve boundary issues and problems for which there is little or no case law, guidelines, or statutes. This area of boundary changes caused by natural forces is a fruitful area for surveyors. The following discussions are just an introduction to a very complex area of practice.

Most persons consider land as fixed in place, when in fact, land is in perpetual movement. A variety of natural phenomena affect the locations of boundaries and land titles.

Earthquakes, land settlement, and land rising in elevation cause movement in the position of monuments, even the land itself. Except along the ocean, vertical displacement of the Earth's surface does not usually alter the location of title interests. Another exception is where title lines are defined by an elevation line or a contour. Subsidence adjoining the oceans and bodies of water may alter the position of the mean high-water or mean high-tide line.

Horizontal and vertical changes to the landscape may be categorized as follows:

Changes Due to Water

Accretion, avulsion, erosion, and reliction are ways in which land boundaries and the location of land titles may be altered. This area is so complex and widespread that Chapter 9—Riparian and Littoral Boundaries is specifically devoted to these concepts.

Changes Due to Wind

Wind may move soil particles, resulting in accretion and erosion. Often, movement by wind is in conjunction with movement by water, and so is not readily distinguishable or recognizable. One case of wind accretion was decided by the Canadian court system.[11]

Violent windstorms not only wreak general havoc across the landscape but also cause movement of soil and destruction of boundary monumentation and references. Hurricanes, tornadoes, microbursts, and related storms are generally isolated events, but frequent enough that they occur on an annual basis.

Changes Due to Earthquakes

Horizontal movements can drastically change land location. After the numerous San Francisco earthquakes of 1906 and the Anchorage earthquake of 1964 many boundary lines were abruptly changed and disturbed. This includes all areas that are subject to these disturbances. In one instance, a sidewalk was observed with a lateral displacement of about 19 feet. Basically, there are three types of movement: (1) sudden horizontal shifts on each side of a fracture; (2) distortions due to stretching or compressing, much as a rubber band can be elongated or shortened; and (3) slides. The last Alaskan earthquake altered the location of the track of the Alaska Railroad, thereby making measurements based on location useless. No litigation was found in which sudden fracture movements, stretching, or compressing was an issue. Landslides are often a consideration in legal cases.

The following discussion is not based on court decisions but represents what is believed to be the surveyors' best solution to land movement problems. In the event of a sudden earth movement, the same rule should apply as for a sudden change in a river's course; the same land belongs to the same owner. Figure 13.25 shows sudden land shifts along the San Andreas Rift. It is probable that in these instances the same owner should continue to own the same land even though displaced along a line of fracture.

After a violent earthquake, land may be distorted without observable rupture. Thus, after the Long Beach earthquake in California, the distances between monuments in one direction became 1 foot longer per 100 feet. The land did not show cracks or other signs of separation. Surveyors in the area recognize local control and prorate the distances between the nearest found monuments. If nearby monuments are not locatable, proportional measurement is based upon nearby improvements. Proration between distance points develops inequities. Long, straight streets may end up with permanent well-rounded bends. In a situation in which everything has moved nearly the same amount, it is inconceivable that owners should move their improvements back to their former absolute position of latitude and longitude.

In the early part of the nineteenth century, this country experienced one of the strongest earthquakes ever, in the vicinity of New Madrid, Missouri. Since there was little settlement in this area, and few recorded lives lost, it is not recorded as a devastating earthquake. However, studies explain how the Mississippi River flowed

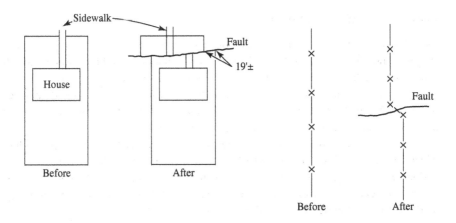

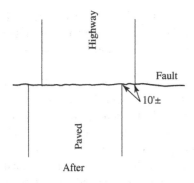

Figure 13.25 Land displacements caused by earthquakes.

backward for three days, and Reelfoot Lake, once nonnavigable, was transformed into a navigable body of water. This issue alone, along with the resulting land title claims, resulted in an 87-page court decision to resolve some of the problems.[12] After the earthquake, the site of the town of New Madrid was under the new location of the Mississippi River. Today, through movement of the location of the river, the site of New Madrid is now on the easterly bank, whereas before the earthquake it was on the westerly bank. Dramatic changes such as this, and their consequences, wreak havoc with land titles and property locations.

Changes Caused by Tsunamis

Tsunamis are caused by other events: earthquakes, underwater volcanoes, underwater landslides, and meteor impact in water. The changes in the landscape can be widespread and major. The 2004 tsunami in the Indian Ocean caused widespread damage, altered shorelines, washed away islands and parts of villages, and resulted in extensive loss of life.

Landslides and Earth Flows

Landslides are due to a variety of causes on slopes that are unstable for one reason or another. They may be in the form of the release of a single boulder, some of which travel great distances, or in the form of a volume of soil or rock material moving downslope. When unstable soils become saturated with water, movement can take place slowly or rapidly, depending on the steepness of the slope, and can result in mass movement of a few to many cubic yards of material.

If slides are created because of negligence, monetary damage is not precluded. Where a slide results from an act of nature, such as an earthquake, the owners undoubtedly own where their bedrock is located. In the Alaskan earthquake of 1964, the shaking caused land to become fluid, and it ran into the bay. In such instances, land boundaries could not flow with the surface material.

Subsidence

Subsidence generally occurs in response to the relieving of support under the land. Withdrawal of groundwater, minerals such as oil and gas, or general erosion of soluble rock may cause the overlying land surface to decrease in elevation. Dramatic effects may come in the form of sudden sinkholes.

Oil removal in Long Beach, California, caused vertical subsidence of almost 27 feet and a horizontal movement from zero to about 8 feet. There are several areas mostly throughout the East including sinkholes in Florida due to a variety of situations. Since the movement is of long duration and imperceptible to the eye, people adjust to it. Streets, curbs, buildings, and improvements all move. It is logical that surface rights should move with the land. Also, it is logical to use proportional measurements in locating surface rights (even for metes and bounds descriptions calling for absolute dimensions).

Earth elongation or shortening is seldom the same in all directions. Thus, land can be stretched 1 foot per 100 north and south and be slightly compressed east and west. In such areas, the surveyor must test original measurements between many monuments spaced in different directions to determine an index of adjustment.

Volcanoes

When volcanoes "blow their tops," abrupt changes to the landscape occur. Ash fall, which can spread for miles, may affect elevations, while flows of lava and other material may also fill in lakes and riverbeds or flow into the ocean and cool, creating land where there was none before the blast. In the 1980 eruption of Mt. St. Helens, within about 10 minutes, the resulting landslide filled 22 kilometers of the North Fork Toutle River Valley with rock debris to an average depth of 45 meters. In places, the deposit is 195 meters thick.

In Hawaii, the leading case concerning ownership of solidified lava due to cooling after flowing into the ocean was decided in 1977.[13]

Glaciers

As glaciers advance, they put tremendous pressure on the landscape, causing it to depress. As they retreat, the land is once again relieved of the pressure and a phenomenon known as glacio-isostatic rebound takes place. One case in Alaska was a friendly suit between a private landowner, who owned the upland, and the State of Alaska, which claimed the bed of the water, over a parcel that was under water as long as the glacier was on top of it but became dry land after the retreat of the ice.[14]

Fire

Fire may burn areas containing trees that are either corner markers or reference monuments. Fire may also burn buildings, destroying party walls, monumentation, and indoor easements, such as passage through a hallway or on a set of stairs.

Fire will also destroy a land records system, as witnessed in a number of cities (Chicago and several Connecticut towns, for example), some town offices, and during the years from 1861 to 1865 a number of courthouses during the Civil War were destroyed giving rise to a change in the system or inception of a completely new recording system.

Fire can also burn heavy brush and dense growth, clearing the way for monument search and recovery, which take place in the Everglades and similar habitats on a fairly regular basis.

Secondary Events

Much of the destruction that occurs is not due to the primary event but stems from the secondary event that follows. For example, earthquakes result in flooding, fire results in erosion, and volcanoes result in landslides and lava flows. In many areas, secondary flooding following hurricanes destroyed many of the original field notes and plats of Colonial records. Coupled with the destruction of the records, many of the field notes and original plats were destroyed (by mother mature).

Even though certain events may seem to be quite rare in certain areas, no place is completely safe from most of the foregoing phenomena. Depending on the category, some areas are more vulnerable than others.

13.15 RECOGNITION OF PAST EVENTS

The older the event, the more difficult it is to recognize after the forces of nature have reshaped the landscape, or if human intervention has taken place in the form of dredging, filling, or grading. Some characteristics are indicative of past events, such as leaning fence posts, power poles, and cemetery markers; cracks in the ground, especially in pavement or cracks in building foundations; cemetery markers rotated on their bases; railroad tracks out of alignment; and marked elevation differences. When things differ from earlier survey measurements, do not quickly assume that the previous work was in error. The earth may have shifted between the two time periods—the older the earlier survey, the more likely that some event may have taken place.

The surveyor should realize that there are several natural ways that boundaries can be altered or land areas created or destroyed. Although there are few decisions that can provide guidance to the surveyor in identifying boundaries that have been altered, any surveyor who accepts the task of locating or positioning a boundary that has been altered or modified by natural phenomena should become versed in what has been written.

Principle 12. *The scrivener should write each description as an individual creation, and each description should assume a continuity of words and definitions.*

Like individuals, there should be no two descriptions exactly the same. If we accept the fact that a description should be unique and should describe a single unique parcel of land, unlike any other parcel in existence, then we understand the purpose of a description.

Principle 13. *When a retracing boundary surveyor encounters a problem, he or she should try to give consideration to all elements of the description, but should be able to explain why specific elements were not adhered to.*

We will assume that the "professional" will endeavor to conduct his or her boundary fieldwork with the highest degree of sophistication. The more elements referred to in the initial boundary description, the more evidence the retracing surveyor will have to identify the original boundary lines. The courts are liberal in interpreting descriptions, and when a problem is encountered, they will look for the "key" to its solution. They will pass this on to the retracing surveyor, even though he or she has no legal authority. They will permit the surveyor to reverse courses, eliminate courses, and just about everything in between, in order to make the description work.[15]

No two boundary descriptions should read the same. Each description should be unique unto itself.

The courts have repeatedly held that the purpose of a land description is to provide the "key" to permit an individual to locate the specific and unique parcel of land on the ground. As such, a description should not take a "cookbook" approach to accomplishing the task. The scrivener should understand the components that are necessary to constitute an adequate and legal description, and then he or she should use those necessary elements to prepare that specific description. The scrivener should avoid using preprinted forms or computer-generated aids; however, the individual who writes the new description should make reference to any previous descriptions and should use newly gathered information to prepare the new description.

NOTES

1. *Earl* v. *Dutour*, 181 Cal. 58 (1919).
2. *The Au Gress Boom Co.* v. *Whitney*, 26 Mich. 42 (1872).
3. *People* v. *Hall*, 8 N.Y.S. 276 (1889).
4. *Oruett* v. *Robinson*, 192 S.W. 537, 108 Tex. 283 (1917).
5. *Woods* v. *Shelby Oil & Gas Co.*, 2 S.W.2d 895, 12 S.W.2s 994 (1929).
6. *Daniel* v. *Williams*, 58 So. 419, 177 Ala. 140 (1912).
7. *Merkel Drainane* v. *Hathaway*, 260 Ill. 186 (1913).
8. *Early* v. *Long*, 42 S 348, 89 Miss. 285 (1906).
9. *Kellogg* v. *Vickory*, 1 Wend. 106 (1828).
10. *Lassus* v. *Gourgott*, 169 La. 577, 125 So. 628 (1929).
11. *Cates* v. *West Tenn. Land Co.*, 127 Tenn. 575 (1913).
12. *State* v. *Zimring*, 58 Haw. 106; 566 P.2d 725 (1977).
13. *Hosinger et al.* v. *State of Alaska*, 642 P.2d 1352 (1982).
14. *Bremner* v. *Bleakley*, 54 O.L.R. (Ont. C.A. 1923)
15. *Martin* v. *Pallin*, 483 S.E. 2d 614 (Ga.). For academic discussion of "Key" see Oconee Land Co. v. *Buchanan*, 686 S.E. 2d 452 (Ga. 2009).

CHAPTER 14

ROLE OF THE SURVEYOR

14.1 INTRODUCTION

The surveyor is involved both in creating original boundaries and in retracing and identifying originally created boundaries. Each of these carries with it significant responsibilities, both technical and professional. When finalizing original boundaries, surveyors must first consider: Were the boundaries established or created in accordance with the technology available and statutes in force at the time? And then: How can these boundaries that were created, identified, and described be recovered?

The following principles are discussed in this chapter:

PRINCIPLE 1. What a boundary is, is a question of law. Where a boundary is, is a question of fact.

PRINCIPLE 2. The creating and retracing boundary surveyors should not give legal advice to a client, either orally or in writing.

PRINCIPLE 3. The creating and retracing surveyors are obligated to keep current on legal opinions rendered by the courts of the various jurisdictions.

PRINCIPLE 4. Only an original survey has permanency of location. A retracement survey is subject to collateral attack by any subsequent surveyor. The original survey creates the lines to be described and then retraced.

PRINCIPLE 5. In a retracement, a surveyor surveys deed lines or description lines created by the original survey.

Brown's Boundary Control and Legal Principles, Eighth Edition.
Donald A. Wilson, C.A. "Tony" Nettleman III, and Walter G. Robillard.
© 2024 John Wiley & Sons, Inc. Published 2024 by John Wiley & Sons, Inc.

PRINCIPLE 6. Unless modified by a written contract, a retracing surveyor's responsibility is to retrace and identify those lines that were created by the original surveyor. No surveyor is vested with the authority to make the determination of property rights. Surveyors report facts and do not advise of property rights.

PRINCIPLE 7. Every parcel of land whose boundaries are surveyed and monumented by a surveyor should conform with the official record that is created. Relationships of possession and use lines to deed lines should be depicted on a plat or indicated in any report furnished to the client.

PRINCIPLE 8. It is the obligation of the surveyor to inform a client of the documents that are needed for the conduct of a boundary survey. The survey should be based on an adequate and satisfactory description from a document or documents when possible.

PRINCIPLE 9. A surveyor should examine sufficient documents, either called for directly or implied, in the client's conveyance or description. This includes obtaining necessary maps, plats, drawings, deeds, and references to adjacent parcels from private, public, and other sources.

PRINCIPLE 10. A surveyor has no authority to determine property ownership or property rights. The surveyor locates boundaries in accordance with and in relation to legal descriptions.

PRINCIPLE 11. A land surveyor locates boundary lines according to the description in the deed and then relates lines of possession that do not agree with these lines and reports the facts to the client, preferably in writing.

PRINCIPLE 12. It is the obligation of the surveyor to search for those corners and monuments called for either directly or implied in the description, and then to report on the evidence found pertaining to them.

PRINCIPLE 13. Because original lines surveyed and marked by the creating surveyor control over all other elements in a deed, the surveyor must determine whether the line of possession recovered is the best remaining evidence of the original line marked.

PRINCIPLE 14. The retracing surveyor locates boundaries in accordance with the best evidence available.

PRINCIPLE 15. Each surveyor should identify the monuments set to facilitate future recovery and to minimize problems.

PRINCIPLE 16. The final product of a land survey should be a plat delivered to the client in conformance with the minimum technical standards of the state as well as a professional opinion as to the identification of the boundaries of the parcel of land. At a minimum, each plat should show all monuments found or set; the basis of bearings; bearings and lengths of all lines, boundaries, and measurements to objects; all visible encroachments; and easements according to agreed requirements.

PRINCIPLE 17. The boundary surveyor is liable for those damages resulting directly from facts not in agreement with the certifications, and the surveyor may be liable for failure to do what an ordinary prudent surveyor should do under the same or similar circumstances.

PRINCIPLE 18. If a surveyor has knowledge—actual or implied—as to how a survey will be used, responsibility will attach for a failure to conduct research and to collect adequate information covering the parcel. In some circumstances, this may even extend to adjacent parcels.

PRINCIPLE 19. A surveyor will assume the responsibility and liability for using the substandard or inaccurate work of other surveyors. Assumption of the risks will attach to the entire substandard work, not only the portion in error.

14.2 FUNCTION OF THE SURVEYOR

The role of the surveyor depends on what he or she is asked to do. The surveyor in boundary issues performs two functions, which are aptly identified in the Florida decision *Rivers* v. *Lozeau*.[1] The theme of surveying is that surveyors, by tradition and education, should be expert measurers. Measurement includes not only keeping current with the latest techniques, but also understanding the historic techniques on which many descriptions are based. It must, by necessity, include those areas on the edges or fringes of surveying: calculations or computations. But under the requirements of today's professional and legal worlds, surveyors are being asked to write land descriptions, interpret descriptions written by others, and evaluate maps, plats, and evidence of wide variety.

> In a retracement, all evidence a surveyor recovers is subject to collateral attack by other surveyors.

No longer is the modern surveyor able to enjoy the seclusion of the woods and fields. The surveyor has been thrust into a rapidly developing world, which is becoming smaller and smaller in terms of parcel size, but larger and larger in terms of the knowledge required and resulting legal ramifications.

14.3 OPINIONS OF FACT AND APPLICATIONS OF LAW

Principle 1.   *What a boundary is, is a question of law. Where a boundary is, is a question of fact.*

This first principle has been enumerated by many judges in many states. Through their authorities, both actual and presumed, judges will determine what weight to give in describing what the boundaries are: Is the fence the boundary, or is the deed description, or will the fence line control? In most jurisdictions, this occurs to the point of declaring deeds void for lack of a valid description.

A. C. Mulford explains how the surveyor fits into this scenario: "He is considered preeminently a measurer of land ... But in the vast majority of cases the actual measuring of land forms the smaller portion of his duties. His hardest work is often, to use a colloquial phrase, to 'Find the land to be surveyed.'"[2]

In another paragraph, he identifies, and distinguishes, the role of the surveyor, as he sees it:

" ... It would perhaps be well to consider ... the relationship between the surveyor and the one who may perhaps be called his *co-worker—the lawyer*." (Emphasis added.)

In Mulford's opinion, this places the surveyor on the team and makes him or her an integral part of any boundary problem.

Principle 2. *The creating and retracing boundary surveyors should not give legal advice to a client, either orally or in writing.*

As surveyors gain more experience in boundary creation and the resulting retracements, they increase their technical and legal knowledge to the point that they may become more proficient than an attorney. This situation may lead the surveyor to want to commence giving legal advice. This practice is discouraged in that if the advice is taken and relied on and it is wrong, the surveyor could be accused of practicing law without a license and be held responsible for any damages and possible reprimand by the registration board.

Boundary surveyors must be very careful to understand that their responsibility is in addressing opinions of fact, whereas the courts address the application of the law. These questions of fact require two groups: witnesses to deliver the facts to the jury and a jury to weigh these facts, as presented, and then evaluate and pass judgment.

Surveyors create the original evidence, then recover and interpret it. Lawyers argue this evidence when it is questioned.

The court interprets the law and applies the law to the facts. Usually, this is done after the attorneys present verbal arguments or written briefs for the benefit of the court. Land surveyors gather facts of the boundaries and, as such, are charged professionally with the responsibility to gather necessary facts adequately and independently, evaluate them correctly, arrive at conclusions of fact, and, if permitted by the court, draw conclusions of law.

Principle 3. *The creating and retracing surveyors are obligated to keep current on legal opinions rendered by the courts of the various jurisdictions.*

This principle may seem to conflict with Principle 2, but each is a separate requirement for the professional. Whereas the professional must possess a greater depth of knowledge to conduct retracement work, in that retracements enter into the legal areas of surveying more than into the technical areas. Most of the minimum standards address the creation of boundaries and not retracement of original boundaries. The retracement area usually looks to legal decisions for guidance.

Principle 4. *Only an original survey has permanency of location. A retracement survey is subject to collateral attack by any subsequent surveyor. The original survey creates the lines to be described and then retraced.*

Starting with the early federal case law, both federal courts and state courts have given precedence that the original survey must control and by law it is technically and legally correct.

A. C. Mulford said it in 1912, when he wrote that the surveyor and the lawyer were a team, with each person doing his (or her) part—the lawyer practicing law, and the surveyor doing the measurements and testifying. The system prohibits attorneys from testifying. It does require surveyors to understand the law, but they should not and cannot make legal opinions to the clients and the courts. Many surveyors will find that they have a better grasp of legal knowledge than the attorneys do, but surveyors can only serve as advisors.

As mentioned, if a surveyor gives a legal opinion to a client, who relies on that opinion, and the opinion turns out to be erroneous, the surveyor may be open to the charge of practicing law without a license.

14.4 ESTABLISHMENT OF BOUNDARIES

All state governments, and the federal government, have identified procedures for locating boundaries. Some procedures are historical, some statutory in nature, whereas others are based on cases or law court decisions. Once township and section lines are located under federal law, no entity has legislative or judicial authority to relocate or reposition section lines or section corners. States can only interpret the evidence as to the "most probable location."

All states have adopted or devised methods for establishing land or property boundaries by court decision or under court direction. With or without the parties' approval, the court using either a county surveyor or a registered surveyor, may direct the location of a disputed, uncertain, or unascertainable boundary line. No matter who conducts the survey, the work must be accomplished in accordance with any statutes in force, and this work is always subject to appeal, unless all parties agree not to do so.

Once the court signs an order identifying these court-ordered boundaries, they become binding on those persons who were named as parties and their successors in interest, but not on others. These corners and lines may now be referred to as *property lines*. If the court or the surveyor does not follow the "letter of the law" in setting these lines and corners, they may be disputed at a later date. In any statutory proceeding, verification of boundary (property) lines requires the government body to do the following:

1. Apply for the survey and usually deposit money.
2. Notify all parties in interest and adjoiners.
3. Perform the survey in conformance with the law.
4. Obtain necessary approval by the statutory officers or court.
5. Provide an opportunity to appeal with a time limit identified.

14.5 ESTABLISHMENT IN LOUISIANA

The legal process to locate lines is unique, and is used in many areas and identified by many names. In Quebec, Canada, France, and Louisiana, it is called *process verbal.* Louisiana has used it to survey unsurveyed lands obtained from the United States. The surveyor general for each parish was nominated by the governor and approved by the state senate. The statutes then related how the lands would be surveyed, marked, and recorded, and the necessary plats prepared. The final survey is then approved by the parish in which the lands are situated.

In Colorado, state statutes, Article 50, Title 38, have identified both (1) how aliquot portions of a section will be located and (2) how disputed boundaries may be settled. The first may be done by private surveyors and the second by county surveyors.

PRIVATE SURVEYS

14.6 RESPONSIBILITY AND AUTHORITY OF THE SURVEYOR

> A surveyor's work can be questioned either in contract or in tort.

In the United States, every state, commonwealth, possession, and territory now requires registration or licensure of surveyors. Some boards are joint boards with engineers and/or architects, and some are separate boards for land surveyors only. Licensing permits a surveyor in a particular state to do those functions that the law defines as surveying. No two states identify surveying the same way. The law prohibits all others from performing these functions under penalty. Other states are helpless to prohibit nonsurveyors from performing, but can control those who are registered.

In recent years, the US Department of the Interior has initiated a federal examination for surveyors with the emphasis on the federal lands. Some states, including the metes and bounds states, are now requiring this certification in order to qualify for any type of survey work with their respective agencies.

Most states now have, in effect, minimum technical standards to govern the standard of work that one would expect from a registered or licensed practitioner. Since each state, commonwealth, and territory has its own definitions and requirements, it is impossible to explain and discuss them all. This book concerns those basic principles that are germane to all registered surveyors.

Principle 5.   *In a retracement, a surveyor surveys deed lines or description lines created by the original survey.*

Most boundary issues occur in retracements, and not in the original surveys. This is basically because retracing surveyors are placed under a greater burden of, first, knowing what evidence to look for and, then, once the evidence is found, determining what weight to give it and being able to support their opinions. Many surveyors are qualified in the fieldwork but lack the communication skills to impart that knowledge to third parties.

Under the principles of retracements, no subsequent surveyor has authority to change or modify any of the originally created lines or corners. The agency is prohibited from changing any originally created lines or corners if to do so would jeopardize any prior rights (bona fide) that were granted in any prior survey. This can only be accomplished through either a dependent or an independent resurvey.

> The surveyor should not let the client dictate how to perform a retracement. A retracement should be the independent work of the surveyor.

Principle 6.   *Unless modified by a written contract, a retracing surveyor's responsibility is to retrace and identify those lines that were created by the original surveyor. No surveyor is vested with the authority to make the determination of property rights. Surveyors report facts and do not advise of property rights.*

Perhaps, this one principle is one of the most difficult for a knowledgeable surveyor to understand and apply. Granted, the qualified boundary surveyor must know the law of retracements, which is steeped in case law, common law, and in some instances statute law, but this knowledge can only be applied in conducting retracement and not in advising clients. There have been instances where the surveyor not only conducted the retracement, but also advised the client as to their legal rights and positions, even to the point of helping them draft legal documents and helping the client to represent himself or herself pro se.

This warning is not given lightly in that many professions are quite jealous of their responsibilities, and if the advice were wrong, giving legal advice is not included in the definition of surveying.

14.7 BASIS OF A BOUNDARY SURVEY

Principle 7. *Every parcel of land whose boundaries are surveyed and monumented by a surveyor should conform with the official record that is created. Relationships of possession and use lines to deed lines should be depicted on a plat or indicated in any report furnished to the client.*

In performing a retracement, the quality of the retracement can only be as accurate as the original survey that created the original boundaries. The more detailed and explicit the original record, the more accurate will be the retracement survey.

A surveyor often can limit future liability or future problems with clients by first having a firm understanding of what the client actually wants the survey to show. It is not logical that clients want surveys of what they possess. The basic question usually is: Am I on the land described in my deed?

Unless authorized in writing to do otherwise, a surveyor should endeavor to locate the client's deed and then show any conflicts with the lines described. There may be times when the surveyor is directed to survey possession lines or other lines that conflict with the deed. A surveyor must understand that surveying to or from corners, points, or lines shown by the client provides no defense in questions of liability.

When a survey is complete and has been provided to the client, the surveyor should indicate the source of the materials used and any conflicts, in possession or title, with the client's title or deed.

Principle 8. *It is the obligation of the surveyor to inform a client of the documents that are needed for the conduct of a boundary survey. The survey should be based on an adequate and satisfactory description from a document or documents when possible.*

The final boundary retracement survey can be only as accurate and as definite as the documents used in its location. The surveyor must determine who will provide the necessary documents. Simply having the client provide the necessary documents will not provide a defense to the surveyor should these documents prove to be inaccurate or deficient. The surveyor is obligated to conduct the survey using the best evidence available. The surveyor may find that the documents provided are unsatisfactory to accomplish the job, and thus additional time must be expended to get those necessary documents.

14.8 HOW MUCH RESEARCH?

Principle 9. *A surveyor should examine sufficient documents, either called for directly or implied, in the client's conveyance or description. This includes obtaining necessary maps, plats, drawings, deeds, and references to adjacent parcels from private, public, and other sources.*

There is no advice that anyone can give a surveyor as to how much research should be conducted. The major causes of disagreement among surveyors are (1) the failure to locate and identify sufficient documents from which sufficient information can be gleaned about the area to be surveyed; (2) insufficient and/or inadequate field search of information, including corners, monuments, and information called for; and (3) faulty or incorrect interpretation of all types and forms of evidence, including documentary evidence. Generally, state standards of practice will dictate how much research should be done. With title insurance surveys, the ALTA-NSPS instructions are quite detailed and offer reliable guidelines.

As discussed in Section 3.6, some courts have emphasized that titles should be taken back to their origin. Some state standards require going back to the origin of the description. When the surveyor, or researcher, fails to do this, the following critical items remain unknown:

1. The differences that exist between the current and the original descriptions, and whether any mistakes or omissions have occurred between the two.
2. The surrounding circumstances at the time the description was first compiled, which is an interpretation requirement self-imposed by the court system.
3. A base year for the conversion of magnetic bearings due to changes in declination.
4. Whether the parcel is the result of a sequential, or a simultaneous, creation.
5. Easements and encumbrances described in the early part of the title but not appearing in later documents.
6. Early references to surveys and plans.
7. Whether any agreements or other documents appear back in time that affect either the title or the boundaries, or both.

Documenting and memorializing what you researched is just as important as the research job itself. This task can be as simple as keeping an Excel spreadsheet with the grantor, grantee, document number (or book/page), and URL link to the image. Alternatively, some surveyors chose to create visualizations of the chain-of-title, such as mind maps, which can be used later in their surveyor reports to more clearly explain the sequence of conveyances. At no time should the researcher say he examined or used *all* of the documents.

Often times, the surveyor concludes the record research process by simply compiling a list of conveyances. A more prudent final step in the record research process is the creation of a working sketch. This working sketch creates a visualization of the critical deeds, easements, leases, and other conveyances. Then these conveyances are overlaid with each other. A prudent practice is to use a software program such as AutoCAD to plot the conveyances and then export that file to PDF. The PDF export allows each conveyance to be turned on or off, made transparent, or colorized. The result of this working sketch is the ability to analyze gaps, overlaps, scrivener's errors, or other problems within your project area before field work has begun.

14.9 OWNERSHIP

Principle 10.   A surveyor has no authority to determine property ownership or property rights. The surveyor locates boundaries in accordance with and in relation to legal descriptions.

In the United States, no state has ceded the authority to assess and determine property ownership and/or property rights to the federal government. Many other countries have vested this authority in their surveyors, but only after education, apprenticeship, and examination. In the United States, the surveyor's main responsibility is to locate boundaries in accordance with a written description or descriptions and then to report any conflicts or encroachments.

14.10 ENCROACHMENTS

Principle 11.   A land surveyor locates boundary lines according to the description in the deed and then relates lines of possession that do not agree with these lines and reports the facts to the client, preferably in writing.

Since surveyors are "expert measurers," they should give their clients measurements and not legal opinions. Some surveyors and companies show such words as "encroachment" and "trespass" on plats. The surveyor's responsibility is to show facts. Black[3] defines *encroachment* as " ... upon a street or highway is a fixture, such as a wall or fence, which illegally intrudes or invades the highway ... "

To really understand the word, we should examine the verb. Once again referring to Black, *encroach* means "to enter by gradual steps or stealth into the possession or rights of another; to trespass or intrude."[4]

The surveyor lacks knowledge of how the possession originated. The involved landowner may have gone into possession under a license or permission that the surveyor does not know about. The surveyor can only show the relationship of the deed lines to possession lines; these are facts. The surveyor *does not* know whether this was done with permission or not.

Unwritten rights of possession and title relative to encroachments are very complicated and involved issues. The surveyor should be careful to keep the survey aspects separated from the legal aspects when it comes to encroachments. A surveyor should not attempt to give advice as to the legality of possession, use, and encroachments. However, a surveyor can give information as to the strength and validity of the evidence of possession.

14.11 SEARCHING FOR MONUMENTS

The principle that called-for monuments in land conveyances control the location of the parcel and the lines is solid in US Survey Law. The search for the true and correct location of these monuments must continue until the surveyor proves them or decides that further searching is impractical.

Principle 12. *It is the obligation of the surveyor to search for those corners and monuments called for either directly or implied in the description, and then to report on the evidence found pertaining to them.*

One will find that much litigation and serious disagreement have resulted from surveyors' failures (1) to understand what it is that should be done in the course of a survey, (2) to find the original monuments and corners called for in documents, and (3) to find a sufficient amount of the original evidence on which to base an adequate retracement or resurvey. Simply because a previous surveyor failed to find the original evidence or located the original corners and monuments incorrectly does not relieve the current surveyor from determining what should be present and then searching for this evidence. One surveyor cannot predicate these decisions on a previous surveyor's acceptance of corners and monuments.

> Reliance of a retracing surveyor on a previous surveyor's work does not relieve the retracing surveyor from liability in the event the previous surveyor's work is in error.

A surveyor should not rely on the number of surveyors who used a corner point, but reliance should be placed on the quality of the independent determination of the corner or monuments. Once a surveyor undertakes a retracement or a resurvey of a boundary, it is obligatory that a diligent, thorough, and complete search of all evidence of the original survey be completed so that the position of the original corners and their monuments can be ascertained. If the original monuments are not or cannot be found, this should be explained. Failure to find the positions of the original monuments and corner positions can lead to unnecessary litigation and possible professional liability.

14.12 POSSESSION MARKING ORIGINAL SURVEY LINES

Principle 13. *Because original lines surveyed and marked by the creating surveyor control over all other elements in a deed, the surveyor must determine whether the line of possession recovered is the best remaining evidence of the original line marked.*

A basic principle that a land surveyor must understand is the difference between possession and title. The official title that a client may have described in a deed or other document may or may not be what is actually being possessed. Lines of possession may originate near, at, or even before a survey and the preparation of the document that gives title. When a surveyor attempts to identify an original surveyed line, it is obligatory that the relationship of the current possession line be analyzed to determine its relationship, both in time and in location, to the original survey that created the parcel.

The main objective of the creating surveyor is to create boundaries that are sufficiently marked and described to be able to recover these lines in the future. The main objective of the retracing surveyor is to recover these original lines with a sufficient degree of certainty to ensure their location.

At times, the surveyor may have to consider what the best available evidence is to prove the location of the original surveyed lines. It is here that possession may be the solution. The surveyor must not use possession as the sole criterion unless a complete analysis of the possession evidence is made. This should include, but not be limited to, the ages of all fences, the ring count over blazes on trees, and the collection of testimony from landowners.

14.13 EVIDENCE

Principle 14. The retracing surveyor locates boundaries in accordance with the best evidence available.

Because the identification of a land parcel is dependent on evidence, there may be isolated times when the surveyor will find that the title to the owner is good, but the property defies location according to the evidence in the documents. Thus, the person owns the property, but it cannot be located. It is at times like this that the surveyor must ascertain what the best available evidence is to locate the parcel. The surveyor is not charged with the responsibility of qualifying title but only with the quality of description.

The retracing surveyor is totally dependent upon the quality of the description that was created by the original surveyor.

Simply because a surveyor cannot find evidence of the parcel on the ground does not make the conveyance void. It is in these difficult circumstances that the surveyor is obligated to use the best evidence available to attempt to locate the parcel, to the best of his or her professional ability. *Best evidence* that is referred to by attorneys in a legal sense should not be confused with the best evidence in a survey sense. The Texas court emphasized this point, by stating that "titles to land are not to fail merely because old markers may have disappeared or because it may be difficult to trace footsteps of the surveyor."

14.14 SETTING MONUMENTS

Principle 15. Each surveyor should identify the monuments set to facilitate future recovery and to minimize problems.

In the everyday work of surveying, one of the most frustrating problems that a surveyor encounters is finding monuments of a survey that are unidentified as to their origin. Today, many states have attempted to rectify this problem by requiring all registered surveyors to place their registration numbers on the monuments set on surveys. However, this does not help the thousands of monuments that were set before the laws were enacted. All surveyors should endeavor to leave an adequate path of identification for future surveyors to follow. This encourages a surveyor to make personal contact and obtain information that may eliminate problems and keep disputes to a minimum.

14.15 PLATS

Principle 16. *The final product of a land survey should be a plat delivered to the client in conformance with the minimum technical standards of the state as well as a professional opinion as to the identification of the boundaries of the parcel of land. At a minimum, each plat should show all monuments found or set; the basis of bearings; bearings and lengths of all lines, boundaries, and measurements to objects; all visible encroachments; and easements according to agreed requirements.*

The liability of a professional land surveyor is determined in two ways: through comparison of the survey product with the state board's minimum technical standards and through comparison of the surveyor's methodology in the current matter with the acts of a fictitious reasonable surveyor.

First, each work product of the surveyor such as a plat or map should be compared to the minimum technical standards (sometimes known as standards of practice). If the plat or map includes the required elements, then the minimum technical standards have been met. These standards are authored by the state legislature (sometimes known as acts) and/or the state board of land surveying registration (sometimes known as rules).

In Florida, Florida Statute 472 includes general expectations of professional surveyors and mappers such as honesty and confidentiality. On the other hand, Florida Administrative Code 5J-17 includes several pages of checklists for both general (all) surveys and boundary surveys. Here, 5J-17 is titled *standards of practice*, but the same statute was formerly titled *minimum technical standards*. While some items in this checklist may seem trivial (i.e., labeling all points, either on the map or in the legend) other checklist items are essential to performing a quality survey (finding or setting all boundary corners).

Second, under common law, a land surveyor is negligent if he fails to adhere to the same standard of care that well-qualified professionals acting under similar circumstances would. The actions of a fictional professional surveyor are determined through consultation of land surveying textbooks, journal articles, continuing education presentations, and other industry treatises. If this reasonableness question arises in court, expert testimony from another professional surveyor is often required.

No matter what state the surveyor is practicing in, or the type of survey map being published, the surveyor should prepare all work to the state minimum technical standards and to the common law standards of the profession.

14.16 LIABILITY

Principle 17.   *The boundary surveyor is liable for those damages resulting directly from facts not in agreement with the certifications, and the surveyor may be liable for failure to do what an ordinary prudent surveyor should do under the same or similar circumstances.*

This principle not only applies to reported facts but to actions taken in the course of the retracement.

This principle actually addresses two areas: (1) certification and (2) the prudent surveyor standard. A surveyor's certificate as to the work performed is no place for "puffing." Simple, clear statements about the information that was used, the work that was done, and the conclusions that were made are all that is necessary. Any statement beyond this may result in liability.

A surveyor's work will be judged on what was contracted for or on the written standards in the local surveying community. If there are no written standards from which to judge the surveyor, the courts will look at the fictional surveyor they call the *ordinary prudent surveyor.* This person does not exist except in the law. The ordinary prudent surveyor is a person who can be expected to survey correctly, with only those errors that can be expected in the ordinary course of conducting a survey.

Liability will result when a practicing surveyor does less than the ordinary prudent surveyor or less than what the minimum technical standards require. To avoid liability, the surveyor in practice should do more than the minimum technical standards require and be a better surveyor than the ordinary prudent surveyor.

Principle 18.   *If a surveyor has knowledge—actual or implied—as to how a survey will be used, responsibility will attach for a failure to conduct research and to collect adequate information covering the parcel. In some circumstances, this may even extend to adjacent parcels.*

Many surveyors in business will be surprised to learn that clients will be less than honest with them when it comes to informing the surveyor of certain information that is needed to perform the work. This might be information about the true story of monuments or about what the survey of a certain parcel is going to be used for. It matters little whether the client informs the surveyor about the use to which the parcel will be put. The ultimate use of the property should not govern the quality of the survey.

Principle 19.   *A surveyor will assume the responsibility and liability for using the substandard or inaccurate work of other surveyors. Assumption of the risks will attach to the entire substandard work, not only the portion in error.*

If a surveyor accepts the work of other surveyors or alters the work, including plats, of other surveyors, liability cannot be passed on or diverted. There will be times when a client may discharge a surveyor and request notes, plats, and all information concerning the project. Simply turning over the information will not relieve the creating surveyor from liability. For decades, questions have been raised as to what the client is entitled to. To limit liability or reduce possible conflicts with clients, the surveyor should never undertake a project without first having a written contract to identify what the client will be entitled to in the event of premature termination or even at the time of the final product. Liability can possibly attach to both parties in the event that the original surveyor made errors and a subsequent surveyor used the erroneous product in a future project.

Many minimum standards will not permit a registered surveyor who is employed by a firm that has numerous crews to continue work that was started by other employees. If the surveyor is required to do so, then it is recommended that a sufficient number of checks be made to verify the quality of the earlier work.

> It is recommended that no surveyor rely on the work of a prior surveyor without first completing sufficient checks to determine its correctness.

14.17 CONCLUSION

In this final section, more questions than answers may be presented, but they will be food for thought and for professional enrichment.

A surveyor will, in all probability, encounter more problem areas when retracing old boundaries than when creating new boundaries. Questions arise as to what to do when one finds a well-established, accepted monument that the new resurvey indicates is incorrect and at the wrong position. Should it be left alone and used, should it be removed, or should a new monument be placed and the old one left? A book can never provide the correct answer for any particular situation. The authors can only relate what they would do. What should a surveyor do when monuments are found that can be proved to have been established inconsistent with the law or the proper rules of survey? The surveyor will find that there are times when more questions exist than answers.

Perhaps one of the most difficult and perplexing problems that a retracing surveyor will encounter is what one group has chosen to call "pin cushion corners." We accept the fact that there can be a corner without a monument and a monument without a corner. However, the problem arises when there is a proliferation of monuments at or near a single corner, none of which were referenced in prior writings. In order to accept any monument in place, that monument must be called for in documents that are referenced in the chain of title. Then the monument called for must be found as described in the writings. Simply finding a monument, in place, places no obligation on the retracing surveyor to accept without other creditable evidence to support its acceptance.

Many surveyors have used or will use monuments that purport to have been the perpetuation of the original corner, but no evidence can be assembled to make the necessary proof to a degree that makes the surveyor comfortable accepting it. What should the surveyor do? A surveyor may determine that time, rather than a correct survey, will ultimately cure any survey or boundary problem. A surveyor may become disgruntled when a judge or jury renders a decision that is totally inconsistent with the evidence or the proper rules of surveying, but we cannot advise as to how such frustration can be eliminated.

These are the questions that a professional should expect. Unfortunately, though, there may be no immediate answers. A book can only set the tone and pattern for learning. As final thoughts, there are certain principles of boundary establishment and recovery that we believe surveyors should understand and follow. These include the following:

1. Courts cannot establish boundaries where they wish, regardless of the evidence and testimony. There must be some foundation in evidence for a court to locate a boundary.
2. Any legal establishment of a boundary will control only those persons named as parties.
3. Boundaries may be altered by unwritten legal means.
4. A retracement that is conducted for a client should be predicated on law: the law that created the original lines and the law providing for the retracement of the lines.
5. If all surveyors were to interpret boundary evidence in the same manner, there would be no conflicts of location.

Most of the everyday work of a surveyor is routine and without conflicts or problems. Much of the routine work consists of fundamental measurements and unquestionable evidence of the location of a parcel and its corners and lines. There are times when a surveyor will spend untold hours, days, and possibly even weeks proving the location of a parcel, only to have all this work questioned by someone who is probably less qualified and less capable and who probably shortcuts the survey process. It is at times like this that we hope the surveyor will reach to the bookshelf and take down an old friend like this book.

The basic philosophy of any textbook on land boundaries can be summarized in two key points, intended as both final rules and final philosophy for the student as well as the experienced surveyor.

> Original surveys create boundaries, and as such the law holds that they are without error and unassailable, even by the courts.

> A retracement of an original boundary is predicated on the evidence recovered and is always open to collateral attack by others, even though the attackers have never conducted their own retracement.

We have added a chapter on ethics and moral responsibilities for the practicing surveyor and, more particularly, for the surveying student in order to provide guidance and direction for bringing the professional into the twenty-first century. We hope that the people who read this chapter will have an appreciation for the beauty of their chosen profession.

NOTES

1. *Rivers* v. *Lozeau*, 539 So.2d 1147 (Fla. Dist. Ct. App. 1989).
2. A. C. Mulford, *Boundaries and Landmarks* (New York: Van Nostrand & Co., 1912).
3. *Black's Law Dictionary*, (St. Paul, MN: West Publishing Co., 1968), 620.
4. Ibid. Page 620.

CHAPTER 15

THE ETHICS AND MORAL RESPONSIBILITIES OF BOUNDARY CREATION AND RETRACEMENTS

15.1 INTRODUCTION

The final chapter of this book transcends the technical and delves into some of the more esoteric areas of the surveying profession: ethics, morality, and responsibilities. These traits are among those that will follow the professional throughout his or her entire career. There may be a few other esoteric traits that are necessary. In over a dozen instances, A. C. Mulford mentioned two traits that surveyors must have:[1]

> "The watchwords of the surveyor are 'patience and common sense.'"

Boundaries are personal in nature. All living things—humans, animals, fish, and some plants—recognize some form of boundary. The landowner, the surveyor, the attorney, and the courts all look at boundaries and approach them in different ways. All parties become involved at the point of the very creation of boundaries—from the preparation of the description of the parcel identifying the boundaries to the possession and then to the actual boundaries created. If problems are encountered early in the possession or at a remote future date, an attorney may be consulted and a surveyor employed to explain why such a circumstance happened. This may come from

Brown's Boundary Control and Legal Principles, Eighth Edition.
Donald A. Wilson, C.A. "Tony" Nettleman III, and Walter G. Robillard.
© 2024 John Wiley & Sons, Inc. Published 2024 by John Wiley & Sons, Inc.

litigating the boundaries and perhaps even appealing the results of the litigation to the peaceful settlement by all parties.

Few professionals consider or realize that boundaries may be created either in writing or by unwritten means, and that either form may be questioned after many intervening years. The positive location of any boundary is determined by the recovery and the identification of the original evidence that was created, described, later recovered, and then interpreted by the surveyor, possibly argued by the attorneys, and, maybe, determined by a jury.

15.2   THE PHILOSOPHY OF BOUNDARIES

As stated in earlier chapters, *boundaries are created*, and any parcel's boundaries do not exist until they are created by any one of several methods. These recognized methods are as follows:

1. By running on the ground and creating a paper trail
2. By words
3. By law

Once a boundary is created, the future responsibility involves finding that original boundary through a process the courts call *retracement,* or "following the footsteps." An amalgamation of the numerous principles that are involved in this little-understood process is as follows:

PRINCIPLE 1. The retracing surveyor should avoid retracing the same line or lines between two different landowners.

PRINCIPLE 2. If the creating surveyor or the retracing surveyor is hesitant about his or her ability to perform the requested services, he or she should decline the project.

PRINCIPLE 3. In a retracement, it is better to have an imprecise survey where the land actually exists, than it is to have precise measurements where the land does not exist at all.[2]

PRINCIPLE 4. The surveyor should strive to help resolve any boundary dispute before it becomes a litigious matter.

PRINCIPLE 5. The surveyor who creates the original boundaries should conduct the work to a degree of technical precision so as to provide the retracing surveyor with sufficient evidence for recovery purposes, regardless of the changes in technology.

PRINCIPLE 6. The creating surveyor should leave sufficient credible evidence, both field and record, that physically and legally describes the boundary lines that he or she creates for future recovery.

PRINCIPLE 7. The creating surveyor creates the boundaries with the technology and words in use at that time; the retracing surveyor retraces the same boundaries with the words and technology used during his or her time.

PRINCIPLE 8. Modern surveyors must ascertain whether they are creating surveyors or retracing surveyors of the boundaries. The requirements and the responsibilities are different.

PRINCIPLE 9. The creating surveyor should not use substandard or unapproved methods in creating boundaries.

PRINCIPLE 10. The creating surveyor should create any boundaries in accordance with the existing laws in effect at the time of the survey.

PRINCIPLE 11. The original surveyor should conduct the original survey with the principle in mind that the boundaries being created may have to be retraced and possibly restored in the future.

PRINCIPLE 12. Before attempting any boundary project, the surveyor should have an understanding of the differences between original surveys and retracements.

PRINCIPLE 13. An original boundary line and corners, once created and described, can never be re-created or redescribed; they can only be retraced in relation to the original survey.

PRINCIPLE 14. The principles of an original boundary creation are distinctly different from those of a boundary retracement.

PRINCIPLE 15. The principles of original boundary creation are more technical than legal.

PRINCIPLE 16. The principles of boundary retracement are more legal than technical.

PRINCIPLE 17. In conducting a retracement, the retracing surveyor should obtain not only the most current description of the lines but also the original description that identified the original boundaries, and then determine the relationship between the two.

PRINCIPLE 18. The retracing surveyor should understand how the priority of calls can aid in the retracement process and should realize that, in some states, not applying these principles could lead to a claim of negligence per se.

PRINCIPLE 19. The retracing surveyor should leave sufficient tangible and field evidence supported by written understandable evidence for the retracement of the client's boundaries.

PRINCIPLE 20. In a boundary retracement, the retracing surveyor should not retrace the common boundary for and between two different clients.

15.3   APPLYING THE PRINCIPLES TO CREATING AND RETRACING BOUNDARIES

This section expands on the principles identified in the preceding list with appropriate discussion to aid the student in being able to understand the reasoning.

Principle 1.   *The retracing surveyor should avoid retracing the same line or lines between two different landowners.*

This principle touches on a possible ethical situation. A retraced boundary is predicated on called for evidence and found evidence. If the two contiguous landowners present the retracing surveyor with different evidence from which to work, a situation may present itself with the surveyor having to defend two locations that are dependent on different documents. The problem is, which one will the surveyor defend?

Principle 2.   *If the creating surveyor or the retracing surveyor is hesitant about his or her ability to perform the requested services, he or she should decline the project.*

Like every profession, surveying has its "Jack of all trades." This principle requires surveyors to examine their ability to conduct the work to the standard required. This includes not only the technical capabilities but also the extent of the knowledge required for a creditable completion. Retracing ancient lines requires people of different capabilities. Much of the work borders on the legal areas of the profession, rather than the technical.

Principle 3.   *In a retracement, it is better to have an imprecise survey where the land actually exists, than it is to have precise measurements where the land does not exist at all.*

This principle requires a person of "patience and honesty." These two attributes require a degree of technical skill, but also the ability to recognize how to work with the law and associated conflicting evidence.

Principle 4.   *The surveyor should strive to help resolve any boundary dispute before it becomes a litigious matter.*

One of the great books of all time stated, *Blessed are the peace makers.* This follows comments made by Justice Thomas Cooley, who said, "When a surveyor enters a neighborhood, chaos follows." Clients engage surveyors because they have faith in their abilities to perform the needed services. Surveyors as professionals usually maintain networks in the surveying community. Many problems can be solved by exchanging information between the two parties.

Principle 5.   *The surveyor who creates the original boundaries should conduct the work to a degree of technical precision so as to provide the retracing surveyor with sufficient evidence for recovery purposes, regardless of the changes in technology.*

For all practical purposes, the creating surveyor determines the quality of the original boundaries. Like the DNA of a person, the surveyor's experience, knowledge, technical capabilities, personality, and human traits are combined to create the boundary lines. Unfortunately, the lines created have certain attributes in that they are invisible and can only be seen in legal terms. Both the lines and the terminal points have only legal status and are not physical until a monument is set at the corner and the line is marked by the act of a human being. If the creating surveyor is not technically proficient, if the equipment is inferior, if the surveyor uses unapproved methodology, if the surveyor is remiss in documenting his or her work—if, if, if—then the evidence of the boundaries created is to an inferior standard and the record left for future surveyors will cause future problems. The case of *Rivers* v. *Lozeau*[3] has been referenced several times for different reasons; in that case, the court emphasized a very important point:

> "[T]he surveyor can, in the first instance, lay out or establish boundary lines within an original division of a tract of land which has theretofore existed as one unit or parcel. In performing this function, he is known as the 'original surveyor' and when his survey results in a property description used by the owner to transfer title to property, that survey has a certain special authority in that the monuments set by the original surveyor on the ground control over discrepancies within the total parcel description and, more importantly, control over all subsequent surveys attempting to locate the same line."

That is why it is vitally important for the surveyor to do good work when conducting an original survey.

Principle 6. *The creating surveyor should leave sufficient credible evidence, both field and record, that physically and legally describes the boundary lines that he or she creates for future recovery.*

One of the weakest elements in any boundary description is the description prepared by the creating surveyor. The law provides that if a bearing tree is referenced to a corner at the time the original survey was performed, *you find the tree; you find the corner*. Little did the creating surveyor realize that his failure to adequately describe the tree's diameter or species would lead to litigation. His estimate of a 26-inch hickory when in reality it was a 20-inch yellow poplar became a critical piece of evidence when the monument disappeared. An additional 3 or 4 minutes' effort at the time of the original work would have prevented costly litigation 50 years later. Many of the original sizes were estimations. Today, the courts look on these estimated distances as absolute. A court in Wisconsin discredited the testimony of a lay witness who saw the monument and two original bearing trees over a registered surveyor who estimated a distance of 22 poles because the surveyor was an expert at "estimating distances."[4]

Principle 7. *The creating surveyor creates the boundaries with the technology and words in use at that time; the retracing surveyor retraces the same boundaries with the words and technology used during his or her time.*

When the authors commenced their careers, the standard equipment for creating surveys were the compass, the transit, and the chain. Today, the standard equipment used by surveyors is the total stations and the global positioning system (GPS) unit. In some states, the compass is prohibited for surveys; thus, that leaves the modern surveyor with a burden to "find the footsteps." This can be a technical dilemma. Description bearings created by compass observation cannot be converted to total station or even GPS bearings. Neither of these instruments can read bearings.

Principle 8. *Modern surveyors must ascertain whether they are creating surveyors or retracing surveyors of the boundaries. The requirements and the responsibilities are different.*

Today's surveyor has to be a diverse individual. A client asks that the 40-acre parcel he owns be subdivided into 40 new 1-acre lots. First, the 40-acre parcel's boundaries must be determined. This is a retracement. The modern surveyor must retrace the original boundaries, which may be 100 years old. Reliance is placed on the original description. This is a classic retracement. Now, after these exterior boundaries are determined, the new 40 parcels must be created. This is an original survey, and new boundaries are being created. New principles apply. Lines must be run, monuments set, and new descriptions prepared.

Principle 9. *The creating surveyor should not use substandard or unapproved methods in creating boundaries.*

In the early surveys, there were no standards for the creation of new parcels. The exception were the federal surveys, which were conducted under federal laws. Today, the states have this requirement under control in that most, if not all, states have enacted *minimum standards* for creating new boundaries. It should be noted that the original federal laws, the ones created in the 1800s, are still in effect today.

Principle 10. *The creating surveyor should create any boundaries in accordance with the existing laws in effect at the time of the survey.*

There are several areas of law that surveyors should understand, apply, and respect when it comes to surveying and retracing boundaries. The major type is statute law, but the surveyor should also understand common or case law, and then administrative law as well. To effectively work in the surveying world of today, all of these areas of law must be understood, followed, and respected.

Statute laws must be followed. These could range from the acceptance of a state plane coordinate system to mandatory continuing education requirements.

Case law also sets standards that must be followed in surveying. Some states identify the controlling elements of conflicting property descriptions, the control of bearings over distances, or, presumably, early descriptions that were slope distances. Although not codified, these case precedents must be followed. This also includes understanding the validity and the application of the priority of calls.[5]

Administrative law identifies the relationship of the surveyor to society through registration requirements.

Principle 11. *The original surveyor should conduct the original survey with the principle in mind that the boundaries being created may have to be retraced and possibly restored in the future.*

This is a difficult principle to understand in that we seldom plan for the future with boundaries. Placing inadequate monuments that are easily destroyed or describing the monumentation in abbreviated or ambiguous words can be harbingers of disaster.

Principle 12. *Before attempting any boundary project, the surveyor should have an understanding of the differences between original surveys and retracements.*

The original creating surveyor should be qualified to conduct the creation of the boundaries that are technically sufficient, while the retracing surveyor should have legal knowledge; knowledge of both historic and modern technology, which includes the ability and qualifications to conduct the historic research and the field investigations; and an understanding of the law of evidence, as well as having a unique ability to communicate this information in writing and other forms of communication to the client and the courts.

Principle 13. *An original boundary line and corners, once created and described, can never be re-created or redescribed; they can only be retraced in relation to the original survey.*

Since the mid-1800s, courts from the US Supreme Court to the lesser courts have held the validity of the original surveys, to the point that the Supreme Court decision in *Cragin* v. *Powell*[6] held that the original surveys are unassailable through the courts. This principle applies to those boundaries created for private parties as well.

Principle 14. *The principles of an original boundary creation are distinctly different from those of a boundary retracement.*

The creation principles have greater reliance on technology. Retracement gives greater weight to law.

In applying these two principles, one should realize the technology area changes exponentially while the legal area changes very slowly.

Principles 15 and 16. *The principles of boundary creation are more technical than legal. The principles of boundary retracement are more legal than technical. The creating surveyor uses the technology and knowledge of the day to create the lines, and, upon creation of the boundary lines, the retracing surveyor must understand and apply the rule of evidence and law of real property to identify those original lines.*

These two principles make the distinction that *two different individuals may have to be employed to conduct these two different aspects of the survey.*

Principle 17.   *In conducting a retracement, the retracing surveyor should obtain not only the most current description of the lines but also the original description that identified the original boundaries, and then determine the relationship between the two.*

Both the original description that created the boundaries and the most current description become important. First, the original description is the one that created the parcel, but the most current one is the one through which the landowners claim possession. The older the description, the more problems that may exist. The secret is the "quality" of the original description and the resultant lines. For example, the original description was created in 1798. It contained calls for identifiable lines and several definitive corners. Throughout the years, this description has been used to convey the land. Although this description was over 200 years old, by applying the elements of survey law and fact, a sufficient number of those description elements could be found to locate the parcel on the ground with a high degree of certainty, or, as the court said, "the preponderance of the evidence."

Principle 18.   *The retracing surveyor should understand how the priority of calls can aid in the retracement process and should realize that, in some states, not applying these principles could lead to a claim of negligence per se.*

The retracing surveyor is given much latitude in analyzing evidence to prove the location of historic boundaries. The first concept the surveyor should know is that the standard of proof to apply is *the preponderance of evidence.*

This is the least demanding—in fact, a court would prefer to give weight to evidence rather than using proportioned measurements.

In a Virginia decision, the court, in identifying the standard of responsibility for the retracing surveyor, wrote as follows[7]:

Land surveyors, like other professionals, are governed by certain Standards which have ripened into certain rules of law. In the absence of evidence of contrary intent, a distinct order of preference governs inconsistencies in the description of land:

1. Natural monuments or landmarks;

2. Artificial monuments and established lines, marked and surveyed;

3. Adjacent boundaries or lines of adjoining tracts;

4. Calls for courses and distances;

5. Designation of quantity.

The foregoing rule is not inflexible .... The retracing surveyor was negligent as a matter of law.

Such exceptions fix binding standards which are not left to the exercise of professional judgment. *Being rules of law, they are not subject to expert opinions."* ... (Emphasis added.)

Principle 19. *The retracing surveyor should leave sufficient tangible and field evidence supported by written understandable evidence for the retracement of the client's boundaries.*

Principle 20. *In a boundary retracement, the retracing surveyor should not retrace the common boundary for and between two different clients.*

One of the most tempting situations in a retracement situation is when a surveyor is engaged to conduct a project for one client and locates a line common to an adjoiner. At a later date, either the original owner or a subsequent owner of the adjacent parcel engages the same surveyor to survey the same line, under the pretext that this surveyor already has the information and can do the job more cheaply, or some other excuse. Based on other evidence, the line may be repositioned in a different location. *This will lead to trouble in the retracement.* Who gets the support from the surveyor's work? Which line is correct? There may also be a question of an ethics violation. Similar situations have led to revocation of the surveyor's license in several states.

Each individual client is the reason we exist as a profession. If we had no clients to serve, we would have no services to sell. Thus, whatever we do in the conduct of any work, we should leave our clients happy and with a record of their parcel of land so that they will be able to occupy their "little acre" with the feeling of security and tranquility that comes from knowing "my description is good, my boundaries are secure, and all is right with the world I live in."

15.4 FINAL COMMENTS

Through this book, we hope to provide its readers with a foundation for the process of creating boundaries and suggestions for retracing those same boundaries or boundaries created by other surveyors. This book cannot answer all of the questions one may have, but we hope it will open doors and inspire its readers to look further for answers to their questions.

Authors like to quote from proven sources. Perhaps no other publication is more suitable for reference here than the final chapter of Mulford's *Boundaries and Landmarks*.[8] We would like to leave the student surveyor as well as the "grizzled" old-timer with the thoughts that follow:

... The problems of boundary lie at the foundation of all surveying, for one must know where a line is before he can measure it, and the solution of these problems call[s] for the same powers of accurate observation and of consecutive and logical thought that are demanded for successful work in any branch of modern science. It is needless to say that the successful surveyor must be accurate in his instrument work and in his computation; yet, if he would really succeed, he must go beyond this. He must add to this the patience to collect all the evidence which can be found bearing upon the case in hand, together with the ability to weigh this evidence to a nicety and to determine clearly the course pointed out by the balance of probability. If, in addition, he possesses enough imagination to cast pleasant lights across the desert of dry details, he should be successful indeed.

The watchwords of the surveyor are patience and common sense.

It is intended to give a little light to men thrown, for perhaps the first time, on their own responsibility or brought face to face with problems which they have not met in their previous experience. For the solution of these problems no general rules can be laid down; each man must work out his own salvation. All that I hope to do is to give a few suggestions from hard-won personal knowledge which may make the road a little easier.

Curiously enough the Surveyor is isolated in his calling, and therein lie his responsibility and temptations. The lawyer comes nearest to understanding the work, yet of actual details of a survey most lawyers are woefully ignorant. The business man who can judge to a hair the fulfillment of a contract has no eye for the shortened line of shifted landmark. To the skilled accountant of the bank the traverse sheet is a closed book. Dishonesty in ordinary business life cannot long be hid and errors in accounts quickly come to light, but the false or faulty survey may pass unchallenged through the years, for a few but the Surveyor himself are qualified to judge it. I maintain that in the hands of the Surveyor to the exceptional degree lie the honor of the generations past and the welfare of generations to come; In his keeping is the Doomsday Book of his community and who shall know if he is false to his trust? Therefore I believe that to every Surveyor who values his honor and has a full sense of his duty the fear of error is a perpetual shadow that darkens the sunlight.

Yet It seems to me that a man of an active mind and high ideals the profession is singular suited; for to the reasonable certainty of a modest income must be added the intellectual satisfaction of problems solved, a sense of knowledge and power increasing with the years, the respect of the community, the consciousness of responsibility met and work well done. It is a profession for men [and women too] who believe that a man is measured by his work, not by his purse, and to such I commend it.

NOTES

1. Mulford, A.C., *Boundaries and Landmarks*. New York: D. Van Nostrand Company, 1912.
2. Ibid.
3. *Rivers v. Lozeau*, 59 So.2d 1147 (Fl. 1989).
4. *Fehrman v. Bissell Lumber Co.*, 204 N.W. 582 (Wisc. 1925).
5. *Spainhour v. B. Aubrey Huffman & Associates, Ltd.*, 237 Va. 340, 377 S.E.2d 615 (Va. 1989).
6. *Cragin v. Powell*, 128 U.S.691 (La. 1888).
7. *Spainhour v. B. Aubrey Huffman & Associates, Ltd.*, 237 Va. 340, 377 S.E.2d 615 (Va. 1989).
8. Mulford, A.C.; ibid.

GLOSSARY OF TERMS

A

About. Approximately; with some approach to exactness, in respect to quantity; all around; on every side; in the immediate neighborhood of. About often indicates a lack of knowledge and should be avoided in descriptions. A deed to land being about 30 feet wide was insufficient to pass title (Tinelite v. Simmott, 5 N.Y.S. 439). About has been interpreted to mean "nearly," "more or less," "in close correspondence to," and "in the immediate neighborhood," depending on the contents of the remainder of the description. "About north" was interpreted as astronomical north when nothing else limited the word (Shipp v. Miller, 15 U.S. 316). An indefinite word to be avoided without other qualifying phrases, as "about 10 feet to a fence."

Abstract. A bare or brief statement of facts written in abbreviated words; a statement of the important parts of a deed, trust deed, or other legal instrument; colloquialism for "abstract of title."

Abstract of title. A compilation of abstracts of deeds, trust deeds, and other pertinent data that affect the title to a piece of real property, all bound together in chronological order. It is a form of title evidence made for the purpose of title examination.

Abstracter. The person who takes the information pertaining to a title in its full form and puts all facts into an abbreviated form called an abstract or an abstract of title.

Abut. To reach or adjoin. In old law, the sides of a property adjoined, whereas the ends were said to abut.

Abuttals. Boundaries or buttings of land; adjoiners.

Access rights. In acquiring freeway, turnpike, or throughway lands, ingress and egress of the abutter, called access rights, are often denied.

Accession to real property. Title to real property can be acquired by accession, which is the addition to property by growth, increase, or labor. Land gradually

Brown's Boundary Control and Legal Principles, Eighth Edition.
Donald A. Wilson, C.A. "Tony" Nettleman III, and Walter G. Robillard.
© 2024 John Wiley & Sons, Inc. Published 2024 by John Wiley & Sons, Inc.

deposited on the bank of a stream by imperceptible means becomes the land of the upland owner by accession.

Accessories to corners.   Physical objects adjacent to corners that have a measured or known relationship to the corner and can be used to identify original corners; these include memorials, bearing trees, mounds, pits, rocks, and banks. Accessories are often considered as part of a corner monument itself rather than as "aiding in a secondary way."

Accretion.   Increase by external addition; enlargement; the act of growing to a thing. Where, from natural causes, land forms by imperceptible degrees upon the bank of a river, stream, lake, or tidewater, either by accumulation of material or recession of water, the process is called accretion, and the end result is called accretions. The process of land formation is also called alluvion, and the land itself is called alluvium.

Accuracy.   Nearness to truth. Precision is the nearness of readings to one another, which may not be accurate, that is, measuring with a long tape.

Acknowledgment.   A formal declaration before some competent public officer declaring it to be his or her act or deed; usually declared before a notary public.

Acquiescence.   Passive compliance or satisfaction; acquiescence in a fence is not avowed consent, nor is it open discontent.

Acre.   Ten square chains, or 160 square poles, but converted to 43,560 square feet.

Acre right.   Early New England term; the share of a citizen of a town in the common lands. It varied from town to town.

Adjacent.   "To be near," "close or contiguous," "in the neighborhood or vicinity of but not necessarily touching," or "adjoining or continuous to" are the common meanings of adjacent. More often used to mean "adjoiner," and if so, adjoiner should be used in preference to adjacent.

Adjoiner.   Adjoin means to be in contact with; hence, the adjoiner is the land in contact with the instant property. When speaking, it is often used to mean the written deed of the adjoiner.

Adjoining.   The word adjoining in a description of a premise conveyed means "next to" or "in contact with" and excludes the idea of intervening space (Yard v. Ocean Beach Assn., 24 A 729).

Admeasurement.   Ascertainment by measure.

Adverse possession.   A method of acquisition of title by possession for a statutory period under certain conditions.

Affidavit.   A sworn statement made before a notary public or other authorized person. The one who makes the statement is called the affiant.

Aforesaid.   Before or already said.

Agreement deed.   A written instrument executed by two or more parties in which the parties agree to do or not to do certain things pertaining to the transfer of land or adjustment of land boundaries.

Alienation.   The transfer of property and possession of land from one person to another.

Aliquot.   In mathematics, a number that divides a larger number into parts without a remainder; 2 and 4 are aliquot parts of 8. As applied to trusts in law, it is treated as fractional, such as ¼ interest. In sectionalized lands, it is a fraction of a whole, as ¼ of a section.

Alluvion.   Where, from natural causes, land forms by imperceptible degrees upon the bank of a body of water, navigable or not navigable, either by accumulation of material or by the recession of water; such land, called alluvium, belongs to the owner of the bank, subject to any existing right-of-way over the bank. The process of land formation is called alluvion or accretion.

Along.   "Along a line" means on and in the direction of the line; it implies motion. "Along the road" means along the centerline or thread of the road unless qualified as, for example, "along the east side line of the road." "Along a line" may be changing in direction by curves or angles. Avoid "with a line," "by a line," or "on a line" where "along a line" is meant. The term along may mean "on"; thus, "along the shore" means "on" the shore and includes the shore (Church v. Meeker, 34 Conn. 421). "With a line," meaning "along a line," is commonly used in the eastern United States.

Ambiguity.   Not clear in meaning. A patent ambiguity is evident and obvious. A latent ambiguity is not apparent on the face of a document.

Ambit.   A boundary line.

Ambitus.   In Roman law, a path worn by going around; a space of at least 2½ feet in width between neighboring houses.

Ancient writings.   Documents over 30 years of age and bearing on their faces evidence of age and authenticity, also coming from proper custody. They are usually presumed true without express proof.

Angle.   The figure formed by the joining in a point of two lines, or the space bounded on two sides by such lines.

Angle, deflection.   The horizontal angle measured from the prolongation of the preceding line of the following line. Right is to the right in the direction of travel.

Apogean tides.   Monthly tides occurring when the moon is farthest from Earth.

Approximate.   "Approximately," "a little more than," "not quite," "not more than" are all terms of safety and precaution. Approximate often denotes uncertainty of dimensions to a greater degree than "more or less" and to a lesser degree than "about." Near to correctness; nearly exact; not perfectly accurate. Reasonable knowledge of dimensions is indicated by the word approximate (it really means you don't know).

Appurtenances.   A word employed in deeds, leases, and so on, for the purpose of including any easements or other rights used or enjoyed with the real property, which are considered to be so much a part of the property that they pass to the grantee automatically under the deed conveying the real property. From appertain, "to belong to."

Appurtenant (easement).   An easement that was created to benefit other land owned by the holder of the easement (see also In gross).

Aqua.   Water.

Arbitrary map.   An office "subdivision" or map made by a title company, assessor, or others for their own convenience in locating property in an area in which all the descriptions are by metes and bounds. On this subdivision, the "lots" are given arbitrary numbers. The deeds and other instruments affecting these lots are posted to what is called an arbitrary account.

Arm of the sea.   Extends as far inland as the water of fresh rivers is pushed backward by the incoming tide. That inland portion of the sea with a tide.

Arpent.   A French measure of land mentioned in the Domesday Book. One hundred perches of 18 feet or about 1 acre.

Assessor's maps.   Maps made for the purpose of assessing land for tax purposes; maps may not accurately indicate conveyance rights.

Assigns.   At law, to transfer or make over to others, especially for the benefit of creditors.

Astronomic.   Of or pertaining to astronomy. Directions determined from the stars are astronomic directions. Unless otherwise stated in a deed, bearings are assumed to be on an astronomical basis.

At.   Absence of motion; at a house may be near or by the house. The word at, when applied to the place or location of an object, is not treated as definitely locative. It denotes nearness or proximity and is less definite than "in," "by," or "on." A boundary of land described as at a road, without qualifying terms, means "to the center of the road." "To," "at," and "by" a stream, unqualified, mean to the limit of private ownership. The word at may have elasticity of meaning, depending upon how it is used. Other locative words are preferred in deeds.

Augmenting easement.   An easement lying outside the parcel being conveyed but of benefit to the parcel.

Avulsion.   The sudden and perceptible separation of land by violent action of water. A stream suddenly adopting a new channel and dividing a parcel into two parcels. Title usually follows the old channel; however, the statute of limitations may specify how soon a person must reclaim the portion cut off. In California, it is one year after the necessity to act.

Azimuth.   The way; the direction. In surveying, azimuth is a direction measured clockwise from a given meridian. The army uses north as the meridian; the geodetic system uses south. If azimuths are used in a description, the meridian must be defined.

Azimuth, astronomic.   The clockwise angle from astronomic north or south (determined by a star observation).

Azimuth, geodetic.   The clockwise angle from geodetic south (may differ from astronomic south a maximum of about $0°019$ and usually not more than a few seconds due to the deflection of gravity). The Lambert or Mercator grid systems use geodetic bearings or azimuths, not astronomic.

B

Backlands.   Lands back from a river, lake, or highway. Indefinite in precise location.

Backside.   English: A yard behind a house and belonging to the house.

Baldio.   Spanish: Wastelands. Unappropriated public domain, not set apart for municipalities (Texas).

Bank of a stream, right or left. When one is facing downstream, the right side is the right bank and the left side is the left bank.

Bar. A bank of sand, gravel, or other material in water, usually forming an obstruction to navigation and usually below average high-water elevation.

Barleycorn. One-third of an inch.

Baseline, sectionalized land. A parallel of latitude or approximately a parallel of latitude running through an arbitrary point chosen as the starting point for sectionalized land within a given area.

Batture. A term used to denote a bed of sand, stone, or rock rising toward the surface of a body of water. As used in Louisiana, it is sometimes applied to portions of the bed of the Mississippi River that are exposed at low water and covered at high water. It is frequently used to mean alluvion as "alluvion or batture" (the process of forming accretions). It has been used to denote the bed between low- and high-water marks in a river.

Bayou. Channel through swamp or marsh; abandoned channels of the parent stream; a secondary channel for floodwater. Used in the South. Of French origin.

Beach. Used in conjunction with boundary lines, the word beach may mean the sea side or the land side of the shore. The meaning intended in a deed depends on other words used in the deed and on surrounding circumstances. In the absence of qualifying terms, beach often conveys to the limits of private ownership, thus giving the greatest advantage to the buyer; that is, "to the beach" means to the mean high-water line unless there are statutes to the contrary.

Beacon. European term for "corner."

Bearing, astronomic. A bearing based on north or south as determined by a star observation.

Bearing, grid. If the Lambert or Mercator grid is meant, it is based on geodetic north as determined at the central meridian. All grid norths within the same zone are parallel with one another. Grid north is not necessarily on an astronomical basis at the central meridian (the difference between astronomic and geodetic is usually insignificant).

Bearing, magnetic. A bearing based on magnetic north.

Bearing, true. A bearing based on an astronomical observation. In early sectionalized land surveys, "true north" was determined by a magnetic north observation corrected to true north by subtracting or adding the declination. Because of confusion in meaning, it should be avoided in new deeds.

Bearing tree. A tree marked or called for to witness the position of a corner (usually a section corner). It is usually blazed with the corner number inscribed.

Beat. In southern states, a political subdivision corresponding to a township or parish in other states.

Bed of water. Varies in meaning but usually means the area below the high-water mark of a stream or lake, or below the mean high-tide line of the ocean or bay. It implies the area of land supporting the water above it.

Being. Existent; being usually denotes a secondary call as "to the northeast corner of Brown's land, being also a 2-inch iron pipe." The 2-inch iron pipe is usually the secondary or informative call, whereas Brown's corner (if senior) is normally the superior call. A "being clause" may be a controlling call, but not necessarily so.

Being clause.   The being clause of a deed denotes the origin of the history of the present deed, such as "being the same land conveyed to Brown in Book 1237, page 672, of Official Records." If a change is made in the wording of a deed, there should be inserted a being clause. Reference to a being clause generally does not operate to enlarge or restrict a particular and sufficient description of the land conveyed (26 C.J.S. 372).

Bench mark.   A monument or point whose elevation is known above a given datum (usually mean sea level). In surveying, it is seldom used to designate a property corner monument, although a property corner monument may also be used as an elevation reference. Webster's New International Dictionary says, "a point of reference from which measurements of any sort may be made."

Bienes.   Spanish: Real and personal property.

Bienes comunes.   Spanish: Common property, as air, water, sea, and beaches.

Biens.   French: Real and personal property. English: Property except estates of inheritance and freehold.

Bisection of a line.   In a property description, bisection of a line is not the cutting of the line into equal parts. Where midpoint is meant, use midpoint.

Blaze.   A mark on a tree caused by cutting off the bark and a portion of the live wood, usually at breast height with a flat scar.

Block.   A square or portion of a city enclosed by streets, whether it is occupied by buildings or composed of vacant lots. In addition, blocks are often enclosed by the boundary of the subdivision. In 318 Mo. 192, it must be surrounded by at least three streets as marked on the ground (not merely platted).

Blunder.   A mistake, not an error. An error is a small residual error of measurement. A blunder is a misreading (e.g. 96 for 69).

Border.   Border is synonymous with boundary; the outer part or edge of anything.

Boundary, land.   Usually, the line of demarcation between adjoining land parcels as determined by legal descriptions. Land boundaries can be marked by monuments, fences, hedges, and so on, or not at all.

Bounded.   Land sold and described as "bounded" by a highway is construed to extend to the centerline of the highway unless other limits are given. A parcel of land bounded by the land of Gretchen Brown is adjacent to and touching the land of Gretchen Brown; "bounded by" implies no intervening space left.

Bounds.   The lines by which different parcels of land are divided. "Butts and bounds" or "butted and bounded" are phrases sometimes used to introduce the boundaries of land. "Buttal" means along the ends of the land. A monument is sometimes referred to as a bound.

Bounds description.   Land described by calling for the adjoiners as "bounded on the north by the land of Leo Orstrom; bounded on the south by Cedar Street; bounded on the east by the Androscoggin River"; and so on.

Boxing the compass.   Naming the points and quarter points of the compass in order, clockwise around the circle, beginning with north.

Branch of the sea.   See Arm of the sea.

But. "That which follows is an exception to that which has gone before" is the common meaning of but in a sentence. But may mean "except," "except that," "on the contrary," or "yet." Because but has so many meanings, it is better to use except in a deed when "except" is meant.

Butts and bounds. Same as metes and bounds. To introduce a description, butted and bounded is sometimes used.

By. In a deed, "by a road" is construed as including the land to the center of the street, but "by the east side of a road" means "along the east side" and not "along the centerline." "To," "on," or "by" a stream means to the limits of the grantor's land. By has many meanings and is generally used to indicate the limits of the grantor's land. By has many meanings and is usually used in deeds to mean "along," "over," or "through" as "by 5th Street." It can mean "near," "close," "within reach," or "toward" (east by north).

By estimation. Same as "more or less" in conveyances. Not measured exactly.

Byroad. New Jersey: A road not laid out but used by people. A road of little importance other than ingress and egress, yet a public road.

C

Cadastral map. Strictly, a map for the purposes of making a cadastre. A cadastre is an official register used to apportion taxes. Hence, it is a map to show the value and relationship of lands for taxing purposes.

Cairn. An artificial mound of rocks whose purpose is to designate or aid in identifying a point.

Call. Within a deed, a call is the designation of visible natural objects, monuments, course, distance, or other matter of description as limits of the boundaries. Locative calls are particular or specific calls exactly locating a point or line. Descriptive calls are general or directory calls that merely direct attention to the neighborhood in which the specific calls are to be found.

Camballeria. Spanish: A quantity of land of variable size in different locations. An allotment of land to a horse soldier as a result of conquest.

Camino. Spanish: Road or highway.

Cardinal direction. Either north, east, south, or west; sometimes used to include all if in the plural form.

Carucata. A quantity of land used for taxation. The land that can be tilled by one plow in a year and a day; varies from 60 to 120 acres, depending on the area.

Center of section. The point formed by lines connecting opposite quarter corners in a section of land. It is also called the center quarter corner.

Centerline of a section. The line connecting opposite quarter corners. A corner point fixed by law.

Centerline of a street. Usually applies to the center of a street prior to widening or closing; that is, the centerline of the original street midway between the sides. To avoid ambiguous conditions, if the street has been narrowed or widened on one side or unequally, the centerline would be defined such as "the centerline as existing on 11/22/62."

Central angle or delta.   The angle subtended by the arc of a portion of a circle. "Through an angle" or "through a central angle" is clear in meaning when the central angle of a course along an arc is described.

Chain.   An instrument used to measure land, which was 33 feet long with 50 links. Later, chains were 66 feet long with 100 links. As a unit of measure, 1 chain equals 66 feet.

Chain of title.   A chronological list of documents that comprise the record history of title of a specific parcel of real estate.

Chainmaker.   A person who assembles a chain of title.

Chamber surveys.   Pennsylvania: False and fraudulent pretense of surveys of public lands by surveyors. At times may be called "paper surveys" or "office surveys."

Chord.   The shortest distance between two points located in a given curve.

Chord, long.   The shortest distance between the beginning and end of a curve (usually, a circular curve).

Clear title.   Good title; one free from encumbrances.

Closing corner.   A corner (usually marked by a monument) that indicates where the new line intersects a previously established land boundary. The point at which a section line closes (intersects) on a prior rancho land grant, correction line, or water boundary. Closing corner monuments are not considered as fixed in position but may be adjusted by a later surveyor to the line closed on. They are an exception to the rule that "wherever an original monument is set, its position is unalterable."

Cloud of title.   A claim or encumbrance on a title to land that may or may not be valid.

Codicil.   A testamentary disposition subsequent to a will, and by which the will is altered, explained, added to, subtracted from, or confirmed by way of republication, but in no case totally revoked (Black's Law Dictionary).

Coincident with.   Constant and continuous contact along a line.

Color of title.   If a claim to a parcel of real property is based on some written instrument, although a defective one, the person is said to have color of title. The title appears good, but in reality it is not.

Common law.   Common law represents the determination of what is right and wrong as found by the courts to which various cases have been submitted by parties to legal actions (see Unwritten law). Statute law does not repeal common law; it merely inactivates it as long as the statute law exists.

Concave.   The inside of a curve; toward the center of the circle.

Conditions.   Restrictions created by a qualification annexed to the estate by the grantor of a deed, upon breach of which the estate is defeated and reverts to the grantor.

Confusion, area of.   Intermixture of the land of two or more owners; ownership of a given area of land is confused.

Consideration.   The inducement, either money or other consideration, that moves a party to enter into a contract.

Constructive notice.   Notice implied or imputed by law as the notice of a deed that has been recorded in a grantee–grantor index. One cannot deny the matter contained in a constructive notice.

Contiguous. Varies in meaning from "in close proximity" to "near, although not in contact to," "touching or bounded," or "traversed by."

Contour. An imaginary line on the ground, all points of which are at the same elevation above or below a specified elevation datum.

Convex. Outside of a curve; away from the center of a curve.

Conveyance. Embraces every instrument in writing by which any estate or interest in real property is created, aliened, mortgaged, or encumbered, or by which the title to any real property may be affected, except wills.

Conveyance, mesne. A title in the chain of title located somewhere between the original conveyance and the present titleholder; not the original or present title but an intermediate conveyance.

Conveyed. All the land transferred in fee title. "That land conveyed to Brown" would include the lot described plus fee rights to adjoining streets.

Coordinates. Of equal rank; in harmony with; not subordinate.

Corn land. In England, grain land.

Corner. The point of intersection of two courses in a land description. The point of change of direction of a land boundary. It may or may not be marked by a corner monument or by a bound. Has also been referred to in descriptions as "a point."

Corporeal. Things that have a material existence; a house is corporeal. The rent collected from its occupation is incorporeal.

Correction line. In the public land surveys, East–west lines usually surveyed 24 miles apart, which were corrected for the curvature of Earth and against which the township lines and section lines terminated.

Course. Course, as used in surveying, usually includes both bearing and distance. "Course and distance" where "bearing and distance" is meant is a common error. Because, when a ship is set on a "course," a bearing is implied, the word course is sometimes used in land description utilizing that meaning. Black's Law Dictionary defines course as "the direction of a line with reference to a meridian." When used as "the third course in the description," it means the third line described and includes bearing, distance, and all objects called for.

Covenant. A word used in deeds for the purpose of creating restrictions; imports an agreement on the part of the grantee to make, or to refrain from making, some specified use of the land conveyed.

Crown lands. Land belonging to the reigning sovereign. Refers to areas of the United States as of a particular date that belonged to a reigning sovereign; crown lands passed to the United States or the colonies upon independence.

Curtesy. A husband's interest in his wife's estate upon her death.

Curves. Most curves used in land descriptions are circular; that is, they have a constant radius. Other mathematical curves sometimes used are spiral (radius changes from infinity to a constant by a mathematical formula), parabolic (used in vertical curves), elliptical, and contour (curve that follows a constant elevation on the ground and usually has a changing radius).

Curves, circular. (1) Centerline of a curve is the midpoint along the arc of the curve and is not the "center of the circle" describing the curve. (2) Parallel curves are concentric. (3) Radius of a curve stops at the arc of the curve. A description intending to extend beyond the arc should state "and on the prolongation of the radius." (4) Compound curves are tangent at the point of compounding (changing of radius). (5) Reverse curves are tangent at the point of reversal. (6) Tangent curves have a common tangent where the curves meet.

Curves, spiral. A variable radius curve; a transition curve between a tangent and a circular curve.

Custom. "Custom is another law." "Custom should be certain, an uncertain custom is null." Custom in surveying often creates common law for locating properties.

D

Date. Derived from datum, meaning "given"; that is, the day, month, and year.

Datum. Any position or element to which others are determined. In Latin, a thing given. If there is more than one reference datum, the plural is datums, not data; for example, "there have been several sea level datums used."

Datum, elevation. Usually, mean sea level as determined by hourly readings over an 18.6- or 19-year average. Others used are mean low water, mean lower low water, mean high water, and mean higher high water. An assumed datum for elevation is sometimes used.

Datum, geodetic. Datum that has five fixed quantities: (1) latitude of the initial point; (2) longitude of the initial point; (3) azimuth of a line from the initial point; (4) the largest diameter of the ellipsoid of reference; (5) the smallest diameter of the ellipsoid of reference.

Datum, tidal. Any elevation datum referred to a phase of the tide.

Datum plane. A surface used as a reference from which to reckon heights or depths. Note: If sea level plane is used, the plane is ellipsoid in shape. The word plane is misused.

Declination. A term that has several meanings, depending on the contents of a sentence. The declination of the sun (or star) is the angle above or below the celestial equator. Magnetic declination is the angle between true north (astronomic) and magnetic north. The act of deviating; turning aside; swerving; in surveying, the angle of deviation from a reference datum.

Decree. A judgment by the court in a legal proceeding.

Decree of distribution. The judicial decision made by a probate court determining who is legally entitled to the real and personal property of a decedent.

Dedication. The appropriation of land or easement for the use of the public and acceptance on the behalf of the public.

Deeds. Act; action, thing done. At law, a deed is evidence in writing of an executed and delivered contract, usually for sale of land. As pertaining to land, its purpose is to define location and title to land. Several types exist. (1) Grant deed: conveys the fee title of the land described and owned by the grantee. If at a later date the grantor acquires a better title to the land conveyed, the grantee immediately acquires the better title without formal documents (after rights). In some states, by law, the grantor warrants the deed against acts of his own volition. (2) Quitclaim deed: passes on to the grantee whatever title the grantor has at the time at which the transaction is consummated. It carries no after-rights; that is, if the grantor acquires a better title at a later date, it is not passed on to the grantee. The deed carries no warranties on the part of the grantor. (3) Agreement deed: an agreement between owners to fix a disputed boundary line. (4) Warranty deed: conveys fee title to the land described to the grantee and, in addition, guarantees the grantor to make good the title if it is found lacking. Probably derived its name from English livery of seisin.

Defeasible title. A title that may be made void but is not yet a nullity.

Degree of curve. Along railroads, the degree of curve is the central angle of a curve subtended by a 100-foot chord on the curve. Along highways, the degree of curve is usually, but not always, defined as the central angle subtended by a 100-foot arc of the curve.

Delta. Angle at the center of a circular curve.

Demise. To convey an estate for life; to lease.

Dependent resurvey. A retracement and reestablishment of the lines of the original survey in their true original positions according to the best available evidence of the positions of the original corners (see also Independent resurvey).

Depesas. Spanish: Land reserved for commons.

Description, land. The exact location of a parcel of property stated in terms of lot, block, and tract, or by metes and bounds.

Deviation of compass. Deviations caused by local attraction, especially as pertaining to a ship's compass when deflected by metal.

Discrepancy. A difference between results of comparable measurements.

Distance between points. Distance between points is always assumed to be the shortest possible horizontal distance unless otherwise specified.

Diurnal. Having a period or cycle of approximately one tidal day. A diurnal tide has only one high and one low tide in a tidal day.

Domesday. (also Doomsday Book.) A book containing a minute and accurate (by the standards of that time) survey of the lands of England made in 1081 to 1086.

Dominant tenement. The land that is benefited by an easement (see also Servient tenement).

Dominium. In old English law, ownership. Sovereignty or dominion.

Donation lands. Lands granted by the sovereign to a person as a gift, particularly to soldiers of the Revolutionary War.

Dower. A wife's interest in her husband's estate upon his death.

Due. Where monuments or other deed terms do not limit the calls, Due north means "astronomical north." A deed reading "thence N 65°109240 W along said southwest line (Cuyama Rancho) as shown on said record of survey map a distance of 2,877.60 feet; thence due north 13,295.04 feet to the true point of beginning" was interpreted as though "due" meant "astronomical." The deflection method, whereby an angle of 65°109240 was turned from the Rancho, was rejected. The "due north" was determined by observation on Polaris and resulted in a difference of approximately 11 feet east and west (Richfield Oil Corp. v. Crawford, 249 Pac. 2d 600). "Due north" as originally used meant "true north" as determined by a declination correction to a magnetic reading. The word has become ambiguous in meaning because of careless usage. If astronomical north is meant in a deed, use "astronomical north," but not "true north."

E

Easement.   An interest in land created by grant or agreement that confers a right upon owners to some profit, benefit, dominion, or lawful use of or over the estate of another; it is distinct from ownership of soil. Example: an easement for road purposes.

Elder survey.   A term more common in the colonies, meaning the older of two surveys.

Ell.   A measure of length (yard).

Eminent domain.   The right or power of government to take private property for public use on paying the owner a just compensation.

Encroach.   Intrude; trespass; to gradually take possession or rights of another; a building encroaches on another's title rights.

Entryman.   US public land laws; one who makes an entry of land.

Equatorial tide.   Tides occurring semimonthly as a result of the moon being over the equator.

Error.   The difference between a measured value and the true value; of smaller magnitude; not a mistake or a blunder. Sometimes the true value is not known and cannot be determined exactly as in a linear measurement. The angles of a triangle must add to 180°; hence, angular error can often be compared to a true value.

Error, index.   A previous surveyor's work that is found to be in error by a relatively constant amount. For example, the distance between monuments is found to be 0.10 feet too long in each 100 feet.

Error of closure.   The error of closure of a survey on itself. A small error of closure denotes consistency (or precision), which may or may not be accurate. Measuring with a long tape can yield consistent but not accurate results.

Escheat.   The lapsing or reverting of land to the state. In feudal law, the reversion of the land to the original grantor or lord.

Escrow.   A grant may be deposited by the grantor with a third person, to be delivered on performance of a condition, and, on the delivery by the depositary, it will take effect. While in the possession of the third person, and subject to condition, it is called an escrow.

Estadal.   Spanish: Measure of 16 square varas.

Estop.   To bar, to stop, to impede.

Estoppel.   A preclusion in law that prevents a person from alleging or denying a fact in consequence of his or her own previous acts.

Et al.   "And others," "and another" (et alii or et alius).

Et con.   "And husband."

Et seq.   "And following."

Et ux.   "And wife" (et uxor).

Evidence.   That which furnishes any mode of proof. Grounds for belief.

Evidence aliunde.   Evidence outside a conveyance document.

Examiner.   A person who analyzes a chain of title to land and passes on validity of various instruments and then renders an opinion.

Exception, excepting.   An exception withdraws a part of the thing described as granted, which would pass but for the excepting clause. The word except means "not included." "Lot 12, excepting the east 30 feet" clearly conveys that portion of lot 12 lying westerly of the east 30 feet. "Lot 12 and lot 13, except the east 30 feet" is not clear since the exception might apply to either one lot or both. "Lot 12 and lot 13, except the east 30 feet of lot 13" is better. Often used interchangeably with reserving in legal descriptions.

Exception doubled. Double exceptions should always be avoided. "Lot A except the east 50 feet, except the south 50 feet" conveys two meanings: (1) "lot A, except the easterly 50 feet of all of lot A and except the south 50 feet of all of lot A," or (2) "lot A, except the east 50 feet is reserved by the grantor." In a double exception, the second exception may refer to the exception or to the lot.

Extrinsic evidence. Evidence of matter not contained in the writings but offered to clear up an ambiguity found to exist when applying the description to the ground.

Eyott. A small island arising from a river.

F

Fathom. Six feet.

Fee. An estate of inheritance in land.

Fee simple. An estate of inheritance in land without qualifications or restrictions as to the persons who may inherit it as heirs. Also called an absolute fee or fee title.

Fee tail. An estate of inheritance given to a person and the heirs of his or her body.

Fence. A barrier between two properties; a hedge, a rock wall, a row of trees, a wire, or a wooden structure.

Feud. An estate in land held by a superior (king, lord, or proprietor) on condition of rendering him or her service; an inheritable right if the conditions are rendered.

Flat. Area too shallow for navigation, as tidal flats.

Following the footsteps. A term of philosophy rather than procedure. Means to follow the evidence and, if possible, the methodology of the original surveyor left by the original surveyor rather than following the exact placement on the ground. Does not equate to finding the original lines, but means finding the evidence of the original lines. Finding the evidence is an absolute. Following the exact original lines is not.

Fractional lot. A portion of a section not subdivided in the regular manner; may be more than or less than the smallest division (40 acres). It is meaningless to refer to a lot in a subdivision, other than government sections, as being a fractional lot.

Fractional sections. A section reduced in size due to a land grant, body of water, and so on.

Free boundary. One that is not limited by a call for a monument as "thence N 12° E, 120 feet." "Thence N 12° E, 210 feet to Brown's south line" is not a free boundary.

Freehold. A term of land ownership. Derived from the Old English, during which time only freemen could own land.

Front. The portion of a land parcel adjoining a street; a corner lot has two possible fronts.

Furlong. One-eighth of a mile (40 poles).

G

GIS.   Geographic information system. A generic term for a land data system that can include a multitude of information about a parcel or parcels of land as to uses and ownership. Usually does not include information about land boundaries.

Gore.   A small triangular piece of land often thought of as an omitted area or area of confusion. A special type of hiatus.

GPS.   Global Positioning System. A method of determining the length, direction, and elevations of lines based on Earth-orbiting satellites. All results are calculated rather than actually running in the field using conventional surveying equipment.

Grant.   The transfer of real property by deed; to bestow; to confer.

Grant deed.   See Deeds.

Grantee.   The person to whom a grant is made; the one who acquires a property.

Grantor.   The person by whom a grant is made; the one who transfers a property.

Great pond.   In Florida, Maine, Massachusetts, and New Hampshire (and possibly other states), a natural pond (lake) having an area of more than 10 acres.

Guide meridians.   True north lines usually run at 24-mile intervals along a standard parallel. Guide meridians terminate in the next northerly standard parallel. Very few were run in the 48 states.

H

Habendum.   That portion of a deed beginning with "To have and to hold." The clause that defines the extent of ownership granted.

Habendum clause.   That part of a deed that limits and defines the rights that the grantee is to have in the property conveyed.

Hack.   A mark on a tree made by cutting out a V-notch well into the live wood.

Half.   Half of a parcel of land, except for federal sectionalized lands; usually indicates half of the area of the parcel. The direction of the dividing line between the two halves may be uncertain.

Hand.   Four inches; measure used for the height of horses.

Heirs and assigns as used in deeds.   Unless the words "and his heirs" are used, the estate conveyed is only for the life of the grantee (estate for life). And his heirs is not necessary in most states because of statutes abolishing the necessity. And assigns is included to take care of corporations, trustees, and so on, who cannot have heirs.

Hereditaments.   Things capable of being inherited: real, personal, heirlooms, and furniture.

Hiatus.   An opening; a gap; a space. A gap between two deeds.

Hide.   Old English: As much land as could be worked with one plow; 60 to 100 acres.

High water.   The maximum height reached by a rising tide.

Higher high water.   The higher of the two high waters of any tidal day.

Higher low water.   The higher of the two low waters of any tidal day.

High-water line.   The intersection of the plane of high water with the shore. Often used to mean the intersection of the plane of mean high water with the shore.

High-water mark. The line the water impresses on the soil by covering it for sufficient periods of time to deprive it of vegetation (Raide v. Dollar, 203 P 469). In the absence of any statement or law to the contrary, it must be construed to mean "ordinary" high-water mark (Rondell v. Fay, 32 Cal. 354).

Hub. A square stake, usually 2 3 2 inches, driven flush with the ground and with tack.

Huebras. Spanish: A measure of land equivalent to what a pair of oxen can plow in a day.

Hui. Hawaii: An association of persons holding the ownership of land, usually as tenants in common.

I

In. Varies in meaning depending on context and may mean "for," "on," "within," and so on. In real property law, used to denote facts of title, as in possession, entitled, invested with title, and so on.

In gross (easement). An easement that exists independently of other land and does not benefit any other land. It is a mere personal interest in or right to use the land of another (see also Appurtenant).

Inadmissible. Under established rules of law, whatever cannot be admitted or received, such as parol evidence to contradict a written document.

Incumbrances. Includes taxes, assessments, and all liens upon real property.

Indefeasible. An estate that cannot be defeated or made void.

Independent resurvey. An establishment of new section lines, and often new township lines, independent of and without reference to the corners of the original survey (see also Dependent resurvey).

Intent. The true meaning. The intent of a deed is determined from the written words only.

Interest. A right, claim, title, or legal share in something.

Intersection. The space occupied by two streets at the point where they cross each other. Point of intersection usually means the point at which the two centerlines intersect. In accident cases, the point of intersection can mean where the cars meet.

Into. To the inside of; within; usually follows the verb expressing motion.

J

Jacob's staff. A single pole (staff) used for mounting a compass; used in place of a tripod.

Joint tenants. Two or more persons owning an estate in equal shares created by a single transfer. Upon the death of a joint tenant, the surviving joint tenant takes the entire property and nothing passes to the heirs of the deceased.

Jurisprudence. The philosophy of law.

Juxtaposition of numbers. Figures used in a description that have differing units, as "thence easterly along the north line of lot 21, 211 feet" can easily be interpreted to mean 21,211 feet instead of lot 21 a distance of 211 feet." Insertion of a phrase as "thence easterly along the north line of lot 21, a distance of 211 feet" is better.

K

Kelp-shore.   Along the ocean, it is the area between high and low water.

Kill.   In early New York (1 N.Y. 96) conveyances, it designates the bed of a river (Dutch origin).

Kuleana.   Hawaiian term for a small area of land awarded in fee by the king.

L

Lace.   One pole (16 ½ feet) of land, as used in Cornwall.

Latent.   That which does not appear on the face of a thing; a latent ambiguity in a deed is not apparent from reading the deed.

League.   Distance measure varying from 2.4 to 4.6 statute miles. Marine league equals 3 geographical miles. Spanish league is 5000 varas, or about 2.63 miles; the square league is then 4428 acres, more or less, depending on the state law.

License.   Permission to do an act which, without permission, would be illegal, such as permission to trespass.

Line.   A lineal measure of 1 ½ or 11/8 inch, depending on the locality. The boundary or line of division between two estates.

Line tree.   A tree existing on a surveyed line and marked with two hacks or notches.

Link.   One-hundredth of a chain; 7.92 inches; a link in a Gunter's chain measuring 7.92 inches.

Lis pendens.   A notice of a pending suit. A person acquiring property after a lis pendens has been recorded takes the property subject to the decree of court that may be rendered.

Littoral.   Ownership of land bordering on the shore of a sea or lake. Often used interchangeably with riparian, the latter being applied to the ownership of land on the banks or shore of a watercourse.

Locative calls.   See Call. In descriptions of land, locative calls are specific calls to determine location. In harmonizing conflicting calls in a deed or survey of lands, courts will ascertain which calls are locative and which are merely directory, and conform the lines to the locative calls. Directory calls are those that merely direct to a neighborhood where the definite calls may be found, whereas locative calls fix boundaries (143 Tenn. 667).

Lost corner.   A point of a survey whose position cannot be determined beyond reasonable doubt, either from traces of the original marks or from acceptable evidence or testimony that bears on the original position, and whose location can be restored by reference to one or more interdependent corners (see also Obliterated corner).

Lot excludes street.   "However clear it may appear that the owner of a lot holds title to the center of the adjoining street, subject to the public easement, and that the boundary of the lot is technically, therefore, the center of the street, in view of the fact that the owner of such lot or land has not right to the possession or occupancy of any portion of such street, the word 'lot' as generally and customarily used does not include that portion of the street" (Earl v. Dutour, 181 C 58). Lot measurements are from the side of the street, whereas title usually extends to the center of the street.

Lot line.   The line shown on the map creating the lot. A lot line is permanent and does not change with street openings.

Lower low water.   The lower of the two low waters of any given tidal day.

Low-water mark.   Low-water mark is the line to which a body of water receded, under ordinary conditions, at its lowest stage.

M

Magnetic meridian.   The north direction a magnetic compass needle points at any given location. Magnetic declination is the deviation of the direction of the needle from astronomical north.

Main channel.   The middle, deepest, or best navigable channel (state boundary lines are sometimes along the main channel, which may not be halfway between the banks).

Mean high water.   The mean height of tidal high waters at a particular station for 18.6 or 19 years. Along rivers, it has various meanings and is usually a gradient (decreases in elevation downstream) boundary. A practical definition is: "where vegetation ceases."

Mean higher high water.   The average height of the higher high waters over a 19-year period.

Mean low water.   The average height of the low waters over a 19-year period.

Mean lower low water.   The average height of lower low waters over a 19-year period.

Mean sea level.   The average height of the surface of the sea for all stages of the tide over a 19-year period, usually determined from hourly readings (see also Mean tide level).

Mean tide level.   Same as half tide level, or the plane halfway between mean high water and mean low water; not to be confused with mean sea level.

Meander line.   A traverse of a body of water for the purpose of determining the size and location of the body of water. For riparian owners, meander lines do not represent the boundary line; the body of water, where it exists, represents the true boundary line. When meander lines are nonriparian, they may become land boundary lines.

Memorial.   A durable article deposited in the ground at the corner to perpetuate the corner's position.

Meridian line.   Any line run astronomical north and south. As meridian lines converge at the North Pole, no two meridians are parallel. Practically, within the limits of a property survey, all lines shown as north or south are considered parallel.

Mesne conveyance.   An intermediate conveyance between the first grantee and the present holder.

Metes and bounds.   As commonly understood, a description of real property that is not described by reference to a lot or block shown on a map but is defined by starting at a known point and describing, in sequence, the lines forming the boundaries of the property.

Mile.   One thousand paces (2000 steps) of the Roman soldier.

Mile, nautical.   The US nautical mile is also termed a sea mile or geographical mile and is considered a minute of arc along the equator, or 6076.10333 feet, for purposes of the Submerged Lands Act.

Mile, statute.   5280 feet.

Moiety.   A part; a fraction of a thing, particularly when applied to interests held in real property.

Monuments.   As pertaining to land surveying, tangible landmarks indicating land boundaries. (1) Physical monument: an existing feature such as a stone, stake, tree, hill, ocean, river, or lake, but not the unmarked line of an adjoiner. (2) Natural monument: a naturally occurring object such as a lake, river, tree, boulder, or hill. Although the courts sometimes refer to a record monument (land of adjoiner) as a type of natural monument, such a broad meaning is excluded in these pages. (3) Artificial monument: a human-made object such as a stake, fence, or set stone. (4) Record monument: an adjoiner property called for in a deed, such as a street or particular parcel of land. Frequently, the boundary line of the adjoiner is referred to as the record monument; actually, the entire property, rather than the line, is the monument. Physical monuments may or may not mark a record monument. In court reports, record monuments are often referred to as natural monuments, but such a meaning is excluded in these pages. A record monument is often referred to as an adjoiner. Usually, a record monument has senior standing, although not always. (5) Legal monument: any monument controlling in a legal description. It is often limited in meaning so as to be synonymous with record monument.

More or less.   The words more or less in their ordinary use are to be taken as words of caution, denoting some uncertainty in the mind of one using them and a desire not to misrepresent. When used in connection with quantity and distance, more or less are words of safety and precaution, intended merely to cover some slight or unimportant inaccuracy (Russo v. Corideo, 102 Conn. 663). When "125 feet more or less to the point of beginning" is used in a deed, the more or less indicates that the 125 feet is an informative term, whereas "to the point of beginning" is the controlling term. "About 12 acres more or less" is indefinite and should be avoided, since the word about has a very broad meaning.

Morgen.   New York: A measure of land equal to about 2 acres. Dutch in origin.

Mortgage.   Mort (death) and gage (pledge); an interest in land created by a written document to provide security for the payment of a debt.

Mother Hubbard clause.   Often of the form, "all abutting strips of land owned by the grantor, whether owned of record or by virtue of the Statute of Limitations." In most states, this allows tacking on of adverse or possession rights. If only the written title is conveyed, an interruption of possession occurs and the time necessary for possession must start again.

N

Navigable.   Capable of being navigated. Rivers or other bodies of water used, or capable of being used, in their ordinary condition, as highways. Two types are recognized; navigable in fact and navigable at law.

Neap tides.   Tides of decreased range occurring semimonthly as a result of the moon being in quadrature (moon and sun 90° to one another from Earth). Neap high water is the average height of high waters of the neap tides.

Near.   A relative term, as applied to space, without positive precise or legal meaning. Its meaning depends on other, more locative terms.

Neighborhood.   A place near; locality.

Nook of land.   English: 12 ½ acres.

Normal.   Normal to a line is 90° to the line. Normal to a curve is a radial line.

North.   "Though the word north, as used in the descriptive call of a deed, may be controlled or qualified in its meaning by other words of description used with it, yet when it is not qualified by other words, it must be construed as meaning due north" (96 Cal. 505). Due north means geographical or astronomical north, not geodetic north. True North differs from magnetic north, which varies from place to place and over time due to local magnetic anomalies. True North is in relation to Earth's axis and not the magnetic poles.

Northerly.   Where nothing is given to limit the exact direction, northerly means due north. Directional calls as northerly are often given in deeds to avoid ambiguity as "thence westerly along Jones Creek to the thread of Merrimack River; thence northerly along Merrimack River; etc."

Northwest Territory.   Northwest of the Ohio River and east of the Mississippi.

O

Obliterated corner.   One for which no visible evidence of the work of the original surveyor remains, the location of which may be shown by competent evidence (see also Lost corner).

Occupancy.   A mode of acquiring property that belongs to nobody; taking possession with the intent of acquiring ownership of a thing that belongs to nobody.

Of.   A term denoting origin, source, descent, and the like. Of course, descendant of kings; a distance of 400 feet, and so on.

On.   Upon; near to; contiguous to; at or in contact with the upper surface of a thing; alongside of; in; at the time of.

Opinion.   An opinion of an attorney as to the marketability of a land title as determined from a review of an abstract of title.

Ordinary tides.   Usually, the mean of all tides, as mean high tide or ordinary high tide.

Original survey.   A survey by which boundaries are created. A parcel has only one original survey.

P

Parallel.   Parallel lines are two straight lines that are an equal distance apart. Parallel curves are always concentric curves. East–west lines are parallel. Technically, north–south lines converge at the poles and cannot be parallel; however, in a legal description or on a map, where two lines are shown with the same bearing, it is implied that the lines are parallel. The same bearing on different maps does not imply parallelism. On township plats, parallel lines may have different bearings due to convergence toward the North Pole. A line is parallel with, not to, another line. To is directional, whereas with is in association with. By mathematical definition, parallel lines are straight lines, but in common speech about boundaries, the words are often used to represent lines that are not straight lines but photographs of each other, and courts, in passing on questions of boundaries, often use them in the latter sense.

Parallel of latitude.   Any line that is run east and west and that is at every point at right angles to the meridian. A parallel of latitude is a curved line on the face of Earth; however, within the limits of a boundary as shown on a map, parallels of latitude are considered straight lines. Where large tracts are surveyed and where the latitude is far north, curvature considerations are necessary.

Parcel.   (noun) Generally refers to a piece of land that cannot be designated by lot number. (verb) To divide an estate.

Parol evidence.   Evidence gathered by testimony of witnesses.

Patent.   (adj.) Open; manifest; evident, as a patent ambiguity. (noun) The title conveyed by the government describing land disposed of by the government; a quitclaim from the government in the form of a patent.

Peonia.   Spanish: A portion of land given to a Spanish soldier in America.

Perch.   A term of measurement that today equals 16½ feet, or 25 links of 7.92 inches, or ¼ of a Gunter's chain, or ½ of a Leybourn chain. A distance set by English Parliamentary statute in the 1600s. Also equals a pole or a rod.

Place lands.   Lands granted to railroads and located on each side of the railroad. The location became fixed upon the adoption of a centerline. Indemnity lands were selected in lieu of place lands that were already granted for other purposes.

Plaintiff.   The person who complains and brings action at law on a charge against another.

Plat.   Pertaining to flat, as a plat of flat ground or to plat (flatten) another person. As used by surveyors, it is a flat map showing land boundaries usually without cultural features such as drainage or hills. Some states have platting acts (statute laws regulating how land must be mapped prior to a sale). The word is not subject to a precise definition and often is used interchangeably with map, chart, or plot.

Plot.   The act of placing survey data on a plat or map. Often used as plot plan or plot, meaning the plat itself. Preferred usage is to use the word plot only as the act of placing data on a plat or map.

Point.   Point in a boundary is the extremity of a line. "To a point" in a description is often meaningless, since the end of a line is a point. If the point is to be referred to later, "to point A" or "to point 1" gives an easy later reference.

Portion.   In legal descriptions, a part of a whole, as "that portion of lot 3 described as follows." A portion of a parcel without a following detail is indeterminate.

Position.   The place occupied by a point on the surface of Earth. Data that define the location (position) of a point on Earth's surface as by coordinates, by reference to a plat, or by metes and bounds.

Precision.   See Accuracy.

Prescription.   The acquisition of an easement by adverse use under claim of right for the statutory period required by law. Adverse possession as applied to easements.

Presumption.   A rule of law that courts and judges shall draw a particular inference from a particular fact, or from particular evidence, unless and until the truth of such inference is disproved (Black's Law Dictionary).

Prima facie.   "At first sight"; "on the face of it"; "presumably."

Principal meridian.   A meridional line running through an arbitrary point chosen as a starting point for all sectionalized land within a given area.

Prolong.   To extend. A line is prolonged, but a curve is continued. Prolongation of a curve is the extension of the tangent to the curve.

Property.   The ownership of a thing is the right of one or more persons to possess and use it to the exclusion of others. The thing of which there may be ownership is called property.

Prorate.   To divide, share, or distribute proportionally. To prorate excess.

Proration.   A method of distributing discovered excess or deficiency between parties having equal rights or proportionate rights to the excess or deficiency. A method of calibrating the tape of a recent surveyor against that of the original surveyor.

Pueblo.   People; pueblo lands (granted to the people as to a city or town).

Q

Quadrature of the moon.   Position of the moon when its longitude differs by 90° from the longitude of the sun.

Quasi.   "As if"; "almost as it were."

Quiet title.   An action at law to remove an adverse claim or cloud on the title of property.

Quitclaim deed.   A deed in the nature of a release containing words of conveyance as well as release. It conveys any interest that the maker may have in the property described without any representations or liability of any kind as to title conveyed or encumbrances that may exist thereon. A patent from the government is a quitclaim deed.

R

Radial or radial bearing.   A radial line is any straight line extending from the center of a defined circle to the circle's circumference. A radial bearing is the direction of a given radial line. On plats, the word radial next to a bearing indicates that that line is radial from the center of the given circle.

Radian.   A unit of circular measure equal to 57°17944.80.

Radius.   A line extending from the center of a circle to the circumference of the circle.

Rancho.   Spanish: In the West, land grant suitable for grazing. Literally, it is a collection of men or their dwellings.

Range.   A term commonly used in the colonies (especially in New England) to designate a tier of lots within a township (not a federal government township). Later adopted in federal sectionalized lands to designate a tier of townships, said tier extending in an east–west direction.

Real property.   Land and generally whatever is erected, growing, or affixed to the land.

Rear.   A deed reading "thence running to the rear of said land" does not always mean that the land extends to the rear line, but may mean "toward the rear" (Moran v. Lezotte, 19 N.W. 757).

Reliction.   An increase of the land by the permanent withdrawal of the sea, river, or lake.

Remainder.   An interest in real property that does not give the right of possession until the rights of the person in possession have been terminated either by that person's death or by lapse of time. One example is when A conveys real property to B, reserving in the deed a life estate to A. A (life tenant) has the right of possession and enjoyment of the property for his lifetime, and at his death, B (remainder-man) acquires this right of possession, thereby giving him the entire interest in the property. During A's lifetime, B's interest is a remainder.

Res judicata.   Res ("thing") settled in court (judicata); a thing judicially acted upon.

Reserving.   When a thing granted is taken back, it is reserved. Easements are usually reserved, as "reserving a 20-foot easement for road purposes, etc." Reserving is used when a new encumbrance is being created. A reservation creates some right or privilege for the grantor over the land described as granted.

Restricted lands.   Lands granted to Indians and restricted by Congress to protect the Indians from their supposed incompetence.

Restrictions.   Provisions in a deed that limit the use of the land.

Retracement.   Retracement is the first phase or step in conducting a resurvey. It is the office and field process conducted to find evidence of any original survey that still remains on the ground. It includes recovering, identifying, and evaluating the remains of the original evidence. It does not include replacing lost corners. It is a process of recovering evidence, not creating new evidence.

Reversion.   The estate or interest that will revert to, or be returned to, the grantor in a deed should restrictions be violated or the term of the conveyance end.

Revert.   To revert or go back to the former owner or his or her assigns.

Right-of-way.   Right to use or cross over property of another.

Riparian.   Of or on the bank; related to or belonging to the bank of a river. Also now used along the seashore.

Riparian rights.   The right that an owner of land bordering on a river has in the water flowing in the river or the land underneath the river. Also, the rights of a person owning land bordering on a body of water in or to its banks, bed, water, or travel on the water.

Rood.   A term of area equated to old English deeds. Equals ¼ of an acre, or 40 square rods.

Running.   In a deed, running along a line adds nothing; a verbose word.

S

Said.   Refers to one previously mentioned with the same name.

Saline land.   Land having salt deposits. Salt spring lands were granted to 14 states; others received none. In early times, salt was a necessary commodity.

Scrivener. A professional writer, deed author.

Seal. An impression upon wax or some other substance. In current practice, the word seal is made in lieu of an actual seal.

Searcher. A person who assembles all the facts concerning the title to real estate for submission to a title examiner.

Seisin. Possession of real property under a claim of a freehold estate.

Servient tenement. The land burdened by an easement (see also Dominant tenement).

Shepard's Citations. Reference material by which a court decision may be examined for its present status and its relation to other, pertinent, cases.

Shore. Land lying between high- and low-water marks. Often used to include the beach, as "to the seashore." Usually referred to the ocean, although river shoreline is used. Not a word with a precise meaning and is modified by the context of a sentence.

Site. A parcel of ground set apart for a specific use. The word itself does not necessarily imply definite boundaries.

Slope rights. Adjacent to highways, the right to extend fills or cuts beyond the side lines of the road easement as dedicated.

So-called. What the public in general, not the particular parties, have to say about the premises referred to.

Spiral. Used to change the curvature of a curve gradually from a given radius to a straight line, or vice versa. As many types of spirals exist, unless the type is defined, ambiguity results.

Spite fence. A fence maintained for the purpose of annoying owner of adjoining land.

Spring tides. Tides of increased range occurring semimonthly as a result of the moon being full or new.

Squatter. A person who settles on another's land, especially public lands, without legal authority.

Squatter's rights. In some states a mere squatter on a land without written title can acquire an adverse right. In some instances squatters had first right to purchase land. Exceedingly variable in meaning.

Standard parallel. Same as correction line. Standard parallels were run as east–west lines in sectionalized land surveys at various intervals from 24 to 96 miles apart.

Statute of Frauds. Statutory requirement that certain contracts be in writing in order to be enforceable. Most such statutes are patterned after the English statute of 1677, originally called the Act for Prevention of Frauds and Perjuryes.

Statute law. Statute law consists of laws passed by proper legislative bodies. A statute generally repeals all earlier laws and inactivates conflicting common law.

Stone bound. A stone, a stone mound, or cut stone used to mark a land boundary corner.

Subdivision. A tract of land divided, by means of a map or plat, into lots, or lots and blocks, for the purpose of resale, generally for residential or agricultural purposes.

Subject to. Refers to something already existing, as "subject to an easement."

Survey.   A term that has several multiple and possible ambiguous meanings. (noun) A product. A survey map, plat, or plan, or "my survey." The physical evidence of the verb "to survey." (verb) To conduct a survey. A process of conducting measurements to create evidence or to gather evidence on the ground from which your survey (noun) will be created. (adjective) Used to describe another word. Such as "survey plat," "survey notes," "survey evidence," or "survey work." *Original survey*: One of the three types of surveys. An original survey does not ascertain boundaries but creates them. *Dependent resurvey (survey)*: The second phase follows a retracement, in which the creator of the original boundaries conducts a retracement of the original boundaries to ascertain the evidence remaining and then proceeds to locate lost corners according to the rules of surveying. The lost corners have the same legal significance as the original corners they replace. *Independent resurvey*: A second place after conducting a retracement in which none of the original evidence is used to locate lost corners and lines, the original evidence is discarded, and new lines and corners are created.

Syllabus.   An abstract heading an adjudged case; the syllabus constitutes no part of the opinion of the court; it is the opinion of the reporter.

T

Tangent.   A tangent to a curve is a line that touches the curve at one point and also is at right angles to the radial line at the point of contact with the curve.

Tax sale.   An official sale of lands by the state for the nonpayment of taxes assessed on them.

Tenements.   At common law, included lands or other inheritances capable of being held in freehold and rents.

Tenure.   The method or system of holding lands or tenements.

Thalweg.   The deepest part of the channel of a river or stream.

Thence.   From that place; the following course is continuous from the one before it.

There.   In or at that place.

Thereon.   A deed that granted a passageway and specified that the grantor must fence the same but reserved the right "to erect gates thereon" implied the right to erect gates either on the side or across the passageway (Gossett v. Chandler, 264 S.W. 853, 204 Ky. 402).

Thread.   The thread of a road is a line midway between the side lines. The thread of a stream is the line midway between banks or the line equidistant from the edge of the water on the two sides of the stream at the ordinary stage of the water. Consult each state's law for a definition within a given state.

Tidal.   For water to be tidal, it does not necessarily have to be salty; it must have tides that regularly flow and reflow.

Tidal current.   Currents caused by tide-producing forces of the moon and sun are tidal currents. Nontidal currents include the permanent currents in the general circulatory system of the sea as well as temporary currents arising from meteorological conditions.

Tide. The periodic rising and falling of the water that results from the gravitational attraction of the moon and sun acting upon the rotating Earth. Horizontal movement of the water resulting from the same causes, although sometimes referred to as tide, should be called tidal current.

Tidelands. Lands covered and uncovered by the ordinary tides (Rondell v. Fay, 32 Cal. 354). The term tidelands has been used loosely with an enlarged meaning to include lands near the tidelands.

Tie. A survey connection from a point of known position to a point whose position is desired.

Tie points. Offset monuments set by the city engineer to mark street lines.

Tied. As used in surveying, monuments are tied together by measurements. A property corner is tied to offset monuments or to other property corners.

Title insurance. Insurance against loss due to any defect or hazard insured against in a policy of title insurance.

Title policy. A policy insuring the title to real property, issued for the protection of persons acquiring interests in real property either as owner, lender, or lessee; it insures against forgery, incompetents, insanities, and other matters that are not shown by public records; it insures the actual title to property as distinguished from the record title, such as is guaranteed in a guarantee of title. It usually does not insure the location.

Title search. The checking or reviewing of all documents affecting the ownership of a piece of property.

To. "To," "on," "by", "at," and "along" a road carry title to the centerline unless otherwise qualified. To implies contact. To does not always include an object, as "to a certain property" does not include the property. But "to a stone" usually means "to the center of the stone." To is directional, as "90° to (not with)" or "at right angles to." To is a word of exclusion rather than inclusion. If you go to an object, you exclude other objects. If you go to a house, you do not necessarily go in the house.

To wit. "That is to say"; "namely."

Torrens system. A system for the registration of land titles. Introduced into South Australia in 1858 by Sir Robert Torrens.

Township. Township is a nearly square area of land usually containing 36 sections of land (25 sections in parts of Ohio). It had variable meaning in the colonies.

Traverse. A sequence of field measurements (length and directions) of lines between points on Earth and used to determine positions of points. A closed traverse ends at the beginning point. Also used in the office as "running a traverse" or meaning performing the calculations necessary to determine whether a field closed traverse mathematically closes.

Trees, ownership of line trees. Trees whose trunks are wholly on the land of one owner belong exclusively to that owner, although their roots grow into the land of another. Trees whose trunks stand partly on the land of two or more coterminous owners belong to them in common. This varies in some states.

Trespass. A legal, not a surveying, term. Denotes unauthorized use of another's land.

True. Correct; honest; conformable to fact (see North).

True North. True North differs from magnetic north, which varies from place to place and over time due to local magnetic anomalies. True North is in relation to Earth's axis and not the magnetic poles.

Trust deed. A written instrument by which a borrower (trustor) conveys his or her land to another (trustee) for the benefit of the lender (beneficiary) as security for the repayment of the money lent. In the event of a failure of the trustor to repay the money, the trustee conducts a foreclosure sale of the real property.

U

Unwritten law. The law not promulgated and recorded but which is, nevertheless, observed and administered in the courts of the country. It has no certain repository but is collected from reports of the decisions of the courts and from the treatises of learned persons.

Uplands. Lands bordering on waters.

V

Vacate. To annul; to render an act void, as "to vacate an easement."

Vara. The measure of land used in connection with Spanish land grants. It is generally accepted as being about 33 inches, but its definition varies from jurisdiction to jurisdiction.

Vest. To give title to or to pass ownership to property.

W

Warrant. A promise that facts are as they are represented to be.

West. A curved line following a latitude. Practically, it is at right angles to north, since most descriptions are too small in area to have noticeable curvature.

Westa. Half a hide of land, or 60 acres.

With. Shows an association, as "parallel with," not "parallel to." With a line, meaning along a line, is commonly used in the eastern states.

Words of exclusion. To, from, by, between, and on are words of exclusion unless there is something in the phrase that makes it apparent that the words were used in a different sense. "To a stone mound," "on Brown's land," and "by the river" exclude other terms.

Writ. An order issued by a court requiring the performance of an act or giving authority to have it done.

INDEX

Note: Page numbers followed by "*f*" and "*t*" indicate figures and tables, respectively.

Brown's Boundary Control and Legal Principles, Eighth Edition.
Donald A. Wilson, C.A. "Tony" Nettleman III, and Walter G. Robillard.
© 2024 John Wiley & Sons, Inc. Published 2024 by John Wiley & Sons, Inc.